Handbook of Spatial Point-Pattern Analysis in Ecology

CHAPMAN & HALL/CRC
APPLIED ENVIRONMENTAL STATISTICS

Handbook of Spatial Point-Pattern Analysis in Ecology

Thorsten Wiegand

Helmholtz Centre for Environmental Research - UFZ
Leipzig, Germany

Kirk A. Moloney

Iowa State University
Ames, Iowa, USA

CRC Press
Taylor & Francis Group
Boca Raton London New York

CRC Press is an imprint of the
Taylor & Francis Group, an **informa** business
A CHAPMAN & HALL BOOK

CRC Press
Taylor & Francis Group
6000 Broken Sound Parkway NW, Suite 300
Boca Raton, FL 33487-2742

First issued in paperback 2020

© 2014 by Taylor & Francis Group, LLC
CRC Press is an imprint of Taylor & Francis Group, an Informa business

No claim to original U.S. Government works

Version Date: 20130904

ISBN 13: 978-0-367-57623-3 (pbk)
ISBN 13: 978-1-4200-8254-8 (hbk)

Library of Congress Cataloging-in-Publication Data

Wiegand, Thorsten.
    Handbook of spatial point-pattern analysis in ecology / Thorsten Wiegand and Kirk A.
Moloney.
        pages cm. -- (Chapman & Hall/CRDC applied environmental statistics)
    "A CRC title."
    Includes bibliographical references and index.
    ISBN 978-1-4200-8254-8 (hardcover : alk. paper)
    1. Ecology--Mathematical models. 2. Statistics. I. Moloney, Kirk Adams, 1952- II.
Title.

QH541.15.M34W54 2014
577.01'51--dc23   2013032825

Visit the Taylor & Francis Web site at
http://www.taylorandfrancis.com

and the CRC Press Web site at
http://www.crcpress.com

For Kathi, Alin, and Carlitos

—**TW**

Martha, Josh, Kendra, Bud, and Bette: Many thanks

for accompanying me along the way

—**KAM**

Contents

Preface

P.1 Why Another Book about Spatial Pattern Analysis?

During the last few decades, there has been increasing interest in the study of spatial patterns in ecology. Ecologists study spatial pattern to better understand the processes that may have caused observed patterns and to test spatially related ecological theories. One approach for doing this is to characterize spatial patterns as accurately as possible, using appropriate statistical techniques. The primary goal then is to derive hypotheses on the nature of the underlying processes producing the pattern, which can then be tested in the field. Another approach is to use ecological theory to derive specific hypotheses about spatial patterns that are then tested using observed data from appropriate systems. The ultimate aim of these efforts is to infer the existence of underlying processes and to identify the spatial scale at which they are operating.

The most basic class of spatial data in ecology is given by the *location of ecological objects* within a known observation window. The set of coordinates of the ecological objects of interest are commonly referred to as "point-pattern data." With only one type of object, we have a univariate point pattern. If there are *a priori* two types of objects (e.g., two different species of trees), we have a bivariate point pattern. And if there are many types of objects (e.g., many different tree species), we have a multivariate point pattern. In some cases, we have additional information about the data objects called *marks*. For example, we may have a qualitative mark, such as the condition of the object (e.g., surviving vs. dead trees), or a quantitative mark, such as the diameter of a gopher mound. The framework for analyzing the spatial structure of marked or unmarked point-pattern data is provided by *point-pattern analysis*, which encompasses a rich array of statistical tools for analyzing the spatial distribution of discrete points.

There has been a nearly exponential increase in the number of published studies during the last two decades that have conducted ecological analyses of point patterns; the same is true for the number of citations those articles received (Figure P.1). However, our personal observation has also been that the application of spatial point-pattern analysis in an ecological context has generally fallen short of its potential; a somewhat curious observation since appropriate statistical methods for analyzing spatial point patterns have been available for several decades (e.g., Ripley 1981; Diggle 1983; Cressie 1993; Stoyan and Stoyan 1994; Møller and Waagepetersen 2004; Illian et al.

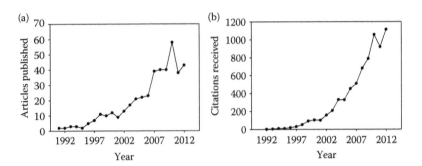

FIGURE P.1

Timeline of point-pattern analysis in the ecological literature. Number of papers published (a) and citations (b) per year, determined by conducting a literature search on ISI's Web of Science (http://apps.webofknowledge.com) using the keyword search terms of "[(Ripley* OR "point pattern") and (Ecolo* OR Biolo* OR Forest* OR Vegetation*)]" for the time span 1945 to 31 December 2012. There were 425 papers, which were cited 7127 times. The average number of citations per paper was 16.77, with an h-index of 44.

2008). This suggested to us that there is a growing need for a book focusing on the application of point-pattern analysis to ecological problems. The *main objective of this book* is thus to provide a synthesis of the techniques of point-pattern analysis that are useful for addressing ecological problems, presented within an ecological framework that guides ecologists through the variety of available methods for different data types and aides in the interpretation of the results obtained by point-pattern analysis. By the same token, our objective is not to present a comprehensive synthesis of all the point-pattern analyses that have been conducted in an ecological context or all the current techniques of spatial point-pattern analysis. Readers who are primarily interested in statistical techniques and point-process modeling, but are not interested in the ecological framework, might prefer to consult the recent books of Diggle (2003), Møller and Waagepetersen (2004), Illian et al. (2008), Baddeley et al. (2006), and Gelfand et al. (2010).

Of particular importance to the successful application of point-pattern analysis is the appropriate use of *null models* and, in a more advanced context, *point-process models*. In the context of point-pattern analysis, a null model is used for detecting "patterns" in the data by randomizing certain aspects of the data while holding other aspects constant. Point-process models are more general models that produce spatial patterns of points using a known set of rules, some of which are stochastic. The patterns produced by a null model or point-process model can be used to test ecological hypotheses of spatial dependence when applied correctly. However, the appropriate use of these models is often missing in ecological papers using spatial point-pattern analysis, although there has been some recent progress (e.g., Goreaud and Pélissier 2004). Usually, a standard null model, such as complete spatial randomness, is used, even though it is often not the correct model for the question being addressed. This can lead to significant misinterpretation of data. Exacerbating this problem is the fact that methods presented in statistical publications are often illustrated with a limited number of "tame" examples, which do not show real-world problems in all their complexity and can only provide guidance for ecologists familiar with advanced statistics.

In contrast to the well-developed body of theory for point patterns, suitable software that can deal with "real-world" data sets has only recently become readily available (e.g., spatstat in R or *Programita*). Most of the earlier software packages that include point-pattern analysis only accommodate a limited range of point-pattern data structures and provide only a limited range of null models. These limitations prohibit the inclusive study of more complex, real-world situations. Additionally, most software also assumes that ecological objects can be approximated as points, suppressing information about the size and shape of the objects being studied. While this may be appropriate in some situations, it may be limiting in others.

Based on many years of teaching experience and collaborative research in the field of ecological point-pattern analysis, we have developed *Programita*, a continuously updated software package. It has been tailored to accommodate the needs of "real-world" applications in ecology and has been developed in response to our own research questions and to requests of colleagues and students who have approached us with their specific research problems in mind, rather than from a motivation to include all existing methods. Most of the examples presented in the book are in fact analyzed using *Programita*. The software and a manual can be accessed at a website accompanying this book using the links www.thorsten-wiegand.de/point-pattern-book.html or point-pattern-book.kamoloney.org. In addition to *Programita*, we would also like to point out the growing availability of statistical packages written in R that are designed to analyze point patterns (e.g., spatstat; Baddeley and Turner 2005, 2006). Although the R learning curve is quite steep, it provides the opportunity to extend methods of point-pattern analysis beyond its current bounds for the more adventurous statistical programmer.

P.1.1 Ecological Viewpoint

This book presents statistical techniques from a somewhat atypical vantage point, as our intended audience is primarily empirical ecologists interested in understanding how to analyze and interpret the information contained in ecological patterns, not theoreticians. For ecologists, the ultimate goal, although difficult to attain, is to infer process from pattern. *The focus of the book is therefore on analysis*, presenting methods that can be used to extract information hidden in spatial point-pattern data that may point to the underlying processes. We will only briefly deal with the construction and fitting of point processes (called synthesis by Illian et al. 2008), a lively and rapidly developing branch of modern spatial statistics (e.g., Møller and Waagepetersen 2004).

In this book, we present only point processes and null models that have proven their immediate utility for broader ecological applications, such as cluster processes. Readers interested in a more in-depth treatment of point-process models and model fitting may refer to the more specialized literature. The underlying reason for not focusing on statistical point-process modeling is that ecology already comes with a mature and well-developed theory for constructing dynamic, spatially explicit, individual-based models (Grimm and Railsback 2005; Railsback and Grimm 2012). These types of model incorporate the actual processes that give rise to observed pattern and can produce spatial point-pattern data relevant for addressing ecological issues (e.g., Wiegand et al. 1998; Jeltsch et al. 1999; Brown et al. 2011).

We hope that this book will stimulate the combined and interactive use of point-pattern analysis and (individual-based) simulation models. While working in the 1990s on the analysis of an individual-based, spatially explicit simulation model (Wiegand et al. 1998), we discovered the potential of combining point-pattern analysis and (individual-based) simulation models. This is one of the reasons we became interested, in the first place, in the study of spatial patterns in dynamic ecological systems.

P.1.2 Audience

In 2004, we wrote an introductory review on point-pattern analysis (Wiegand and Moloney 2004) for an ecological audience in which we also introduced our point-pattern analysis software *Programita*. An interesting and somewhat unexpected aspect of the review paper has been the number of requests for *Programita* that have come from a variety of disciplines outside ecology (e.g., archaeology, analysis of spatial patterns in brain tissue, social sciences, etc.). Indeed, there is no point-pattern analysis methodology specific to ecology; the methods are universal and can be applied in a variety of fields, given that they have similar data structures. We, therefore, expect that this book will have a much broader audience than just ecologists. Readers from outside ecology are encouraged to consider examples from their own discipline, taking advantage of the underlying structural similarities. Nonetheless, the primary readership of the book is intended to be ecologists interested in analyzing spatial point-pattern data who may be intimidated by more advanced theoretical literature. The book and the software will allow advanced students, research workers, practitioners, and professionals to perform appropriate analyses with their point-pattern data.

We hope that this book will be a source for teaching graduate students spatial point-pattern analysis, because it not only offers techniques, but also embeds point-pattern analysis into an ecological framework with a comprehensive selection of real-world examples. In Chapter 5, we provide suggestions on how to use the book as a textbook for teaching.

P.2 This Book: What It Is and What It Is Not

Excellent textbooks that introduce the theory of statistical point-pattern analysis in depth to a broader readership are available (e.g., Ripley 1981; Cressie 1993; Stoyan and Stoyan 1994; Diggle 2003; Møller and Waagepetersen 2004; Illian et al. 2008). This book is intended to complement these reference works, not compete. Much of the statistical literature on spatial point-pattern analysis, and even articles in scientific journals with material highly relevant for ecologists (e.g., Waagepetersen and Guan 2009), is fairly technical and inaccessible for nonspecialized readers. The excellent book by Illian et al. (2008) is written in a relatively nontechnical manner and presents mainly mathematical–statistical facts, but it will still exceed the mathematical capacities of most practicing ecologists. We, therefore, decided to avoid mathematical formalism as much as possible; instead, we try to explain the underlying ideas and principles from the perspective of practicing ecologists; ideas that for the nonstatistician are too often hidden behind a formalized mathematical language. However, this book includes equations. They form a part of the universal scientific language and are indispensable tools of spatial statistics that allow concise and precise communication. We cannot completely circumvent their use even when avoiding formal mathematical language and must, therefore, expect that our readers have received training in basic mathematics and statistics. Most of the equations are concentrated in Sections 3.1 through 3.3, which present important background information on the estimators of the different summary statistics, and in Chapter 4, where more complex point-process models are presented. Nevertheless, the core of the book comprises ecological examples of the application of point-pattern analysis.

As stated above, this book is written primarily for empirical ecologists who are interested in extracting information that is hidden in their spatial point-pattern data and may point to the key underlying processes producing the pattern. Given this objective, our book will provide an ecological framework that guides ecologists unfamiliar with advanced statistics in their selection of available methods. Special emphasis is given to the formulation of appropriate null models and point processes for describing the features of point patterns and testing ecological hypotheses of spatial dependence, the essential ingredient in the successful application of point-pattern analysis in an ecological context.

Chapter 1 puts the *analysis of spatial patterns in an ecological context* and provides insights into why this approach to understanding ecological systems may be helpful for ecologists. In this chapter, we justify why an entire book on spatial point patterns is needed, provide some examples on how analysis of spatial patterns has interacted with ecological theory, and discuss the pattern-process link, which has often been disputed.

Chapter 2 presents the *fundamentals of point-pattern analysis*. After a brief summary of the typical steps of a point-pattern analysis (Section 2.1), we dedicate Sections 2.2 through 2.6 to the following five basic issues that have to be considered in each point-pattern analysis:

1. Data structures
2. Summary statistics
3. Null models and point processes
4. Methods to compare data and null models
5. Dealing with heterogeneous patterns

Chapter 3 deals with *more technical aspects of point-pattern analysis* and presents a detailed treatment of the estimators of summary statistics of practical utility for ecologists (Section 3.1), analysis of replicate patterns (Section 3.2), and independent superposition of point processes (Section 3.3). Section 3.4 presents a toolbox of different techniques and methods that are especially useful in a variety of applications, such as analysis of point patterns in irregular observation windows (Section 3.4.2) or pattern reconstruction (Section 3.4.3).

We place special emphasis on the separate treatment of the different data structures that are used in ecology. Treating data types separately is necessary because each type of data corresponds to a different subset of questions and requires a different set of methods. Chapter 4 is, therefore, dedicated to presenting examples of the analysis of specific data structures. Within each subsection, we present the typical ecological questions related to specific data structures and provide a collection of examples. In providing the examples, we present the application of specific null models and point processes corresponding to real-world applications. Finally, Chapter 5 provides suggestions regarding the use of this book for teaching.

We have attached to each chapter a *star code* that helps the reader to assess the level of the chapter. A single star ($*$) indicates that the corresponding chapter presents basic material that each reader should study carefully. This represents, in general, the background material relevant for any application of point-pattern analysis. Two stars ($**$) indicate standard material relevant for readers who are interested in specific analyses (e.g., analysis of a certain data type). Finally, three stars ($***$) indicate more technical material, such as the discussion of the advantages and disadvantages of a certain estimator or edge correction method, which is often not of vital interest for standard users.

Acknowledgments

The authors thank the numerous students, collaborators, and *Programita* users for their stimulus to develop many of the methods presented here and for convincing us of the need to write this book. Thanks to Mauricio Aguayo, Martin Aguiar, Martín Almiron, Gerardo Azócar, Robi Bagchi, Willie Batista, Fernando Biganzoli, Klaus Birkhofer, Paula Blanco, Doug Bruggeman, Justin Calabrese, Chechu Camarero, Antonio Castilla, Wirong Chanthorn, Pablo Cipriotti, Roger Cousens, Carlos De Angelo, Marcelino De la Cruz, Brent Danielson, Hansjörg Dietz, Bruno Djossa, Fernando González-Taboada, Alex Fajardo, Rusty Feagin, Cani Fedriani, Birgit Felinks, Andres Fuentes Ramirez, Jana Förster, Valéria Forni Martins, Fernando Garelli, Stephan Getzin, Urs Giesselmann, Pamela Graff, Nimal Gunatilleke, Savit Gunatilleke, Juan Gurevitz, Florian Hartig, Fangliang He, Bob Howe, Steven Hubbell, Andreas Huth, Hans Jacquemyn, Florian Jeltsch, Raja Kanagaraj, Martin Kazmierczak, Daniel Kissling, Ben Klaas, Stephanie Kramer-Schadt, Guoyu Lan, Michael Lawes, Yiching Lin, Elena Lobo, Ramiro Pablo López, Marcella Lunazzi, Fernando Maestre, Isa Martínez, Leandro Mastrantonio, Felix May, Stephanie Melles, Katrin Meyer, Erika Mudrak, Nacho Mundo, Aris Moustakas, Jose Paruelo, Aníbal Pauchard, George Perry, Marion Pfeifer, Ruwan Punchi-Manage, Simon Queenborough, Josep Raventós, Drew Rayburn, Corinna Riginos, Juanjo Robledo-Arnuncio, Javi Rodríguez-Pérez, Michael Lawes, Paul Ramsay, David Roshier, Katja Schiffers, Nicolas Schtickzelle, Guochun Shen, Nate Swenson, Solana Tabeni, Cheng-Han Tsai, Alfredo Valido, Claudia Vega, Eduardo Velazquez, Dario Vezzani, Xugao Wang, David Watson, Kerstin Wiegand, Jan Wild, Chris Wills, Amie Wolf, Hong Yu, Yan Zhu, and many others. The authors would also like to thank Dietrich Stoyan for his outstanding work that, in many respects, provided the backbone for this book.

The foundation for the book all started when the Graduate School "Alberto Soriano"–Faculty of Agronomy–University Buenos Aires asked TW in 1998 to give a course on *"Patrones espaciales en ecología: modelos y análisis"* (Spatial patterns in ecology: models and analysis). In the seven courses that followed between 1999 and 2010, TW received continuous motivation based on the needs of students to analyze their spatial data, which resulted in the development of the software *Programita* and finally this book. *A mis alumnos, muchas gracias para todo, el asado y la música ...*

KAM began exploring the techniques described in this book at Cornell University as a postdoc in the mid-1980s while working in the lab of Simon Levin, to whom he is extremely grateful. The field has advanced a great deal since that time, and it has been exciting to participate. As with TW, an interest in producing a book introducing the techniques of point-pattern analysis to practicing ecologists was inspired, in part, by conducting short courses on spatial analysis and modeling at the University of Potsdam in Germany, the ETH in Zürich, Switzerland, and at the Universidad de Concepción in Chile as well as teaching point-pattern analysis in KAM's course on landscape ecology at Iowa State University.

Work on the book for KAM would not have been possible without finding the time needed to focus on writing. There were several opportunities provided in this regard. KAM would like to thank the Velux Foundation, Peter Edwards, and Hansjörg Dietz for making possible his stay at the ETH in Zürich during 2006. It was during that time the initial outline and proposal for the book were developed. A summer spent in Potsdam, Germany in 2010 as a Mary Curie Fellow on the FEMMES project also provided time away from the fray to work on the book. KAM is grateful to Florian Jeltsch and Frank Schurr and the support from the European Union through Marie Curie Transfer of Knowledge Project FEMMES (MTKDCT-2006-042261) for making the stay in Potsdam possible. Ultimately, none of this would have been possible without the guidance of Lyman Benson and Edwin "Jonesy" Phillips at Pomona College, Hub Vogelmann and Ian Worley at the University of Vermont, and Janis Antonovics and Bill Schlesinger at Duke University, who helped KAM navigate the perilous shoals along the way to academic adulthood.

Sometimes the request to collaborate changes the course of things, such as the invitation by Nimal and Savit Gunatilleke that made TW aware of their exciting data from the Sinharaja Plot and the entire network of the Center for Tropical Forest Science. This collaboration finally resulted in the ERC advanced grant 233066 SPATIODIVERSITY to TW that supported studies that developed many of the methods presented here. We used the resulting data set in several of the example analyses in this book. The Sinharaja Plot research was supported by the Center for Tropical Forest Science of the Smithsonian Tropical Research Institute and the Arnold Arboretum of Harvard University, the John D. and Catherine T. MacArthur Foundation (94-29503 and 98-55295), the National Science Foundation, USA (0090311), and the National Institute for Environmental Studies, Japan.

Our special thanks go to Steven Hubbell and Robin Foster for their visionary and "crazy" idea to establish 50 ha fully mapped plots of tropical forests in the early 1980s, which stimulated much of the development of point-pattern methods presented here, and to Steven Hubbell, Robin Foster, Rick Condit, and Stuart Davis for their effort in maintaining these plots over decades within the global network of the CTFS Center for Tropical Forest Science of the Smithsonian Tropical Research Institute. We analyzed this data set in many of the examples presented in the book. The BCI forest dynamics research project was made possible by National Science Foundation grants to Stephen P. Hubbell: DEB-0640386, DEB-0425651, DEB-0346488, DEB-0129874, DEB-00753102, DEB-9909347, DEB-9615226, DEB-9615226, DEB-9405933, DEB-9221033, DEB-9100058, DEB-8906869, DEB-8605042, DEB-8206992, DEB-7922197, support from the Center for Tropical Forest Science, the Smithsonian Tropical

Research Institute, the John D. and Catherine T. MacArthur Foundation, the Mellon Foundation, the Small World Institute Fund, and numerous private individuals and through the hard work of over 100 people from 10 countries over the past two decades. We are indebted to Willie Batista and Marcella Lunazzi for providing the data used in Figures 2.26 and 2.28, Martín Aguiar and Pablo Cipriotto for the data analyzed in Figures 3.36 through 3.38, and Josep Raventós for the data analyzed in Figure 3.41. Special thanks to Paul Ramsay for his critical review that forced us to abandon the originally planned fractal structure of the book, and to David Grubbs for his continuous support, patience, and pressure to finally complete the book.

Authors

Dr. Thorsten Wiegand is a theoretical ecologist with more than 20 years of research experience in question-driven research in biodiversity and conservation. He studied physics in Marburg, Germany, where he earned a PhD in 1992. However, during his diploma and PhD work, he moved to the field of theoretical ecology, where he has since remained, first as a postdoc and later as a senior scientist at the Helmholtz-Centre for Environmental Research—UFZ in Leipzig, Germany. He works in close collaboration with field ecologists to conduct model-data integration and synthesis. His research centers on the investigation of the role of species interactions, spatial processes and spatial structures for population and community dynamics, community assembly, and biodiversity. The overarching methodological themes of his research are development of advanced computational methods of (1) spatial point-pattern analysis in complex large data sets, (2) dynamic, stochastic, and spatially explicit simulation models that incorporate real-world ecological complexity, and (3) statistical inference to rigorously confront the patterns in the data detected in (1) with the models developed in (2) to achieve optimal model-data integration. His study systems include plant communities such as tropical forests in Sri Lanka and Panama, treeline dynamics in Europe, semiarid shrub and grasslands in South Africa, the Patagonian steppe in Argentina, and endangered animal species such as brown bears (*Ursus arctos*), Iberian lynx (*Lynx pardinus*), European lynx (*Lynx lynx*), and tigers (*Panthera tigris*).

Dr. Kirk A. Moloney is a professor of plant ecology at Iowa State University in Ames, Iowa. He has been working in the field of ecology for the last 27 years, specializing in the study of the spatial dynamics of populations and communities. His current focus is on the ecology and evolution of exotic species, working primarily in the southwestern deserts of North America, an amazing region of this diverse world. He earned a PhD from Duke University in 1986.

1

⋆ *Application of Spatial Statistics in Ecology*

One could say that ecology is a science driven by an interest in processes that are inherently spatial by nature. We strive to explain why organisms appear in one location and not in another. We characterize change in community composition along environmental gradients and explore the interplay between extinction and colonization in maintaining spatially distributed metapopulations. Indeed, one of the classic textbooks in ecology originally published in the 1970s by Charles Krebs (Krebs 1978) is *Ecology: The Experimental Analysis of Distribution and Abundance*. The title nicely illustrates the point that many questions in ecology arise from the observation that organisms and the ecological processes that influence them vary in space.

Although there has always been a background awareness of the importance of spatial relationships in ecology, until the mid-1980s most ecological research, with the exception of a few pioneering studies, avoided the explicit consideration of space. Ecological field experiments were generally designed to remove spatial "contamination" through the use of techniques such as randomization and block designs. Empirical studies were situated so that individual samples were obtained from what were perceived to be spatially homogeneous sites and comparisons among sites were designed to contrast different idealized ecological settings, minimizing the variability produced by uncontrolled spatial heterogeneity. Even theories such as island biogeography, which invoked spatial relationships, only contained an implicit, not explicit, consideration of space in their formulation.

During the 1980s, there was a fundamental shift in the ecological community toward an explicit consideration of spatial relationships (Galiano 1982; Sterner et al. 1986; Getis and Franklin 1987; Kenkel 1988). This was driven by a number of factors. First, desktop computers were quickly being integrated into ecological research, providing an easily used tool capable of visualizing, modeling, and analyzing complex spatial relationships. Aerial photographs and images from satellites became readily available, providing ecologists with an overview of ecological patterns never before available. And, perhaps most importantly, ecology had matured to a point where it needed to break free of nonspatial thinking to advance the science. With the increasing integration of a spatial component into ecological thinking, it was necessary to develop the tools required for characterizing and analyzing spatial relationships. Fortunately for ecologists, there was an ongoing renaissance in spatial statistics that could be incorporated into ecological studies.

1.1 Analysis of Spatial Patterns

The statistical analysis of spatial patterns comes in three basic forms, depending on the nature of the underlying data (Figure 1.1). (1) *Quantitative data* are those that are continuously distributed in space. For example, within a terrestrial ecosystem soil nitrogen can be measured at any point in space; abundance of a plant species can be determined in quadrats placed at any location. These types of data are generally analyzed using the techniques of geostatistics (e.g., Journel and Huijbregts 1978; Webster and Oliver 2007), the origins of which are found in the very rich and well-developed literature of time-series analysis. (2) *Categorical data*, which are characterized by defined regions in space, are referred to by spatial statisticians as lattice data. In an ecological setting, the most common form of data falling into this category are maps, where bounded regions in space are classified into discrete categories, such as vegetation type, landscape category, soil type, and so on. Availability of remotely sensed data has augmented the use of this type of data, greatly. Various techniques are used in the field of landscape ecology for analyzing map data, with many supplied by the well-known package FRAGSTATS (McGarigal and Marks 1995) and a variety of Geographic Information Software (GIS) packages (e.g., ArcGIS, Grass). (3) *Spatial point-pattern* data are about the location of objects in space. In ecology, these could be trees in a forest, gopher mounds in a prairie, insect galls distributed

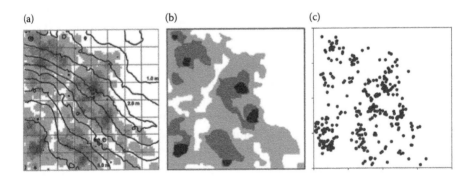

(a)   (b)   (c)

FIGURE 1.1

Three forms of spatial data representing the production of gopher mounds in a tallgrass prairie. (a) Quantitative data are indicated in grayscale and represent expected rates of gopher mound production for each m² area in the study site; higher expected rates of gopher mound production are represented by darker shades. (b) Categorical data are represented by a map segmented into areas of mound-building activity corresponding to four rates of activity: high, medium, low, and 0. (c) Point-pattern data show the location of mounds produced over a discrete period of time. (Modified after Klaas, B.A., K.A. Moloney, and B.J. Danielson. 2000. *Ecography* 23: 246–256.)

among goldenrods in an old field, nest locations, and so on. The questions of interest with these kinds of data often revolve around assessing the potential ecological causes (i.e., processes) behind the observed distribution of the objects in space (i.e., patterns). There are two lines of research that have developed around this question. First, *species distribution or habitat suitability models* (Elith and Leathwick 2009) examine environmental factors that determine the presence/absence or abundance of ecological objects in space. In this case, the assumption is that the environment is heterogeneous and environmental covariates can explain important characteristics of the spatial distribution of objects in space. Note that possible interactions between objects (especially autocorrelation) pose a problem to this type of analysis (e.g., Dormann et al. 2007). In the second line of research, *point-pattern analysis* usually assumes that the environment is homogeneous but it investigates possible small-scale interactions between points. Here, the heterogeneity of the environment poses a problem to the analysis (e.g., Baddeley et al. 2000; Wiegand and Moloney 2004; Wiegand et al. 2007a; Law et al. 2009).

Initially, during its historical development, point-pattern analysis only provided an assessment of whether or not a pattern was completely, spatially random (CSR), a notion that will be defined later in the book. The two potential alternatives to CSR were that the pattern was clustered or regularly distributed, both having implications regarding the processes responsible for producing the pattern. More recently, the approach to analyzing spatial point patterns has become much more sophisticated due to a number of important developments. One extension, in particular, is the consideration of the structure of point patterns across a range of spatial scales, rather than simply characterizing the overall pattern. Other important developments involve techniques that allow the analysis of bivariate, multivariate, and marked patterns. Now, there are also methods being developed for the analysis of heterogeneous patterns, using covariates, which reconciles to some extent the dichotomy of species distribution modeling and spatial point-pattern analysis mentioned above.

Although all three forms of spatial data are important in ecological research, we will focus on the analysis of spatial point patterns in the remainder of the book. There are several reasons for this. First, a more voluminous book would be necessary to consider all three forms of spatial analysis. Second, we feel that there is a need to present a modern approach to the analysis of spatial point processes targeted specifically at practicing ecologists. Point-pattern analysis in the ecological sciences is increasingly employed (Figure P.1) and one striking aspect of this literature is that, while well intentioned, it is often misapplied. Finally, while earlier studies in point-pattern analysis utilize mostly univariate techniques for unmarked patterns, a wealth of methods has recently developed around the different types of data available to the ecologist (Table 2.1), meriting a book with an in-depth presentation from an ecological point of view.

1.1.1 Spatial Patterns and Ecological Theory

As indicated above, until the mid-1980s, most ecological theory avoided the explicit consideration of space. One reason for this was that differential and difference equations, the historically preferred tools of theoretical ecologists, quickly become intractable when space is considered explicitly. However, another, perhaps equally important reason for avoiding a consideration of spatial relationships was that there were an overwhelming number of fundamental problems to be addressed that did not require an explicit consideration of space, for example, the dynamics of single species, predator–prey interactions, and host–parasitoid systems. (The interested reader can find a good introduction to this nonspatial theory in the books by May (1976) and Yodzis (1989).)

With the maturation of theoretical ecology, interest in spatial phenomena increased. In plant ecology, for example, this interest was fueled by increasing evidence that spatial patterns and processes play an important role in the assembly, temporal dynamics, and internal functioning of plant communities (e.g., Watt 1947; Pacala 1997; Tilman and Kareiva 1997; Bolker and Pacala 1999; Murrell et al. 2001). It was also well established that most plant species are not randomly distributed, but instead are often locally aggregated or overdispersed at one or several spatial scales (Watt 1947; Pielou 1977; Kenkel 1988; Condit et al. 2000). These observations continue to stimulate a large body of theoretical work investigating the relevance of the spatial arrangement of individuals for the coexistence of species and the maintenance of biodiversity (e.g., Hurtt and Pacala 1995; Pacala and Levin 1997; Chesson 2000a; Murrell et al. 2001) and give rise to the need for further development of techniques for characterizing and interpreting spatial pattern in the distribution of organisms.

An interesting illustration of the influence of a consideration of spatial patterns on the development of ecological theory is provided by the case of competition in plants. In a classic article, Kenkel (1988) tested the idea that intraspecific competition may lead to spatial regularity in the distribution of individuals. While the presence of a regular pattern may point to the presence of competitive interactions, he argued that more convincing evidence would be obtained through careful mapping of the fates of individuals over time. As a consequence, he used maps of the spatial distributions of living and dead individuals to test for spatial pattern. In this case, the appropriate null model is the random mortality hypothesis, that is, no spatial pattern in mortality. If the data are compatible with the random mortality hypothesis, we cannot infer the presence of competition from the data. However, a rejection of the null model would indicate either a clustered or hyperdispersed pattern of mortality, which, according to Kenkel (1988), would suggest either scramble or contest competition, respectively. Under scramble competition (in which a limited resource is partitioned equally among contestants so that no competitor obtains the amount it needs) one expects that dense clumps of conspecifics may die, with more isolated individuals surviving. This should result in a pattern where surviving and dead individuals are segregated. However, contest competition (in which the resources are unequally partitioned so that some competitors obtain more than others) will lead to the clustering of dead individuals around individuals obtaining a greater share of resources. In his study, Kenkel (1988) found a two-phase pattern suggesting scramble competition early in stand development, with dead trees clustered together, and contest competition at a later stage, resulting in hyperdispersion among living trees. These results motivated the subsequent development of a general model of self-thinning that explains how competition shapes the spatial structure of plant communities over time, especially forests (Stoll et al. 1994; Adler 1996; Kenkel et al. 1997; Mast and Veblen 1999).

A more recent example also illustrates how spatial pattern analysis can be used to test ecological hypotheses. The spatial segregation hypothesis of Pacala (1997) argues that intraspecific aggregation due to dispersal limitation (Hubbell 2001) leads to aggregated, intraspecific patterns and segregated interspecific patterns. As a consequence, the average plant in a community will compete mostly with con-specific plants at a local scale (Pacala 1997). The resulting segregation of species prevents (or retards) competitive exclusion and contributes to coexistence (Pacala 1997; Murrell et al. 2001). A recent study by Raventós et al. (2010) found support for the segregation hypothesis based on point-pattern analysis. They studied a long-term data set of fully mapped seedling emergence and subsequent survival after experimental fires in a Mediterranean gorse shrubland. They used these data to test whether the observed spatial patterns were consistent with the segregation hypothesis, examining community dynamics from early seedling emergence to the establishment of a mature community. The initial pattern of seedling emergence showed a pattern in agreement with the segregation hypothesis; seedlings of individual species were aggregated and the clusters of individual species were spatially segregated (De Luis et al. 2008). The interesting question, however, was to determine whether the initial segregation did indeed prevent interspecific competition from becoming dominant. To address this, Raventós et al. (2010) employed a suite of point-pattern analysis techniques to characterize the key spatial interactions related to observed patterns of mortality. They determined the probability of mortality as a function of the distance to conspecific versus heterospecific plants and found that mortality was controlled almost entirely by intraspecific interactions. Dead plants were aggregated together, but segregated from surviving plants, indicating two-sided scramble competition. Additionally, spatial interactions were density-dependent and changed their sign over the course of time from positive to negative when plants grew to maturity.

These two examples demonstrate that the attempt to understand the potentially complex interactions among processes and mechanisms that may cause spatial patterns can benefit from a tight integration between ecological theory and analysis of spatial patterns. One approach toward doing so would be to analyze observed spatial patterns to characterize their structure in as much detail as possible (e.g., identify the degree to which there is intraspecific clustering and interspecific segregation), and subsequently use this characterization to derive specific hypotheses and theories about the underlying processes (e.g., the spatial segregation hypothesis; Pacala 1997). Alternatively, one may use general theory to derive specific hypotheses about the development of spatial patterns for a given ecological system. Predicted patterns are then confronted with spatial data to infer the existence of underlying processes and identify the scale at which those processes are operating.

The *underlying assumption of point-pattern analysis* is that spatial patterns may conserve an imprint of past processes, constituting an "ecological archive" from which we may recover information on the underlying processes (Wiegand et al. 2003; Grimm et al. 2005; McIntire and Fajardo 2009). If spatial interactions and mechanisms are indeed important for the dynamics, assembly and diversity of an ecological system we would expect that spatial patterns should conserve a signature of the underlying processes that operated in generating the patterns (Moloney 1993). Because ecological processes and systems are dynamic in nature, this approach would ideally require several "snap-shot" patterns. Analysis of two snap shots has been carried out in plant communities for the analysis of recruitment and mortality, but longer series of snapshots are rare. For example, the establishment of a worldwide network of megaplots (25–50 ha) of tropical and temperate forests (http://www.ctfs.si.edu), one of the most expensive data collecting and monitoring enterprises in ecology, is driven by this idea. Within these plots, every tree with a diameter at breast height larger than 1 cm is mapped, identified, and monitored every 5 years. In some plots, this has been going on for more than 30 years. Hubbell et al. (2001) summarized the original premise for creating and monitoring the first of these plots at Barro Colorado Island (BCI) in Panama by stating "whatever coexistence mechanisms were operating in the BCI forest, they should leave a spatial signature that could be detected by making explicit maps of individual tree locations in the BCI forest, p. 860." Figure 1.2 provides an example of a snapshot of a fully mapped mega-plot in a tropical forest. These patterns contain information on the internal organization of the system, but in a coded form. The difficult task is to decode this information into ecological understanding and point-pattern analysis provides a sophisticated tool for doing so.

1.1.2 The Pattern-Process Link

Although spatial patterns may contain abundant information on the internal organization of ecological systems, decoding the signal from spatial patterns is challenging. Potential problems are caused by the following well-known phenomena (McIntire and Fajardo 2009): (1) substantially different processes may create the same spatial patterns (i.e., the problem of equifinality; Bertalanffy 1968; Levin 1992; Barot et al. 1999; Wiegand et al. 2003);

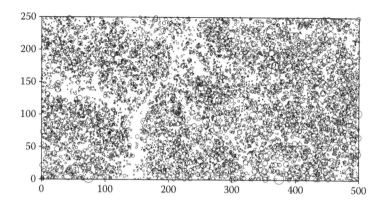

FIGURE 1.2
Trees from a fully mapped forest plot, Sinharaja, Sri Lanka. The figure presents a 250×500 m subsection of the original 500×500 m plot with the circles representing all trees having a diameter at breast height (dbh) greater than 1 cm. The diameters of the circles are proportional to dbh.

(2) causality may not be straightforward because several dynamic processes may interact and modify the spatial patterns in a complex manner (Peterson and Squiers 1995; Dovčiak et al. 2001; Wiegand et al. 2009); (3) well-defined, nonrandom processes can produce patterns indistinguishable from apparent random assembly (Cale et al. 1989; Molofsky et al. 2002); and (4) processes may also be the result of specific spatial patterns rather than being the source of the pattern (Stoll and Prati 2001; Getzin et al. 2008).

Because of these problems, a perspective has prevailed that suggests that analyses of spatial patterns *per se* might be insufficient to identify the processes underlying observed ecological patterns and subsequent experimental manipulations would be required to disentangle the relationships. However, McIntire and Fajardo (2009) recently argued that even if these phenomena would have elements of truth, their inverses are not necessarily false: "a single process *can* create a single precise pattern (e.g., Schurr et al. 2004; Fajardo and McIntire 2007), nonrandom processes *can* create highly structured patterns (e.g., Molofsky et al. 2002; Broquet et al. 2006), and the impact of pattern on process may not act at the same scales as the impact of process on pattern and so will be discernible" (p. 47). Clearly, while the link between process and pattern is usually not perfect and may be confounded by other processes acting at diverse scales, strong spatial signals of biological organization may exist in many data sets. This should allow, at least in principle, the uncovering of the link between process and pattern, enhancing our ability to understand and predict the appropriate relationships.

We argue that much of the uncertainty about the ability of spatial point patterns to uncover the underlying processes, especially regarding points (1) and (3) above, is a consequence of (i) an incomplete description of the properties of the observed spatial patterns, (ii) insufficient focus and development of ecological hypotheses, or (iii) mistakes in the technical implementation of null models. Indeed, in ecological applications it is widespread to use just one or occasionally two summary statistics (e.g., Ripley's K) to describe the properties of the observed spatial patterns. Wiegand et al. (2013a) systematically tested the performance of combinations of commonly used summary statistics and found that only one summary statistic was required to describe patterns with near random structures, but four or five to capture the properties of patterns with more complex spatial structures. Thus, the current practice in ecology of using only one or two summary statistics runs the risk of not detecting the essential characteristics of more complex patterns.

Additionally, the majority of older studies have used only very simple null models, such as complete spatial randomness, to determine whether a pattern is clustered or regular (hyperdispersed). In evaluating the literature for our 2004 review paper (Wiegand and Moloney 2004) we were surprised by the lack of awareness concerning the importance of null models and the dearth of precise formulations of ecological questions. One of the intentions in writing the 2004 review, as indicated in the title "Rings, circles, and null-models for point pattern analysis in ecology," was to indicate the availability of a much broader range of null models and point process models that could be used to conduct better and more specific analyses. Fortunately, during the last several years there has been an increase in the use of more sophisticated null models and more precise statements of ecological hypotheses in studies utilizing point-pattern analysis. In fact, these studies often incorporate multiple *a priori* hypotheses (Schurr et al. 2004). This is clearly a positive trend, as precise ecological hypotheses and detailed characterizations of the properties of spatial patterns are needed for successful inference regarding the underlying processes producing ecological pattern and for evaluating the impact of spatial patterns on ecological processes.

As noted by McIntire and Fajardo (2009), the application of point-pattern analysis often has the problem of contending with complex chains of causality, where several dynamic processes interact and modify the spatial patterns in a complicated manner. Another problem arises from the enormous effort required to repeatedly census larger plots. As a consequence, point-pattern analysis usually employs the analysis of static snapshot patterns (but see, e.g., Wiegand et al. 1998; Felinks et al. 2008; Raventós et al. 2010) despite the fact that these patterns are the result of dynamic temporal processes. Because of this limitation, we may ultimately only be able to understand the significance of static patterns within a dynamic framework.

A formalized protocol for integrating ecological theory and analysis of spatial patterns within a dynamic perspective is to *use pattern-oriented modeling* (Wiegand et al. 2003, 2004; Grimm et al. 2005; Grimm and Railsback 2005, 2012). In this approach, the spatial patterns of ecological objects and other (nonspatial) characteristics of an ecological system, which are important for the scientific question being asked, are quantified by means of summary statistics (called "patterns"; Grimm et al. 2005). Next, general theory and data on the specific system are used to construct dynamic, spatially explicit simulation models that incorporate different processes and mechanisms hypothesized to generate the observed patterns. A prerequisite common to these models is that they need to be able to produce exactly the same type of (spatial) data as observed. Then, either competing models are simulated with many different parameterizations that represent a systematic sample of the parameter space (e.g., Wiegand et al. 2004) or optimization techniques are applied to find parameterizations that minimize the difference between observed and simulated patterns (Komuro et al. 2006; Duboz et al. 2010; Martinez et al. 2011; Hartig et al. 2011). Comparing the summary statistics of the observed data with those emerging from the simulation models allows the identification of those model versions and parameterizations that are compatible with the observations (Wiegand et al. 2004; Grimm et al. 2005; Martinez et al. 2011). If a given model version is not able to provide a match for all of the summary statistics regarded to be important, the model is still deficient and needs to be modified. Conversely, if the model matches all summary statistics, one may simplify it. Thus, the quantification of spatial patterns plays an integral part in the general research program in science, the explanation of observed patterns.

An early example of the pattern-oriented approach was provided by Jeltsch et al. (1999), who examined whether a tree population of the southern Kalahari savanna was in a phase of decline, increase, or equilibrium with respect to tree abundance. The point pattern used in this study consisted of six, replicate, spatial distributions of trees within an area of 50 ha digitized from aerial photographs. They used the *L*-function to compare observed point patterns with point patterns produced by an individual-based simulation model. A more recent example is the study of Martinez et al. (2011), which used the intensity function of an observed point pattern in model selection and parameterization. Their study applied approximate Bayesian techniques to obtain parameterizations that minimized the difference between observed and simulated patterns. Another interesting study by Calabrese et al. (2010) provides an example for a direct pattern-process link. They developed a minimalistic, spatially explicit, cellular automaton simulation model to study savanna dynamics. Each cell had two possible states, either being a tree or grass covered. By using an analytical approximation based on product densities (see Section 2.3.2), they could explicitly derive the pair-correlation function of the spatial tree pattern and link it to the underlying processes included in their model.

2

Fundamentals of Point-Pattern Analysis

This chapter is the core of the book and provides the conceptual frame-work for point-pattern analysis in ecology. We begin by briefly explaining the basic steps of point-pattern analysis in Section 2.1. This is followed by the introduction of the basic data types of interest in ecology (Section 2.2) and the summary statistics adapted to explore each of these data types (Section 2.3). We then describe the different ways that null models can be used to test ecological hypotheses or conduct exploratory analyses (Section 2.4), followed by an introduction of the general methods used to compare data and the output of point-process models (Section 2.5). We end the chapter with a discussion of the methods used to detect and deal with heterogeneities (Section 2.6).

⋆ 2.1 Fundamental Steps of Point-Pattern Analyses

2.1.1 Basic Ecological Considerations

2.1.1.1 What Is an Ecological Point Pattern?

An ecological point pattern consists of a set of ecological objects that can be characterized by their locations in space. These may be either real objects, or objects that are the result of a numerical simulation. In classical point-pattern analysis, it is assumed that ecological objects can be approximated by points and have no size or shape. The types of objects that might be explored include the locations of individual organisms, structures built by organisms—burrows, mounds, middens, seed caches—or key aspects of the environment that influence ecological interactions—water holes, lakes, "safe sites" (e.g., Cutler et al. 2008). Generally, ecological point patterns are distributed in two-dimensional (2-D) space, although analyses can also be conducted on one-dimensional (1-D) linear systems, for example, invasive species distributed along ditches (Maheu-Giroux and de Blois 2007), or in three-dimensional (3-D) systems, for example, spider webs within shrub canopies, although we currently know of no examples of ecological point-pattern analysis in three dimensions (although, see Wiegand et al. 2008 for an ecological example in a 12-D space).

The motivation for point-pattern analysis centers primarily on an underlying interest in spatially explicit ecological processes and the distribution of ecological objects resulting from those processes. For example, individual trees competing for resources might lead to a hyperdispersed pattern or localized dispersal might lead to a clustered pattern in the distribution of individuals. These tendencies might not be obvious through casual inspection. However, an appropriately designed analysis might reveal a connection between the observable pattern and the ecological processes leading to or resulting from that pattern.

In the simplest form of point-pattern analysis, all of the points are of the same type and the only characteristic distinguishing among them is their location, an example being the location of gopher mounds in a grassland (Figure 2.1a). In this case, we can apply *a univariate point-pattern analysis*. However, for many ecological questions we need to consider the relationship between different types of ecological objects, where some characteristics, in addition to position, differentiate among points. This produces a more complex situation and a careful consideration of the underlying hypotheses and types of objects involved is required in determining the appropriate form of analysis.

In some cases, the interesting question relates to the relationship between two or more types of objects. For example, we might want to determine whether two species of plants are distributed independently of one another (Figure 2.1b). In this case, we would most likely employ *a bivariate point-pattern*

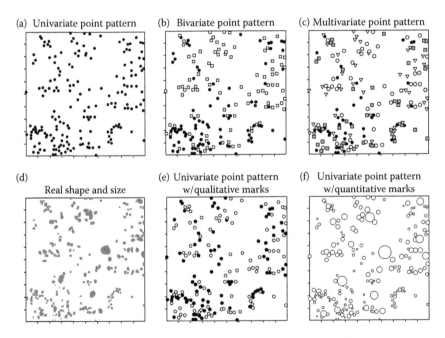

FIGURE 2.1
Categories of point patterns typically encountered in ecology.

analysis (Figure 2.1b). This can be extended to a *multivariate point-pattern* analysis when there are more than two types of objects (Figure 2.1c). Other questions, such as facilitation in plants that vary greatly in size, may require conserving the *real shape and size* of the objects (Figure 2.1d). In many real-world situations, the locations of the points may also be influenced by the environment; for example, the resources for a given tree species may be heterogeneously distributed in space and some locations may sustain a higher density of trees than others. In this case, the data sets are supplemented by *mapped covariates* (such as elevation, slope, or wetness), which can be used to characterize elements of environmental heterogeneity that may be, in some way, responsible for producing the observed point pattern.

   In other cases, the objects may be of the same underlying type, but differ in some important trait that is of ecological interest. For example, we might be interested in trees of the same species that can be categorized as alive or dead (a *qualitative mark*; Figure 2.1e) and ask whether dead trees died in clumps, as expected by scramble competition. However, the trees may also be characterized by their size (a *quantitative mark*; Figure 2.1f) and we may ask whether the size of trees shows a correlation with the distance between trees. In the latter two cases, the locations of objects are augmented by *marks*, which represent characteristics that differentiate among points of the same underlying type, and the set of ecological objects is represented as a marked point pattern. These examples demonstrate that the potential types of point-pattern data are manifold. Indeed, each data type requires a different set of methods to be appropriately analyzed. One of the objectives of this book is to address this diversity in data types. However, before we proceed with describing the basic steps of point-pattern analysis, we will briefly outline the prerequisite steps of assembling and processing point-pattern data so that they can be properly analyzed.

2.1.1.2 Assembling Ecological Point-Pattern Data for Subsequent Analysis

Prior to analyzing an ecological point pattern, data must be collected and assembled in a form suitable for analysis. Historically, the use of point-pattern analysis in ecology had the simple goal of determining the number of trees in a forest through extrapolation (Illian et al. 2008, their Section 1.3.1) or determining whether an observed pattern differed significantly from a random spatial distribution (Upton and Fingleton 1985; Cressie 1993; Dale 1999). Two approaches to the analysis were generally applied, the so-called *plot and plotless methods*. It is also important to note that the approach currently used in most cases, that is, fully mapped plots, was not the standard approach used in earlier studies, due to the technical difficulties of mapping large areas and the lack of statistical methods to analyze fully mapped data sets.

   In the plotless methods, various nearest-neighbor approaches to data collection and analysis are employed. The simplest involve locating individual objects (e.g., trees) at random and measuring the distance to the nearest object. Other, more complex forms of plotless methods have also been developed, for example, distance from a randomly selected object to its second nearest neighbor, distance between a randomly selected point and the nearest object, and so on. The interested reader is referred to the textbooks of Upton and Fingleton (1985) and Cressie (1993) for details on these methods. In plot-based methods of point-pattern analysis, data are generally collected in randomly distributed quadrats by tallying the number of ecological objects occurring in each quadrat. The data are then analyzed using approaches such as χ^2 analysis, assuming that a *completely spatially random (CSR)* distribution would produce a Poisson distribution in the sample. (For details on these and other plot-based methods, see the treatment presented in the book by Dale [1999].)

In both the plot and plotless methods, the overall goal is to determine whether the distribution of ecological objects differs significantly from a CSR distribution. If a pattern departs from being random, it does so either by being a clustered distribution or a hyperdispersed (regular) distribution. In a *clustered* distribution, objects tend to lie closer to one another than would be expected with a CSR distribution, whereas for a *hyperdispersed* distribution objects tend to be farther apart than expected with a CSR distribution. Note that both the plot and plotless methods ignore the important question of scale, as well as additional information that may exist for each point (i.e., marks).

More refined approaches to point-pattern analysis, which we will focus on in this book, have been developed during the last several decades and are increasingly employed in ecological studies. One of the major reasons for their development is that they provide a more general assessment of pattern, which may be quite complex due to processes operating at multiple scales. An example might be a species that has a strong tendency to be hyperdispersed at a local scale, due to strong competitive interactions among individuals, but is influenced by dispersal limitation leading to clustering at a broader scale. In this example, the distributional pattern of individuals would be regular at short distances (local scales), but clustered at intermediate distances. The traditional methods of point-pattern analysis briefly described above, however, can only provide a single assessment of the nature of the pattern. They can only answer the simple statistical question, "is the point pattern significantly different from CSR?" The answer to that question can depend greatly upon the sampling scheme used to assess the pattern. Plotless approaches tend to focus primarily on local scale interactions, whereas plot-based methods are quite sensitive to the size and shape of the plots used in sampling. In the example above, we might decide that the pattern is random, clustered, or hyperdispersed depending on the plot size and shape used in sampling.

Recognizing the limitations of the traditional approaches to point-pattern analysis, modern approaches to analysis have been designed to assess the structure of point patterns across a range of scales. One cost of this approach is more rigorous requirements with respect to data collection. Generally, all ecological objects of interest in the observation window must be included in the analysis. What this means in a practical sense is two things. The area from which the sample data is to be obtained (the observation window) must first be clearly delineated and then data (i.e., coordinates, and possibly marks and environmental covariates) must be collected within the entire observation window for all ecological objects of interest.

An important step in assembling point-pattern data is the *selection of the appropriate observation window* (i.e., sample area) for the scientific questions being asked. Usually, it is only feasible to fully map an observation window of limited size, which will then encompass only a part of a pattern that extends well beyond the window. For selection of the location of the sample area and its size, the scientific questions must be considered. For example, if the broader scale influence of the environment on the distribution of ecological objects is of primary interest, but not the possible interactions among objects, the observation window must be sufficiently large to encompass the typical variability in the environmental conditions. Conversely, if potential interactions among objects are the focus of the analysis, one would most likely select observation windows that minimize environmental variability and the size of the plot would generally be somewhat smaller. As a rule of thumb, the spatial structures to be analyzed should be repeated multiple times within the observation window to provide sufficient information for their characterization. For example, if the pattern shows clusters with typical diameters of say 50 m then the minimum width of the observation window should be a few hundred meters, at minimum. An observation window which encompasses just two or three clusters would be certainly too small to study the characteristics of clusters. Only the spatial structure of the pattern within clusters could be studied.

Generally, a rectangular area is used as an observation window, although it is possible to use areas delineated by complex borders to exclude heterogeneities. It is even possible to exclude areas totally enclosed within a larger observation window from the analysis, if necessary. Of course, the more complex the geometry of the area included in the analysis, the more complex the analysis, as we shall see later. To minimize edge effects (see Section 2.3.2.5), observation windows should not be too narrow. Long narrow windows are not suitable for point-pattern analysis because too many points will potentially have unobserved neighbors outside the window.

Another consideration in determining the size of the observation window is the minimum number of points of a pattern that can be sampled (and therefore the minimum window size) and result in a successful analysis. Clearly, if the observation window contains too few points, a meaningful analysis will become difficult because the variance in the estimator of the summary statistics becomes too large. *We recommend as a rule of thumb using sample sizes of more than 70 points for each pattern analyzed.* It may be possible to analyze fewer points (e.g., >50), but only under exceptional circumstances.

A more systematic approach to this issue is presented in Illian et al. (2008, p. 266), where they consider both the characteristics of the pattern (e.g., degree of clustering and overall point density) and the shape of the observation window. Here, the aim is to determine a window size that ensures that the variance in the estimators of the summary statistics is smaller than a predefined value. Two examples presented by Illian et al. (2008) indicate that, in one case, a minimum of 100 points should be used to give an acceptable relative error rate of 10% in the estimation of the intensity λ (their example 4.15, p. 266) and, in a second case, a minimum of 70 points should be used to produce an acceptable standard deviation of 0.1 in the estimation of the pair-correlation function (their example 4.16, p. 267).

Once an observation window is defined, the next step is to *collect the data* for the analysis. This requires locating all of the ecological objects of interest within the observation window and determining their coordinates (x and y locations for a 2-D sample). Additionally, other characteristics of the ecological objects such as type (e.g., species) or marks, which can be quantitative (e.g., size) or qualitative (e.g., surviving vs. dead), may be determined and included in subsequent analysis. Various approaches can be employed in determining the coordinates of the ecological objects within the observation window. A grid may be laid out in the field, and positions of ecological objects within the grid can be determined by measuring the distance and direction from the nearest node of the grid. The locations of objects can also be determined using survey equipment or GPS devices. A third approach to data collection is to determine the locations of objects indirectly from aerial photographs or satellite images, potentially allowing a very broad area to be sampled (e.g., Nelson et al. 2002; Moustakas et al. 2006; Gil et al. 2013). In all cases, it is critical to sample as completely as possible.

2.1.1.3 Homogeneity as an Assumption

Most point-pattern analyses conducted by ecologists make the assumption of homogeneity. This means that the environmental conditions and processes influencing the appearance of a point are the same everywhere within the observation window. As a consequence, the statistical properties of *homogeneous* point patterns are the same everywhere within the observation window. Or, in other words, the properties of the pattern are invariant under translation. Note that the term *stationary* is often used instead of homogeneous (see, e.g., Illian et al. 2008, p. 38). While stationary point patterns have properties that are invariant under translation, *isotropic* point patterns have properties that are invariant under rotations around the origin. Anisotropy is often driven by strong directionality in environmental conditions such as prevailing winds or steep slopes, which make processes dependent on direction (Haase 2001).

If a pattern is homogeneous, one may define a *typical point* of the pattern and develop summary statistics, which characterize the specific point configuration in the neighborhood of the typical point. Examples of commonly used summary statistics are the mean number of points within a certain distance of the typical point or the probability that the nearest neighbor is within a certain distance. Because of this convenient property, most methods for the analysis of spatial point patterns have been developed for homogeneous patterns. A majority of studies begin by attempting to define an observation window that meets the criterion of homogeneity and obvious heterogeneities, such as rock outcrops, changes in soil type or other clearly discernable boundaries, are excluded. However, heterogeneities in the point pattern are often difficult to detect without conducting appropriate tests. If the departure from homogeneity is slight then there may be no need to adjust the analysis to account for the heterogeneity in the pattern. However, in many cases, ignoring underlying spatial heterogeneity in the pattern and/or processes producing the pattern may produce a biased analysis that is quite misleading. In this case, as will be described in Section 2.6, a number of approaches can be used to test and account for the underlying heterogeneity.

Heterogeneity occurs in ecological point patterns primarily through the occurrence of heterogeneous environmental conditions in space; some areas are more suitable for a given species than others. Additionally, environmental conditions may change the relative importance of ecological processes and the characteristics of the resulting spatial pattern may differ in space (e.g., Couteron et al. 2003; Getzin et al. 2008). Another source of pattern heterogeneity could be produced by dispersal limitation, which may cause some areas of suitable habitat to be unoccupied in the observation window. Similar effects may arise by species actively expanding or contracting their distributions, for example, invasive species or species at risk of extinction. In all of these cases, heterogeneous distributions can result, even in the absence of environmental heterogeneity, since pattern formation can be sensitive to initial conditions, expansion dynamics (invasive species), and demographic processes (declining species), when species are undergoing significant shifts in their broad-scale distributional patterns.

Whether due to environmental heterogeneity or shifting distributional patterns, spatial patterns may show considerable "heterogeneity," especially when looking at larger scales. However, point-pattern analysis deals predominantly with the small-scale correlation structure of spatial patterns, that is, it exploits the characteristics of local point configurations, which are mainly influenced by interactions among ecological objects rather than by environmental determinants or transient dynamics. One may naively assume that presence of larger-scale heterogeneity may not influence the results of point-pattern analyses that deal with the small-scale correlation structure of spatial patterns. However, as discussed in more detail in Sections 2.6.3.1 and 2.6.3.3, cross-scale effects may be strong and without appropriate adjustments the analysis will produce flawed results. These issues are currently being explored in research focused on the development of methods for the analysis of heterogeneous point patterns (e.g., Baddeley et al. 2000; Hahn et al.

2003; Baddeley et al. 2005; Waagepetersen and Guan 2009). Although this is a young, highly technical, and rapidly developing research field, a number of relatively simple and practical methods exist that can be employed to analyze certain aspects of heterogeneity in a manner that is relevant for most practical applications in ecology (Section 2.6.3).

2.1.1.4 Dimensionality as an Assumption

One of the general assumptions of point-pattern analysis is that the objects being studied are composed of dimensionless points. Clearly, this is not the case for ecological objects, which nearly always have a 2- or 3-D structure. However, if the spatial scales at which interactions between objects occur are generally greater than the scale of the underlying objects, little bias is introduced by reducing the scale of the objects being analyzed to being dimensionless points. In some cases, however, this may not be so and one may need to analyze the spatial structure of *objects of finite size and real shape*. While such analyses are usually not within the scope of point-pattern analysis (this field is covered by the more general field of stochastic geometry; e.g., Stoyan et al. 1995), several methods developed for point-pattern analysis can be generalized to account for objects of finite size and shape (Wiegand et al. 2006). We consider these approaches initially in Section 2.2.4, and in more detail in Section 3.1.8.

2.1.2 Basic Steps in Conducting a Point-Pattern Analysis

2.1.2.1 Determination of Data Type

After the research question has been specified, the first step in conducting a point-pattern analysis is to determine the most appropriate data type to use in responding to the ecological questions at hand. Do we simply need to characterize the data by location, or do we need to include marks? Do the data consist of one type of data object, two types, or more? Our choice of data type, of course, may be constrained by the nature of the data, if it has already been collected. Although the choice of data type to use seems to be a rather obvious task, there are, in some cases, subtle differences that need to be considered before proceeding. For example, there are several types of pattern that comprise two types of points, but may require completely different sets of summary statistics and null models for an appropriate analysis (e.g., Goreaud and Pélissier 2003).

2.1.2.2 Selection of Appropriate Summary Statistics and Their Scales of Analysis

Once we have decided on the appropriate data types to be used in an analysis, we need to determine the most appropriate metrics (*summary statistics*) for characterizing the data. Some examples of summary statistics used in the analysis of point patterns are the commonly employed Ripley's K-function, the pair-correlation function, and the nearest-neighbor distribution function. In essence, *summary statistics quantify the statistical properties of spatial point patterns and "provide a brief and concise description of point patterns using numbers, functions or diagrams"* (Illian et al. 2008, p. 179). Of course, the summary statistics used to characterize a particular point pattern will depend on the type of data and the hypotheses being explored. We will introduce the fundamentals of summary statistics in Section 2.3, and then provide more detailed information regarding summary statistics of general importance for ecological applications in the respective sections of Chapter 4 dedicated to each particular data type. Note that most textbooks use the term "summary statistic" (e.g., Diggle 2003), whereas Illian et al. (2008) prefer the term "summary characteristic." We decided to use the term "summary statistic" because it appeared to be, for us, a more intuitive phrase for a general audience than "summary characteristic."

In general, the utility of summary statistics is that they depict the mean tendency of the data or the variability around the mean behavior, which can be tested against an appropriate null model for significance. As discussed above, traditional approaches to point-pattern analysis were independent of spatial scale and the summary statistics employed were simple scalar values characterizing the general tendency of the data. In the modern approach, a pattern is generally assessed at multiple scales, producing an array of statistics, instead of a single scalar value. For example, the popular K-function is related to the mean number of points within distance r of the typical point of the pattern (Ripley 1976). In this case, the summary statistics characterizing the pattern consist of an array of values associated with all distances r considered in the analysis. Because of the scale dependence of modern summary statistics, for many analyses we may have to first answer a key question before summarizing the data: *What are the appropriate scales of analysis?*

A range of scales from very short distances to very long ones may be assessed in a point-pattern analysis (Figure 2.2a). At shorter scales, analysis is technically constrained by the resolution of the measurements used in characterizing the data, often referred to as the grain of the data. It makes no sense to assess scales at distances shorter than the smallest distance capable of being measured, even though we could do so. Additionally, the shortest scale considered in an analysis is generally greater than the resolution of the data, as there are other considerations that are of equal importance. In most cases, it makes no sense to assess scales that are so short that the physical nature of the underlying objects prevents the co-occurrence of two objects (although in some cases there can be exceptions to this rule; see Section 3.1.8). Also, the choice of shortest scale to use in assessing a pattern may depend on how many objects are included in the data set relative to the size of the observation window. This is analogous to a consideration of sample size in traditional statistics.

At the opposite end of the spectrum, a decision has to be made as to the longest scale to analyze (Figure 2.2b,c). The rule of thumb in the analysis of

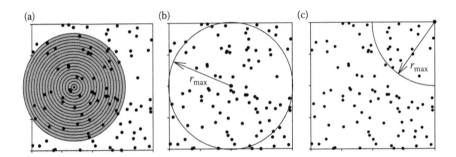

FIGURE 2.2
Range of spatial scales analyzed. (a) Rings around a point of a pattern representing distance classes (or spatial scales r) to be analyzed, starting with the shortest distance available, continuing by multiples of that distance up to a maximum distance of r_{max}. The scale of r_{max} is generally constrained to be half of the length of the shortest side of the observation window. (b) *Best case scenario* for r_{max}: the entire circle of a point located at the geographic center will be within the observation window. (c) *Worst case scenario* for r_{max}: only one-fourth of the circle of a point located in a corner will be located within the observation window.

spatial pattern is often that the *upper limit of the analysis is constrained to be half the length of the shortest dimension of the observation window*. The reasoning for this cutoff comes from time-series analysis, based on the need to be able to sample at least a full wavelength for frequencies that are to be resolved. This logic does not apply directly to point-pattern analysis. However, at scales greater than half the length of the smallest dimension of the observation window, a large fraction of the possible comparisons from an arbitrary location would fall outside of the observation window. This has the potential for introducing serious bias in the resulting analysis. Under the best-case scenario, a location being analyzed would be located at the center of the plot and, at a scale r set to half the length of the shortest side of the observation window, the entire disk of radius r surrounding the location would be located within the observation window (Figure 2.2b). Under the worst-case scenario, with the location at a corner, only one-fourth of the disk would be within the observation window (Figure 2.2c). This has two consequences. First, estimates of the summary statistics become unreliable if too much of the area that should be sampled falls outside of the mapped area, even with the application of standard "edge correction" techniques (see Section 3.1.1). Second, points not actually mapped (i.e., points located just outside the observation window) may influence the observed point configuration. If the proportion of observed versus nonobserved points at a given scale becomes too small, the true effects of spatial interactions cannot be accurately determined. As a consequence, it is a generally accepted practice to limit spatial scales at the upper end to less than half of the smallest linear dimension in the observation window. In practice, it is often better to limit the upper scale even more.

Once the upper and lower limits of the analysis are defined, the number of spatial scales to be assessed between these bounds needs to be determined.

Generally, all multiples of the smallest scale less than or equal to the upper bound are included in the analysis (Figure 2.2a). The net result is that an estimate of the spatial structure is obtained at each scale analyzed. This provides a quantitative description of the point pattern that must be further evaluated to determine how the results are to be interpreted. This is done by developing suitable null models against which the observed pattern can be compared.

2.1.2.3 Selection of Appropriate Null Models

As described above, a spatial point pattern is composed of a set of locations of ecological objects in space, which may carry additional information characterizing the objects (called marks). Generally, the distribution of points is assumed to be generated by some form of stochastic mechanism (Diggle 2003). *Point processes* are mathematical models characterizing these stochastic mechanisms. An important use of point processes is to generate stochastic realizations of point patterns with known properties. Comparison and/or fitting of point processes to observed patterns allows for detailed characterization of their statistical properties. Point processes thus provide methods to explore the nature of observed point patterns. For example, a very simple point process, for a univariate point pattern, posits that objects are distributed in space at random, with an equal probability of occurring anywhere. It also posits that the location where a point occurs is independent of the locations of other objects already present. Statisticians refer to the properties of objects (points) produced in this way as being identically and independently distributed (i.i.d.) and, if strictly followed, a process with these properties will produce a *completely spatially random (CSR)* pattern.

Point processes can be used as null models for assessing point-pattern data. *Null models* in ecology are usually viewed as pattern-generating models that randomize certain aspects of the data while holding others fixed. The objective of the null model is to create patterns that are expected in the absence of a particular ecological mechanism of interest (Gotelli and Graves 1996). A null model in this role thus functions as a standard statistical null hypothesis for detecting pattern. The null model is used to produce a pattern that is tested against observed data to determine whether there is structure in the data that does not exist in the null model. Null models in this context are thus employed to confirm that an observed pattern departs from the null model, the inference being that the data contain more structure than the null model. For example, CSR, which represents the complete randomization of a univariate pattern, is the simplest point process that can be used to assess univariate point patterns. The point process in this case serves as a benchmark for determining whether there is any spatial structure in the observed spatial pattern.

Using point processes in the manner described above is generally done in an *exploratory context*, where the objective is to identify the statistical properties of the observed point pattern, as much as is possible (Wiegand et al. 2009). The goal is to gather information on the biological organization present in a spatial point pattern that can be used to generate hypotheses on the underlying processes. Once these hypotheses are generated, point-process models can then be extended beyond the classical use of null models by constructing them to create patterns that are expected in the presence (not absence) of particular mechanisms, such as clustering of offspring around parents or trends in density related to environmental gradients. In this case, a point-process model is constructed to represent a hypothesized process. The task here is to use a *confirmatory approach* by developing *a point-process model that produces patterns that cannot be distinguished from the observed pattern*, rather than showing that the patterns are statistically different. As we will see in the next section, these two complementary modes of point-pattern analysis require different methods to compare observed and simulated data.

2.1.2.4 Comparison of Data and Null Models

Once a point process for assessing an observed pattern is developed, the patterns it generates can be compared to the observed pattern. There are generally two steps in this comparison. First, as with most statistical analyses, a test is constructed to determine whether or not the observed pattern differs significantly from the point-process model. The test involves calculation of one or more summary statistics characterizing the output of the point-process model and the observed pattern. If the observed pattern differs significantly in one or more summary statistics from the expectation under the point process, we can then assess the direction that the observed pattern deviates from the expectation in the second step.

In essence, a point process acts in this context as a dividing line between two tendencies leading away from the expectation. If the point process corresponds to the simplest possible randomization of the given data type (e.g., CSR for univariate patterns), we will refer to the line drawn in the sand between these two tendencies as the *fundamental division*. The fundamental division is used to determine whether the data contain a signal that can be distinguished from pure stochastic effects for the given data type with the observed sample size. On one side are patterns that deviate by having points that occur, on average, closer to one another than would be expected for a purely random point process. For univariate patterns, this is often referred to as a *clustered pattern*. On the other side are patterns that deviate by having points that occur, on average, farther apart from one another than would be expected. In this case, the pattern is referred to as being *hyperdispersed*.

In many cases, null models arising from the fundamental division are too simple to characterize the complex spatial structure of point patterns in an ecologically meaningful way. They may also disregard important information that is available for assessing the point pattern. For example, if we analyze a bivariate pattern and know that one of the patterns has a habitat dependency, we should not use the standard test for independence, which is the fundamental division for bivariate patterns. This is because the test for independence may create too much variability in the null model by randomizing the pattern globally and not accounting for the underlying habitat association. However, constraining the randomization of this pattern with the known habitat association may reveal a small-scale departure from independence.

Another case where the use of the fundamental division is not very informative is the situation of a pattern that contains a complex cluster structure, with more than one critical scale of clustering (e.g., Wiegand et al. 2007c, 2009). In this case, it may already be evident from visualization that the pattern is not random. Subsequently, we may be interested in extracting more precise information on the critical scales of clustering, since they may conserve information on the underlying processes (e.g., Wiegand et al. 2009). Such analyses require the use of more sophisticated point-process models, with known properties, that serve as a "point of reference" or *benchmark*. The parameters of these models, if they fit the data well, can be used to succinctly summarize the properties of the point pattern.

In other cases, we may strive to test specific ecological hypotheses and their affects on the development of the spatial structure of a point pattern. In this case, we need to implement these hypotheses by selecting specific point processes that are equivalent to the ecological process hypothesized (e.g., Shen et al. 2009). Thus, the critical issue in hypothesis testing and characterizing complex spatial patterns in detail is to pick the appropriate point process to test against the observed pattern. This is relatively simple for univariate point patterns, but becomes more difficult for marked point patterns and for bi- and multivariate point patterns, as we shall see.

Once a point process is chosen for an analysis, we use it to assess whether or not the observed pattern deviates from the null expectation and, if it does, we then determine in which direction it deviates. Generally, *Monte Carlo simulation* is used to determine whether a pattern is significantly different from a point process used as a null model. Multiple simulations based on the point process are conducted, generating simulation envelopes by calculating the same summary statistics for each simulation as used in assessing the observed pattern. If the observed pattern falls outside of the 95% simulation envelopes produced through the Monte Carlo study, this is used as evidence that the observed pattern significantly departs from the null model (the expected pattern). Note that more refined approaches to test goodness of fit, which consider all scales analyzed simultaneously, are also possible (e.g., Loosmore and Ford 2006; Grabarnik et al. 2011). The comparison of the null model to the observed pattern generally indicates how the observed pattern deviates from the null model. This is important information that can be used to refine the point process applied in assessing the observed pattern, if desired. Exploring deviations of observed patterns from the expectations produced by point processes will be especially effective if several summary statistics are compared with those of the observed pattern, as this tests different aspects of the observed pattern against the expectation (Wiegand et al. 2013a).

In concluding this section, we would like to emphasize the point that the concept of a fundamental division is key to the analysis of more complex types of point pattern, since testing against the fundamental division is generally the first step in exploratory data analysis. If the pattern does not depart from a pattern generated by chance (i.e., realizations of the fundamental division), then there is usually not much point in conducting further analyses. Some exceptions to this rule are cases where the fundamental division disregards preexisting knowledge of some component influencing the point pattern (see above). Also, confrontation of data by the fundamental division, although exploratory in nature, in many cases produces deep insight into the structure hidden in the data and helps to formulate hypotheses for conducting field research to further elucidate the processes underlying observed patterns. Alternatively, assessing the relationship between the observed pattern and the fundamental division helps in the selection of more refined point processes that will allow a more detailed description of the spatial structure hidden in the data or provide a mechanism for testing alternative hypotheses. In a later section, we will expand this concept to encompass the different data types treated in this book.

BOX 2.1   SUMMARY OF THE TYPICAL STEPS IN A POINT-PATTERN ANALYSIS

1. PRIOR TO ANALYSIS

After the research question has been specified, the first step is to compile information on possible drivers of the spatial structures of the system of interest, to determine the appropriate data types for analysis, and to critically inspect a visualization of the pattern based on the ecological information available. Next, the point-pattern *data type* must be determined and it must be decided whether the pattern is, in good approximation, *homogeneous* or if it is *heterogeneous*. In the latter case, one may need to devise techniques to address the heterogeneity.

2. SELECTION OF APPROPRIATE SUMMARY STATISTICS

In a second step, appropriate *summary statistics* need to be selected and calculated to concisely summarize the properties of the spatial structure of the observed spatial point pattern. We recommend the use of several summary statistics simultaneously.

3. SELECTION OF APPROPRIATE NULL MODELS AND POINT-PROCESS MODELS

Selection of an appropriate *null model* is at the heart of point-pattern analysis. A null model is used to test against data to determine whether there is structure in the data that does not exist in the null model. Consequently, a first step is to find out whether the data contain a signal that can be distinguished from pure stochastic effects. The *fundamental division* is the corresponding null model, which represents stochastic patterns that arise through randomization of the data, conditional on a given data type and sample size. In contrast, more general *point-process models* may be used to represent more complex spatial structures or a hypothesized process. The task is to characterize the complex structure of the patterns in more detail. Thus, a null model or point-process model may serve two purposes related to exploratory versus confirmatory analysis.

- Within an *exploratory context*, the researcher wants to describe the statistical properties of the observed point pattern as closely as possible to gather information on biological organization present in a spatial point pattern that can be used to generate hypotheses on the underlying processes. Exploratory analysis may also use refined null models with known properties, such as cluster processes, as a benchmark to summarize the properties of the point pattern.
- Within a *confirmatory context*, the null model is the point-process implementation of an ecological hypothesis (or several alternative hypotheses) about the stochastic properties of the observed point pattern. Comparison of the summary statistics of the data with those of the simulated patterns (step 4) allows a determination of the hypothesis that receives the most support from the data.

4. COMPARISON OF DATA AND NULL MODELS

The fourth step involves *comparison of observed data and null models*, that is, the observed and simulated patterns are compared to test whether the hypothesis holds. For this purpose, summary statistics are used as test statistics and are calculated for both the observed pattern and the simulated patterns. Statistical tests are then used to evaluate the match between the test statistics of the observed and simulated patterns or to evaluate the support for different hypotheses depending on the given data.

★ 2.2 Data Types

A point pattern in ecology consists of the coordinates of ecological objects within a given observation window. In addition to coordinates, we may have information that characterizes objects by traits such as stem diameter or condition (e.g., alive or dead). This information is called a *mark*, and a data set comprising coordinates with marks is referred to as a *marked point pattern*. If no marks are present, we call the pattern an *unmarked point pattern*. Marks can be either *quantitative* (such as the diameter of a stem) or *qualitative* (i.e., categorical such as dead vs. surviving). As a result of these considerations, different types of data may be used in a point-pattern analysis (Figure 2.3), each requiring the use of different summary statistics and analytical methods. Because of the tight linkage between data type and methods, it is important to begin with a classification of data types before going into greater detail.

Here, we provide a hierarchical classification of the general categories of point-pattern data types as summarized in Table 2.1. We also briefly describe the *fundamental division* associated with each data type. The fundamental division is basically the pattern produced by the simplest null model for a given data type—the simplest model being one that produces a pattern that can be considered a full randomization of the points, conditioned on the type of point data and the number of points. The fundamental division consequently provides a dividing line between two alternative possibilities for significant spatial relationships with opposing tendencies, for example,

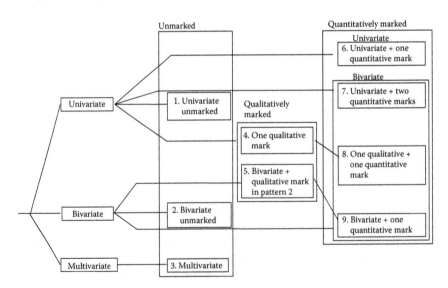

FIGURE 2.3
Scheme presenting the data types for point patterns treated in the book.

TABLE 2.1

Data Types of Point Patterns Treated in the Book and the Corresponding Fundamental Divisions

	One Type of Point	Two Types of Points	More than Two Types of Point
(a) Unmarked patterns	Data type 1	Data type 2	Data type 3 (> 2 types of points)[a]
Fundamental division	Random pattern (CSR)	Independent patterns	Neutral pattern
Case I	Aggregation	Attraction	Accumulator
Case II	Hyperdispersion	Repulsion or segregation	Repeller
Example	Adult acacia trees in a savanna	Juvenile trees relative to adult trees	All adult trees categorized by species in a forest
(b) Qualitative marks	—	Data type 4	Data type 5 (3 types of points)[b]
Fundamental division	—	Random labeling	Random labeling of the qualitatively marked pattern
Case I	—	Several cases	Mark occurs more likely close to points of focal pattern
Case II	—	Several cases	Mark occurs less likely close to points of focal pattern
Example	—	Surviving and dead seedlings	Surviving and dead seedlings relative to adult trees
(c) Quantitative marks	Data type 6 (one type of point)	Data types 8 and 9 (two types of points)[a]	Data type 7 (one type of point, two qualitative marks)[a]
Fundamental division	Independent marking	Independent marking between two types of points	Independent marking between two types of marks
Case I	Positive correlation or stimulation	Positive correlation or facilitation	Positive correlation
Case II	Negative correlation or inhibition	Negative correlation or inhibition	Negative correlation
Example	Trees of different size in a forest	Sizes of surviving and dead trees (type 8) or sizes of trees of two different species (type 9)	Abundance of two orchid species on host trees
(d) Objects	Data type 10 (one type of object)	Data type 10 (two types of objects)	
Fundamental division	Randomly placed objects	Independent patterns	
Case I	Aggregation	Attraction	
Case II	Hyperdispersion	Repulsion or segregation	
Example	Shrubs of one species	Two species of shrubs	

Note: The data types refer to the cases in the scheme of Figure 2.3.

[a] These cases did not yet reach the level of more general theories; there exist only a very few examples.

[b] Three types of points, a focal pattern *f* and a qualitatively marked pattern.

clustered versus hyperdispersed for univariate patterns and attraction versus repulsion for bivariate patterns.

Of practical relevance in ecology are four general categories of data, all of which can be further subdivided (Figure 2.3): *unmarked point patterns*, which involve only the locations of objects of possibly different types; point patterns that include *qualitative marks*, as well as location; point patterns that include *quantitative marks*, as well as location; and "point patterns" that cannot be adequately analyzed as true points but require consideration of *real shape and finite size*. The latter case represents the situation where an appropriate analysis cannot ignore the fact that the ecological objects of interest are 2-D by nature, that is, the general simplifying assumption that treats ecological objects as being zero dimensioned points would lead to a strongly biased interpretation. This case falls a bit outside of what is traditionally considered to be standard point-pattern analysis, but some of the tools of point-pattern analysis can be appropriately used to interpret relationships within this additional data type.

Another consideration occurs if environmental heterogeneity is present and has a significant impact on the spatial structure of the point pattern. Each of the four basic data types may still be analyzed, but environmental covariates must be incorporated into the analysis. In this case, the point process producing the fundamental division still randomizes the pattern of a given data structure in a way that produces no spatial relationships among objects, but the randomization of the objects or the marks is conditioned on the covariate.

2.2.1 Unmarked Point Patterns

Unmarked point patterns are those that are appropriately studied by simply noting the locations of objects and their type, in the case where more than one type of object is being considered. For example, we may want to characterize the spatial relationships among gopher mounds as a way of understanding behavioral relationships within gopher colonies (e.g., Klaas et al. 2000). Or, we may want to examine the spatial relationships between two or more plant species to determine whether the distributional pattern is consistent with either the process of facilitation or competition. From these examples, it is evident that unmarked point patterns may comprise one to many types of ecological objects. However, the type must be an *a priori* label (e.g., different species), and not an *a posteriori* label (such as surviving vs. dead), which was created by a process that acted *a posteriori* over an existing pattern. How we proceed in analyzing these types of data depends to a large extent on the number of object types involved. In general, the analyses applied can be subdivided among three basic categories: (i) *univariate* point patterns; (ii) *bivariate* point patterns; and (iii) *multivariate* point patterns. Univariate point patterns are the most studied in ecology and a broad range of methods is available to examine their properties. The methods available for bi- and multivariate point patterns are much more restricted, but in many cases they can be developed in a manner analogous to the methods used in analyzing univariate patterns.

2.2.1.1 Univariate Point Patterns (Data Type 1)

Univariate point patterns involve only one type of ecological object, for example, individuals of a single species or ant mounds produced during a set period of time (Figure 2.1a). The relevant ecological questions for these types of data generally revolve around developing an understanding of the basic processes responsible for producing observed spatial patterns. In this case, the fundamental division can be represented by a point process that distributes objects with equal probability to any location within the observation window, independent of the location of any other object. The resulting pattern is often referred to as *complete spatial randomness* (CSR) or, more formally, as a homogeneous Poisson process, and it represents the fundamental division between patterns that are aggregated or hyperdispersed (Table 2.1; data type 1).

2.2.1.2 Bivariate Point Patterns (Data Type 2)

Bivariate point patterns involve two distinct types of objects, such as two different species or two different life history stages that were *created a priori by a different set of processes* (Figure 2.1b). The relevant ecological questions for these types of data generally involve detecting and understanding *possible interactions* between the two types of objects. In this case, the fundamental division is given by *independence* of the two patterns, which separates the two alternative cases of attraction and repulsion (or segregation). Analysis of bivariate point patterns is relatively rare in ecology compared with univariate analysis, although a basic interest in ecology is to reveal the role of interspecific interactions in allowing the coexistence of species (e.g., He and Duncan 2000; Koukoulas and Blackburn 2005).

Unfortunately, inappropriate point processes are often used to represent the fundamental division of independence. For example, many studies have used CSR for both component patterns to test for independence. However, independence needs to be represented by a point process that maintains the univariate spatial structure of the two component patterns, but breaks apart possible spatial dependence between the two patterns. The reason for this is that the general interest is in the relationship between the two patterns and not in the spatial structure of the composite pattern. In practice, it is not easy to derive a point process for the fundamental division producing independence between two patterns, where the spatial structure of each individual pattern is not altered. One possibility is to employ a toroidal shift (Lotwick and Silverman 1982), which produces patterns that approximately maintain the univariate spatial structure of the two, univariate component

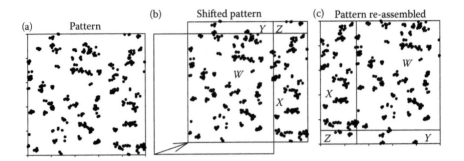

FIGURE 2.4
Toroidal shift used in the implementation of the independence null model for bivariate patterns. (a) Original pattern within a 1000 × 1000 observation window. (b) Pattern shifted by adding a value of 300 to all *x*-coordinates and a value of 100 to all *y*-coordinates. (c) Rearrangement
of the shifted pattern, following toroidal geometry. The areas *X*, *Y*, and *Z*, which now lie outside the observation window, are reallocated following the rules of toroidal geometry. Points
lying within zone *X* in the shifted pattern are relocated within the observation window by
subtracting a value of 1000 from the *x*-coordinate. Points lying within zone *Y* are relocated by
subtracting 1000 from the *y*-coordinate. Points in zone *Z* are relocated by subtracting a value of
1000 from both the *x*- and *y*-coordinates.

patterns (Figure 2.4). As a null model, one of the two component patterns
is shifted in its entirety by first adding a fixed random vector (dx, dy) to the
coordinates (x, y) of each point in the pattern (Figure 2.4b), and then reassembling the shifted pattern by wrapping it on a torus (Figure 2.4c). The
latter step is necessary because the initial shift of position will move some
points outside of the original observation window. Using the geometry of a
torus moves the displaced points back into the observation window during
reassembly. More complex, but superior techniques for generating repre-
sentations of independence exist. These use the nonparametric technique of
pattern reconstruction (Section 3.4.3.7) or fit parametric point processes to
the data (see Section 2.5.2).

2.2.1.3 Multivariate Point Patterns (Data Type 3)

Multivariate point patterns involve several distinct types of objects, such as
different species, that are created *a priori* by different sets of processes (Figure
2.1c). The relevant ecological questions for these types of data generally
involve detecting and understanding *nonrandom spatial structures in diversity*.
A basic issue that could be explored in this context is, for example, to determine whether species tend to form intraspecific clusters with interspecific
segregation (a spatial pattern hypothesized by the segregation hypothesis to
promote coexistence; Pacala 1997), or whether species tend to be well mixed
throughout a plot. A basic goal in analyzing multivariate point patterns is to
examine the number of species to be found in a subarea *A* of the observation

window W. This is related to the species-area relationship (SAR), one of the fundamental summary statistics of α diversity (Shimatani and Kubota 2004). To represent the "detectability" of an individual species s in calculating the SAR, so-called spherical contact distributions are used. They yield the probability that an individual of species s is located within distance r of an arbitrary test location within W Shimatani and Kubota (2004). Another basic goal in analyzing multivariate point patterns would be to examine patterns of β diversity (i.e., the amount of change in species richness as a function of spatial displacement). This can be done, for example, by estimating the expected similarity in species composition of two small plots, which are separated by distance r (i.e., the distance-decay hypothesis; Nekola and White 1999). To represent species turnover for calculating β diversity, the so-called mark connection functions can be used. They yield the probability that two individuals separated by a distance r belong to the same species (e.g., Chave and Leigh 2002; Condit et al. 2002) and are related to the spatially explicit Simpson index (Shimatani 2001). Such spatial structures in diversity are likely to be sensitive to key spatial processes and mechanisms such as dispersal limitation, species interactions, habitat association or Janzen–Connell effects, making them a powerful tool for testing mechanistic hypotheses on species coexistence.

Traditionally, diversity data are collected within sampling plots at different locations. For such data, local diversity can be compared among locations in a nonspatial way by using *diversity indices* such as species richness, the Shannon index, or the Simpson index. Similarity in species composition between plots as a function of their geographic locations can also be calculated using several measures of dissimilarity such as the Sørensen index, the Jaccard index, or the Bray–Curtis index. However, as outlined by Shimatani (2001) and Shimatani and Kubota (2004), continuous data of fully mapped plots contain substantially more information than data from separate, disjunct sampling plots because the information on the precise locations of individuals within sampling plots is lost. Methods of pattern analysis could, in principle, be applied to these types of data, so that we can take advantage of the information contained in fully mapped multivariate patterns. However, the complexity of such patterns has hindered the development of a theory of truly multivariate summary statistics (but see Podani and Czaran 1997; Shimatani 2001; Shimatani and Kubota 2004; Wiegand et al. 2007b; Pommerening et al. 2011).

In practice, applications of point-pattern analysis to multispecies patterns have been mostly limited to an examination of pairwise interactions among species (e.g., Kubota et al. 2007; Lieberman and Lieberman 2007; Wiegand et al. 2007a; Perry et al. 2009; Wang et al. 2010). The few attempts to develop summary statistics for multivariate patterns have been, in essence, logical extensions of summary statistics for uni- or bivariate patterns, as we shall see when we delve into this topic in greater detail in Sections 3.1.5 and 4.3.

Difficulties in developing a general framework for point-pattern analysis of multivariate patterns arise, not only with respect to summary statistics that characterize the different aspects of spatial structure in diversity, but also in the formulation of the fundamental division and appropriate null models. Two generalizations of the null model of independence may be used, depending on the nature of the summary statistic. First, if the summary statistic is based on the sum of bivariate summary statistics, spatial structure in diversity can be best analyzed from the viewpoint of a focal species. In this case, the fundamental division would require the locations of individuals of all species, other than the focal species, to be fixed in space. In contrast, the locations of the individuals of the focal species are randomized in such a way that the univariate spatial structure of the pattern is conserved, while any potential interdependence with the other patterns is removed (Wiegand et al. 2007b; Section 4.3.3).

The second generalization of the null model arises for summary statistics that are not centered on a focal species, but may average univariate point-pattern summary statistics over all species, or incorporate more traditional grid-based statistics, such as the species–area relationship (Shen et al. 2009). In the latter example, the number of species is counted in quadrats. In this case, a suitable fundamental division is the *"null community,"* where all species are randomized, conserving certain aspects of their univariate spatial structures. In a study by Shen et al. (2009), which followed this approach, "null communities" were constructed based on point-process models that conserved some aspects of the observed univariate spatial structures attributable to the action of processes, such as habitat association or dispersal limitation. This allowed the identification of the mechanisms that produced the largest contribution to the observed multivariate patterns, at least as conditioned on the summary statistic used (Wang et al. 2011; Section 4.3.4).

2.2.2 Qualitatively Marked Point Patterns

Qualitatively marked patterns carry a qualitative (discrete or categorical) mark that is descriptive of their state, such as surviving versus dead seedlings. What differentiates qualitatively marked point patterns from bivariate or multivariate point patterns (e.g., trees of two or more species) is that the latter are composed of different objects that were *created a priori*, whereas a qualitative mark is produced by a process *acting a posteriori* over a given univariate (unmarked) pattern (Figure 2.1b vs. 2.1e). In essence, this means that we are defining qualitative marks here as something created conditional on a given, preexisting pattern; as an example, a point pattern exists (e.g., trees in a forest) and we are interested in the characteristics of the point process that generates the marks (e.g., surviving vs. dead) over the existing pattern (Figure 2.1).

This approach contrasts with the one taken by Illian et al. (2008), who consider bi- and multivariate point patterns to lie within the framework of marked point patterns. They distinguish among quantitative marks (e.g., size), *a posteriori* qualitative marks (e.g., surviving vs. dead), and *a priori* qualitative marks (e.g., different species) (Illian et al. 2008: p. 296). While the classification used by Illian et al. (2008) is correct from a more technical point of view (e.g., a species identifier can be considered a mark that characterizes a tree in the forest), it obscures a fundamental difference in ecological questions related to *a posteriori* and *a priori* marking and would make the ecological classification of questions and methods more difficult. We find it simpler, and more ecologically meaningful, to distinguish between unmarked (uni, bi-, and multivariate) and qualitatively marked point patterns and thereby follow the guidance of Goreaud and Pélissier (2003, Table 2.1) to avoid "misinterpretation of biotic interactions."

2.2.2.1 Random Labeling (Data Type 4)

The simplest case to consider for patterns consisting of qualitative marks is one with two categories that differentiate among points of a pattern, for example, alive versus dead. The relevant ecological questions for this data type generally try to develop an understanding of the basic processes responsible for producing the marks. In the example of surviving versus dead trees, the interest is often to determine whether mortality events occur in a spatially correlated way over a given univariate pattern of trees. The fundamental division for qualitatively marked patterns can be represented by a point process that shuffles the marks (e.g., dead) with equal probability among all of the objects of the pattern (e.g., tree), independent of the location of any other object, while conserving the number of marks in each category that characterize the observed pattern. The marked point process for this case is referred to as *random labeling* (Table 2.1 and Figure 2.3; Data Type 4). Note the fundamental difference between this case and the tests used to explore the independence of bivariate patterns. The test for independence of bivariate patterns relocates the points of pattern 2 potentially anywhere within the study area (but conditional on the spatial structure of pattern 2), whereas random labeling shuffles marks over the points of the univariate pattern, which is fixed in space. It is interesting to note that most applications of random labeling in the ecological literature have explored the random mortality hypothesis presented in Kenkel (1988).

There are several ways in which the qualitative marks of a univariate pattern may be correlated. For example, the mark "dead" may be clustered within the pattern leading to the segregation of the points with the mark "dead" from the points carrying the mark "surviving" (as predicted by scramble competition in the case of dead vs. live trees; Kenkel 1988), or we may have the attraction of disparate marks (dead trees are found in the vicinity of living trees as would be expected with contest competition; Kenkel 1988). In general, a variety of test statistics are needed to capture the different ways in which a qualitatively marked pattern can depart from the random labeling null model (Jacquemyn et al. 2010). The different cases and the correspondingly appropriate summary statistics are presented in more detail in Section 4.4.1.

2.2.2.2 *Trivariate Random Labeling (Data Type 5)*

Trivariate random labeling is a natural extension of data type 4 (i.e., qualitatively marked patterns). The relevant ecological questions for this data type investigate the influence of an additional (antecedent) pattern on the processes responsible for producing the marks of a qualitatively marked pattern. For example, Biganzoli et al. (2009) investigated whether fire-caused mortality of the shrub species *Eupatorium buniifolium* was higher in the vicinity of its competitor shrub *Baccharis dracunculifolia*. This was a natural hypothesis because the short-lived shrub *B. dracunculifolia* is readily killed by fire, but establishes abundantly from seed in recently burned sites (i.e., a "killer strategy"), whereas the long-lived shrub *E. buniifolium* has the ability to survive and resprout vigorously after fire. Killing its neighbors could therefore be a strategy of *B. dracunculifolia* to escape from competition of nearby *E. buniifolium* shrubs. This example involves an antecedent pattern that is hypothesized to exert an influence on a qualitatively marked pattern. It is called "trivariate" because it involves three types of points: the antecedent pattern (the killer species *B. dracunculifolia* in the example) and the two types of points of the qualitatively marked pattern (i.e., surviving and dead individuals of the resprouter species *E. buniifolium*). The fundamental division for trivariate random labeling is *random labeling* of the qualitatively marked pattern (Table 2.1 and Figure 2.3, data type 5), which represents the case of no influence of the antecedent pattern on the distribution of marks. This null model separates cases of positive and negative impacts of the antecedent pattern on the qualitative marking. The appropriate summary statistic for this data type is a mark connection function (see Section 2.3.6), which provides the conditional probability that a typical *E. buniifolium* shrub located at distance *r* from a *B. dracunculifolia* shrub is killed by fire. Further ecological examples of this data type are given in De la Cruz et al. (2008), Biganzoli et al. (2009), Jacquemyn et al. (2010), and Raventós et al. (2010).

2.2.3 Point Patterns with Quantitative Marks

The size of trees, number of fruits produced by shrubs, or the number of epiphytic orchids located on a host plant are examples for quantitative marks. The relevant ecological questions for these types of data generally revolve around whether the values of the marks are spatially correlated. Quantitatively marked patterns are rarely analyzed in the literature and, when they are, usually only the simplest case of univariate patterns with one attached mark is considered (but see Raventós et al. 2010; Wiegand et al. 2013b). However, a plethora of potential data types is possible for quantitative marks, because one or more quantitative marks may be attached to several of the data types discussed so far (Figure 2.3). There is also no theoretical restriction to the number of different marks a given pattern may carry. However, in this book we will restrict our presentation to the theory of univariate quantitative marks as developed by Illian et al. (2008, their Section 5.3.3) and will develop analogous theories for patterns with two marks, which may be of general interest in ecological applications. However, we will leave the theory of multivariate marks for future research. In the following, we briefly introduce the four cases for quantitatively marked patterns considered in this book.

2.2.3.1 Univariate Quantitatively Marked Pattern (Data Type 6)

One quantitative mark (e.g., size) attached to a univariate pattern (e.g., trees of a given species) is the simplest case of a quantitatively marked pattern. The ecological questions related to this data type explore issues concerning distance-dependent correlations in the marks. The fundamental division for this data type can be represented by a point process that shuffles the given marks with equal probability to any object of the univariate pattern, independent of the location of any other object. The resulting null model is given by *independent marking*, which acts to separate several forms of spatial structure in the marks. For example, we may have inhibition in cases where trees lying close together are smaller than average and, if trees close together are larger than average, this may indicate some form of mutual stimulation among nearby trees. However, the marks may also show positive and negative correlations. In the case of a positive correlation at small distance r, marks tend to have similar magnitudes; for example, larger ant mounds are closely associated with larger mounds and smaller mounds are associated with one another. Conversely, in the case of negative correlations we may have some form of inhibition among nearby points; for example, a mound found in the vicinity of a large mound tends to be small and vice versa. Similar to random labeling, several test statistics are used to describe the different possible departures from independent marking (see Sections 3.1.7.1 and 3.1.7.2).

2.2.3.2 Bivariate Quantitatively Marked Patterns (Data Types 7, 8, and 9)

More complex situations arise for quantitatively marked patterns, if they require bivariate summary statistics to be appropriately analyzed

(Figure 2.5). The simplest, bivariate generalization of the case of a univariate pattern with one quantitative mark (i.e., data type 6) is the case a univariate pattern with two quantitative marks (i.e., data type 7). The ecological questions related with this data type generally consider distance-dependent correlations between the two types of marks. For example, the univariate pattern could be composed of host trees, which are occupied by two different species of orchids. The two quantitative marks would be the abundance of the two orchids on a given tree (Figure 2.5a). A suitable fundamental division for this data type shuffles one of the marks randomly over the univariate pattern but leaves the other mark unchanged. The fundamental division, in this situation, is the case of no spatial correlation between the two types of marks, which differentiates between the cases of positive or negative inter-mark correlation.

A second, bivariate generalization of the data type of a univariate pattern with one quantitative mark (i.e., data type 6) is the case where the points of the univariate pattern also have one qualitative mark (i.e., data type 8; Figure 2.5b). The ecological questions related with this data type generally consider the distance-dependent correlation between the marks of the two types of points. For example, if the qualitatively marked pattern would be given by dead and surviving trees and the quantitative mark would be the size of the trees, we may ask if the difference in size between a dead tree located near a surviving tree is on average greater than expected for all pairs of surviving and dead trees present in the plot. This would be expected if dead trees were suppressed by neighboring trees. The data type in this case can

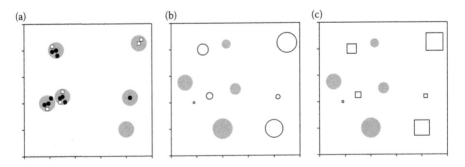

FIGURE 2.5
Bivariate quantitatively marked patterns. (a) Data type 7. One univariate pattern (e.g., host trees, indicated by the gray disks) which is augmented by two types of quantitative marks that may represent orchids of two different species which are hosted by the trees (represented by the closed disks and the white quadrats). (b) Data type 8. One qualitative mark (e.g., surviving vs. dead, indicated by gray and white color) and one quantitative mark (size, indicated by the size of the circles). (c) Data type 9. A bivariate pattern (e.g., two different tree species, indicated by gray circles and white quadrats) and one quantitative mark (size, indicated by the size of the circles and quadrats).

be randomized in two ways, depending on whether the quantitative or the qualitative mark is shuffled over all points. If the focus of the analysis is on the correlation structure within the quantitative mark, the suitable fundamental division would be *independent marking,* where the quantitative mark (e.g., size) is shuffled over all points of the pattern. However, if the interest focuses more on differences arising because of the qualitative mark, the suitable fundamental division would be *random labeling* where the quantitative mark (e.g., size) is fixed and the qualitative mark (e.g., dead) is shuffled over all points of the pattern.

Finally, a third generalization of the data type of a univariate pattern with one quantitative mark (i.e., data type 6) is the case where a bivariate pattern carries one quantitative mark (i.e., data type 9; Figure 2.5c). An example of this data type is given by two competing tree species for which we know the size. Analogous to data type 8, the ecological questions related with this data type ask about distance-dependent correlations between the marks of the two types of points. However, the randomization of this data type requires *independent marking of only one of the two component patterns,* while holding the mark of the other "antecedent" pattern constant. Shuffling the mark over all points of the bivariate pattern would not be appropriate because the two patterns are *a priori* of a different nature and may show quite different mark distributions. For example, to test whether the proximity of a particular tree species influences the size of trees of a second species, we can use a bivariate r-mark correlation function (introduced in Section 2.3.7) that gives us the mean size of trees of the second species that are located at distance r away from individuals of first species. This is analogous to trivariate random labeling but uses a quantitative mark instead of a qualitative mark.

More complex quantitatively marked patterns—Complex marked data types are possible. One interesting example is given by a case of genetic parental analysis, which is used to estimate pollen dispersal kernels and has important applications in ecology (Niggemann et al. 2012; Section 3.2.2.3). For such studies, one harvests seeds from trees, which are then known to be the female parent, and uses genetic analysis to determine the pollen donor (i.e., male parent) from among all potential pollen donors (i.e., all male trees in the study area). In this case, there is a complex data type that involves female trees i, male trees j, and marks m_{ij}, which are the number of seeds of female parent i fathered by male j. This data set can be used *to nonparametrically estimate the pollen dispersal kernel.* The basic interest in analyzing these data is to find out whether pollen dispersal occurs in a spatially correlated manner (i.e., pollen of nearby males has a higher probability of reaching a female than pollen from males located further away). The fundamental division is *random mating,* in which the marks m_{ij} for a fixed female parent i are randomly shuffled over all potential male parents j. This process is then independently repeated for each potential female parent i. This example shows that marked patterns of direct interest for ecology may show quite complex structures and that there is no theoretical restriction on the number of qualitative and quantitative marks a pattern may carry. Recent research even generalizes the concept of marked point patterns to functional marks (e.g., Comas et al. 2011). In this case, the attached mark is not a single number (e.g., size), but a function (e.g., annual growth increment during the last 10 years).

2.2.4 Objects of Finite Size and Real Shape (Data Type 10)

The approximation of objects as points may not be appropriate in some analyses and one may need to analyze the spatial structure of *objects of finite size and real shape* (Figure 2.1d). While such analyses are usually not within the scope of point-pattern analysis (this field is covered by the more general field of stochastic geometry; e.g., Stoyan et al. 1995) it has important applications in ecology. Because several methods developed for point patterns can be directly generalized for objects of finite size and shape (Wiegand et al. 2006), we include them here. In univariate analyses, the fundamental division is given by *completely randomly distributed objects*, as opposed to aggregation or hyperdispersion, and in bivariate analyses, the fundamental division is given by *independently distributed patterns* as opposed to attraction or repulsion (segregation). Note that the framework of marked point patterns can consider size and measures of shape of an ecological object as a quantitative mark. The size of objects is also sometimes considered in the null model within the framework of uni- or bivariate point patterns (i.e., hard- and soft-core processes producing hyperdispersion as described in Section 2.3.2.3).

2.2.5 Covariates

Although the fundamental objective in point-pattern analysis is to investigate small-scale spatial interactions between objects, in many cases the locations and marks of objects are also impacted by extrinsic environmental factors. If environmental conditions are spatially correlated, the resulting point patterns may show spatial correlation in the locations or marks even without intrinsic spatial interactions among the objects of study. In cases where we can model the impact of the environment on the locations or marks of the objects, we may be able to modify the fundamental division to account for the underlying influence of the environment.

For univariate patterns influenced by extrinsic factors, we need to model how the probability of occurrence of an object of the pattern depends on the environment. An appropriate framework for this is to apply species distribution or habitat modeling (Elith and Leathwick 2009), in which maps of environmental factors are statistically related to the presence,

presence/absence, or abundance of ecological objects in space. The fundamental division will then distribute the points independently of each other, conditioned on the probability of occurrence of an object as influenced by extrinsic environmental factors. The resulting null model is generally referred to as a *heterogeneous Poisson process* and will be introduced later in the book (Sections 2.6.1.2, 2.6.3.3, and Box 2.3).

Assessment of the independence of bivariate patterns where one or two of the component patterns is influenced by an environmental covariate is complicated. The complication arises because the inference must be conditional on the univariate spatial structures of the two component patterns. One approach is to leave a focal pattern unchanged and use the technique of pattern reconstruction (see Section 3.4.3.7; Jacquemyn et al. 2012b) for randomizing the second pattern, additionally conditioning on the intensity function $\lambda(x)$ of the second pattern, which may be driven by environmental factors (Wiegand et al. 2013a). In this way, both the probability of occurrence, as affected by environmental covariates, and the small-scale spatial structure of the second pattern are conserved. Or, in other words, the larger scale spatial structure of the second pattern is conserved, but the typical small-scale structures appear at somewhat displaced locations compared with the observed pattern. A similar approach can be used for multivariate patterns, for example, in an analysis where the goal is to analyze the average species richness in neighborhoods centered on the individuals of a focal species (i.e., the individual–species relationship, ISAR; see Section 3.1.5.2). In this case, a suitable approach for generating realizations of the fundamental division is to leave all other patterns unchanged and to use pattern reconstruction, as in the bivariate case, for the focal pattern, conditioned on the probability of occurrence as affected by environmental covariates. This is a nonparametric extension of the parametric approach presented in Shen et al. (2009), where a heterogeneous Thomas process is fitted to the data of each species.

A similar strategy can be pursued for marked patterns in which the process that distributes the mark may be influenced by the environment (e.g., Raventós et al. 2012). For example, the probability of mortality may not only depend on interactions with neighboring trees, but may also depend on local environmental factors, such as the amount of moisture available. In this case, the task is to model how overall tree mortality depends on localized environmental variables. The probability of mortality can be calculated, for example, by using logistic regression based on the environmental variables, but the potentially confounding effect of spatial autocorrelation (i.e., interactions among points) must be factored out. In this case, a point process for which the probability of mortality is given by the results of logistic regression provides a suitable fundamental division.

⋆ 2.3 Summary Statistics

After defining one or several hypotheses to be tested and placing the data in the appropriate format, the first step in the analysis of a point pattern is to characterize the pattern using the appropriate summary statistics. *Summary statistics* quantify the statistical properties of spatial point patterns and are the key for exploring their spatial structure. Dissimilar data types require, in general, different summary statistics. Additionally, spatial point patterns may exhibit complex spatial structures at multiple scales. As such, it cannot be expected that one summary statistic will always suffice in characterizing a pattern. For example, it is well known that several, substantially different point processes (e.g., cluster processes) yield identical pair-correlation functions, but differ in their nearest-neighbor statistics (e.g., Diggle 2003; Wiegand et al. 2007c). To investigate this essential issue further, Wiegand et al. (2013a) conducted a detailed analysis of how well frequently used summary statistics, or a combination of them, capture the spatial structure of univariate patterns. They found that, in general, four to five summary statistics were required to capture the spatial structure of "real world" univariate patterns. Their findings revealed that the current practice in ecology of using only one or two summary statistics risks missing the detection of the essential characteristics of more complex patterns, which would be unmasked if a more diverse set of summary statistics were employed.

In this section, we present the underlying ideas of the basic (univariate) summary statistics commonly used in point-pattern analysis. We also present mark-connection functions and mark-correlation functions, which are used in analyzing marked point patterns, and explain the principal idea of how the basic univariate summary statistics can be expanded to consider marks. However, because summary statistics adapted to bi- and multivariate patterns often arise as a natural extension of univariate summary statistics, we discuss this more complex topic later in Section 3.1 and the respective sections of Chapter 4 reserved for the different data types.

Before we discuss the summary statistics, a few general definitions and considerations need to be introduced. We also note that some of the more technical details required for calculating univariate summary statistics, such as estimators and edge correction, will be introduced in a later section of the book (Section 3.1).

Locations of individual objects will be referred to in vector format as x_i, where x_i is the location of object i:

$$
\begin{aligned}
x_i &= (x_i) && \text{in 1-d} \\
x_i &= (x_i, y_i) && \text{in 2-d} \\
x_i &= (x_i, y_i, z_i) && \text{in 3-d}
\end{aligned}
\tag{2.1}
$$

etc.

We will focus primarily on the 2-d case because it is the most appropriate for the majority of ecological studies. For a particular analysis, the number of objects and the total area of the observation window are generally fixed. *Total area* of the observation window will be denoted as A and the *number of objects* for a univariate pattern as n. Notation for the number of objects for bivariate and multivariate analyses will be indicated by a subscript where appropriate.

Illian et al. (2008) classify summary statistics in two ways; they may not only be *numerical* or *functional*, but also *location-* or *point-related*. Point-related summary statistics describe the spatial structure of the pattern from the viewpoint of the points of the pattern, often the typical point, and summarize properties of the neighborhood of the points. This viewpoint is especially amenable to ecological analyses of plants because point-related summary statistics represent the "plant's-eye view" of a community (Turkington and Harper 1979; Law et al. 2009). In contrast, location-related summary statistics are evaluated from test locations, which are placed within the observation window independently of the points in the pattern.

Numerical summary statistics are indices that characterize the spatial pattern with a single value. For example, the intensity λ, which is the mean number of points per unit area, is a location-related numerical summary statistic. Another well-known, point-related, numerical summary statistic is the mean distance m_D to the nearest neighbor. The classical Clark–Evans index CE (Clark and Evans 1954) divides m_D by the corresponding expectation of m_D under CSR. It indicates hyperdispersion if the nearest neighbor is further away than expected by CSR (i.e., $CE > 1$) and aggregation if $CE < 1$. Earlier work in ecology and forestry, in particular, used indices as summary statistics (Pielou 1959; Pommerening and Stoyan 2006). Classical examples of location-related indices are the index of dispersion and Pielou's index of nonrandomness (Pielou 1959), both of which have also been used to determine whether a pattern is hyperdispersed or aggregated (Illian et al. 2008: p. 195f).

Modern point-pattern analysis uses mostly *functional summary statistics*, which characterize a pattern as a function of scale. Important point-related, functional summary statistics are, for example, Ripley's K-function $K(r)$, which is proportional to the mean number of points within distance r from the typical point, and the distribution function $D(r)$ of distances r to the nearest neighbor. Both characterize the neighborhood of the typical point. An important location-related, functional summary statistic is the spherical contact distribution $H_s(r)$, which is the distribution of distances r from an "arbitrary" test location to the nearest point of the pattern. It is conceptually related to the nearest-neighbor function $D(r)$, but characterizes the holes in the pattern, whereas $D(r)$ characterizes clusters.

Summary statistics can be further subdivided into first-order statistics (e.g., the intensity function), second-order statistics (e.g., the K-function), higher-order statistics (e.g., the $T(r)$ function), nearest-neighbor statistics (e.g., the distribution function $D(r)$ of the distances to the nearest neighbor), and morphological functions (e.g., the spherical contact distribution $H_s(r)$). In the following, we present important examples of each of these for the basic univariate case.

2.3.1 First-Order Statistics

2.3.1.1 The Intensity Function

The intensity λ of a univariate point pattern is the mean number of points per unit area. From a theoretical point of view, the intensity is a *first-order statistic*. This can be understood when referring to an important theoretical quantity, the so-called *product density*. In general, the kth-order product density $\rho^{(k)}$ describes the frequency of possible configurations of k points in space (Illian et al. 2008, p. 32). For $k = 1$, the first-order product density $\rho^{(1)}(x)$ is obtained. An interpretation of $\rho^{(1)}(x)$ is as follows: if we have a disk with infinitesimal area dx and center x then the probability that a point occurs within the disk yields $\rho^{(1)}(x)dx$ (Figure 2.6a). The first-order product density $\rho^{(1)}(x)$ therefore characterizes the point density around location x and is commonly called the *intensity function* $\lambda(x)$. Because it characterizes the configuration of an individual point ($k = 1$), it is referred to as a "first-order" statistic. In general, the intensity function will depend on the location x because the local point density will vary with location. Note that the intensity function is a location-related summary statistic.

Homogeneous point patterns are characterized by a constant intensity function, whereas (first-order) heterogeneous patterns may show a spatial

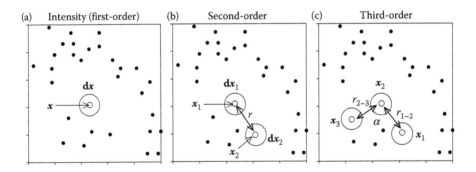

FIGURE 2.6
Product densities. (a) First-order product density represents the probability of a point being located within a randomly placed disk of infinitesimal area dx. (b) Second-order product density represents the probability of two points being located in two randomly placed disks separated by distance r. (c) Third-order product density represents the probability of three points being located in three randomly placed disks characterized by two distances (r_{1-2} and r_{2-3}) and one angle (α).

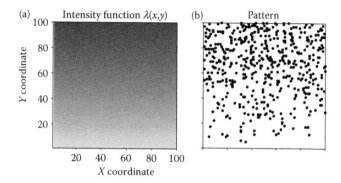

FIGURE 2.7
The intensity function and its effect on random point distributions. (a) The intensity function $\lambda(x, y) = 0.01y$ for a linear gradient in the probability of point occurrence. Darker shading corresponds to an increase in the intensity. (b) A point pattern created with the intensity function shown in (a) through application of a heterogeneous Poisson process (Section 2.6.3.4, Box 2.3).

trend or some other form of spatially nonconstant intensity function. For homogeneous patterns, an estimate $\hat{\lambda}$ of the intensity is the number of points n within the observation window W divided by the area of the window A, that is, $\hat{\lambda} = n/A$. For a homogeneous, univariate point pattern, the intensity function can be used to determine the expected number of points at any sampling scale, that is, $n = \hat{\lambda} A$, where A can take any value (e.g., Figure 2.9a).

Heterogeneous point patterns can occur due to various factors, such as spatial trends in the background environment or nonstationary interactions among points (i.e., the interactions depend on location within the observation window). In the first case, the local point density is a function of spatial location, that is, $\lambda(x)$ is required instead of λ, the latter being sufficient for the homogeneous case. Figure 2.7a shows a simple intensity function $\lambda(x) = 0.01y$ produced by a linear gradient in the y-direction. Figure 2.7b shows a point pattern generated through a heterogeneous Poisson process model parameterized with the intensity function $\lambda(x)$ (see Section 2.6.1.2 for an explanation of how this type of modeling is conducted).

2.3.2 Second-Order Statistics

Unlike the case for first-order statistics of point processes, where there is only one relevant summary statistic, that is, the intensity function, there are several possible second-order statistics. The form that a second-order statistic takes depends on the nature of the comparisons to be made and the hypotheses being considered; it will differ between univariate, bivariate, multivariate, and marked point processes. Here, we will briefly outline the general characteristics of second-order statistics for the basic, univariate case.

Second-order statistics are based on the spatial relationships of pairs of points. In this case, the "second-order product density" $\rho^{(2)}(x_1, x_2)$ depends on two locations x_1 and x_2. If we consider two disjoint disks of infinitesimally small size dx_1 and dx_2, centered on locations x_1 and x_2, respectively (Figure 2.6b), then $\rho^{(2)}(x_1, x_2)\, dx_1\, dx_2$ approximates the probability that one point of a given point process occurs within the first disk and a second point occurs within the second disk. For homogeneous and isotropic point patterns (i.e., the statistical properties of the pattern are the same over the entire observation window and are independent of the direction considered), the product density depends only on the distance $r = \|x_1 - x_2\|$ between the two points x_1 and x_2. This is a convenient property, which greatly simplifies the interpretation and estimation of second-order statistics. For a univariate, CSR (i.e., homogeneous Poisson) process, the second-order product density can be easily calculated. Because the intensity of CSR is λ, the probability that one point of a CSR pattern occurs within the disk dx_1 yields λdx_1 and the probability that a second point occurs within the second disk dx_2 yields λdx_2. Because the points are independently placed under CSR, we find $\rho^{(2)}(x_1, x_2)\, dx_1\, dx_2 = \lambda^2\, dx_1\, dx_2$ and therefore $\rho^{(2)}(x_1, x_2) = \lambda^2$.

2.3.2.1 Pair-Correlation Function and the O-Ring Statistic

The *pair-correlation function* $g(r)$ is closely related to the second-order product density. For homogeneous patterns it is the second-order product density normalized by dividing by λ^2, producing $g(r) = \lambda^{-2}\, \rho^{(2)}(r)$. The advantage of normalizing is that, for CSR, the pair-correlation function yields a value of $g(r) = 1$, independent of the intensity λ of the pattern. If the pattern shows a tendency toward aggregation, there will be, on average, more close-by points at smaller distances r than expected by CSR. Thus, the probability $\rho^{(2)}(x_1, x_2)\, dx_1\, dx_2$ that one point of a given point process occurs within the first disk and a second point occurs within the second disk at small distance r away from the first disk is greater than expected under CSR. Thus, we will observe for small distances r that $g(r) > 1$. Conversely, if the pattern shows a tendency toward hyperdispersion, there will be on average fewer close-by points at smaller distances r than expected by CSR. We will therefore observe for small distances r that $g(r) < 1$. Thus, the pair-correlation function is a summary statistic that provides the dividing line between hyperdispersion and aggregation in a simple way (i.e., $g(r) = 1$).

A summary statistic that is closely related to the pair-correlation function is the *O-ring statistic* $O(r)$, since $O(r) = \lambda g(r)$ (Table 2.2; Figure 2.8a) (Wiegand and Moloney 2004). It also has a very intuitive interpretation for homogeneous patterns that is especially useful for ecological applications; it is the expected density of points at distance r from the typical point of the pattern. Because of this property, it is sometimes called the neighborhood density

TABLE 2.2

Relationships among Second-Order Statistics and Other Summary Statistics, Employing a Distance-Based Approach under Conditions of Homogeneity and Isotropy

Metric	Symbol	Relationships	Value under CSR
Ripley's K-function	$K(r)$	$K(r) = \int_{t=0}^{r} g(t)(2\pi t)dt$	πr^2
Pair-correlation function	$g(r)$	$g(r) = \dfrac{dK(r)}{dr} \Big/ 2\pi r$	1
K2-function	$K2(r)$	$K2(r) = \dfrac{dg(r)}{dr}$	0
Product density function	$\rho^{(2)}(r)$	$\rho^{(2)}(r) = \lambda^2 g(r)$	λ^2
O-ring	$O(r)$	$O(r) = \lambda g(r)$	λ
L_1-function*	$L_1(r)$	$L_1(r) = \sqrt{\pi^{-1} K(r)}$	r
L_2-function	$L_2(r)$	$L_2(r) = \sqrt{\pi^{-1} K(r)} - r$	0
Cumulative distribution function of distances to nearest neighbor	$D(r)$	$\lambda K(r) = \sum_{k=1}^{\infty} D^k(r)$	$1 - \exp(-\lambda \pi r^2)$
Distribution function of distances to kth neighbor	$d^k(r)$	$\lambda g(r) 2\pi r = \sum_{k=1}^{\infty} d^k(r)$	$\dfrac{2(\lambda \pi r^2)^k}{r(k-1)!} e^{-\lambda \pi r^2 \dagger}$
Mean distance to kth neighbor	$nn(k)$	$nn(k) = \sum_{r=1}^{\infty} r d^k(r)$	$\dfrac{1}{\sqrt{\lambda}} \dfrac{k(2k)!}{(2^k k!)^2}$ $\approx \dfrac{1}{\sqrt{\pi \lambda}} k^{1/2 \ddagger}$
Spherical contact distribution	$H_s(r)$		$1 - \exp(-\lambda \pi r^2)$

Note: The parameter λ, when it occurs in an equation, is the intensity of the point process.
* The L_1-function is sometimes used in the statistical literature; we use the L_2 function in this book referring to it as the L-function.
† Illian, J. et al. (2008, p. 76).
‡ Thompson (1956).

function (Perry et al. 2006). In a sense, the O-ring statistic is a point-related analogue to the location-related intensity function $\lambda(x)$. In the theoretical literature it has also been referred to as the Palm intensity function (e.g., Tanaka et al. 2008).

2.3.2.2 Ripley's K-Function

Up to now, the most popular second-order summary statistic in the ecological literature has been *Ripley's K-function* $K(r)$. The idea behind the K-function

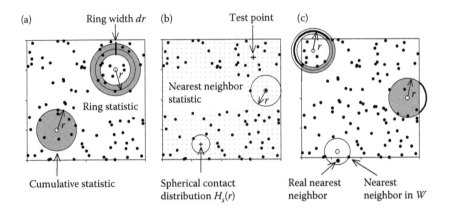

FIGURE 2.8
Schematic representation of basic summary statistics and edge correction methods. (a) *Point centered statistics.* Cumulative second-order statistics, such as Ripley's K-function, are based on determining the number of points expected within a disk (neighborhood) of radius r around the typical point (small open circle). Ring statistics, such as the O-ring or pair-correlation function, are based on the expected density of points within a ring of radius r and width dr located at distance r from the typical point of the pattern. (b) *Nearest-neighbor statistics.* Nearest-neighbor statistics are based on the distance of the closest neighboring point to a typical point of the pattern. This concept can be expanded to include statistics characterizing distance to the kth nearest neighbor. The spherical contact distribution is conceptually similar, but is based on the distance of the closest neighboring point to a test point (gray crosses) instead of the typical point. (c) *Edge correction methods.* Edge correction compensates for unrecorded points in the incomplete rings or disks (ring on top left or middle right) when calculating second-order statistics. In the case of nearest-neighbor statistics, the nearest neighbor may potentially be located outside the observation window.

is, in fact, quite straightforward (Figure 2.8a). It is based on determining the expected number of points occurring within a distance r of the typical point, which is given by the quantity $\lambda K(r)$ where λ is the intensity of the pattern. The expectation for a CSR process is that there should be $\lambda \pi r^2$ points within r units of the typical point, since this is the area of a disk of radius r around the point multiplied by the intensity of the process (Figure 2.9a). The influence of the intensity λ is removed in calculating the K-function, so that the expected value under CSR yields $K(r) = \lambda^{-1} (\lambda \pi r^2)$ or, more simply, $K(r) = \pi r^2$. The K-function is, in fact, the cumulative version of the pair-correlation function, that is,

$$K(r) = \int_{t=0}^{r} g(t)(2\pi t)dt \qquad (2.2)$$

This can be understood intuitively: the quantity $\lambda g(r)$ gives the density of points within a ring with radius r and width dr, centered at the typical point of the pattern (Figure 2.8a). Thus, multiplying the neighborhood density $\lambda g(r)$ by the area of the ring $2\pi r \, dr$ yields the number of points within the ring with

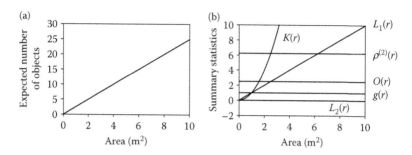

FIGURE 2.9
Expected values under CSR. (a) Relationship between sample area and number of objects expected in the sample for a CSR process with $\lambda = 2.5$; the slope of the line is equal to the intensity of the underlying CSR process. (b) Graph of the expectation for univariate statistics under CSR. Values for $O(r)$ and $\rho^{(2)}(r)$ are based on $\lambda = 2.5$ and would differ for different intensity values.

radius r, and adding up all nonoverlapping rings up to radius r and width dr yields the total number of points within distance r of the typical point.

A major disadvantage of the K-function is that its expectation under CSR increases at the rate of r^2. Because of this, a transformation is often applied to produce what is known as the L-function (Besag 1977). There are two variants of the L-function:

$$L_1(r) = \sqrt{\pi^{-1}K(r)} \tag{2.3}$$

$$L_2(r) = L_1(r) - r \tag{2.4}$$

Under CSR, $L_1(r) = r$ and $L_2(r) = 0$ (Table 2.2). Spatial statisticians often prefer to use the variant $L_1(r)$ of the L-function, which increases linearly with r because it emphasizes the cumulative nature of the K-function-based summary statistics (Figure 2.8a). In contrast, most ecological studies utilize the $L_2(r)$ variant, as the expectation (always equal to zero for a CSR pattern) is easier to plot and assess against the observed pattern. In all cases, a value at distance r that is greater than the expected value indicates a tendency toward clustering, whereas a value less than the expected value indicates a tendency toward hyperdispersion. Assessing whether or not a value differs significantly from the expectation will naturally depend on developing a suitable significance test (Section 2.5).

2.3.2.3 Differences between the K- and Pair-Correlation Functions

A major problem with the K-function and its L-transformed relatives can be traced to the cumulative nature of the statistic (Figure 2.9b). Effects at shorter scales obscure the effects at broader scales. Although this is not apparent in the case of $L_2(r)$ when plotting the expectation under CSR (Figure 2.9b) we shall see that it is a problem with this statistic when analyzing real patterns, which can lead to problems of interpretation (see Figure 2.10). Because of this, the pair correlation and related O-ring function potentially provide a better assessment of pattern across scales,

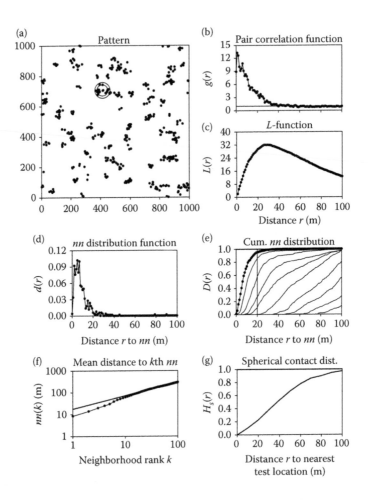

FIGURE 2.10
Summary statistics for a clustered point pattern. (a) Pattern comprising 500 points within a 1000×1000 m^2 observation window. The pattern was created by a Thomas process (Box 2.5) composed of 100 randomly distributed clusters with an approximate radius of 20 m. Eighty-six percent of the points have a distance less than 20 m from the cluster center and each cluster consists of 5 points, on average. Panels (b) through (g) present the results of several summary statistics applied to the pattern in (a). Panel (e) presents a sequence of 10 distribution functions $D^k(r)$ of the distances to the kth neighbor with $k = 1, 2, 4, 6, 8, 12, 16, 20, 25,$ and 30, where $D^1(r)$ is indicated by solid dots.

particularly in the context of an exploratory analysis. In fact, Dietrich Stoyan, a spatial statistician, has been actively promoting the use of the pair-correlation function, as opposed to the *K*-family of statistics. This issue has only recently been recognized in the ecological literature (Condit et al. 2000; Wiegand and Moloney 2004; Schurr et al. 2004; Perry et al. 2006); as a consequence the *K*-function family of statistics still predominates in most published studies.

The easiest way to get a sense of the differences between the pair correlation and *L*-functions is by way of example. Figure 2.10a provides a map of a clustered point pattern produced by a Thomas process, a pattern composed of clusters of points distributed at random within the observation window (Box 2.5). The map is accompanied by the statistical analysis of the pattern, using a variety of summary statistics, some of which remain to be introduced (Figure 2.10b–g). *What is the information about the pattern that is provided by the pair-correlation function?* The pair-correlation function $g(r)$ is the expected point density at distance r from a typical point of the pattern, relative to the intensity λ of the pattern. For the fundamental division of univariate patterns (i.e., a random pattern) the value of the pair-correlation function yields $g(r) = 1$. A value of $g(r) > 1$ within a small neighborhood r indicates that the typical point configuration contains more points at distance r from the typical point than expected for a completely random pattern; that is, the pattern is clustered at this spatial scale. The reverse is true for $g(r) < 1$. Numerical estimates of the pair-correlation function based on the pattern in Figure 2.10a yield a value of approximately $g(r) = 13$ at a neighborhood radius of 2 m (Figure 2.10b). We can interpret this to mean that the average neighborhood density 2 m away from the typical point is 13 times higher than that expected under a random pattern.

The shape of the estimated pair-correlation function also reveals information regarding the range of clustering. Point densities at increasingly larger distances from the typical point in a locally clustered pattern will usually decline and the indication of clustering disappears as the pair-correlation estimates approach a value of 1. (If the pair-correlation function does not decline to a value of 1 at larger scales r we most likely have a heterogeneous pattern.) In our example, the values of the pair-correlation estimates decline rapidly with increasing scale r, and clustering disappears roughly after a distance of 35–40 m (Figure 2.10b). In fact, if we define cluster size to be the mean radius of a cluster, the pair-correlation function approaches a value of 1 at a distance r corresponding to the diameter of the clusters (i.e., the largest distance between two points within the cluster). In the example, 40 m is the average diameter of the clusters produced by the point process used to construct the pattern. Later we will see that the shape of the estimated pair-correlation function contains even more information that can be used to reveal cluster characteristics, such as the typical size of a cluster, the number of clusters, and the average number of points within a cluster (see Section 4.1.4).

In contrast to the pair-correlation function, *the shape of the estimated L-function is difficult to interpret* (Figure 2.10c). The first observation is that the estimated values for the L-function are clearly greater than zero for all scales r shown in Figure 2.10c. This is in sharp contrast to the results of the pair-correlation analysis, which indicates that clustering disappears at distances of approximately 40 m away from the typical point. The reason for this apparent contradiction is that the L-function counts all points within distance r from the typical point (Figure 2.8a) and compares the result to the expectation of a completely random spatial pattern, whereas the pair-correlation function looks only at those points located at distance r away from the typical point (Figure 2.8a). Clearly, high neighborhood densities at small scales influence the values of the L-function at larger scales (but not the values of the pair-correlation function). This is the "memory" of the K-function (Wiegand and Moloney 2004; Section 2.6.3.1). In essence, what occurs is that the L-function increases in value as long as there is still, on average, some clustering at increasing scales. Eventually, the magnitude of the L-function declines at a scale beyond which there is no more local clustering. This can be seen when comparing the plot of the estimated pair-correlation function (Figure 2.10b) with that of the estimated L-function (Figure 2.10c). The maximum value of the L-function occurs approximately at a distance of $r = 30$ m, which is approximately the scale at which the pair correlation starts to drop to low values. Thus, the L-function has the largest values at scales where clustering disappears, not at the scales where clustering is the strongest. This particular behavior is somewhat counterintuitive and bears the danger of being misinterpreted. A related problem is that the numerical values of the L-function are relatively small where clustering is strongest (up to say 10 m in Figure 2.10c) and large at scales where clustering begins to disappear. This may lead to underestimation of small-scale clustering when using the K- or L-function.

If we now investigate a pattern containing points that are hyperdispersed at a scale of 20 m (Figure 2.11a), we find that numerical estimates of the pair-correlation function yield a value of 0 for distances smaller than 20 m (Figure 2.11b). This indicates that the pattern shown in Figure 2.11a does not contain interpoint distances smaller than 20 m. Such patterns are referred to as *"hard core"* patterns because they behave like a pattern produced by randomly placing nonoverlapping disks with radius R within the observational window, where the centers of the disks are treated as the point pattern. Indeed, the pair-correlation function approximates a step function with a value of 0 for distances at scales below twice the hard-core distance $R = 10$ m (i.e., the minimum distance between two disk centers) and approximates a value of 1 for larger distances (Figure 2.11b). If the nonoverlapping disks have a range of sizes, the centers of smaller disks would come closer to each other than the centers of larger disks. In this case, the pair-correlation function would be zero up to twice the minimal radius of the disks and then increase monotonically to a value of 1. The latter value would be reached at

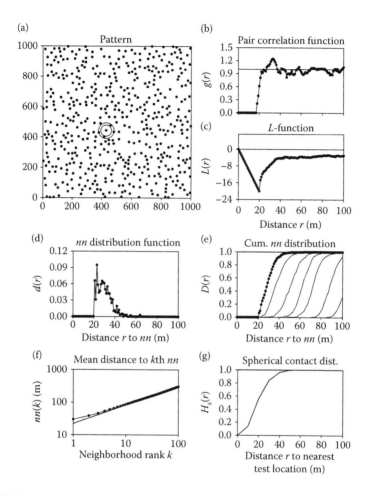

FIGURE 2.11
Summary statistics for a hyperdispersed point pattern. (a) Pattern comprising 500 points within a 1000×1000 m² observation window. The pattern was created by a hard-core process with random points being distributed in sequence within the observation window. Each new point was accepted if it was located at least 20 m away from all other points already placed. Panels (b) through (g) present the results of several summary statistics applied to the pattern in (a). Panel (e) presents a sequence of 10 distribution functions $D^k(r)$ of the distances to the kth neighbor with $k = 1, 2, 4, 6, 8, 12, 16, 20, 25$, and 30, where $D^1(r)$ is indicated by solid dots.

a distance of twice the radius of the largest disks. Such patterns are referred to as "*soft core*."

Similar to what we observed in analyzing the clustered pattern (Figure 2.10c), we also find that the shape of the estimated L-function is difficult to interpret for a hyperdispersed pattern (Figure 2.11c). The first observation in our example is that the L-function decreases linearly up to distances of 20 m. Such a behavior is typical of a hard-core pattern and arises because of the definition of the L_2-function as $L_1(r) - r$. Clearly, the hard-core pattern has no neighbor up to double the hard-core distance of 20 m and therefore we find $K(r) = 0$ and $L_2(r) = -r$. The second observation is that the estimated values for the L-function are clearly smaller than zero for all scales r shown in Figure 2.11c. This is in sharp contrast to the results of the pair-correlation analysis, which indicates that regularity disappears at distances of approximately 20 m away from the typical point. The reason for this apparent contradiction is again the "memory" of the K-function (Wiegand and Moloney 2004). In essence, what occurs for the hard-core pattern is that the typical point will always contain fewer neighbors within distance r than a random pattern because the closer distances are missing. As a consequence, the estimates of the L-function will be negative. Because of the memory effect and the nonintuitive nature of the L-function, we recommend the use of the pair-correlation function under most circumstances.

2.3.2.4 The K2-Function

The $K2$-function developed by Schiffers et al. (2008) is a straightforward extension of the pair-correlation function, which can be used to reveal patterns of hyperdispersion and aggregation in uni- and bivariate patterns, despite the presence of spatial variation in the intensity function. The $K2$-function, in essence, mitigates the problem of virtual aggregation that impacts the pair-correlation and the K-functions when the intensity function is heterogeneous (see Section 2.6.3.1). Heuristically, the $K2$-function can be interpreted as being the first derivative of the pair-correlation function, which basically characterizes change in neighborhood density over a small range of distances (i.e., from $r - \Delta r$ to $r + \Delta r$):

$$\hat{K}2(r) = \frac{\hat{g}(r + \Delta r) - \hat{g}(r - \Delta r)}{2\Delta r} \tag{2.5}$$

The underlying rationale of the $K2$-function is that heterogeneity, due to a nonconstant intensity function, introduces a rather minor change in the magnitude of the values of the pair correlation, relative to the effect of local interactions among points. In Section 2.6.2.2, we will explicitly explore how heterogeneity in the intensity function influences the shape of the pair-correlation function. For example, "patch heterogeneity," where the pattern is restricted to some part of the observation window, yields an elevated and almost constant value of $g(r)$ at smaller distances r, which then declines smoothly at larger distances r (Figure 2.22). A similar shape results from a smooth gradient in the intensity function. However, point interactions that cause clustering (Figure 2.10b) or hyperdispersion (Figure 2.11b) often yield more rapid changes in the neighborhood density compared with the changes introduced by patch or gradient-type heterogeneity. This means that the shape of the $K2$-function should be largely independent of heterogeneity in the intensity function, therefore allowing the assessment of regularity or clustering, even in the presence of broader scale heterogeneity.

Figure 2.12 presents two examples that illustrate how virtual aggregation impacts the *L*-function, the pair-correlation function and the *K*2-function. To this end, we expanded the original 500×500 m^2 observation windows of two homogeneous patterns (with respect to the intensity function) by attaching an empty window of the same size to each (small panels to the left and is right of Figure 2.12a,b), yielding two extreme cases of patch heterogeneity (i.e., the pattern is only located in one-half of the observation window and is void of points in the other). Comparison of the values of the summary statistics of the homogeneous patterns (within the 500×500 m^2 observation window) with those of the enlarged 500×1000 m^2 observation window allows us to assess the impact of this type of heterogeneity on the shape of the summary statistics.

First, we consider the behavior of the summary statistics for the original homogeneous patterns (i.e., the summary statistics were calculated only for the 500×500 m^2 observation windows in Figure 2.12a,b). The results are shown as gray disks in Figure 2.12c–h. Note that a negative value of *K*2 indicates clustering, since neighborhood density declines with increasing distance under clustering, whereas a positive value indicates hyperdispersion. As a consequence, the *K*2-function of the clustered pattern exhibits significant negative values for distances in the interval of $r = 3$–20 m, that is, the values drop below the simulation envelopes for the CSR null model (Figure 2.12g). This indicates that the change in the pair-correlation function at this distance interval was greater than expected. To correctly interpret this result, we have to compare the *K*2-function to the pair-correlation function, which has high values at distances of 1–3 m with relatively little change in magnitude (Figure 2.12e). Considering the results for the *K*2- and pair-correlation function together, we conclude that the pattern shows clustering at distances up to approximately 20 m, as confirmed by the simulation envelopes of the pair-correlation function. In contrast, for the hyperdispersed pattern the *K*2-function lies outside of the simulation envelopes for the distance interval of 8–11 m, where it has positive values (Figure 2.12 h). This indicates that the neighborhood density sharply increases at distances from 8 to 11 m. Considering this, in conjunction with the fact that the pair-correlation function exhibits low values at distances below 8 m (Figure 2.12f), we can interpret the results as indicating that hyperdispersion occurs at distances below 8 m.

After analyzing the original homogeneous patterns, we now estimate the summary statistics of the corresponding heterogeneous patterns (i.e., the summary statistics were calculated for the enlarged 1000×500 m^2 observation windows shown as small panels next to Figure 2.12a,b). The resulting summary statistics are shown as black disks in Figure 2.12c–h. Comparing the shape of the *K*2-function for the heterogeneous pattern (black dots in Figure 2.12g) with the shape for the corresponding homogeneous pattern

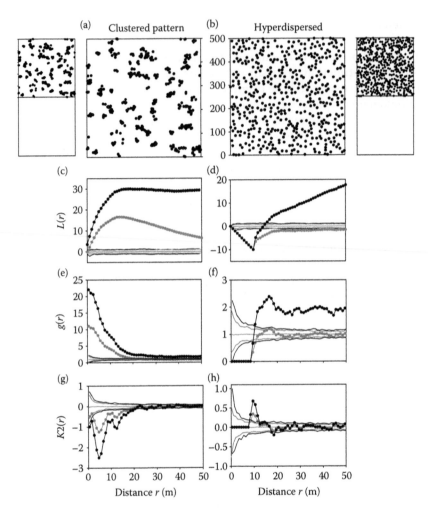

FIGURE 2.12
K2-function compared to the pair-correlation and *L*-functions. (a) *Clustered pattern* of 500 points generated by a Thomas process with parameters producing $\rho A = 100$ clusters with an approximate radius of $2\sigma = 10$ m. (b) *Homogeneous pattern* of 500 points created by a hard-core process with a separation distance of 10 m. (c) and (d) *L*-function applied to the clustered and homogeneous patterns, respectively. (e) and (f) Pair-correlation function applied to the clustered and homogeneous patterns, respectively. (g) and (h) *K2*-function applied to the clustered and homogeneous patterns, respectively. Observation windows shown to the left of panel (a) and the right of panel (b) contain the same patterns as the main panels with an additional 500 × 500 m² area void of points attached to the bottom. In all analyses, results for the 500 × 500 m² homogeneous patterns are shown with gray circles and results for the enlarged window are shown with black circles. Results for the corresponding CSR model, with associated simulation envelopes, are shown as lines with no symbols.

(gray disks), we see that the K2-function is relatively invariant to strong heterogeneity in the intensity function. More importantly, the scales at which there are significant departures from the CSR null model are basically the same when comparing curves for the heterogeneous and homogeneous patterns for the K2-function, although the magnitude of the values that lie outside the simulation envelopes are somewhat different between the two patterns (Figure 2.12g,h). Thus, the assessment of the scales of the significant effects by the K2-function is largely unaffected by heterogeneity in the intensity function. This is in stark contrast to the pair-correlation and L-functions (Figure 2.12e and c, respectively), which exhibit a strong response due to virtual aggregation. This result is essentially the same for the hyperdispersed pattern (Figure 2.12h).

Schiffers et al. (2008) also apply the K2-function to several other examples of heterogeneous patterns with known properties, including a real example of molehill patterns, and obtain essentially the same results as in our examples. These findings indicate that the use of the K2-function, in concert with the pair correlation function and the CSR null model, provides a more accurate assessment of the scales of clustering and hyperdispersion for patterns produced with a spatially varying intensity function than would be possible using either the L- or pair-correlation function.

2.3.2.5 Edge Correction and Estimation of Second-Order Statistics

Estimation of second-order statistics requires methods to avoid or correct for the bias produced by being near the edge of the observation window. For example, a naive way to estimate the K-function for a set of points in a pattern would be to visit each point i, count the number of additional points within distance r of point i, and calculate the average over all points i, irrespective of the locations of the points. However, the problem with this approach is that, for some points i, the distance to the nearest border of the observation window will be shorter than the distance r at which $K(r)$ should be evaluated (Figure 2.8c). This means that some of the disks with radius r centered at point i are not completely inside the observation window. Because points outside the observation window are not recorded, they will be missed by this estimation method and a downward bias will be produced in the estimator. The bias in the overall estimator will be greater as the distance r becomes larger. An extreme example of this form of bias occurs when point i is located at a corner of the observation window (Figure 2.2c), where only one-quarter of the disk with radius r centered on point i is located within the observation window. Also, at the maximum distance r usually considered in the standard analysis (i.e., half of the shortest side of the observation window W; Figure 2.2), only a point i located exactly in the center of the observation window will have its entire disk located fully within W (Figure 2.2b). The same problem occurs for the pair-correlation function; in this case we may have incomplete rings (Figure 2.8c). It is clear that we must consider some form of edge correction to minimize the biases introduced by edge effects into second-order point-pattern analyses.

Five principal possibilities exist for edge correction. First, we may ignore the problem and conduct *no edge correction*. This may be an option if the bias is small. Indeed, if the observation window is large compared to the range of distances r considered, only a few points i would be affected and we can ignore the bias as a first approximation (Figure 2.13a). Later in the book we will calculate the expected bias for rectangular observation windows analytically (see Section 3.1.2.1). We shall see that in most cases in ecology the size

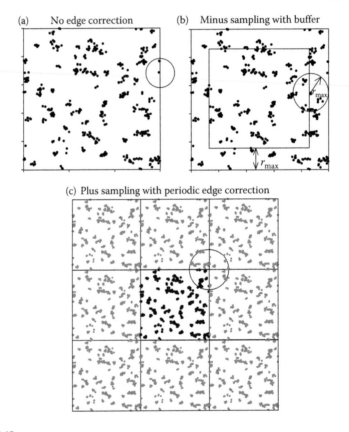

(a)   No edge correction   (b)   Minus sampling with buffer

(c) Plus sampling with periodic edge correction

FIGURE 2.13
Strategies for edge correction. (a) *No edge correction.* The neighborhood of points close to the border cannot be correctly determined because points outside the observation window are not recorded. (b) *Minus sampling.* Estimation of summary statistics is based only on focal points within an inner window (rectangle) for which all neighborhoods up to a maximum distance of r_{max} can be correctly determined. (c) *Plus sampling.* All observed points, shown as black circles within the central observation window, are used as focal points for estimation of summary statistics. Unobserved points are "reconstructed" by replicating and translating the observed pattern outside of the observation window, shown as the eight regions with gray circles surrounding the central observation window. The reconstructed points are used in subsequent analyses to complete the partial neighborhoods of observed points surrounding the focal points.

of the observation window W produces a significant limitation and some form of edge correction will be required.

The second approach to edge correction is to avoid bias by using only focal points i that have complete rings or disks within W for estimation. This method is called *"minus sampling"* (Figure 2.13b) and has the disadvantage that the available information is only partially used, especially for larger distances r. A third class of methods reconstructs the missing points outside W based on the knowledge of the points inside W. Here, the method of *"plus-sampling"* is used, because (reconstructed) points outside W are counted. One example is periodic edge correction, where the point pattern in W is enlarged by copying the pattern in W and repeating it at the borders of the observed pattern (Illian et al. 2008, p. 184; Figure 2.13c).

The fourth method of edge correction is to adjust the estimate for each focal point i by accounting for the proportion of the disk or ring falling outside the observation window. In practice, each point pair i–j is weighted individually with a correction factor w_{ij} based on the relationship between r and the distance of the focal point i from the border of the observation window. One commonly used weighting factor is provided by dividing the circumference of the full disk centered on i with radius r and passing through point j by the circumference of the disk lying inside the sample domain (i.e., *Ripley edge correction*; Figure 2.8c). For example, if only one-third of the circumference of the disk falls within the sample area, the point pair i–j is given a weight of 3, thus extrapolating the observed number of points within incomplete disks of radius r around focal points to that expected for a full disk. This edge correction method is widely used in ecology. Haase (1995) and Goreaud and Pélissier (1999) provide formulas for the correction factor w_{ij} used in an ecological context.

Finally, we may calculate the expected bias directly for a given distance r and multiply the naive estimator with a correction factor (i.e., *Ohser edge correction*). Note that in this approach the weighting factor does not depend on the individual point pair and is the same for all point pairs i–j separated by distance r. The last two approaches to edge correction are the most commonly used in the current application of point-pattern analysis. We will present an in-depth treatment of these methods in Section 3.1.1.

2.3.3 Higher-Order Statistics

Higher-order statistics are theoretically possible, but for reasons of practicality are rarely used. For example, the third-order product density would involve the consideration of three points simultaneously, and would represent the probability that there is one point in each of three small disks $\mathbf{d}x_1$, $\mathbf{d}x_2$, and $\mathbf{d}x_3$ (Figure 2.6c). In the case of homogeneous patterns, calculation of the third-order product density would involve two distances and one angle (Figure 2.6c). As a result, $\rho^{(3)}(x_1, x_2, x_3)$ is usually not applied in point-pattern analysis because it is considered too complex (Illian et al. 2008, p. 245). Sometimes, however, integrated or aggregated statistics are used when considering the statistical properties of more than two points at a time. One example is the *T*-function introduced in Schladitz and Baddeley (2000), which is a generalization of the *K*-function. While the *K*-function basically counts the number of points within distance *r* of the typical point of the pattern, the *T*-function assesses triplets of points (the typical point and an additional pair of points). The third-order statistics may be useful in testing point-process models parameterized from data or in assessing the quality of pattern reconstruction (e.g., Tscheschel and Stoyan 2006). They are suitable in this context because they are fundamentally different from the second-order or nearest-neighbor statistics used in model fitting or reconstruction, providing a relatively independent source of valida-tion. However, because the higher-order statistics are difficult to interpret and are rarely used in applications concerning ecological problems, we will not consider these methods in detail.

2.3.4 Nearest-Neighbor Statistics

Nearest-neighbor statistics provide a useful complement to the second-order statistics just described. They characterize, for the most part, the small-scale correlation structure of point patterns by assessing the mean statistical prop-erties of the distance separating the typical point and its nearest or *k*th near-est neighbor (Figure 2.8b). The difference between the nearest-neighbor and the second-order approaches is in some ways subtle, but it is also quite sig-nificant. The second-order approaches assess the average number of points within a neighborhood, whereas the nearest-neighbor approaches assess the probability that a neighborhood is empty or contains $k - 1$ neighbors.

2.3.4.1 Distribution Functions of Nearest-Neighbor Distances

The function $D(r)$ represents the (cumulative) distribution of the distances *r* to the nearest neighbor, measured from the typical point of a pattern (Figures 2.8b, 2.10e, 2.11e). Values of this function range from zero (scales at or below which no nearest neighbors have been encountered in the pat-tern) to one (scales at or above which all points have a nearest neighbor). The function $d(r)$ is the corresponding noncumulative distribution (Figures 2.10d and 2.11d), which represents the probability that the nearest neigh-bor of the typical point is located at distance *r*. Note that the estimation of the noncumulative distribution function $d(r)$ is typically based on only a few points, and is often not sufficient to produce smooth distributions. Therefore, $d(r)$ is rarely used (but see Illian et al. 2008, p. 211f). Instead of the nearest neighbor, one may also analyze distances to the *k*th near-est neighbor. The resulting summary statistic is the distribution function $D^k(r)$ of the distance *r* to the *k*th nearest neighbor. In this terminology, the nearest-neighbor distribution function is given by $D^1(r)$, but the super-script is usually omitted for simplicity of notation. Note that the symbol

$G(y)$ is sometimes used for the nearest-neighbor distribution function (e.g., Diggle 2003).

If the distance of a focal point i to the nearest border is shorter than the distance to its nearest neighbor within W, the true nearest neighbor might be located outside the observation window and, consequently, the nearest-neighbor distance r_i of point i cannot be correctly determined (Figure 2.8c). Thus, there may be a need for edge correction, as discussed with respect to second-order statistics (Section 2.3.2.5). However, in contrast to second-order statistics, the error of not considering the points outside W is less severe. One reason for this is that the true nearest neighbor may indeed be located inside W, even if the focal point i is located relatively close to the border. This may happen if the intensity is relatively high and for clustered patterns where the nearest neighbor tends to be close by. Another reason is that in cases where the true nearest neighbor is located outside W, there is a high probability that the second, third, or fourth-nearest neighbor may be located inside W and serve as a reasonable approximation for the distance r_i. Thus, estimation of the nearest-neighbor distribution function $D(r)$ without edge correction provides a reasonably good approximation of the true $D(r)$, if the number of points in the pattern is relatively high, and/or if the pattern is clustered. In cases where the $D(r)$ of the observed pattern is assessed relative to simulation envelopes of a null model, edge correction may not be necessary because the $D(r)$ of the observed pattern, as well as those of the simulated patterns, are subject to the same small biases that will cancel out in the comparison.

In addition to the minus sampling and reconstruction methods discussed in the previous chapter on second-order statistics (Section 2.3.2.5), an elegant edge correction method for nearest-neighbor statistics does not reconstruct points, but uses only points i for which the nearest neighbor can be correctly determined (i.e., *Hanisch edge correction*). These are the points for which the nearest-neighbor distance (measured within the observation window) is shorter than the distance to the boundary (Figure 2.8c). To compensate for the omission of some points, the remaining points are given weights in the analysis. If the observed distance of point i to the nearest neighbor is large, it is downweighted in the analysis because points with a large nearest-neighbor distance in W are rare (Illian et al. 2008, p. 187). Details of this edge correction method are presented in Section 3.1.3.1.

The value of the nearest-neighbor distribution function $D(r)$ for CSR can be calculated based on the Poisson distribution. To do this, $D(r)$ must be reinterpreted as being the inverse of the conditional probability that there is no point within a disk of radius r around the typical point of the pattern. Under a random pattern with intensity λ, the probability of having no point within an area $A = \pi r^2$ is given by the Poisson distribution $P(k = 0, A) = \exp(-\lambda A) = \exp(-\lambda \pi r^2)$. Thus, we have $D(r) = 1 - \exp(-\lambda \pi r^2)$ for a random pattern. If the pattern is clustered, the nearest neighbor will be, on average, located closer than under CSR and $D(r) > 1 - \exp(-\lambda \pi r^2)$ (Figure 2.10e), whereas $D(r) < 1 - \exp(-\lambda \pi r^2)$ for a hyperdispersed pattern (Figure 2.11e).

Nearest-neighbor statistics are "short-sighted" and sense only the immediate neighborhood of the typical point, thus providing information about local cluster structures, whereas all points within distance r contribute equally to the estimator of the K-function (Figure 2.8a). Thus, $D(r)$ only provides information on the local neighborhood of points and does not describe any behavior at larger distances. Because of this, $D(r)$ measures different properties of a pattern than the second-order statistics $g(r)$ and $K(r)$. This may be desirable for specific ecological questions that focus on immediate neighborhood relationships, such as the distribution of seedlings around parent trees (Getzin et al. 2008). The distribution functions of distances to the nearest neighbor for the clustered-pattern example are shown in Figure 2.10d,e. In most cases (i.e., 95%), the nearest neighbor is located less than 20 m away (vertical line in Figure 2.10e). This indicates that the size of the typical cluster is approximately 20 m. Figure 2.10e shows $D(r)$ (dotted line) together with a sequence of distribution functions $D^k(r)$ of the distance y to the kth nearest neighbor. The $D^k(r)$ show that the typical cluster does not comprise many points, since only 58% and 28% of the 4th and 6th nearest neighbors, respectively, are located within 20 m. Indeed, the mean number of points per cluster of the underlying Thomas process (Box 2.5) was five. Thus, using a sequence of $D^k(r)$s allows additional inference on the local cluster structure of a point pattern.

The typical point in the hard-core pattern shown in Figure 2.11a has no nearest neighbor within 20 m, consequently we find $D^1(r) = 0$ for $r < 20$. Figure 2.11d shows that 95% of all points of the hard-core pattern have their nearest neighbor within 42 m. A comparison of Figures 2.10d and 2.11d shows that the kth nearest-neighbor distribution functions of the clustered versus hyperdispersed patterns have substantially different shapes. Whereas some points of the clustered pattern have their kth neighbor at large distances (e.g., the 4th or 6th nearest neighbor), which causes a "tail" in the distribution of $D^k(r)$, the range of distances over which the kth nearest neighbor occurs is substantially smaller for the hyperdispersed pattern. If the neighborhood rank k is slightly larger than the number of points in a particular cluster, the kth nearest neighbor is located in the nearest cluster, which may be in some cases far away, thereby creating the long tail in $D^k(r)$. In contrast, the hyperdispersed pattern has fewer large gaps than the clustered pattern, which reduces the possible distances to the kth neighbor.

2.3.4.2 Mean Distance to kth Neighbor

The distribution functions $D^k(r)$ of the distances r to the kth nearest neighbor provide information as to how the distances to the kth neighbor vary for a given point pattern and thus captures the variability around some mean behavior. If we look at the first nearest neighbor (i.e., $k = 1$) the mean is given by the mean distance to the nearest neighbor, which is a measure that has been frequently used to characterize univariate point patterns. An alternative way to look at nearest-neighbor statistics is to generalize this approach and characterize the pattern by the mean distances $nn(k)$ to the kth neighbors. This summary statistic has the convenient property of approximating a power law (Hubbell et al. 2008; Table 2.2) and thus may have the potential to reveal scale-independent features of a pattern. Figure 2.10f shows the mean distances $nn(k)$ to the kth neighbors as a function of the neighbor rank k. We calculated $nn(k)$ without edge correction, only considering the point pairs present in the observation window. This may lead to an overestimation of $nn(k)$ for larger neighborhood ranks because an unobserved point outside the observation window may be, in fact, the kth nearest neighbor. The data can be fit well with a power law $nn(k) = ak^e$ (solid line in Figure 2.10f) with parameters $a = 17.7 \pm 0.14$ and $e = 0.61 \pm 0.002$. We find some departure from the power law up to the 8th nearest neighbor. This, however, is only apparent on a log–log plot (Figure 2.10f). Note that a random pattern has an exponent of $e = 0.5$ (Table 2.2). The hyperdispersed hard-core pattern shown in Figure 2.11a can also be fit with a power law $nn(k) = ak^e$ (solid line in Figure 2.11f) with parameters $a = 22.1 \pm 0.14$ and $e = 0.57 \pm 0.001$.

2.3.5 Morphological Functions

Morphological functions offer an alternative way of characterizing point patterns (Mecke and Stoyan 2005). The idea of morphological functions applied to point patterns is a relatively simple, geometrical concept. Each point of a pattern is enlarged to a disk of radius r. As a result, a pattern of potentially overlapping disks is obtained (Figure 2.14). The topology or morphology of this pattern is then analyzed for different values of r. Of special interest is the fraction of the observation window covered by the union of all disks, which is represented in Illian et al. (2008, p. 200) by the symbol $A_A(r)$. Somewhat surprisingly on a first view, $A_A(r)$ gives the probability that the first neighbor to an arbitrary "test point" in W is within distance r. This metric is also known as the spherical contact distribution function $H_s(r)$. Additional morphological summary statistics may be derived from $A_A(r)$ (see Mecke and Stoyan

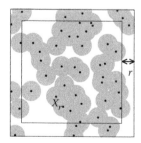

FIGURE 2.14
The spherical contact distribution $H_s(r)$ as a morphological function. $H_s(r)$ is the proportion of the observation window covered by the union X_r of all (possibly overlapping) disks of radius r centered on the points of a pattern. The points in this example are shown as the small dark circles and the disks are the gray shaded areas.

2005; Illian et al. 2008, p. 200ff). We present in the following only the spherical contact distribution function because of its analogy to nearest-neighbor statistics and because it is an especially useful complement to other summary statistics in point-pattern analysis (Wiegand et al. 2013a).

2.3.5.1 Spherical Contact Distribution

The most important location-related summary statistic in point-pattern analysis, after the intensity function, is the spherical contact distribution function $H_s(r)$. $H_s(r)$ is the distribution of the distances r from an "arbitrary" test location to the nearest point of the pattern (Figure 2.8b). The test locations may be points that are randomly distributed or they may be constructed from the nodes of a grid. In general, the test locations should "sufficiently" cover the observation window to allow good resolution of $H_s(r)$. One interpretation of $H_s(r)$ is that it is a statistic that basically characterizes the empty space in a pattern. For this reason it is sometimes referred to as the empty space distribution function, using the symbol $F(x)$, rather than $H_s(r)$ (e.g., Diggle 2003). The spherical contact distribution function provides important information about the pattern, complementing the point-related summary statistics. Indeed, $H_s(r)$ is most closely related to the distribution function of nearest-neighbor distances $D(r)$.

The numerical value of $H_s(r)$ under CSR is the same as that for the distribution function of nearest-neighbor distances $D(r)$, that is, $H_s(r) = 1 - \exp(-\lambda \pi r^2)$. However, if the pattern shows clustering, the distances from an "arbitrary" test location to the nearest point of the pattern will be, on average, larger than under CSR because a clustered pattern contains larger areas of empty space than a random pattern. Conversely, if the pattern is hyperdispersed the holes are smaller and the distance to the nearest point from an arbitrary test location will be shorter than under CSR.

Figure 2.10g shows the spherical contact distribution $H_s(r)$ for the clustered pattern in 2.10a. It reveals large gaps in the pattern. While 95% of all points have their nearest neighbor within 20 m (Figure 2.10e), only 22% of all test points have a point in the clustered pattern within 20 m (Figure 2.10h) and 5% of all test points have no neighbor within 90 m. This indicates that some of the gaps have diameters of up to 90 m, a diagnosis that is confirmed by visual inspection of the pattern. Figure 2.11g shows the spherical contact distribution $H_s(r)$ for the hard-core pattern in 2.11a. In contrast to the clustered pattern, there are no large gaps in this pattern and 95% of all test locations have the nearest point of the pattern within 40 m.

2.3.6 Mark Connection Functions for Qualitatively Marked Patterns

Mark connection functions are summary statistics adapted to analyze qualitatively marked patterns, where a univariate pattern (e.g., trees of a given species) carries one qualitative mark (e.g., surviving vs. dead). In most cases, the interest in analyzing this type of pattern is in characterizing the process that distributed the mark over the (antecedent) univariate pattern, not in the univariate pattern itself. Qualitatively marked patterns can be analyzed with bivariate pair-correlation functions or K-functions; however, these analyses still contain information on the underlying univariate pattern. Adapted summary statistics therefore condition on the underlying spatial structure of the univariate pattern, removing its influence on the analysis.

The idea of mark connection functions is simple: for two points l and m of a pattern separated by distance r, the mark connection function $p_{lm}(r)$ gives the probability that the first point is of type l and the second of type m. If the pattern contains only type l and type m points, the sum of the four possible mark connection functions equals one, that is, $p_{ll}(r) + p_{ml}(r) + p_{lm}(r) + p_{mm}(r) = 1$. If the process distributes the marks randomly over the points (i.e., *random labeling*), the mark connection functions yield $p_{lm}(r) = p_l\, p_m$ where p_l and p_m are the proportions of type l and m points in the pattern, respectively. However, if the process that distributed the mark l (e.g., burned shrub) is positively autocorrelated at some distance r (e.g., shrubs at distance r from a burned shrub have a higher probability of also being burned), the mark connection function $p_{ll}(r)$ yields $p_{ll}(r) > p_l\, p_l$ and it will be above the simulation envelopes of the random labeling null model that assumes spatially uncorrelated marks.

Mark connection functions are closely related to pair-correlation functions and product densities, as follows: If we consider two disjoint disks of infinitesimally small size dx_i and dx_j, centered on locations x_i and x_j, respectively, then the quantity $\rho_{lm}(r)dx_i\, dx_j$ approximates the probability that a type l point occurs within the first disk and a type m point within the second disk. Additionally, we can introduce a quantity $\rho_{l+m,l+m}(r)dx_i\, dx_j$ (where the subscript $l + m$ indicates the combined pattern of type l and type m points), which approximates the probability that a type l or a type m point occurs within the first disk and a type l or type m point occurs within the second disk. Thus, the quantity $\rho_{lm}(r)/\rho_{l+m,l+m}(r)$ yields the probability that the first point is of type l and the second of type m. Given this relationship, we find that

$$p_{lm}(r) = \frac{\rho_{lm}(r)}{\rho_{l+m,l+m}(r)} = \frac{\lambda_l \lambda_m g_{lm}(r)}{\lambda_{l+m}^2 g_{l+m,l+m}(r)} = p_l p_m \frac{g_{lm}(r)}{g_{l+m,l+m}(r)}. \tag{2.6}$$

Dividing in Equation 2.6 by the product density of the joined pattern $\rho_{l+m,l+m}(r)$, therefore, adjusts the result to account for situations where one point of the pattern is located at location x_i and another point is at location x_j. Equation 2.6 also demonstrates that the expectation of the mark connection function under random labeling yields $p_{lm}(r) = p_l\, p_m$, because in this case $g_{lm}(r) = g_{l+m,l+m}(r)$.

An estimator of the mark connection function $p_{lm}(r)$ would visit all pairs of points that are separated by distance r and determine the proportion of those pairs in which the first point is a type l point and the second point a type m point. This procedure can be interpreted as calculating the mean value of a "test function" over all pairs of points separated by distance r. The value of the test function is one if the first point is a type l point and the second point a type m point, and zero otherwise. This interpretation provides an analogy to mark correlation functions presented in the next section and embeds the mark connection functions within the broader framework of marked point patterns.

Figure 2.15 provides an example of the use of the mark connection function in analyzing a pattern with known properties. The pattern in the example resulted from the superposition of a random pattern (100 points) with a clustered pattern (500 points), which is inspired by the distribution of seedlings in tropical forests (Wiegand et al. 2009). The clustered pattern was created with a Thomas process and comprised 100 randomly distributed clusters with an approximate radius of 10 m (86% of all points have a distance smaller than 10 m from the cluster center) and each cluster consisted of, on average, 5 points. Additionally, the 600 points in the pattern contained marks of dead "1" (200 points) or surviving "2" (400 points), which were created in a density-dependent way. The probability of a given seedling dying was

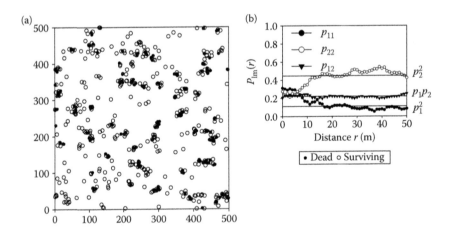

FIGURE 2.15
Mark connection functions. (a) Map of a marked point pattern consisting of dead and surviving seedlings distributed in the same pattern as the Thomas process in Figure 2.10, but with an additional 100 random points. (b) Results of analyses using mark connection functions (symbols) for the pattern shown in (a) with the expectations under no spatial correlation among the marks shown as lines. The notations to the right of the figure indicate the lines representing the respective expectations. Note that $p_{12}(r) = p_{21}(r)$. The example is based on a pattern one would typically see for seedlings in a tropical forest. Dead seedlings are represented by filled disk and a subscript of 1; surviving seedlings by open disks and a subscript of 2. There are 200 dead seedlings in the pattern and 400 that have survived. The probability of a seedling dying was proportional to the number of seedlings occurring within a 10 m neighborhood; more isolated seedlings had a higher probability of surviving.

proportional to the density of seedlings within its 10 m neighborhood (i.e., the approximate cluster radius). Thus, isolated points had a high probability of surviving, whereas points in the center of clusters had a high probability of dying (Figure 2.15a). Figure 2.15b shows the mark connection functions for this pattern. Because the pattern is highly clustered, a seedling lying near another seedling (i.e., within the same cluster) has a higher probability of dying than would be expected under the case of spatially uncorrelated marking. Thus, the mark connection function $p_{11}(r)$ is larger than expected (i.e., $p_{11}(r) > p_1 p_1$) for distances r smaller than the diameter of the clusters (i.e., 20 m), but approximates the expectation at greater distances (Figure 2.15b). Conversely, the more isolated, surviving points have over the same distance interval a lower than expected probability of having a surviving seedling as a neighbor, that is, $p_{22}(r) < p_2 p_2$. Given that the process distributing the mark for dead seedlings was only driven by the neighborhood density, there was no positive or negative association of surviving seedlings to dead seedlings, that is, $p_{21}(r) \approx p_2 p_1$.

2.3.7 Mark Correlation Functions for Quantitatively Marked Patterns

Mark correlation functions are summary statistics adapted for quantitatively marked patterns, where a univariate pattern (e.g., trees of a given species) carries a quantitative mark (e.g., size). As with mark connection functions, the interest in analyzing quantitatively marked patterns is generally on characterizing the process that distributed the mark over the (antecedent) univariate pattern, whereas the characteristics of the univariate pattern itself are usually not of interest.

As with the mark connection functions, *the idea of mark correlation functions is simple*; a (nonnormalized) mark correlation function $c_t(r)$ yields the conditional mean of a test function $t(m_i, m_j)$ calculated from the marks m_i and m_j of two points i and j, respectively, given that they are located distance r apart. For example, if the test function is the size of a tree j located distance r from a focal tree i, the corresponding mark correlation function is the mean size of trees conditioned on those trees that have a neighbor at distance r. To make the mark correlation functions independent of the values of the marks, the function is usually normalized by the mean value of the test function over all pairs of points. In the current example, this would be the (unconditional) mean size of the trees.

Several test functions are used in practice (Illian et al. 2008). For example, the mean value of the mark of the first m_i (or the second m_j) point of a pair of points separated by distance r is called the *r-mark correlation function*, the product of the two marks (i.e., $m_i m_j$) is the *mark correlation function*, and the squared difference of the two marks (i.e., $0.5(m_i - m_j)^2$) is called the *mark variogram*, because of the analogy to variograms used in geostatistics. Figure 2.16a shows a marked pattern based on simulated data. The pattern is the same as that of Figure 2.15a (i.e., a superposition of a cluster pattern

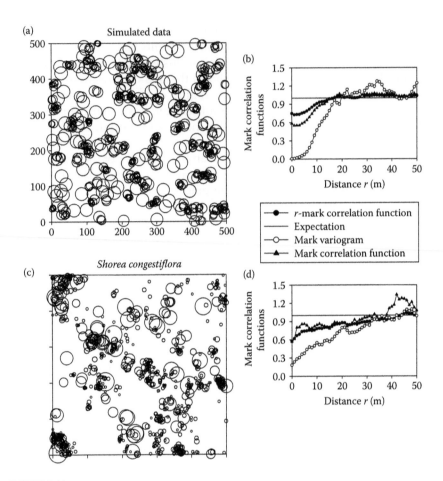

FIGURE 2.16
Mark correlation functions for quantitatively marked patterns. (a) Simulated map of quantitatively marked points representing an independent superposition of 100 random points with the cluster pattern shown in Figure 2.10a, the latter comprising 500 points. The marks attached to the points are inversely proportional to the number of neighbors within a distance of 10 m; that is, marks for isolated points are larger than those for points occurring in clusters. (b) Mark correlation functions calculated from the data shown in panel (a). (c) Map of the species *Shorea congestiflora* in the Sinharaja Plot, Sri Lanka, with dbh as a quantitative mark (Wiegand et al. 2007c). (d) Mark correlation functions calculated from the data shown in panel (c).

with 500 points and 100 random points), and the mark attached to a point is proportional to $1/d^{0.5}$ where d is the number of neighbors within a distance of 10 m. Isolated points are, therefore, larger than points located in the clusters. Under these conditions we expect that points located in the center of a cluster have many neighbors and are therefore smaller. The distribution of sizes in this example is spatially correlated. A tree that has another tree nearby tends to be smaller than a tree that is more isolated. Figure 2.16b shows the results of the *r*-mark correlation function (closed circles), which is the mean size of trees conditioned on those trees that have a neighbor at distance *r*. As expected, the conditional size is smaller than the mean size for distances less than the diameter of a cluster (i.e., 20 m). At larger distances, the conditional size approximates the expectation without spatial correlation. Because trees with more neighbors are smaller, the mark variogram (open circles) shows that trees that are very close to each other have very similar sizes (they are small), and that the size difference increases with increasing distance *r*. Because nearby trees are generally small, the mark correlation function (triangles) shows a stronger response than the *r*-mark correlation function.

Figure 2.16c shows the pattern of the species *Shorea congestiflora* with dbh as a quantitative mark. Wiegand et al. (2007c) analyzed this pattern in detail and found that many small trees were grouped in clusters. They also found that a large tree was sometimes located close to a cluster of small trees. The resulting mark correlation functions produced results similar to those of the simulated pattern shown in Figure 2.16a and indicated cluster diameters of approximately 30–40 m. This is especially true for the mark variogram. The *r*-mark correlation function also shows a response very similar to that of the simulated pattern, but the mark correlation function shows a weaker response. This can be explained by the presence of some larger trees close to clusters of small trees, which produces the larger values in the mark product of nearby trees. Accordingly, the mark variogram of the real data starts with values of 0.2 at very small distances, whereas the mark variogram of the simulated data starts with values of 0. Note that, in the case of mark variograms, lower values represent stronger positive correlation.

2.4 Null Models and Point-Process Models

Null models and point-process models are important tools employed in conducting spatial point-pattern analysis. They can be used to (1) determine whether there is significant spatial structure in a data set, (2) succinctly summarize the properties of spatial structure in the data, and (3) test ecological hypotheses about the factors that may have produced an observed spatial structure. In the following, we will briefly discuss the differences among fundamental divisions, null models, and point-process models, and present different possibilities for hypothesis testing within the framework of point-pattern analysis.

2.4.1 Fundamental Division: The Most Basic Null Model

The fundamental division is a specific type of null model. It corresponds to the simplest possible randomization of a given data type and is used to *determine whether the data contain a signal that can be distinguished from pure stochastic effects* consistent with the given data type and observed sample size. Thus, testing against the fundamental division is usually the first step of an exploratory point-pattern analysis. While testing against the fundamental division is often not very informative for univariate patterns (where departures from CSR are often obvious *a priori* from initial visualization of the data), the fundamental division is the most important null model for other data types such as qualitatively or quantitatively marked point patterns.

For a univariate pattern, the fundamental division corresponds to complete spatial randomness (i.e., CSR), where each point of the pattern is located at a random location independently from every other point within the observation window. The fundamental division in this case can be used to detect aggregation (i.e., the points are on average closer together than expected by chance) or hyperdispersion (i.e., the points are on average further apart than expected by chance) in an observed pattern. However, the fundamental division in general does not correspond to a complete randomization of the data. The randomization often needs to be conditioned on the specific characteristics of the given data type. For example, the fundamental division of a bivariate pattern needs to conserve the univariate structure of the two component patterns and remove only potential interdependence between the two patterns. Thus, only characteristics specific to the bivariate pattern are randomized. This is accomplished by randomizing the relationship between the points of the two component patterns, while keeping the univariate patterns intact. The fundamental division of bivariate patterns constructed in this way is designed to detect attraction (the points of the two patterns are on average closer together than expected by chance) or repulsion (the points of the two patterns are further apart than expected by chance) between the two patterns. Therefore, a complete randomization of one or two of the component patterns (i.e., CSR), although frequently used in earlier studies, does not provide a suitable fundamental division. This is because the null model, when misspecified in this way, also randomizes the univariate component pattern(s) and possible departures from the null model may be caused by univariate effects rather than by bivariate effects. Indeed, if both component patterns show strong clustering, some clusters of the component patterns may accidently overlap and generate pair-correlation functions with the appearance of a positive association. While such pair-correlation functions may not arise by chance for two CSR patterns, they may be commonly observed for clustered patterns. This problem has been recognized for quite a while (e.g., Lotwick and Silverman 1982). However, it should also be noted that while CSR for one or both component patterns does not usually produce a suitable fundamental division, it might be a suitable null model for specific questions (e.g., Wang et al. 2010; Wiegand et al. 2012; Section 4.2.2.1).

The subtle differences between qualitatively marked patterns versus bivariate patterns have also frequently caused confusion in the use of the fundamental division. The constraints imposed by the specific data type need to be considered. For instance, in the case of a qualitatively marked pattern we have one univariate pattern (e.g., shrubs of a given species) that was subsequently marked by a given process (e.g., fire causing mortality of some shrubs). The fundamental division of a pattern qualitatively marked in this way must condition on the observed univariate pattern (i.e., the locations of points are left unchanged) and the mark (i.e., dead vs. surviving) must be randomized. There are many examples in the literature where the authors treat this type of process as a bivariate pattern, randomizing the positions of the points, rather than the marks, as a test against the observed data. This produces a misspecified null model and generates incorrect inferences about the spatial structure of the data.

In essence, the fundamental division forms an important reference point because it provides a theoretical model for patterns that arise through pure chance mechanisms. This is particularly useful if the pattern produced by the fundamental division can be characterized analytically. In some cases, this also allows for the construction of additional summary statistics. An important example is the *L*-function (Besag 1977), which transforms the *K*-function so that the expectation under the fundamental division is $L(r) = 0$. Departures from the fundamental division are, therefore, easily recognized in a plot of the *L*-function. This is more difficult for the *K*-function for which the expectation under the fundamental division increases with the square of the distance *r*, that is, $K(r) = \pi r^2$ (see Section 2.3.2.2). Another example is random labeling, for which knowledge of the theoretical expectation of the pair-correlation and *K*-functions suggests several additional summary statistics to test specific departures from random labeling (see Section 4.4).

2.4.2 Null Models

The fundamental divisions presented in the last section are very specific null models used to determine whether data contain a signal that can be distinguished from pure stochastic effects. However, depending on the scientific question posed, a broad range of null models beyond the fundamental division can be used in point-pattern analysis. Null models in ecology have been mostly associated with the discipline of community ecology (Gotelli and Graves 1996). In this tradition they are often viewed as pattern-generating models designed to create patterns that are expected in the absence of a particular ecological mechanism of interest. The patterns generated are then used to test data suspected of being produced by the very same ecological mechanism (Gotelli and Ellison 2004). The objective is to find out *whether there is structure in the data that does not exist in the null model*. A null model in this role functions as a standard statistical null hypothesis where one wants to prove that it is not correct. Thus, in a manner similar to null hypothesis testing, the task is to confirm that the observed pattern departs from the null model.

2.4.3 Point Process Models

A point pattern is given by a set of locations that are distributed within an observation window due to some form of stochastic mechanism (Diggle 2003). *Point processes* are mathematical models characterizing those stochastic mechanisms. As such, point-process models can be used to simulate spatial point patterns with known properties to be compared with observed patterns. Consequently, point-process models include all null models and fundamental divisions. In contrast to null models, more complex point-process models are often used to "model the data." In this context, the focus is not to prove that the observed pattern contains structure that is not included in the null model. Instead, the task is to confirm that *the observed pattern cannot be distinguished from the patterns produced by the point process.* This approach can be applied to create patterns that are expected in the presence (not absence) of particular mechanisms, such as clustering of offspring around parents or trends in density related to environmental gradients. This approach is also very useful if the objective is to *summarize the statistical properties of the observed point pattern.* In this case, parametric models are fit to the data and the parameters of these models succinctly summarize the properties of the point pattern, if they fit the data well.

While the (nested) classification of fundamental divisions, null models and point-process models provides a useful framework for the different roles of point-process models in ecology, it does not fully acknowledge the different roles point-process models can play in exploratory, confirmatory, and multimodel analysis. The distinction between these forms of analysis is important because the confirmatory versus exploratory versus multimodel approaches have a significantly different focus, requiring slightly different methods, and confounding exploratory and confirmatory approaches may produced incorrect inferences on the data.

2.4.3.1 *Exploratory Context*

Within an *exploratory context*, the goal of the researcher is to describe the properties of the observed point pattern as completely as possible, with the final objective being to derive suitable hypotheses on the underlying processes producing the pattern (Wiegand et al. 2009). Baddeley (2010, p. 347ff) provides a more detailed treatment of exploratory data analysis in the context of point-pattern analysis. An exploratory analysis usually starts with visualization of the data and a closer examination of the intensity function before a test against the fundamental division is conducted.

The latter is used to determine whether the pattern contains nonrandom spatial structure in the interpoint interactions. If such a structure is identified, the researcher may wish to derive a more detailed description of the pattern, which may help identify the nature of the underlying processes. In this case, a point process with known properties is used to serve as a "point of reference" or benchmark. Parametric point processes can be fit to the data and if the fit is satisfying, the parameters succinctly summarize the properties of the point pattern. This is of great value, especially if several observed point patterns, corresponding to different ecological situations, are analyzed and compared. An example for such an analysis is a study by Seidler and Plotkin (2006). They fit a Thomas cluster process (see Box 2.5) to characterize the properties of clustering in the spatial patterns of several tropical tree species and found a strong correlation between cluster size (as measured by the parameter σ of the Thomas process) and the species' mode of dispersal.

Constructing and fitting a point-process model is in some ways the synthesis of the knowledge gained from a thorough exploratory analysis (Illian et al. 2008, p. 445). However, exploratory studies that aim to reveal detailed properties of spatial patterns are frequently dismissed as being "descriptive" and are regarded as less rewarding than studies that test ecological hypotheses. This is a misconception that however still persists, especially for many referees of manuscripts on point-pattern analysis. A precise and detailed description of the properties of a point pattern that reveals biological organization is a prerequisite for deriving specific hypothesis about the underlying processes (McIntire and Fajardo 2009). Consequently, this type of analysis represents an important and challenging part of the scientific process. Clearly, in simple cases, which have dominated most studies of spatial patterns in ecology, description of the pattern is not very challenging and indeed a rather uninformative task. In contrast, we will present examples of complex spatial patterns where finding the "pattern" in the data is challenging. Examples for such complex spatial patterns include the patterns of recruits in tropical forests (e.g., Wiegand et al. 2009), the interaction structure among species in high diversity communities (Wiegand et al. 2007a; Illian et al. 2009), or spatial patterns of diversity in species-rich communities (Wiegand et al. 2007b). We will present these examples in more detail in Sections 4.1.4 and 4.3.

2.4.3.2 Confirmatory Context

Theory or previous studies may suggest a certain hypothesis on the spatial structure embedded in observed point patterns and the researcher may wish to test this hypothesis by means of point-pattern analysis. In this case, the analysis will be conducted in a *confirmatory context* and the point process is the implementation of a (null) hypothesis (or several alternative hypotheses) about the stochastic properties of the observed point pattern. In this case, the null model or point-process model needs to be designed carefully and be based, as much as possible, on ecological considerations to truly reflect the ecological hypothesis. For formal testing of the ecological hypothesis it is convenient to formulate the null model in a manner that creates patterns that are expected in the absence of the particular ecological mechanism of interest. A significant departure of the observed data from the null model is then taken as support for the ecological hypothesis. Because statistical significance is important for this approach, more simulations of the null model are conducted compared to the exploratory approach and specific methods are required to minimize the error rates of the statistical tests.

2.4.3.3 Multimodel Inference and Fitting

In recent studies, the classical approach of testing individual null hypotheses is being broadened to incorporate an approach that tests several alternative hypotheses simultaneously. Here, the aim is to *identify, from among competing hypotheses, the hypothesis that receives the most support from the data*. Placing a hypothesis within the context of several other hypotheses reveals how important that hypothesis is compared to the alternatives (McIntire and Fajardo 2009). These tests usually consider a tradeoff between model fit and model complexity and tackle the question as to whether inclusion of an additional mechanism (or process) within the model can be justified given the data. In this context, multiple point-process models can be viewed as an implementation of alternative pattern generating mechanisms (e.g., clustering and/or environmental heterogeneity) (Shen et al. 2009; Section 4.3.4). Similar to other branches of statistical modeling, comparing alternative hypotheses in a point-pattern context can be done by fitting of parametric models to data. Approaches from information theory or Bayesian statistics are then used to compare the observed pattern with realizations of fitted point-process models that correspond to different hypotheses. This allows the identification of the hypothesis that receives the most support from the data. Such methods, however, are only in their infancy for point-pattern analysis (McIntire and Fajardo 2009) and have not been routinely applied in ecology (but see Tanaka et al. 2008; Shen et al. 2009). The reason for this is related to the fact that likelihood functions are difficult to derive for complex point patterns and suitable simulation techniques need to be developed in concert with approximations of the likelihood function.

2.4.4 Process-Based Models

While point-process models usually do not explicitly describe the processes that created the observed point patterns, a wide class of dynamic process-based models is used in ecology to reveal the processes underlying observed patterns (Grimm et al. 2005). Dynamic process-based models, if they are spatially explicit and individual-based (Grimm and Railsback 2005), can also create point patterns and their output can be analyzed with the techniques presented in this book (e.g., Wiegand et al. 1998; Jeltsch et al. 1999).

Use of individual-based simulation models allows for two types of analyses. First, in a forward analysis, a set of dynamic processes is implemented in the model (e.g., birth, death, dispersal), and the emerging spatial patterns are studied. Execution of simulation experiments, in which the relative importance of processes or certain characteristics of the processes (e.g., dispersal distances) are changed, can provide deep insights into the interplay of different processes in creating spatial patterns. This approach can also be applied to determine how parameters of the processes influence the emerging spatial patterns. For example, Brown et al. (2011) systematically investigated the spatial and nonspatial signals of simulated ecological processes, such as neutral, niche, lottery, Janzen–Connell and heteromyopia models, by comparing first- and second-order measures for the patterns they generated. In another approach, an inverse analysis of the information encoded in spatial point patterns can be used for parameterization and model selection of individual- and process-based models (Wiegand et al. 2004; Grimm et al. 2005; Martínez et al. 2011; Harting et al. 2011).

In the inverse approach, an individual-based and spatially explicit simulation model is constructed that explicitly models the most important processes hypothesized to affect the individuals during their lifetime (i.e., a process-based model). Usually, a suite of alternative models that represent competing hypothesis on the processes that drive the system under study are developed. The objective of the analysis is then to assess which hypothesis is most likely, given a set of observational data. The observational data can be the spatial patterns of the individuals (e.g., trees in a forest), but will also generally comprise a suite of other (nonspatial) characteristics of the system that are important for the scientific question being asked. The task is to find those model parameterizations that match the different observations best for each of the competing hypotheses. For this reason, the approach is called "inverse," because the parameters (usually the model input) are unknown, but the model outputs (i.e., the observational data) are known. Finding the best parameterization is a challenging problem that only recently became possible for more complex simulation models, and can be accomplished by techniques such as rejection filters, Bayesian analyses combined with Markov chain Monte Carlo (MCMC) approaches, or other optimization techniques (Wood 2010; Martínez et al. 2011; Harting et al. 2011). Once the best fit for each hypothesis is determined, the hypothesis that provides the most parsimonious explanation of the data is selected (i.e., balancing fit and model complexity). For example, Martínez et al. (2011) followed this approach and tested a set of alternative individual-based models with data from four contrasting *Pinus uncinata* ecotones in the central Spanish Pyrenees to reveal the minimal subset of processes required for the observed tree-line formation. The spatial patterns used in this study were the intensity functions of different life forms (i.e., seedlings, adults, krummholz) measured along transects across the treeline.

2.5 Methods to Compare Data and Point-Process Models

As described in the previous section, point-process models provide an essential tool for point-pattern analysis and are used to determine whether there is spatial structure in point-pattern data. They are also used to summarize the properties of spatial structure and to test ecological hypotheses. *An essential element in testing ecological hypotheses is comparison of observed patterns with patterns generated by point-process models.* In fact, if a point-process model represents the fundamental division or a suitable null model, it can be used to verify that the observed pattern differs significantly from the patterns produced by the point-process model (Section 2.5.1). However, if parametric point-process models are used to summarize the spatial structure in an observed pattern, a first step is to fit the model to the data. The task here is to determine the unknown parameters of the model (Section 2.5.2). Once the parameters are determined, realizations of the resulting point-process model are used to evaluate model fit.

2.5.1 Evaluate the Match between the Data and a Null Model

Exploratory and confirmatory approaches require comparison of observed patterns and patterns created by a null model to evaluate their match. Summary statistics are used for this purpose as *test statistics*, which compare the characteristics of the observed point patterns with those of the spatial point patterns created by the null model. Theoretically, distribution theory could be used to determine confidence envelopes of a given test statistic under a given null model. However, this approach quickly becomes analytically intractable if more complex null models or edge effects for irregularly shaped observation windows are considered. Therefore, the more practical alternative is to use Monte Carlo simulations to produce realizations of the stochastic point process underlying the specific null model. Because summary statistics involve a substantial reduction of information, we can expect that a given summary statistic captures only certain aspects of the point pattern. Therefore, it is important to conduct comparisons between simulated and observed patterns for several summary statistics, capturing different characteristics of the pattern (Wiegand et al. 2013a).

2.5.1.1 Simulation Envelopes

The most useful procedures (for ecologists) for comparing observed and simulated patterns are those based on functional summary statistics (test statistics) of the data together with simulation envelopes to indicate the range of statistical variation under the null model (Diggle 2003, p. 28). Figure 2.17 provides an example in which the pair-correlation function is used as a test statistic and the null model is a homogeneous Poisson process. Each

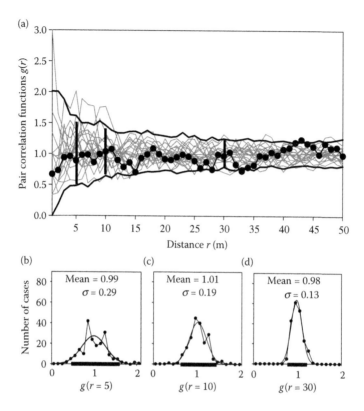

FIGURE 2.17

Construction of simulation envelopes to test for departure from CSR, using the pair-correlation function as an example. (a) Results from application of the pair-correlation function to an observed pattern of 500 points and 199 replicate CSR patterns produced by randomly distributing 500 points within a 1000 × 1000 m² observation window. The line connected by black, filled circles shows the results of the analysis for the observed pattern; whereas, the gray lines show the first 19 simulations of the CSR null model to provide an idea of the variability produced among replicates. The solid black, horizontal lines show the simulation envelopes, which represent 95% simulation envelopes, obtained as the 5th lowest and highest values of the 199 simulations for each scale analyzed, with black vertical bars indicating the range of values falling within the simulation envelops at scales of r = 5, 10, and 30. (b)–(d) Figures showing the distribution of pair-correlation values for the 199 replicates of the null model for scales of r = 5, 10, and 30, respectively. The pair-correlation values from the simulations were fit by a normal distribution, as shown by the black lines, with the black horizontal bars indicating the range of values lying within the simulation envelopes as in panel (a).

simulation of the null model generates a point pattern and for each simulated pattern the test statistic is calculated (gray lines in Figure 2.17a). Because the null model is stochastic, the test statistics of the different simulated patterns do not exactly match one another, but instead exhibit fluctuations around the theoretically expected value (in this case a value of 1). The objective of the simulation envelope approach is to determine whether the test statistic of the observed data falls outside the range of stochastic fluctuations produced by the null model. In the case shown in Figure 2.17, a possible signal of nonrandomness that may be hidden in the data cannot be distinguished from the stochastic fluctuations of the null model.

As suggested earlier, if the number of simulations of the null model is sufficiently high, the distribution of the values of the test statistic produced by the model can be constructed at each spatial scale r. The panels B to D of Figure 2.17 show a few examples in which the distribution of the values of the test statistic can be approximated by a normal distribution. In the example shown, the simulation envelopes are given by the 5th lowest and highest values of the 199 values of the pair-correlation function of the simulated patterns. We used 199 and not 200 simulations because calculation of the error rate is based on the 199 simulations of the null model plus the one observed pattern. To understand this, imagine that the observed pattern does in fact follow the null model. In this case, the risk that the value $\hat{g}(r)$ of the pair-correlation function of the observed pattern lies, by chance alone, outside the bounds of the envelope is 5/200 for lying below and 5/200 for lying above. Thus, the error rate of the two-tailed test is $\alpha = 0.05$. More generally, the simulation envelopes can be defined as the sth highest and lowest values of the test statistic taken from m simulations of the null model, which have an approximate error rate of $\alpha = 2 * s/(1 + m)$ if the test is two tailed and $\alpha = s/(1 + m)$ if the test is one tailed.

If not stated otherwise, *we will use the following conventions in the book regarding the construction and presentation of simulation envelopes*: simulation envelopes will be constructed from the fifth lowest and highest values of a test statistic estimated from 199 simulations of the null model and figures of the results will represent the summary statistic for the observed pattern by closed disks, the simulation envelopes by black lines and the expectation of the simulated point-process model by a bold, gray line.

In Figure 2.17a we can see that the pair-correlation function of the observed pattern lies within the simulation envelope for most distances r and lies slightly outside at only a few distances (i.e., $r = 33, 43$). Thus, the test statistic of the observed pattern is well within the range of stochastic fluctuations of the pair-correlation function for the simulated patterns and we infer that there is no significant departure from the null model. However, it is somewhat more difficult to judge whether there is a significant departure from the null model for values that fall somewhat outside the envelopes. The simulation envelope test is exact for a single spatial scale r, but if we test several scales simultaneously, as is commonly done with this approach, we run the risk of Type I error inflation due to the phenomenon known as "simultaneous inference." For example, if we simultaneously test three scales in our analysis (with $\alpha = 0.05$) the risk that the null model is true, even if the three values of $\hat{g}(r)$ are outside the simulation envelopes, is $1 - (1 - 0.05)^3 = 0.143 > 0.05$. For this reason, the simulation envelope approach has been criticized, especially in combination with cumulative test statistics such as the K-function (Loosmore and Ford 2006).

In spite of these concerns, *simulation envelopes are a valuable tool of exploratory analysis* because the primary concern is not with formal hypothesis testing. The envelopes depict the expected range of values of the summary statistic produced by stochastic effects under the null model and thus contain important information for interpreting the summary statistic of the observed pattern. Note that the simulation envelopes are often called confidence intervals. Because of the problem of simultaneous inference outlined above it is better to not use the term confidence interval, which suggests an exact test, but to use the more neutral term "simulation envelopes." Statisticians also use the term "critical bands." It also must be kept in mind that, when interpreting the results, *slight departures of the observed pattern outside of the simulation envelopes should not be overinterpreted* as being meaningful. The over interpretation of point-pattern data in the literature under these circumstances has been, unfortunately, quite common.

2.5.1.2 Goodness-of-Fit Test

If the objective of a study requires formal hypothesis testing, a Goodness-of-Fit test (GoF) can be used as a complement to analyses based on simulation envelopes (see Diggle 2003, p. 14; Loosmore and Ford 2006; Grabarnik et al. 2011). These tests are not subject to Type I error inflation as described in the previous section. The GoF test collapses the scale-dependent information of a functional test statistic into a single index u_i. The index u_i represents the accumulated deviation of the observed summary statistic from the expected test statistic under the null model, summed up over an appropriate distance interval (r_{min}, r_{max}):

$$u_i = \sum_{r=r_{min}}^{r_{max}} (\hat{H}_i(r) - H(r))^2 \tag{2.7}$$

where $\hat{H}_i(r)$ is the test statistic of the observed pattern $(i = 0)$ and that of the simulated patterns $(i = 1, \ldots m)$, and $H(r)$ is the expected test statistic under the null model. If the expected test statistic $H(r)$ is not known analytically, $H(r)$ can be replaced by

$$\bar{H}_i(r) = \frac{1}{m} \sum_{j \neq i} \hat{H}_j(r) \tag{2.8}$$

which is the average over all test statistics $\hat{H}_i(r)$, excluding the test statistic for index i. Note that $\bar{H}_0(r)$ yields the average over the test statistics of all m simulated patterns and, therefore, provides an unbiased estimate of $H(r)$ under the null model (Diggle 2003, p. 14).

For the GoF test, the u_i are calculated for the observed data ($i = 0$) and for the simulated data ($i = 1...m$) and the rank of u_0 among all u_i is determined. The observed P value of this test is

$$\hat{p} = 1 - \frac{\text{rank}[u_0] - 1}{m + 1} \tag{2.9}$$

For example, if the u_0 computed for the observed pattern was larger than the u_i computed for each of the $m = 99$ simulations of the null model we have rank$[u_i] = 100$ and $\hat{p} = 1 - 99/100 = 0.01$. Details can be found in Diggle (2003), Loosmore and Ford (2006), and Grabarnik et al. (2011).

Note that this approach does not strictly test whether the null model is accepted or rejected. However, it does determine whether the specific index u_0 calculated for the observed pattern for the chosen functional test statistic over the specified distance interval (r_{min}, r_{max}) is within the range of the u_i calculated for the stochastic realizations i of the null model (Loosmore and Ford 2006). This means in practice that the *GoF test is somewhat sensitive to the distance interval selected*. For example, if the departure from the null model occurs only across a small range of scales, for example scales of less than 5 m, but the test is conducted over a much broader interval, say 0–100 m, the ability to detect a true departure from random may be overpowered. For this reason, the distance interval (r_{min}, r_{max}) must be carefully selected prior to the analysis based on ecological arguments as to the range of distances over which departures from the null model are expected to occur. Therefore, the *p*-value alone does not convey the nature of discrepancy between the data and the null model. It should always be used in conjunction with visual inspection of the simulation envelopes and *a priori* hypotheses. No single test statistic should be allowed to override a critical inspection of the plot of the summary statistic and its simulation envelopes (Diggle 2003, pp. 14).

The test statistic u_i is a one-sided test statistic. Similar two-sided test statistics can be constructed based on the standard T-test (Diggle et al. 2007):

$$T_i = \sum_{r=r_{min}}^{r_{max}} \left(\frac{S_i(r) - E(r)}{\sqrt{V(r)}} \right) \tag{2.10}$$

where $S_i(r)$ is the summary statistic for the observed data (for $i = 0$) and the summary statistic of the m simulations of the null model (for $i = 1$ to m), $E(r)$ and $V(r)$ are the mean and variance of the summary statistic of the m simulations of the null model and (r_{min}, r_{max}) is an appropriate distance interval over which departures from the null model are expected. The *p*-value of the test is obtained by estimating T_0 from the data and comparing it to each of the m estimated T_i statistics from the null model. The significance level of the test is given by $(k + 1)/(m + 1)$, where k is the number of simulated T_i greater (or smaller) than T_0 if the departure was positive (or negative).

2.5.2 Fitting Parametric Point Processes to the Data

In many practical applications the goal is to characterize the statistical properties of an observed point pattern as closely as possible. Parametric point processes play an important role in this context because they represent patterns with known (often complex) spatial structures that can be used as a yardstick to better quantify the statistical properties of the observed point pattern. One important example is the Thomas cluster process (Figure 2.10; Box 2.5), the simplest point process that can produce clustered patterns. The cluster properties of this point process can be tuned by two parameters: one parameter determines how many "clusters" are randomly distributed within an observation window and a second parameter determines the average size of the clusters. Each cluster in a Thomas process has a random number of points determined by a Poisson process, and the distances of these points to the center of the cluster are determined by a normal distribution. This point process generates clustered point patterns with known properties that arise from the selected parameters. The question now is how to find the parameters of the Thomas process that will resemble the properties of an observed pattern most closely (i.e., to fit the point process to the data). This is the theme of this section.

There are three main approaches used to fit parametric point processes to data. The first approach is the most convenient. In this case, a summary statistic of the point process can be expressed analytically. This is, for example, possible for the Thomas process where the analytical form of the pair-correlation function and K-function is known. In this case, we only need to estimate the empirical summary statistic of the data and fit it to the analytical expression of the point process. The method of choice for this is the minimum contrast method. The second approach does not rely on summary statistics, but calculates a likelihood function (or approximate pseudolikelihood functions) directly from the raw data. This method is extensively used in regression analysis, but is only possible for a limited range of point processes. In the third and final approach, if a summary statistic cannot be calculated analytically or through a likelihood function, we must rely on simulations to fit the parameters of the point-process model.

2.5.2.1 Minimum Contrast Methods

If the analytical formula of a summary statistic expected for a given point-process model is known, we can conveniently fit the unknown parameters of the model to the observed pattern. In this case, the summary statistic of the observed pattern can be directly compared to the analytical formula. The method of choice for doing this is the minimum contrast method as described in Stoyan and Stoyan (1994), Diggle (2003) and Illian et al. (2008, p. 450f). This method searches for the vector of parameters θ that minimize the difference between the observed summary statistic $\hat{S}$ and the summary statistic $S(\theta)$ for which the dependence on the parameters θ is known analytically (or from simulations). The choice of the summary statistic is important. It should capture a substantial degree of the spatial structure in an observed pattern. Therefore, functional summary statistics will, in general, be used. In this case, the minimum contrast method minimizes the sum of squares

$$\Delta(\boldsymbol{\theta}) = \int_{r_0}^{r_{max}} (\hat{S}(r)^c - S(\boldsymbol{\theta}, r)^c)^2 dr \tag{2.11}$$

with respect to the vector of parameters $\boldsymbol{\theta}$. The values r_0, r_{max}, and c are tuning constants, $\hat{S}(r)$ is the summary statistic of the data and $S(\boldsymbol{\theta}, r)$ is the value of the summary statistic of the point-process model for distance r and parameter vector $\boldsymbol{\theta}$. Note that the minimum contrast method can also be applied for cases in which the analytical solution for $S(\boldsymbol{\theta}, r)$ is not known. In this case, $S(\boldsymbol{\theta}, r)$ may be replaced by an estimate derived from multiple realizations of the underlying point-process model for parameters $\boldsymbol{\theta}$. However, this approach is in general computationally intensive and often not feasible. In practice, the integral of Equation 2.11 is replaced by a sum.

In the ecological literature, the K- or L-function is most commonly used as the summary statistic $S(r)$ (e.g., Batista and Maguire 1998; Plotkin et al. 2000; Potts et al. 2004; Seidler and Plotkin 2006). However, one may also use the pair-correlation function, since its noncumulative property may have certain advantages for model fitting, especially if the point process describes the data only over a limited range of distances. In fact, Wiegand et al. (2007c, 2009) propose the joint use of the pair-correlation function and the L-function.

An immediate question in applying the minimum contrast approach is how to choose appropriate values for the tuning constants r_0, r_{max}, and c. In Equation 2.11, larger values of $S(r)$ are weighted more when $c > 1$ than with a transformation of $c = 1$, whereas a transformation of $c < 1$ weights larger values less. The problem is that the sampling fluctuations increase for the K-function with distance r and the K-function may therefore have a larger influence on the estimation of the parameters at greater distances r. Besag (1977) proposed the transformation $K(r)^{1/2}$, as a mechanism for stabilizing the variance, based on the finding that the variance of $K(r)^{1/2}$ for CSR is approximately independent of distance r. Thus, as a rule of thumb, Diggle (2003) recommends using a value of $c = 0.5$ for the K-function (the same as $c = 1$ for the L-function). However, empirical evidence suggests that a value of $c = 0.25$ would be a better choice for patterns with strong clustering (Diggle 2003), as this reduces the influence of large values of $K(r)$, which are expected under clustering.

The value of r_0 should not be too small when the pair-correlation function is used, because estimates of $g(r)$ at small distances r are often not very accurate, since only a few point pairs exist at shorter distances. This is also indicated by wide simulation envelopes of the pair-correlation function at small values of r (e.g., Figure 2.23d). The maximal distance r_{max}

over which Equation 2.11 is evaluated should not be much larger than the maximal distance over which biological effects (such as clustering) are expected to exist. Slight heterogeneities in the pattern may "contaminate" the summary statistics at larger scales. Also, the value of r_{max} should be smaller than one-fourth of the smallest side of the observation window W to avoid significant bias on the estimator of the summary statistic due to edge effects.

As mentioned earlier, summary statistics involve a substantial reduction of information and use of a single summary statistic may not be sufficient to adequately characterize the underlying point process. In fact, it is well known that several alternative point-process models share the same second-order summary statistics. One clear example of this is provided by the Thomas cluster processes (e.g., Diggle 2003; Tscheschel and Stoyan 2006; Wiegand et al. 2007c). The pair-correlation and K-function of the Thomas cluster process yield exactly the same functional form as does an independent superposition of a Thomas cluster process and a random pattern. It is thus not possible to distinguish between these processes based on second-order statistics alone and other summary statistics, such as the nearest-neighbor distribution function or the spherical contact distribution, need to be explored as well (Stoyan and Stoyan 1994; Diggle 2003; Wiegand et al. 2007c, 2009). This can be done by using the additional summary statistics for model evaluation (e.g., Wiegand et al. 2007c, 2009) or by directly fitting the model to several summary statistics.

As an example of the minimum contrast approach, we fit a Thomas process (Box 2.5) to the data shown in Figure 2.10a. A Thomas process with 100 clusters characterized by a cluster radius of 20 m was used to generate this pattern. The pair-correlation function and the L-function were used in fitting the two parameters σ (the cluster size is 2σ) and ρ ($A\rho$ yields the number of clusters) to the data. To calculate the discrepancy between model and data for a given parameter set $\theta = (\sigma, \rho)$ we replace the integral of Equation 2.11 by a sum and normalize with the total sum of squares of the summary statistics:

$$\Delta_L = \sum_{r=r_0}^{r_{max}} [\hat{L}(r)^c - L(r,\sigma,\rho)^c]^2 \Big/ \sum_{r=r_0}^{r_{max}} [\hat{L}(r)^c]^2$$

$$\Delta_g = \sum_{r=r_0}^{r_{max}} [\hat{g}(r)^c - g(r,\sigma,\rho)^c]^2 \Big/ \sum_{r=r_0}^{r_{max}} [\hat{g}(r)^c]^2 \qquad (2.12)$$

$$\Delta_{L,g} = \sqrt{\Delta_L \Delta_g}$$

We used the tuning constants $r_0 = 1$ m, $r_{max} = 70$ m, and $c = 1$ for the L-function and $c = 0.5$ for the pair-correlation function.

Equation 2.12 was normalized by the sum of squares to obtain error indices with a direct interpretation: the error indices Δ_L and Δ_g give the fraction of the total sum of squares of the empirical g-function and L-function, respectively, which is not explained by the model. To fit the parameters, we minimized the average contrast $\Delta_{L,g}$ of the g- and the L-function. To conduct the fit, we placed a regular 100×100 grid over the parameter space and calculated the error $\Delta_{L,g}$ for each parameterization θ on the grid. The error profile shown in Figure 2.18c has a clear minimum, which indicates that the parameterization can be accurately determined. The fit yields a cluster size of 20 m with 94.4 clusters, which is in fact very close to the parameterization used to generate the pattern (100 clusters). Additionally, the pair-correlation function and the L-function exhibit a good fit (Figure 2.18a,b), as do summary statistics

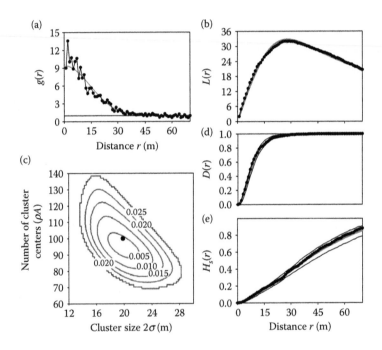

FIGURE 2.18
Minimum contrast method used to obtain best fit of parameter values as applied to the Thomas process in Figure 2.10a. (a) Observed pair-correlation function and best fit. (b) Observed L-function and best fit. (c) Error profile contours for the index $\Delta_{L,g}$ with the parameter values used to generate the pattern shown by the dot. (d) Assessment of model fit using the nearest-neighbor distance function $D(r)$, a function not used in fitting the model. (e) Assessment of model fit using the spherical contact distribution $H_s(r)$, a function not used in fitting the model. Closed disks in panels (a), (b), (d), and (e) are values for the observed data. Solid lines in panels (a) and (b) provide the values associated with the best-fit model. In panel (c), the error index $\Delta_{L,g}$ is an average of the total sum of squares of the empirical g- and L-functions not explained when comparing values for the observed data with those of the model fit for specific values of ρ (y-axis) and σ (x-axis). In panels (d) and (e), simulation envelopes (black lines) and mean values (gray line) were obtained from 199 realizations of the best-fit model, which was obtained using the pair-correlation and L-functions.

not used for fitting the parameters (Figure 2.18d,e). It is also clear, since the Thomas process is stochastic, that a single realization may not exactly fit the parameters used to generate it.

2.5.2.2 Maximum Likelihood Methods and Pseudolikelihoods

The minimum contrast method described in the previous section works well, but only if at least one summary statistic of the point-process model can be calculated analytically. This method, therefore, does not use the raw data but relies on a summary statistic that summarizes certain aspects of the spatial structure of the data. As mentioned earlier, the common practice of using a single summary statistic for model fitting may involve substantial loss of information (Wiegand et al. 2013a). This suggests the use of alternative methods that rely on the raw data. Approaches that use the raw data for model fitting are based on maximum likelihood methods and are widely used in ecology (Hilborn and Mangel 1997; Burnham and Anderson 2002). The *likelihood function* is the probability of observing the data for a given model parameterization. In this framework, the data are fixed and the hypothesis (i.e., the parameters θ) is variable. One can derive statistical methods for parameter estimation, model selection and uncertainty analysis based on the likelihood function. This approach is used extensively for regression modeling (e.g., Burnham and Anderson 2002). However, most current methods of statistical inference work only for models in which the likelihood function is known for each possible model parameterization.

For point processes, it is in general not possible to find the likelihood function and the maximum likelihood method can only be applied for specific classes of point-process models. However, one important point-process model for which likelihood methods can be applied is the heterogeneous Poisson process (Section 2.6.2.9). In this case, the likelihood function can be derived for parametric estimates of the intensity function $\lambda(x)$ (Diggle 2003, p. 104; Møller 2010, p. 320). A natural choice is to use a log-linear model of the form $\lambda(x) = \exp(c_0 + c_1 v_1(x) + \cdots + c_k v_k(x))$ for k environmental variables v_i with coefficients c_i. Recent extensions of this approach also allow inhomogeneous Thomas processes to be fit under the framework of Cox processes (see Section 4.1.5; Waagepetersen and Guan 2009). An implementation of this method is available in the software spatstat (Baddeley and Turner 2005). Ecological examples of this approach are also provided by Shen et al. (2009) and Wang et al. (2011).

In cases in which the likelihood function is not known explicitly, several approximations for deriving parameter estimates are possible, for example, pseudolikelihood methods (Besag 1975; Møller and Waagepetersen 2004, p. 171; Møller 2010, p. 322). These can provide good parameter estimates, although application of pseudolikelihood-based approaches to model selection and uncertainty analysis may not work. There are other alternatives available, such as methods for simulation-based inference for spatial point processes (Møller and Waagepetersen 2004) and computational methods for likelihood-based inference (Møller and Waagepetersen 2007). Note that point-process modeling and fitting their parameters is a highly technical, but lively and rapidly advancing, branch of spatial statistics, which is only loosely related to the core objectives of our book (Ogata and Tanemura 1985; Ripley 1988; Degenhardt 1999; Baddeley and Turner 2000). As such, we will only briefly touch upon those topics here.

2.5.2.3 *Testing Fitted Models*

After a model is fit to the data it should be tested. The first step in this process is to use the fit of the point-process model to generate stochastic realizations of the model output and calculate simulation envelopes for the summary statistic(s) used for fitting the model (Section 2.5.1.1) and perhaps to conduct a GoF test (Section 2.5.1.2). Because the initial model is developed by fitting its parameters to one or two summary statistics of the point pattern, we expect that the resulting analysis should exhibit a good model fit, when assessed by the same statistics. However, the model may still, in fact, not provide a good fit to the data. Therefore, *it is still important to determine whether the point process is suitable for characterizing the pattern.* If the model fits the original summary statistics well, a critical step is to examine the model output with summary statistics that are of a different nature than the summary statistic(s) used to originally fit the model. If second-order statistics, such as the pair-correlation function, were used for model fitting, summary statistics such as the nearest-neighbor distribution functions or, even better, the location-related spherical contact distribution may be used for model testing. An example of this approach is provided in Figure 2.18.

The analyses shown in Figure 2.18 are based on the data presented in Figure 2.10, which were produced by a Thomas process. The observed values for the pair-correlation function and the L-function are quite close to the values that would be produced by the Thomas process, as would be expected since these two sets of statistics were used in fitting the model. However, the nearest neighbor distance function $D(r)$ and spherical contact distribution $H_s(r)$, fit to the data, also match the output of the Thomas process quite well, even though these summary statistics (Figure 2.18d,e) were not used in fitting the model. This result is not particularly surprising since the data were created with a Thomas process, however, in practical applications this is not always the case and the failure of a summary statistic to be fit by the model may yield important information on the characteristics of the observed pattern. An example for this is given in Wiegand et al. (2007c) where a complex cluster process was fit to the pattern data of a topical tree species using the pair-correlation function. However, it is well known that the pair-correlation function often does not fully characterize the spatial properties of a point pattern (e.g., Tscheschel and Stoyan 2006; Wiegand et al. 2013a). Indeed, there are several substantially different Thomas processes which have the same pair-correlation function but differ in their nearest-neighbor distribution functions (Wiegand et al. 2007c). Testing the fitted Thomas process with the nearest-neighbor distribution function $D(r)$ indicated that the observed pattern was likely to be a pattern composed of an independent superposition of a cluster process and a random pattern. This finding had important consequences for the assessment of potential processes that were generating the spatial pattern of the species.

Difficulties may arise if one wishes to validate a fitted model that also includes heterogeneity because the test statistics are affected by spatial heterogeneity as well as by spatial dependence between points. Recent developments in point-process statistics are oriented toward general methods of residual analysis and model diagnostics for such fitted point processes (e.g., Baddeley et al. 2005).

★ 2.6 Dealing with Heterogeneous Patterns

Two types of questions can be asked about spatial point patterns. The first question aims to reveal intrinsic factors and mechanisms, such as competition, dispersal, or facilitation, which may cause nonrandom spatial patterns of organisms (e.g., Wiegand et al. 2009). This is the traditional scope of spatial point-pattern analysis. However, since consistent relationships between species distributions and the physical environment have been observed for a long time (Elith and Leathwick 2009), ecologists are also interested in determining the effects of extrinsic environmental factors on the distribution of organisms (e.g., Kanagaraj et al. 2011a, in the case of tigers *Panthera tigris* or Martínez et al. 2012 in the case of *Pinus unicata*). A general overview of these techniques is provided in the reviews of Guisan and Zimmermann (2000), Guisan and Thuiller (2005), and Elith and Leathwick (2009). While the primary focus of such "species distribution models" is on the larger distributional range of species or populations, environmental heterogeneity may also occur at local spatial scales that are traditionally the domain of spatial point-pattern analysis. For example, the distribution of species in a 50 ha plot of tropical forest at BCI, Panamá is influenced by topographic variables and soil nutrients (e.g., Harms et al. 2001; Comita et al. 2007; John et al. 2007). In cases like this, there is a fundamental ambiguity between clustering and heterogeneity as the causative factor producing spatial pattern. Indeed, it is difficult to distinguish between the two, statistically, without additional biological information (Diggle 2003). This is intuitively clear since both clustering and environmental heterogeneity can generate patterns with locally elevated point densities.

In general, small-scale clustering is more likely to be caused by point–point interactions, whereas aggregation beyond the local neighborhood of an organism is generally attributable to environmental heterogeneity. Environmental heterogeneity, in turn, typically varies along environmental gradients that are often related to topographical features (e.g., Harms et al. 2001; Valencia et al. 2004; Gunatilleke et al. 2006). However, it will be extremely difficult or even impossible to tease apart the two factors with respect to their impact on pattern formation, if the scales at which environmental factors influence a pattern are of the same order of magnitude as the interactions among points. For example, such a situation may arise for seedlings because there may be strong competitive interactions among seedlings operating at the same scale as small-scale environmental heterogeneity that determines the suitability of regeneration sites (Harper 1977). Successful separation of effects in point-pattern analysis will, therefore, only be possible if there is some separation of scales among those effects or if additional information is available (Wang et al. 2010; Wiegand et al. 2012).

Curiously, the extrinsic and intrinsic aspects of spatial pattern formation have generally been explored independently, with the view that interactions between the two are problematic: most techniques of spatial point-pattern analysis assume homogeneous patterns (i.e., no environmental heterogeneity), avoiding a consideration of extrinsic influences on pattern formation, whereas species distribution models generally ignore spatial dependence caused by intrinsic factors or, in some more recent examples, adopt nonstandard techniques to account for spatial dependence (e.g., Dormann et al. 2007; Beale et al. 2010). However, recent developments in point-pattern analysis allow for the exploration of certain classes of inhomogeneous point patterns (Baddeley et al. 2000; Wiegand et al. 2007a; Shen et al. 2009) and hold promise for a reconciliation of these two aspects of spatial analysis. In this chapter, we describe point-pattern methods for dealing with heterogeneous patterns as an approach to reconciling these two components of pattern formation.

We first define the different types of homogeneous and heterogeneous patterns based on the homogeneous Poisson process (Section 2.6.1). Next we provide an overview of methods to detect heterogeneity (Section 2.6.2), and conclude with a discussion of methods to account for heterogeneity (Section 2.6.3). Because most methods dealing with heterogeneous patterns were developed for univariate patterns, we concentrate on the univariate case in this chapter. However, many of the concepts developed here can be translated directly to bi- and multivariate point patterns and in some cases also to marked point patterns. We will present this material in the respective sections of Chapter 4 dealing with those data types.

★ 2.6.1 Classification of Homogeneous and Heterogeneous Patterns

Homogeneous (or stationary) patterns exhibit the same statistical properties over the entire observation window, that is, the chance of observing a specific point configuration at a specific location is independent of location (Illian et al. 2008; Section 2.1.1.3). In this case, differences in point configurations at specific locations result only from random fluctuations that follow the same laws at all locations (Stoyan and Stoyan 1994, p. 191). This property allows us to define the *typical point of the pattern*. The typical point plays an important role in the construction and interpretation of many summary statistics that use neighborhood properties to describe the small-scale correlation structure of the pattern. If the pattern is heterogeneous no such typical point exists.

The two conditions that define a *homogeneous Poisson process* (Box 2.2, case A in Table 2.3) can be used to classify different types of heterogeneous (univariate) point patterns. We use the two conditions (i.e., constant intensity, no interactions among points) as two axes for classification. First, as indicated in the columns of Table 2.3, there are two possibilities for the intensity function. It can be either approximately a constant λ or a function $\lambda(x)$ of location x. Second, as indicated in the rows of Table 2.3, there are three possibilities for point interactions: (1) no point interactions; (2) homogeneous interactions (i.e., the statistical properties of specific point configurations are the same at all locations within the observation window); or (3) heterogeneous interactions (i.e., the statistical properties of typical point configurations are different at different locations in the observation window). This two-way classification results in six different types of patterns, which we present briefly in the following sections.

BOX 2.2   THE HOMOGENEOUS POISSON PROCESS

A homogeneous Poisson process, often referred to as CSR, is characterized by two fundamental properties:

1. The intensity λ of the process (i.e., the mean point density in a unit area) is a constant, which means that the number of points in a study plot of area A follows a Poisson distribution with an expected mean of λA. In this case, the probability of finding a point in an infinitesimally small disk of center x and area dx yields λdx.
2. The points of the process are independently distributed, which means that there is no interaction between the points of the pattern determining their locations.

TABLE 2.3

Categories of Homogeneous and Heterogeneous Univariate, Point Patterns Based on Departures from the Homogeneous Poisson Process

Interaction among Points	Intensity Constant	Intensity Is Function of Location (x, y)
None	(A) Homogeneous Poisson process	(B) Heterogeneous Poisson process
Homogeneous[a]	(C) Homogeneous point process with interactions among points	(D) First-order heterogeneous pattern or dispersal limitation or range expansion/contraction
Heterogeneous[b]	(E) Second-order heterogeneous pattern	(F) First- and second-order heterogeneous patterns

[a] The interactions among points follow the same stochastic rules within the entire observation window.

[b] The stochastic rules governing the interactions among points change within the observation window, for example, repulsion at some places and aggregation at others.

2.6.1.1 Homogeneous Poisson Process (A)

The homogeneous Poisson process (case A in Table 2.3; Box 2.2) represents the case of complete absence of spatial structure in a univariate pattern and is also referred to as *Complete Spatial Randomness (CSR)*. From the theoretical point of view, the homogeneous Poisson process has a central role in point-process statistics. For example, the textbook on spatial point patterns by Illian et al. (2008) dedicates an entire chapter (of the seven) to the homogeneous Poisson process (their Chapter 2). Many of the theoretical concepts of point processes can be analytically explored for the homogeneous Poisson process, and it is a building block for more complex point-process models. However, the homogeneous Poisson process is also fundamental in ecological applications because it represents the fundamental division for univariate patterns. As discussed in Section 2.4.1, the fundamental division is used to determine whether the data contain a signal that can be distinguished from pure stochastic effects for the given data type with the observed sample size. The homogeneous Poisson process is, therefore, a null model used as a reference point for distinguishing between clustering (Figure 2.10) and repulsion (Figure 2.11). Implementation of a homogeneous Poisson process is simple; it requires only the calculation of random coordinates for independently distributed points.

2.6.1.2 Heterogeneous Poisson Process (B)

A heterogeneous Poisson process exists if there are no interactions among points but external factors alter the intensity of points at different locations within the observation window (case B in Table 2.3; Box 2.3; Figure 2.7). This

---

BOX 2.3   THE HETEROGENEOUS POISSON PROCESS

A heterogeneous Poisson process is characterized by two fundamental properties:

1. The intensity of the process depends on location x (i.e., the probability that there is a point in an infinitesimally small disk of center x and area dx is $\lambda(x)dx$).
2. The points are independently scattered, which means that there is no interaction between the points of the pattern.

An implementation of the heterogeneous Poisson process generates points at random coordinates within the entire observation window, but only accepts a point with a probability that is proportional to the intensity function at the location of the point. Thus, a homogeneous Poisson process is independently thinned with a "thinning surface" which is proportional to the intensity function $\lambda(x)$.

---

represents the simplest case of a heterogeneous pattern and is completely determined by the intensity function $\lambda(x)$, that is, the intensity is a function of location. Modeling the intensity function $\lambda(x)$ is an important branch of statistical modeling in ecology with a long tradition (see Section 2.6.3.4 below). Implementation of a heterogeneous Poisson process is simple; point positions are initially distributed by a homogeneous Poisson process and are then accepted or rejected with a probability proportional to the intensity function $\lambda(x)$ depending on their location x. The probability of accepting a point is given by $p(x) = \lambda(x)/\lambda^*$, where λ^* is the maximal value of the intensity function within the observation window. Note that the procedure of rejecting or accepting points based on the function $p(x)$ is called *"independent, location-dependent thinning"* (Illian et al. 2008, p. 119 and p. 365). In this way, the points are independent of each other (guaranteed by the selection of tentative points) but they follow the intensity function $\lambda(x)$ (guaranteed by the thinning). Figure 2.7a shows an example for a simple thinning function and Figure 2.7b is a realization of the corresponding heterogeneous Poisson process.

The heterogeneous Poisson process has important applications in ecology, mostly as a null model that conditions on known environmental dependence (e.g., Shen et al. 2009; Cheng et al. 2012; Lin et al. 2011) or on observed, broader-scale, spatial structure (e.g., Wiegand and Moloney 2004; Perry et al. 2006; Wiegand et al. 2007a). This can be used to reveal small-scale interactions among points, after accounting for the environmental dependence or the broader scale structure of the pattern (Wiegand et al. 2007a; Wang et al.

2010). The heterogeneous Poisson process is especially important for bivariate patterns (Wiegand et al. 2007a, 2012; see Section 4.2.4).

2.6.1.3 Homogeneous Point Processes with Interactions (C)

This is the target class of most applications of point-pattern analysis (case C in Table 2.3). A point may occur with the same probability in any small area within the observation window (i.e., the intensity function is a constant λ), but in contrast to the homogeneous Poisson process there are interactions among the points of the pattern that follow the same stochastic rules at all locations within the observation window (i.e., homogeneous interactions). Examples for homogeneous patterns with interactions that yield aggregation and hyperdispersion are shown in Figures 2.10 and 2.11, respectively. Note that cases A and C in Table 2.3 comprise the homogeneous (or stationary) patterns.

If the same interaction type occurs throughout the observation window (e.g., only clustering), the statistical properties of the pattern are also the same over the entire observation window. In this case, we can summarize the statistical properties of the pattern in terms of properties of the *typical point*. This is the basis for many summary statistics. For example, in Section 2.3 we make extensive use of the typical point of the pattern for defining summary statistics and for interpreting the properties of homogeneous patterns. A large variety of methods exist to characterize homogeneous point patterns.

2.6.1.4 First-Order Heterogeneous Point Processes (D)

A "first-order heterogeneous" point process exits if there are interactions among points and external factors also exert a spatial influence on the intensity of points (case D in Table 2.3). In this case, there are no reciprocal effects between the intensity and point interactions. Conceptually, this type of point process can be modeled by a two-step procedure, in which a homogeneous point process with interactions (case C in Table 2.3) is independently thinned according to $p(x) = \lambda(x)/\lambda^*$ as in case (B). Baddeley et al. (2000) called this type of process "*second-order intensity reweighted stationary.*" In analyzing this type of pattern one can use the so-called "*inhomogeneous second-order summary statistics,*" introduced by Baddeley et al. (2000), to reveal the second-order properties of the underlying homogeneous point process, since the point interactions in this class of pattern are independent of the intensity function. This approach allows for a characterization of the statistical properties of the point interactions, despite the environmental dependence forced on the pattern by the spatially varying intensity function $\lambda(x)$. We will present this technique in Section 2.6.3.5, with technical details provided in Sections 3.1.2.5 to 3.1.2.7.

2.6.1.5 Second-Order Heterogeneous Point Processes (E)

Heterogeneity in the intensity $\lambda(x)$ of a process (as given in types B and D; Table 2.3) is only a special, but nevertheless important case of more general heterogeneity (Table 2.3). Even if the intensity of a process is constant, the process may be heterogeneous if the nature of the point interactions changes in response to environmental conditions. We call such a pattern *"second-order heterogeneous."* For example, the cluster size of a cluster process may depend on location, or a pattern may be hyperdispersed in some areas of the observation window, but aggregated in others. An interesting example of such heterogeneity is provided in Couteron et al. (2003). They studied tiger bush in Niger within a 100×50 m observation window in which all woody plants with a height >0.5 m were mapped. A substantial part of the plot was virtually devoid of woody individuals. However, when looking inside the irregular observation window adapted to the vegetated area, they found that the distribution of woody plants tended to be clumped in some subplots, while being characterized by inhibition in others. The observed shift from clumping to inhibition was consistent, with patterns of water resource limitation occurring within the study plot.

Internal, as well as external mechanisms of population dynamics may also produce patterns with a heterogeneous appearance that are not related to environmental factors, but are related to the spatial effects of population dynamics, such as those caused by dispersal limitation, invasion, or population contraction (Fang 2005). In these cases, areas that are potentially suitable for the species are not occupied because of internal effects related to population dynamics.

2.6.1.6 First- and Second-Order Heterogeneous Point Processes (F)

While thinning of a homogeneous point pattern with a thinning function $p(x)$ substantially expands the range of patterns that can be analyzed, it has to be noted that the condition of second-order, intensity-reweighted stationarity is rather restrictive (Illian et al. 2008, p. 282), because it does not include cases in which the typical spatial structures vary with location (other than being thinned from a homogeneous pattern). For example, hard-core processes in which the hard-core distance varies with location are not second-order, intensity-reweighted stationary. The same is true for cluster processes in which the cluster size varies with location. We also may have direct interactions between the intensity and the point interactions. Thus, a pattern may be heterogeneous both in its intensity and in its second- and higher-order point interactions (case F, Table 2.3). Clearly, this is the most complicated type of pattern to be analyzed and research to find methods for characterizing such patterns is only in its infancy. In Section 2.6.2 we present some methods that allow detection of this type of heterogeneity, but no general methodology exists to deal with this case. As a consequence, we will not deal with these types of patterns any further.

★★ 2.6.2 Testing for Heterogeneity

Because environmental conditions are usually not completely homogeneous, most patterns ecologists encounter are to some extent heterogeneous. *One of the tasks of point-pattern analysis is to determine whether or not environmental dependency contributes substantially to the assembly of an observed spatial pattern.* This knowledge is important in selecting the appropriate analytical techniques, since analysis of heterogeneous patterns with tools designed for homogeneous patterns may produce misleading results (an example is given in Section 2.6.3.3).

In general, it is difficult to determine whether a pattern observed in a specific observation window is heterogeneous, particularly when no additional information is available. Under these circumstances, only certain aspects of heterogeneity can be investigated (Illian et al. 2008). One reason for this is that a given pattern is only one realization of an underlying stochastic point process, and particular realizations of homogeneous processes may have a heterogeneous appearance. Similarly, realizations of heterogeneous point processes may have a homogeneous appearance. This problem is especially true when the range of dependence among points is relatively large compared to the size of the observation window (Figure 2.19). However, in such cases, point-pattern analysis may still reveal the properties within local clusters.

Additional information that can be helpful in determining whether environmental dependency contributes substantially to the spatial structure of a point pattern can be provided by maps of environmental covariates

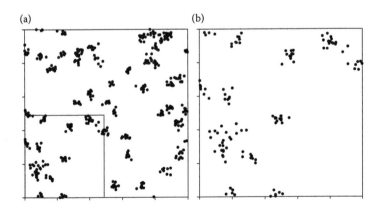

FIGURE 2.19
Homogeneous cluster process. (a) Pattern with 50 randomly distributed clusters that have on average 10 points. (b) Lower left quadrant of panel (a) as outlined. Panel (b) demonstrates that a small sample from a homogeneous pattern can be heterogeneous in appearance; in this case there are no points located on the lower right side of the small window.

that are suspected of influencing the local point density of the pattern. For example, the distribution of tree species in tropical forests on local scales (i.e., <1 km^2) is often influenced by topography or soil nutrients (e.g., Harms et al. 2001; John et al. 2007). Several snapshot or replicate patterns, although rarely available, can be helpful in deciding whether the pattern is heterogeneous. Because spatial patterns at local spatial scales that arise from intrinsic factors, for example, limited dispersal, can also be similar to patterns resulting from extrinsic factors, for example, habitat heterogeneity, any additional biological information on potential interactions among the points of the pattern can be used to help evaluate the evidence for heterogeneity.

In the following, we present two general categories of approaches for testing the various aspects of heterogeneity that can occur in univariate patterns. First, we consider approaches that assess the pattern as observed and do not use additional information on environmental covariates in attempting to identify potential heterogeneities (Sections 2.6.2.1 through 2.6.2.6). In this case, an initial step is often to derive an estimate of the intensity function for the observation window and inspect the results for possible systematic gradients in the point density function (Section 2.6.2.1). This a good starting point, particularly since a nonconstant intensity function is the most frequent type of heterogeneity encountered in practical applications. An alternative approach in diagnosing potential heterogeneities is to use knowledge about the expected behavior of different summary statistics when encountering first-order heterogeneous patterns. In this case, we begin by studying the behavior of the K-, L- and pair-correlation functions under typical cases of first-order heterogeneity (Section 2.6.2.2). We use this knowledge, together with information on the range of point interactions (e.g., direct competition among trees is expected to occur only within 30 m), to devise a simple test, based on CSR, for determining the nature of the heterogeneity (Section 2.6.2.3). Additionally, we can compare second-order summary statistics among different subwindows within the original observation window and look for systematic differences indicating an underlying heterogeneity (Sections 2.6.2.4 and 2.6.2.5).

A second general category of approaches for studying heterogeneity includes a consideration of environmental covariates or categorical habitat types suspected of influencing the intensity function in the analysis (Sections 2.6.2.6 through 2.6.2.9). A natural, first step in applying this type of approach is to investigate whether a point pattern exhibits a significant association with particular covariates, before incorporating the covariates in the analysis of the intensity function. In developing these ideas, we first present methods that can detect the association of categorical (Sections 2.6.2.6 and 2.6.2.7) and continuous (Section 2.6.2.8) covariates with the spatial distribution of points. We then conclude the section by presenting methods that can be used to estimate the intensity function based on the influence of environmental covariates (Section 2.6.2.9).

2.6.2.1 Nonparametric Intensity Estimation

The most apparent type of heterogeneity is a nonconstant intensity function. Therefore, it is natural to begin investigation of potential heterogeneity of a pattern by attempting to estimate the intensity function. The intensity function $\lambda(x)$ can be estimated nonparametrically using *kernel estimation* (Box 2.4). This is particularly appropriate for situations in which no additional information on environmental variables is available, as it allows the detection of potential gradients in the intensity that may point to underlying heterogeneity (i.e., type B, D, or F heterogeneity; Table 2.3). Figure 2.20b shows the results of such an analysis using the simple box kernel, which

---

**BOX 2.4 NONPARAMETRIC KERNEL ESTIMATE
OF THE INTENSITY FUNCTION**

Intuitively, the intensity $\lambda(x)$ at a given point x is the mean number of points within a small disk centered at x, divided by the area of the disk. This is the basic idea for nonparametric kernel estimation. The simplest approach is, therefore, to count all points j within the observation window W that are located within a disk of radius R centered on the focal location x, and divide this number by the area of the disk overlapping W (Diggle 2003, p. 117). The radius R is often referred to as the bandwidth (with symbol h). The so-called "box kernel"

$$k_R^B(d_{xj}) = \begin{cases} 1 & d_{xj} \leq R \\ 0 & \text{otherwise} \end{cases}$$

weights all points j within distance R of location x with a weight of one and all points further away with a weight of zero. The dummy variable d_{xj}, therefore, returns the distance between x and the location of j within W (i.e., $|x - x_j|$). By moving the disk over all locations x in W, the intensity $\lambda(x)$ can be estimated for all locations x within the observation window.

One negative aspect of the box kernel is that it tends to produce rough surfaces, especially in areas of low point density, where the disk shape is often still visible. Smoother estimates can be obtained with kernel functions that weight points less the further away they are from the focal location x. A popular kernel for this purpose is the Epanechnikov kernel (Stoyan and Stoyan 1994). To derive smooth intensity surfaces the Epanechnikov kernel weights each point j at distance d from the focal location x with a kernel function $k_R^E(d_{xj})$:

$$k_R^E(d_{xj}) = \begin{cases} 2\left(1 - \dfrac{d_{xj}^2}{R^2}\right) & d \leq R \\ 0 & \text{otherwise} \end{cases}$$

Note that the formulas for 2-d kernels that are used in estimating the intensity centered on a location x must conform to the following condition: $\int_0^R k_R(r)2\pi r\,dr = \pi R^2$.

generally produces somewhat rugged estimates of the intensity function, whereas the Epanechnikov kernel produces noticeably smoother estimates (Figure 2.20c).

The pattern shown in Figure 2.20a is a realization of a heterogeneous Poisson process with a linear intensity function $\lambda(x, y) = ay$ (shown in Figure 2.7a), where x and y range between 0 and 1 and a is a normalizing constant. The estimates produced by either the box or the Epanechnikov kernel point to a strong spatial trend in the intensity function. Although the trend is obvious in Figure 2.20b and c, spurious spatial structure, which is difficult to distinguish from the true signal in the data, can also be observed in the estimates of the intensity function. This points to a distinct tradeoff between not smoothing enough (small R), which introduces too much spurious small-scale spatial structure related to stochastic effects into the estimate (e.g., Figure 2.20d), and oversmoothing (large R), which dampens the depiction of the trend by averaging over too broad a scale. Clearly, the ideal kernel estimate must find a balance between these two extremes. This is accomplished by choosing the correct scale for smoothing, which removes the stochastic fluctuations in the observed pattern and reveals a close approximation of the true structure of the intensity function.

This tradeoff can be illustrated by considering a random pattern with n points within an observation window W of area A (i.e., the intensity is $\lambda = n/A$), where the probability of finding k points within a circular moving window with radius R is given by the Poisson distribution $P(k, \lambda\pi R^2)$. In this case, $\mu = \lambda\pi R^2$ is the expected number of points within a circular window of radius R, since this is the intensity λ multiplied by the area πR^2. If the radius R is small, the relative variability in the range of possible values k around the expectation μ is large. For the Poisson distribution, the 95% confidence interval (2.5%, 97.5%) for k is given, approximately, by $\mu \pm 2\mu^{0.5}$. If we wish to limit the range of the 95% confidence interval (i.e., width $4\,\mu^{0.5}$) to no greater than half of the expected value μ, we have to solve for $4\,\mu^{0.5} = 0.5\,\mu$. This yields a mean number of points of $\mu = 64$ and a 95% range of (48, 80) (Figure 2.21). Similarly, if we accept a 95% confidence interval that is as large as the

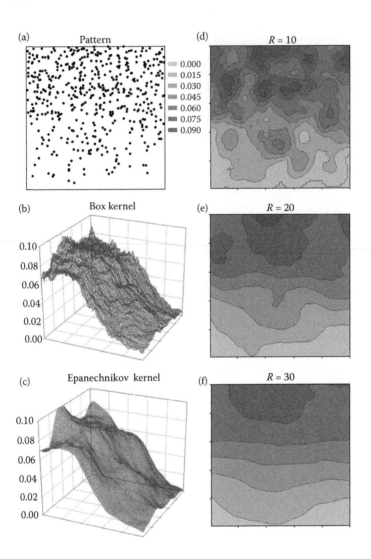

FIGURE 2.20
Nonparametric kernel estimates of the intensity function. (a) Observed pattern, comprising 485 points within a 100×100 m^2 observation window (i.e., $\lambda = 0.0485$). (b) Intensity function $\lambda(x)$ estimated with a box kernel with a bandwidth R equal to 1/5 the width of the observation window (i.e., 20 m). (c) $\lambda(x)$ estimated with an Epanechnikov kernel with same bandwidth used in panel (b). (d–f) Contour plots of $\lambda(x)$ estimated with an Epanechnikov kernel with different bandwidths R.

expected value, we obtain a mean of $\mu = 16$ and a 95% confidence interval of (8, 24) (Figure 2.21).

Once we decide on the amount of variability we are willing to accept, we can determine the radius R of the moving window we need to employ in obtaining our estimate of $\lambda(x)$. For the first example above, in which we accept the variability associated with a mean of $\mu = 64$, we have to solve the

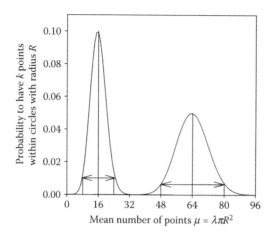

FIGURE 2.21
Tradeoffs due to under- and oversmoothing when obtaining a nonparametric kernel estimate for CSR patterns with different intensities λ (= n/A). Refer to Section 2.6.2.1 for more details on how to interpret this figure.

relationship $\mu = \lambda\pi R^2 = 64$ to estimate the associated radius R of the moving window. This yields a value of $R = 4.5\ \lambda^{-0.5}$. Therefore, for a pattern with 500 points within a 100×100 m observation window, we obtain $R = 20.2$. Thus, the radius R of the moving window should be selected to be proportional to the inverse of the square root of the number of points, that is, $R \sim n^{-0.5}$ (Diggle 1983).

It should be pointed out that the above calculation is only a rough estimation of the ruggedness of the intensity function for a homogeneous Poisson process. Indeed, Illian et al. (2008, p. 115) note that there is no general recipe for determining the optimal value R for calculating the bandwidth. Additional information on the processes underlying the distribution of the points is required to distinguish between spurious smaller-scale spatial structure and meaningful spatial structure in the distribution of $\lambda(x)$. However, even with additional information that helps in determining a suitable bandwidth, one should still consider a range of values of R and visualize the resulting intensity surfaces (e.g., Figure 2.20d–f) in determining the correct scale for smoothing. For example, returning to the pattern shown in Figure 2.20a, there are 485 points within a 100×100 m observation window, resulting in a bandwidth of $R \approx 20$ m if we decide to focus on smoothing with a value of $\mu = 64$ points. The corresponding intensity estimate is shown in Figure 2.20e. A bandwidth of $R \approx 10$ (i.e., a mean of $\mu = 16$) is certainly too small as shown in Figure 2.20d, as it produces spurious small-scale structures that arise only because of the underlying stochasticity. This has no particularly relevant ecological meaning. A bandwidth of $R = 30$ m ($\mu = 138$) produces a smoother estimate (Figure 2.20f), that may already gloss over the ecologically relevant details underlying the pattern.

A related parametric method that can be used to detect spatial trends is to fit trend surfaces to the data using approaches such as regression analysis based on polynomial functions of the coordinates x and y (e.g., Law et al. 2009).

2.6.2.2 Inspect Pair-Correlation and L-Functions

While the estimation of the intensity function, as described in the last section, can provide a first indication of possible nonrandom trends and gradients in a point pattern, it is often not sufficient evidence for concluding that the pattern is heterogeneous. This is especially true when the number of points is relatively low and a large moving window would be required to yield low variability in localized estimates of the intensity function. In this case, the size of the moving window may become much larger than the range of expected point–point interactions, potentially glossing over biologically relevant information. As a consequence, we present an alternative test in this section for diagnosing potential heterogeneity of types B and D (i.e., nonconstant intensity; Table 2.3) that avoids this problem by analyzing the shape of suitably chosen summary statistics, such as the pair-correlation function and the L-function, and comparing the results with the shapes expected for known patterns.

The test for heterogeneity introduced here is based on the fact that the pair-correlation function of a homogeneous point process must eventually approach a value of 1 asymptotically. The L-function, in turn, must eventually approach a value of 0. In other words, second-order effects (i.e., direct interactions among points) will eventually fade away with increasing spatial scale, mirroring a fundamental property of interactions in physics. As a consequence, possible (first-order) heterogeneity is indicated if the value of the pair-correlation function does not approximate a value of 1 for scales clearly larger than the interaction range, and L does not approach a value of 0. However, one should note that this approach toward detecting heterogeneity assumes separation of scales, that is, direct interactions among points operate at smaller distances than typical variations in point density due to environmental heterogeneity.

Before evaluating the shape of the summary statistics as a diagnostic for heterogeneity, we must first determine the expected shape of the pair-correlation function and the L-function under heterogeneity. Following Wiegand et al. (2007a), we derive approximations of the K-function for two simple cases of heterogeneity (gradient and patch type heterogeneity), which allows us to assess the typical shape of the pair-correlation function and the L-function under these conditions. Remember that the univariate K-function can be defined using the quantity $\lambda K(r)$, which is the average number of points within distance r of the typical point of the pattern.

As an example of an analysis of *gradient heterogeneity*, we consider a pattern with n points within a 1×1 unit square observation window that has been produced by a linear gradient in the x-direction, that is, $\lambda(x, y) = \alpha + \beta x$ with $\alpha \geq 0$ and $\beta \geq 0$ (similar to the pattern shown in Figure 2.22a). Because we have a total of n points, the value of β yields $\beta = 2n-2\alpha$ (noting that $n = \int_0^1 \int_0^1 \lambda(x,y)dxdy$). Thus, $\lambda(x, y) = \alpha + 2(n - \alpha)x$. At each position x of the gradient (except for those too close to the border), the expected number of points within distance r is the intensity multiplied by the area πr^2, since departures from the mean intensity in the x-direction cancel out due to the linearity of the gradient. Thus, without consideration of edge effects, we find in a first approximation for scales of $r \ll 1$:

$$K(r) \approx \frac{1}{n} \frac{1}{n} \int\limits_0^1 \int\limits_0^1 \lambda(x,y)[\lambda(x,y)\pi r^2]dxdy, \qquad (2.13)$$

which yields

$$K(r) \approx \left(\frac{4}{3} - \frac{2\alpha}{3n} + \frac{\alpha^2}{3n^2} \right)\pi r^2 \qquad (2.14)$$

and, because $g(r) = (1/2\pi r) \, dK/dr$, we find

$$g(r) \approx \left(\frac{4}{3} - \frac{2\alpha}{3n} + \frac{\alpha^2}{3n^2} \right) \qquad (2.15)$$

Additionally, if the intensity has a value of 0 at $x = 0$, then $\alpha = 0$ and for small values of distance r we find that $K(r) \approx 4/3 \, \pi r^2$, $g(r) \approx 4/3$, and $L(r) \approx [K(r)/\pi]^{0.5}$ $-r = 0.1547 \, r$. Instead, if there is no gradient (i.e., no heterogeneity), then $\beta = 0$ and $\alpha = n$, and consequently $K(r) = \pi r^2$, $L(r) = 0$, and $g(r) = 1$, as expected.

From the above, it can be observed that a gradient in the intensity function leads, in a first approximation, to elevated values of the K- and g-functions at small scales r (relative to the dimensions of the observation window) and a linear increase in the L-function. The maximal effect for a linear gradient (a factor of 4/3) occurs if the intensity is zero at one border (i.e., $\alpha = 0$ in our situation). If the gradient is less pronounced, the effect is smaller (i.e., $1 < g(r) < 4/3$). Figure 2.22a shows a pattern with $n = 100$ points generated with the linear gradient $\lambda(x, y) = \beta y$ for the y-axis. The corresponding pair-correlation function (Figure 2.22c) approximates a value of 4/3 at smaller scales (i.e., $r < 0.25$ which is one-fourth of the width of the observation window). However, note that departures from the expected value occur for larger scales because the approximation of Equation 2.13 does not consider edge effects. Similarly, the L-function shows a linear increase that coincides well with the expectation (Figure 2.22e).

In the second case, we analyze an example of *patch heterogeneity* in which a univariate point pattern with N points within the 1×1 unit square

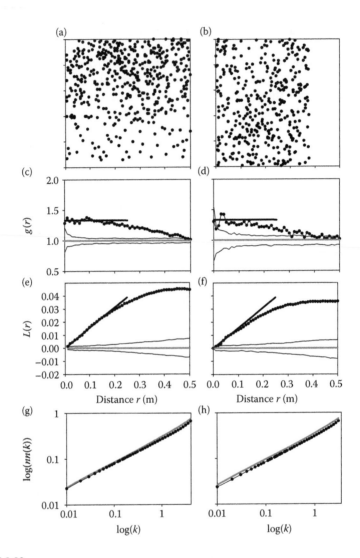

FIGURE 2.22
Influence of simple gradient and patch heterogeneities on the shape of the pair correlation, *L*-, and mean distance to the *k*th nearest-neighbor. (a) Pattern produced by a simple linear gradient. (b) Pattern produced by a simple patch heterogeneity. Panels (c) through (h) line up under the patterns they analyze. In panels (c) through (f), the bold solid lines represent analytical approximations of the functions for the heterogeneous patterns, without consideration of edge correction. Panels (c) through (f) also contain simulation envelopes with mean values for the underlying CSR null model for comparison to the analyses of the heterogeneous patterns. The pair-correlation function typically exhibits a fixed value greater than one at short distances, whereas the *L*-function exhibits a linear increase. This behavior is often indicative of heterogeneity. In the case of a patch heterogeneity, the observed mean distances to the *k*th neighbor (panel h) are always a factor of $-\log(c)/2$ less than the expectation under CSR. For the pattern in panel (b) we observe a difference of 0.058, which is very close to the expectation of 0.062, since $c = 0.75$.

observation window forms an internally homogeneous patch covering a proportion $c = 3/4$ of the study region (Figure 2.22b). There are no points outside the patch and the pattern inside the patch is random (i.e., CSR). The expected intensity under a homogeneous pattern yields $\lambda = 1/n$. However, since the n points are only distributed within a patch of area c, they have a local density of $\lambda_l = (1/c)\,\lambda$ and disks with radius r around the points of the pattern contain, in a rough first approximation, $\lambda K(r) = (1/c)\,\lambda\,\pi r^2$ points. Thus, for smaller scales r we find, approximately, that $K(r) = (1/c)\,\pi r^2$ and $g(r) = (1/c)$. For larger scales r, the disks around the points of the pattern increasingly overlap the area outside the patch. As a consequence, the simple approximation made above overestimates the number of points and the pair-correlation function declines. This is analogous to the problem of edge correction (Section 2.3.2.5).

In the case of patch heterogeneity, we can approximate the expectation of the mean distance $nn(k)$ to the kth neighbor. Recall that the expectation of $nn(k)$ under CSR yields $nn(k) = (k/(\lambda\pi))^{0.5}$ (Table 2.2). If the pattern is a CSR pattern in a patch covering a proportion c of the observation window, the local point density inside the patch yields $\lambda_l = (1/c)\,\lambda$, and the mean distances to the kth neighbor are related to those expected for a CSR pattern with an intensity λ_l. Thus, the observed $nn(k)$ under the patch heterogeneity will yield $nn(k) = (c * k/(\lambda\pi))^{0.5}$. If we plot $nn(k)$ against k on a double-logarithmic scale, we find $\log(nn(k)) = 0.5[-\log(\lambda\pi) + \log(k) + \log(c)]$, whereas the expectation of the corresponding CSR yields $\log(nn(k)) = 0.5[-\log(\lambda\pi) + \log(k)]$. Therefore, in this case, the observed $nn(k)$ and the expectation under CSR differ by an additive term $\log(c)$. Because c is smaller than one, the observed $nn(k)$ will therefore always be below the expected value, with a constant difference of $\log(c)/2$ (Figure 2.22h). However, this is not the case for the gradient heterogeneity (Figure 2.22g).

Summarizing the two basic cases of heterogeneity, that is, gradient and patch, we find that both cause the same characteristic behavior of the pair-correlation function at smaller scales (say smaller than one-fourth of the width of the observation window): the pair-correlation function is larger than one with a constant value $g(r) = g_0$ and the K-function yields $K(r) = g_0\,\pi r^2$. As a consequence, the L-function $L(r) = (K(r)/\pi)^{0.5} - r$ increases linearly: $L(r) = (\sqrt{g_0} - 1)r$ with slope $(\sqrt{g_0} - 1)$. *Thus, a steep linear increase of the L-function at smaller scales, as shown for the gradient case in Figure 2.22e, may be an indicator of larger scale heterogeneity.* This phenomenon is also called "*virtual aggregation*" (Wiegand and Moloney 2004) and is explained in more detail in Section 2.6.3.1. Additionally, in the case of patch heterogeneity, we find that the observed value for $nn(k)$ is always $\log(c)/2$ smaller than the expectation under CSR, where c is the proportion of the study area covered by the patch. The metric $nn(k)$ is thus an excellent tool for verifying this type of heterogeneity.

Note that a linear gradient (with $\alpha = 0$) yields approximately the same pair-correlation function as an internally homogeneous patch covering three-quarters of the study window. Thus, the potential effect of larger gaps in the point pattern exceeds the effects of gradients, if the gaps are larger than one-fourth of the study window (Figure 2.22). The two examples given also show that "pure" gradient heterogeneity (as shown in Figure 2.22a) has a relatively moderate effect on the second-order statistics, and that the effect of empty areas is much larger, if the gaps are larger than one-fourth of the study window.

2.6.2.3 CSR at Scales Larger than the Expected Interaction Range

The findings presented in the previous section illustrate the typical shape of the pair-correlation and *L*-function under heterogeneity and provide indications for identifying potential heterogeneity. However, this approach is not sufficient for a more formal test. It is based only on the shape of the summary statistics and does not consider the degree of stochasticity, which can cause substantial variation in the shape of the summary statistics among realizations of the same point process, especially if the number of points is low. We can devise a relatively simple test to determine whether the intensity is nonconstant (cases B, and D in Table 2.3), based on examining the shape of second-order summary statistics under heterogeneity, as explained in the previous section. The idea is to compare an appropriate summary statistic, such as the pair-correlation function or the *L*-function of the observed data, to that arising from a completely random pattern. The assessment is generally made only at scales larger than the expected range of dependence in the interactions between the points. Thus, departures from randomness at scales larger than the expected interaction range may be due to environmental heterogeneity. In this case, it is useful to have information on the maximal expected interaction ranges for the points of interest.

As an example of this approach, Getzin et al. (2008) assessed heterogeneity in two old-growth, Douglas-fir forests plots on Vancouver Island, Canada. They analyzed the spatial pattern of all mature, adult trees for all species occurring within the plots (Figure 2.23a,c). Their assumption was that the pattern of all mature adult trees should be a good indicator of environmentally driven habitat quality, caused by factors such as rock outcrops or wet drainage sites. The argument was that mature trees have undergone excessive thinning and are expected to have exploited all available sites, with the exception of recently produced canopy gaps arising through the mortality of large trees. Thus, potential heterogeneities caused by environmental habitat factors common to all species should be detectable by contrasting the pattern of all mature trees to random patterns (i.e., CSR). As shown in the previous section, large-scale heterogeneity will cause locally elevated tree densities, leading to a slight increase in the magnitude of the pair-correlation function and a considerable increase in the *L*-function at larger scales. Therefore, it is expected that type B and D heterogeneous patterns will show significant departures from CSR at larger scales beyond the scale of direct tree–tree interactions.

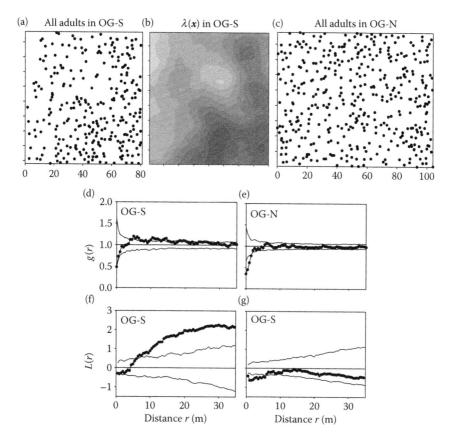

FIGURE 2.23

Analysis of homogeneous and heterogeneous patterns of mature trees (dbh > 15 cm) in two old-growth Douglas-fir forests on Vancouver Island, Canada. (a) Old-growth south plot, exhibiting several forms of environmental heterogeneity. (b) Intensity function for the map in panel (a), estimated using an Epanechnikov kernel with a bandwidth of 15 m; degree of shading is proportional to intensity. (c) Old-growth north plot, exhibiting no environmental heterogeneity. (d) Pair-correlation analysis of the heterogeneous south plot using a null model of CSR. (e) Same as panel (d), but for the homogeneous north plot. (f) Analysis of the heterogeneous south plot with the L-function. (g) Same as panel (f), but for the homogeneous north plot. Environmental heterogeneity in panel (d) is indicated by departure of the L-function from CSR at larger scales. (Modified after Getzin, S. et al. 2008. *Journal of Ecology* 96: 807–820.)

As expected, the spatial patterns for all mature trees within the two plots exhibited a tendency to regularity at small scales (i.e., values of the pair-correlation function below one at small distances *r*; Figure 2.23d,e). Indications of heterogeneity in the old-growth south plot were also observed in the steep increase of the L-function at larger scales (i.e., *r* > 10 m; Figure 2.23f) and an almost constant, but elevated value of the pair-correlation function (Figure 2.23d). In contrast, the pattern for mature trees in the old-growth, north plot showed no indications of heterogeneity (Figure 2.23e,g).

Another example that illustrates the use of homogeneous summary statistics to assess heterogeneities is provided by the longleaf pine data set collected and analyzed by Platt et al. (1988) and presented in Cressie (1993, p. 600, their Table 5.1), which has been repeatedly analyzed in the literature (e.g., Stoyan and Stoyan 1996; Guan 2008; Tanaka et al. 2008). Visual inspection of the data (Figure 2.24a) does not provide a clear answer regarding potential heterogeneities, although higher point densities are apparent in some areas of the plot around the coordinates (50, 100) to (100, 170). The pair-correlation function indicates strong clustering at scales less than 10 m (Figure 2.24b) and values greater than 1 at larger distances (10–60 m), before slowly approaching the expected value of 1 at scales >60 m. This indicates type D heterogeneity. The cumulative *L*-function provides an even clearer diagnosis of heterogeneity

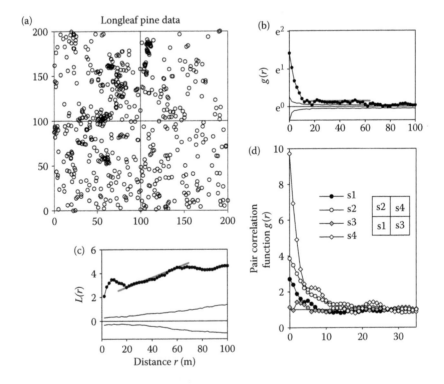

FIGURE 2.24
Longleaf pine data from the Wade Tract Preserve, Florida, USA. (Adapted from Cressie, N. 1993. *Statistics for Spatial Data, Revised Edition.* John Wiley & Sons, New York, p. 600.) (a) Map of longleaf pine data with boundaries of four subplots with tree locations indicated by open circles. (b) Pair-correlation function of the data. (c) *L*-function of the data. (d) Pair-correlation function of the data for the four 100 × 100 m² subplots shown in (a).

(Figure 2.24c). There is a strong signal of aggregation at small scales, but for scales of 20–50 m the *L*-value increases linearly, thereby indicating a heterogeneity gradient as shown in Figure 2.22a. At a scale of 100 m, which is half of the size of the plot, the *L*-function still does not drop to values of 0, indicating a larger scale trend. The test with the CSR null model shows that the elevated values of the pair-correlation function are clearly outside the simulation envelopes for distances smaller than 60 m, and do not approach the expectation of $g(r) = 1$ for distances larger than 60 m (Figure 2.24b). Additionally, the *L*-function indicates aggregation at all scales, which is a clear indication of heterogeneity (Figure 2.24c). Based on these results, we can interpret the pattern in the following way: there is small-scale clustering in some areas of the plot, which causes high values of the pair-correlation function at distances of $r < 10$ m. Some broader scale clustering exists, producing constant values of the pair-correlation function at distances of 10–60 m and a linear increase in the *L*-function. Finally, a weak, larger-scale gradient in the intensity causes the elevated values of the pair-correlation function at distances $r > 60$ m and the constant values of the *L*-function at these scales. In the next section, we will test this hypothesis by exploring subsets of the data.

2.6.2.4 Analyze Pattern in Subwindows

The methods presented in the previous section clearly diagnose heterogeneity in the pattern of the longleaf pine data set; however, additional methods are required to more clearly identify the type of heterogeneity and to determine whether or not second-order heterogeneity may occur in the data set. Methods that compare appropriate summary statistics within smaller subwindows of the original observation window represent a powerful approach for detecting heterogeneities, not only in the intensity, but also in the local point configuration. If the point pattern is homogeneous then the summary statistics of all subplots should be the same, except for stochastic variation. However, if the point pattern is first-order heterogeneous (types B and D Table 2.3), that is, the intensity changes, but the point configurations are the same over the entire observation window, one can use the *O*-ring statistic $O(r) = \lambda g(r)$ to assess whether the patterns within subwindows show different intensities. Also, the intensity-normalized pair-correlation function $g(r)$ can be used to assess whether the second-order structures differ among subplots.

We applied the latter test to the longleaf pine data by calculating the pair-correlation function for four 100×100 m subplots of the data (Figure 2.24a). Indeed, the pair-correlation functions of the four subplots varied substantially in shape. Within the SE subplot (s3), the pair-correlation function exhibits no indication of clustering and yields values close to one. In contrast, the NE subplot (s4) exhibits strong small-scale clustering and the other two subplots (s1 and s2) exhibit intermediate levels of clustering (Figure 2.24d). This diagnosis suggests that the pattern contains some degree of complexity in its spatial heterogeneity, with clustering appearing only in some parts of the plot.

The previous assessment of the longleaf pine data set suggests a complex heterogeneous pattern, although earlier attempts at analyzing the pattern generally assumed homogeneity. Cressie (1993) fit a Thomas process (Box 2.5) to the data, accounting for local-scale clustering. In contrast, Stoyan and Stoyan (1996) suggested a refined point-process model that assumed two different scales of clustering. In a third analysis, Tanaka et al. (2008) fit several, complex-cluster point processes to the data. However, these attempts

---

BOX 2.5   THE THOMAS PROCESS: CLUSTERING WITH ONE CRITICAL SCALE

The Thomas cluster process (see Section 4.1.4) is the simplest point process that creates clustered patterns with one critical scale of clustering. It is defined by the following three postulates:

1. "Parent" events form a homogeneous Poisson process with intensity ρ,
2. Each parent produces a random number of "offspring" S following a Poisson distribution $p_S(\mu)$ with mean $\mu = \lambda/\rho$ where λ is the intensity of the offspring pattern,
3. The positions of the offspring relative to their parents are independently and identically distributed according to a radially symmetric, bivariate normal distribution with variance σ^2. The distribution function for the offspring around a parent is therefore given by

$$h(r,\sigma) = \frac{1}{2\pi\sigma^2}\exp\left(-\frac{1}{2}\frac{r^2}{\sigma^2}\right).$$

The final pattern consists of the offspring only, as the parents are removed from the pattern. The univariate Thomas cluster process is stationary, with intensity $\lambda = \rho\mu$, and isotropic, since the bivariate normal distribution is radially symmetrical. In less technical terms, a number of clusters (equal to $A\rho$) are randomly and independently distributed, each cluster has a random number of points (with mean μ), and the distribution of the distances from the cluster center to the points of the cluster follows the normal distribution $h(r, \sigma)$.

Several summary statistics, such as the pair-correlation function [that is, $g(r, \sigma, \rho) = 1 + (1/\rho)(\exp(-r^2/4\sigma^2))/(4\pi\sigma^2)$], the K-function and the distribution function of nearest-neighbor distances, can be calculated analytically, which allows the process to be fit to a given data set.

may be inappropriate, if in fact the pattern contains an underlying heterogeneity. We, therefore, continue our analysis of this data set by a close inspection of the pattern, guided by the shape of the pair-correlation function and the *L*-function in different subareas (Figure 2.24d). Clustering is virtually absent in subplot s3, most of subplot s4, and part of subplot s1. This suggests that a suitable approach might be to modify the subplots in such a way that one subplot covers an area where clustering occurs and a second subplot covers the area where no clustering is apparent. We, therefore, divided the observation window (visually) into two subplots: a north-western (NW) subplot, which exhibited clustering and an eastern subplot, which exhibited no apparent clustering (Figure 2.25a).

Analysis of the pines located within the north-western subplot produced the surprising result that the pair-correlation function approached the expected value of one at scales greater than 10 m, just after disappearance of the strong signal of clustering (cf. Figure 2.24b and 2.25b). Thus, it is likely that the pattern in the NW part of the study site is homogeneous. A simple Thomas process fit to the data (Box 2.5) yields parameter estimates of 54 clusters in the NW subplot with an approximate radius of 6.3 m and a mean of 4.7 points. The pair-correlation function fit to the Thomas process approximates the pair-correlation fit to the data satisfactorily, although there is some additional small-scale clustering (cf. disks and gray line in Figure 2.25b). Testing the fit with the distribution function $D(r)$ of the nearest-neighbor distances reveals only a small departure from the expected pattern. The observed $D(r)$ is below the expectation of the null model, indicating slightly more isolated trees than expected at distances of 5–11 m (Figure 2.25d). Testing the fit with the spherical contact distribution $H_s(r)$ reveals slightly higher values of $H_s(r)$ than expected by the null model (Figure 2.25f). Because the $H_s(r)$ basically measures the gaps in the pattern, this result means that the observed pattern shows a slight tendency to have gaps that are too small compared with the null model. This is reasonable given that the observed pattern shows somewhat more isolated points and because isolated points appear in gaps.

Interestingly, the pair-correlation function of the eastern subplot (Figure 2.25c) did not exhibit small-scale clustering as found in the NW part of the plot, and the values of the pair correlation were almost constantly elevated, indicating a smooth gradient in point density (similar to Figure 2.22a). As a consequence, no departures from a heterogeneous Poisson null model, based on a nonparametric estimation of the intensity function, were found (Figure 2.25c,e,g). The intensity estimate for the eastern part of the study plot, using a bandwidth of $R = 40$ m (see Box 2.3), results in the heterogeneous Poisson null model as shown in Figure 2.25a. Thus, the longleaf pine data set provides an example of a complex heterogeneous pattern exhibiting small-scale clustering, with a homogeneous intensity function, within one area of the study plot and no clustering, but a gradient in density, in a second area of the plot. One should also note that the information contained in the diameters of

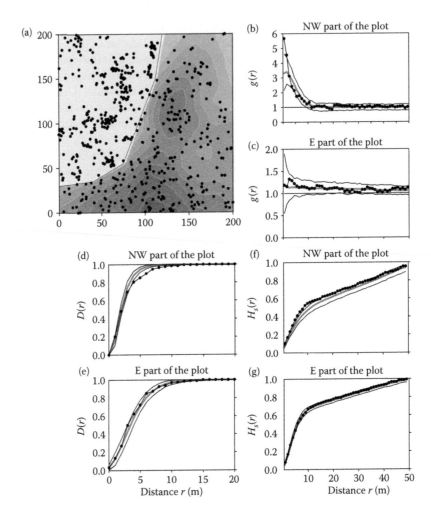

FIGURE 2.25
Analysis of heterogeneity in the longleaf pine data from Cressie (1993). (a) Subdivision of the observation window into two parts through visual assessment. For the northwest (NW) portion we used a Thomas process and for the east (E) portion a heterogeneous Poisson process. Tree locations are indicated by black dots and contours in the E portion represent the nonparametric kernel estimate with bandwidth $R = 40$ m to describe the heterogeneity in this portion. Panels (b), (d), and (f) show the results of the analysis for the NW portion of the plot. We fitted the observed pair-correlation and L-functions with a simple Thomas process and estimated simulation envelopes of the fitted point process. The fit yields parameter estimates of 54 clusters in with an approximate radius of 6.3 m and a mean of 4.7 points. Panels (d) and (f) show results for the summary statistics $H_s(r)$ and $D(r)$ not used for fitting. Panels (c), (e), and (g) show the results of the analysis for the E portion of the plot based on a heterogeneous Poisson process using a nonparametric kernel estimate with bandwidth $R = 40$ m.

the trees reveals additional insight. The clusters within the upper left, NW area of the observation window consisted of trees smaller than those found in the other area.

In another approach using subwindows to assess heterogeneity, Brix et al. (2001) and Couteron et al. (2003) proposed a test of second-order heterogeneity for cases in which the typical point configuration changes in response to some environmental variable (i.e., cases E and F in Table 2.3). This test is based on a partition of the original observation window into smaller subplots. In each of the subplots, a "local" test based on the CSR null model is carried out. For this test, Couteron et al. (2003) used the mean distance to the nearest neighbor or the mean inter-point distance as a test statistic. The local tests then provide maps of p-values that indicate the magnitude of departure from a Poisson process and the direction of the departure (i.e., clustering or hyperdispersion). They defined the p-value as the relative frequency of CSR realizations exceeding the observed value. This allows for an assessment of potential second-order heterogeneity. Finally, their method allows for a global test using the entire set of quadrats partitioning the observation window, which can be used to determine whether the pattern is most likely produced by a heterogeneous Poisson process. The null hypothesis is that the whole point pattern in the plot follows a heterogeneous Poisson process and, therefore, quadrats are expected to yield a set of p-values uniformly distributed on [0,1] (Couteron et al. 2003). This can be tested via a Kolmogorov-Smirnov test. A departure from the null hypothesis toward local clustering (at the subplot scale) would yield a distribution of p-values that is more abundant above 0.5 than below and vice versa, in the case of local inhibition. Brix et al. (2001) and Couteron et al. (2003) also provide a method to determine the optimal size of the subplots. For example, local tests, such as those just described, maintain the number of points in each quadrat and are therefore able to assess spatial relationships between neighbors in spite of larger scale heterogeneity. They can also reveal interactions between density and spatial pattern, which may occur in the presence of resource gradients (Couteron et al. 2003).

2.6.2.5 Divide Observation Window into Approximately Homogeneous Subareas

A lesson from the example of the longleaf pine data set (Figure 2.25) is that division of the observation window into appropriate, possibly irregularly shaped, subwindows can lead to surprising insights into the spatial structure of the pattern and may provide a simpler and more "parsimonious" explanation of the pattern than applying complex point processes within the entire observation window. In many practical situations, a heterogeneous point pattern may be separated into two or more, approximately homogeneous subwindows. This corresponds to a situation similar to "patch heterogeneity" as presented in Section 2.6.2.2. A Poisson-based approach for detecting such simple cases of heterogeneity was proposed by Pélissier and Goreaud (2001). Their idea was to distribute a number of test locations i over the observation window, for example, at the nodes of a regular grid, and count the number of points in disks with radius r placed at those locations [$= n_i(r)$], correcting for edge effects if the disk did not fully overlap the observation window. This is very similar to the kernel estimates of the intensity function described in Section 2.6.2.1.

An example of this approach is shown in Figure 2.26, where test points were located in a regular grid placed at 10 m intervals. The area of the circles in Figure 2.26b is proportional to the number of points found in circles with a radius $R = 20$ m placed at the test points. Figure 2.26c shows a histogram with the frequency distribution of $n_i(r)$. If the pattern shows two internally homogeneous areas with different intensity, we would expect the histogram to show two peaks, composed of a superposition of two Poisson distributions. This is confirmed in Figure 2.26c, where we provide the histogram together with the best fitting superposition of two Poisson distributions. There is one subarea in the window with very low (or even zero) point density yielding $n_i(r) < 4$ points, mostly located at the borders of the observation window, whereas the inner area contains a higher intensity. However, the inner area is not completely homogeneous, resulting in a histogram that has a somewhat wider distribution than a Poisson distribution. A value of $n_i(r) = 5$ can then be used to divide the observation window into two separate subareas, as shown in Figure 2.26b. In conducting an analysis in this way, Pélissier and Goreaud (2001) used spatial interpolation techniques applied to the $n_i(x)$ to determine the contour line dividing regions of different local

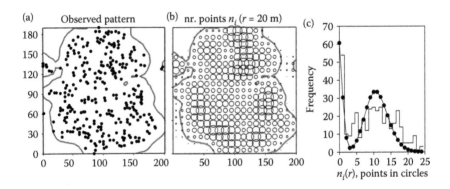

FIGURE 2.26
Delineating homogeneous subareas of a point pattern. (a) Observed pattern of *Syagrus yatay* palms in the National Park El Palmar, Argentina. (b) Test locations on an evenly spaced 10 m grid, indicated by circles proportional in size to $n_i(r)$, the number of points occurring within a 20 m radius. (c) Histogram of $n_i(r)$ values and the best fit with a superposition of two Poisson distributions with mean values of 0.5 and 10.5. The contour line for $n_i(r) = 5$, which represents the separation between the two Poisson distributions in panel (c), is shown in both panels (a) and (b).

density. It should be recognized that this method requires careful consideration of the characteristics of the pattern. For example, we may exclude some areas of an internally homogeneous study area that are accidently void of points, if the pattern is clustered (e.g., Figure 2.19).

2.6.2.6 Chi-Square Test for Association with Categorical and Continuous Habitats

The most convincing tests of heterogeneity are those that can directly show a dependency of the observed pattern on one or more environmental covariates. However, these tests require additional information, such as a complete map of an environmental variable within the observation window. In the following, we present tests that have been most frequently used for this purpose. The χ^2 and torus translation tests are applied in cases where habitat categories are discrete in nature. In contrast, the Berman test applies to situations involving continuous covariates. While these tests can reveal significant dependence of the pattern on environmental variables, it is still often necessary to estimate the intensity function. Thus, in Section 2.6.2.9 we briefly describe methods to estimate the intensity function parametrically based on environmental variables that have demonstrated a significant association with the observed pattern.

A simple test to assess the association of a point pattern with specific categorical habitats is the χ^2 test for goodness of fit (Snedecor and Cochran 1989, p. 77). The adaptation of this test to a data type comprising a point pattern and a categorical habitat map, with c different categories, compares the distribution of the points over the c discrete habitats $i = 1, \ldots, c$ with the distribution expected under the null model of random placement. To this end, the observed numbers of points O_i in a given habitat category i is compared with the expected numbers E_i of points based on the χ^2 statistic

$$\chi^2 = \sum_{i=1}^{c} \frac{(O_i - E_i)^2}{E_i} \tag{2.16}$$

summed over all $i = 1, \ldots, c$ categories. If the null hypothesis holds, the test statistic χ^2 follows the theoretical χ^2 distribution for large samples. If the null hypothesis does not hold, the observed number of points in habitat category i tends to agree poorly with the expectation, and the χ^2 test statistic becomes large. Depending on the degrees of freedom ($= c - 1$) and the value of α (e.g., 0.05), the data are in agreement with the null hypothesis, if the empirical value of the test statistic χ^2 is below the corresponding value of χ^2_α. In the case of random placement, the expectation E_i is given by the number of points (n) multiplied by the proportion of the total area covered by habitat type i. However, in some cases additional information can be used. For example, if the entire community is mapped, one may wish to correct for differences in overall densities among different habitat types (e.g., Harms et al. 2001).

Plotkin et al. (2000) placed the χ^2 test statistic into a Monte Carlo simulation framework. To this end, the n points of the pattern were randomly placed within the observation window using the CSR null model and the proportion Γ_i of points found within habitat type i was determined. This procedure was repeated 1000 times and the 5% simulation envelopes, that is, the 25 lowest and highest values Γ_i of the 1000 simulated patterns, were determined. The null hypothesis of no correlation between habitat type and the placement of points can be rejected at the 5% confidence level if the observed value of Γ_i lies outside the simulation envelopes. Plotkin et al. (2000) showed that this test, based on CSR, is equivalent to the standard χ^2 test. However, the simulation framework allows for the extension to more sophisticated null models.

A critical assumption of the χ^2 test and the corresponding CSR based Γ test is that the point pattern does not show strong clustering, that is, the points of the pattern must be independent. For example, if a pattern is composed of a few clusters, which are coincidently located within patches of habitat i, the Γ test may indicate spurious association, because a random distribution of points may not be able to yield the high Γ_i values that are observed. However, if a null model is used that conserves the observed spatial clustering of the pattern, it would be quite likely that such Γ_i values would be observed. Thus, the CSR null model which does not condition on the observed spatial structure of the pattern will be inappropriate for assessing the Γ_i test statistic. Taking this into account, Plotkin et al. (2000) fit the data with a Thomas process (Box 2.5) and used the realizations of the Thomas process as the null model. Not surprisingly, many of the previously observed associations based on CSR turned out to be a consequence of the violation of the assumption of independence (Plotkin et al. 2000).

It should be noted that the test using a Thomas cluster process is based on a parametric fit that works well only if the observed pattern is appropriately fit by a Thomas process. This may not always be the case, since the Thomas process makes specific assumptions concerning the spatial structure of the pattern. For example, as the Thomas process is based on clustering it does not allow too many points to have a nearest neighbor distance greater than the typical diameter of a cluster. Wiegand et al. (2007c, 2009), however, showed that such situations may actually occur frequently in real-world data sets, where there are often isolated points. In such cases, the Thomas process cannot characterize these "isolated" points and the fit will yield a somewhat larger number of clusters. Because the Thomas process cannot recreate important spatial properties such as isolated points, the associated tests may be biased. A nonparametric alternative to the test based on the parametric Thomas process can be provided by the use of the techniques of pattern reconstruction, which conserve the observed spatial pattern in evaluating the overall spatial structure, thus avoiding some of these problems (see Section 3.4.3).

2.6.2.7 Torus Translation Test for Categorical Habitats

The torus translation test is frequently applied to detect association of spatial patterns to predefined categorical habitat types (e.g., Harms et al. 2001; Valencia et al. 2004, Gunatilleke et al. 2006; Baraloto et al. 2007; Lai et al. 2009; Chuyong et al. 2011). The test was explicitly designed to deal with spatial patterns that do not meet the assumption of independence among points (i.e., they show a clustered pattern), as is generally assumed for tests such as the χ^2 GoF statistic discussed in the previous section. In the torus translation test, the null hypothesis is that the point pattern is independent of the spatial pattern of habitat types. In general, a test of independence must conserve the spatial structure of the point pattern and the habitats, but break up any potential interdependence. This is analogous to a test of independence for two point patterns (e.g., Section 2.2.1.2). Therefore, the question of interest is whether the observed association between the point-pattern and the habitats could be caused by pure coincidence rather than by a connection between the spatial structures of the point pattern and the arrangement of habitats. For example, clusters of a given species may coincidently overlap one habitat. If a test distributed the points randomly over the habitat map, a spurious positive association may result from the reduced variance in the test pattern, because the cluster structure of the point pattern was not conserved. If the pattern is randomized in such a way that it conserves the observed spatial autocorrelation among the points, in some realizations of the null model some clusters may coincidently overlap with the habitat, thereby creating a pattern similar to the observed association, but in other cases the clusters may be coincidently located outside the habitat. As a result, the expected range of values of the test statistic quantifying the association would be much larger than that of the CSR null model.

The solution presented in Harms et al. (2001) to conserve the spatial structure of the point pattern and habitats was in some respects opposite to the approach employing the Thomas process as used by Plotkin et al. (2000). Harms et al. (2001) did not change the observed distribution of individual points, as in Plotkin et al. (2000), but generated simulated null distributions of the habitat map by moving the true habitat map about a 2-D torus. This test goes back to a conditional Monte Carlo test suggested by Lotwick and Silverman (1982, Section 4) that maintains the critical properties of the spatial structure of both the habitats and points, while altering the positions of the habitats. The test then compares the observed relative densities of points within the different (true) habitats with the expected, or null, distributions of the relative densities based on many simulated habitat maps (Harms et al. 2001). Relative densities were calculated based on the density of the focal species within a given habitat type relative to overall stem density of all species within that habitat. A significant positive (negative) association of a given species with a particular habitat occurred if its relative density in the true habitat map was >97.5% (<97.5%) of the values obtained from the translated maps.

The torus translation test depends on predefined habitat categories. For example, Harms et al. (2001) and Gunatilleke et al. (2006) defined habitats based on topography (e.g., elevation, slope), biological considerations, and a priori assumptions, but not statistical analysis. This may weaken the power of the analysis if the habitat categories are not well chosen. An alternative method, deriving a habitat classification based on statistical analysis, was proposed by Legendre et al. (2009). Here multivariate regression trees were used to define habitat types based on environmental covariates and associated species assemblages (see also Lai et al. 2009; Kanagaraj et al. 2011b).

Clearly, a test of independence between habitats and a species pattern must conserve the spatial structure of the point pattern and habitats, but break apart their possible interdependence. The solution of Harms et al. (2001), generating simulated null distributions of the habitat map by moving the true habitat map about a 2-D torus (for torus translation of a point pattern see Figure 2.4), conserves the internal spatial structure of the block of habitat that still overlaps the observation window after translation. However, it also artificially shifts and relocates the parts of the habitat map that do not overlap the original observation window after the translation (see Figure 2.4b,c). The latter may create artificial situations, if the pattern of habitats shows a strong departure from stationarity (e.g., underlying gradients in the distribution). Therefore, a better alternative would be to leave the habitat map unchanged, as is done in Plotkin et al. (2000), and randomize the pattern of points based on the techniques of pattern reconstruction (see Section 3.4.3). This approach allows the production of stochastic replicates of the pattern that result in an array of summary statistics close to those observed for the actual pattern. In this way, the user can define which aspects of the spatial structure should be conserved.

2.6.2.8 Berman Test

The approaches discussed in the last two sections are designed for categorical habitat variables. The torus translation test of Harms et al. (2001) requires a priori definition of discrete habitat types which are often developed from continuous habitat variables, such as elevation, slope, or soil nutrient maps. It may be, therefore, desirable to test directly for association of a spatial point pattern with a continuous environmental variable. This can be accomplished by the Berman test, which goes back to a study motivated by a geological problem (Berman 1986). The question was whether copper ore deposits (representing a point pattern) were spatially associated with linear-shaped geological features (lineaments) visible from satellite images. Berman (1986) reduced the problem to a test of spatial association between a point pattern and a spatial covariate $v(x)$. In the case of this motivating example, the spatial covariate $v(x)$ was the distance of a point x to the nearest lineament, but the test is applicable to any covariate $v(x)$. The test is performed by comparing the observed distribution of the values of a spatial covariate $v(x)$ taken at the locations x_i of the points i of a point pattern and the predicted distribution of the same covariate under a null model that randomizes the points of the pattern. The test statistic $Z_1 = (S - \mu)/\sigma$ introduced by Berman (1986) is based on the sum S of the covariate values $v(x_i)$ at the points x_i of the pattern. The value μ is the predicted mean value of S under the n realizations of the null model and σ^2 is the corresponding variance. The null distribution of this test statistic is approximately the standard normal distribution.

Based on this test statistic one can formulate the null and alternative hypotheses. The null hypothesis H_0 is that the pattern was generated by a homogeneous Poisson process independent of the environmental covariate $v(x)$. The alternative hypothesis H_1 is that the pattern is an inhomogeneous Poisson point process with an intensity function depending on $v(x)$. In a GoF test, significant deviation of the test statistic Z_1 from the null hypothesis H_0 can be assessed by comparing the observed value of Z_1 with the standard normal distribution or by comparing the rank of the observed S within the corresponding values of S for the simulations of the null model. Note that the original Berman test did not take into account the effect of spatial autocorrelation (clustering) in the spatial pattern and, therefore, provides a conservative estimate of habitat association (Berman 1986). Berman (1986) also used torus translations of the point pattern to account for potential clustering. However, the effect of spatial autocorrelation (clustering) in the spatial pattern can be accounted for by generating the spatial patterns of the null model using techniques of pattern reconstruction (Wiegand et al. 2013a; Section 3.4.3.7).

A related test was used in John et al. (2007) to test for associations of tree species to continuous maps of soil nutrients. They based their test on the mean and standard deviation of an index of association between the spatial point pattern and the continuous map of an environmental variable. To this end, the observation window was divided into 20 × 20 m quadrats j, and the number of points within each quadrat $j (= n_j)$ was related to the value X_j of the environmental variable in quadrat j. The mean value of the environmental variable measured at the locations of the points yields

$$\bar{X} = \frac{\sum_j n_j X_j}{\sum_j n_j} \tag{2.17}$$

and the corresponding standard deviation yields

$$SD = \frac{1}{N} \sqrt{N \sum_j n_j X_j^2 - \sum_j (n_j X_j)^2} \tag{2.18}$$

where N is the total number of points. The two test statistics $\bar{X}$ and SD were then calculated for the observed pattern and for 1000 realizations of the fitted Thomas process. If the observed value of $\bar{X}$ was greater than the 25th largest value of $\bar{X}$ calculated for the 1000 simulations, there was a positive association (with a one-tailed error rate $\alpha = 0.025$) between the number of trees and the environmental variable. Conversely, there was a significant negative association if the observed values of $\bar{X}$ were below the 25 smallest value of the test statistics calculated from the simulations. For SD, a one-tailed test with $\alpha = 0.05$ was used (i.e., observed values smaller than the distributions of values expected under the null model) (John et al. 2007).

The tests presented in Sections 2.6.2.7 and 2.6.2.8 for assessing association, use different null models to test for independence. Some use a random pattern (i.e., CSR) as a null model and others try to approximate the observed spatial aggregation structure of the pattern by using a torus shift or Thomas process. It is evident that the only "correct" test is one that conditions on both the spatial structure of the habitat map (or the environmental covariate) and the spatial structure of the point pattern. This can be accomplished by pattern reconstruction, which should, therefore, be used if the objective of the study is to confirm that an environmental dependency exists (e.g., the studies by Harms et al. 2001 or John et al. 2007). However, for other objectives the CSR null model may be completely sufficient and the additional effort of conducting pattern reconstruction may not be justified. For example, if the objective is to conduct variable reduction previous to estimation of a parametric intensity function using statistical regression models (i.e., the next section), we only want to sort out variables that are clearly not associated with the observed point pattern. In this case, the conservative result of the CSR null model is sufficient.

2.6.2.9 Habitat Models

One of the most important ideas in ecology that has developed in attempting to explain the diversity of species is the niche concept (Hutchinson 1957). The basic tenet of the niche concept is that species can coexist only if they explore different resources. As a consequence, since environmental conditions are heterogeneous in space, some areas are more suitable for a given species than others. A rich body of methods has developed in ecology to describe the distribution of organisms in space and to investigate species–environment relationships. These approaches can be modified to help understand the relationship between the distribution of resources and point patterns of ecological objects.

Predictive habitat distribution models (Guisan and Zimmerman 2000; Guisan and Thuiler 2005; Elith and Leathwick 2009), which are probabilistic in nature, are an important tool that can be used in determining the spatial structure of the intensity function $\lambda(x)$. The aim of a habitat model, in this context, is to determine the intensity function $\lambda(x)$ in a parametric way. This contrasts with the tests described in the three previous sections, which are designed to determine whether point patterns have a significant association with a categorical habitat map or environmental covariate. Habitat models relate the information on the presence of the organisms of interest with a set of environmental covariates to predict the probability of finding the organism at a given location. The advantage of the predictive habitat distribution model approach is that the final intensity function can include several covariates and the estimated intensity function can be used in further analyses. Data types from the field used in this kind of analysis that are closely related to point patterns may include presence locations, presence and absence locations, or spatially distributed abundances.

The most obvious parametric model to fit the intensity function for point patterns is the so-called log-linear model (Waagepetersen 2007). The assumption here is that the logarithmic transformation of the intensity function can be modeled as a sum of covariates: $\log(\lambda(x)) = c_0 + c_1 v_1(x) + \cdots + c_n v_n(x)$, where the c_i are coefficients and the $v_i(x)$ are the covariates. The intensity function can then be fit using maximum likelihood estimation (Waagepetersen 2007). This model has been used in a number of ecological studies to determine the intensity function $\lambda(x)$ for the spatial pattern of trees (e.g., Shen et al. 2009; Wang et al. 2011). This approach is closely related to Cox processes, which are discussed in Section 4.1.5.

An alternative method for estimating the intensity function is based on methods borrowed from techniques developed to model resource selection by animals (Manly et al. 2002; Lele and Kaim 2006). In such studies, the basic approach is to determine the resource selection probability function (RSPF), which gives the probability that a particular location x (= resource), as characterized by a combination of environmental covariates $v_i(x)$, will be used by an individual animal (Lele and Keim 2006). In the case of animal resource selection, only a subset of the locations that were visited by the individuals under study is known. This suggests a use-versus-available study design in which the data of visited locations are compared with a subset of the available sites. In the case of plants, the actual locations occupied by a given species represent only a subset of all potentially suitable locations. Unused locations in this case consist not only of unsuitable locations, but also locations that could have been used, but were not. Lele and Keim (2006) and Lele (2009) proposed a sophisticated method for the analysis of use-available patterns based on the logistic form of an RSPF, which provides the absolute probability of use. This approach can be used to determine the intensity function for a data set comprising the locations of individuals of a given species and a set of environmental covariates.

Figure 2.27 shows an example of habitat modeling based on the log-linear model and the resource selection approach proposed by Lele and Keim (2006) and Lele (2009). The pattern includes all 428 living individuals of the species *Ocotea whitei* at the BCI plot (Box 2.6), as determined from a census in 2000 (shown in Figure 2.27c). The environmental variables were all derived

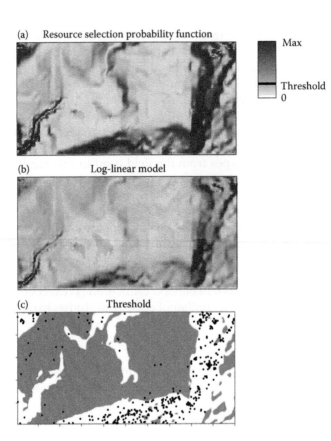

FIGURE 2.27
(See color insert.) Habitat model for *Ocotea whitei* within the BCI plot, Panama, for the 2000 census. (a) Probability of occurrence determined by the logistic, resource-selection probability function (Lele 2009). (b) Thinning surface $p(x) = \lambda(x)/\lambda^*$ for the log-linear model; λ^* is the maximum value of $\lambda(x)$. (c) Categorical map showing areas of high suitability ($p(x) > 0.2$; white areas) versus low suitability ($p(x) \leq 0.2$; gray areas), with points of the pattern as black dots.

from elevation measures (see Kanagaraj et al. 2011b) and included elevation, a topographic wetness index (TWI), vertical distance to the channel network (VDist.Chann; Tarboton 1997; Sørensen et al. 2006), slope, curvature, and aspect. However, the Berman test (Section 2.6.2.8) revealed only a significant association by some of the variables with the pattern (i.e., TWI, VDist.Chann, Slope, Aspect). The variables without significant association were therefore excluded. The resource selection probability function analysis revealed a significant and positive relationship of the probability of use with slope ($P < 0.001$) and a significant, negative relationship with aspect ($P < 0.01$) (Table 2.4a). The model also exhibits a good ability to discriminate (AUC = 0.80), as indicated in an almost binary switch from suitable to nonsuitable habitat (Figure 2.27a). The log-linear model based on the approach of Waagepetersen (2007) yields a similar result with similar coefficients (Table 2.4b; Figure 2.27b).

BOX 2.6   THE BCI FOREST DYNAMICS PLOT

A 50-hectare (500 × 1000 m) permanent tree plot was established in 1980 by Steve Hubbell and Robin Foster in the tropical moist forest of Barro Colorado Island (BCI) in central Panama (9°10′N, 79°51′W) and is maintained by the Smithsonian Tropical Research Institute. Censuses have been carried out in 1981–1983, 1985, 1990, 1995, 2000, 2005, and 2010. In each census, all free-standing woody stems at least 10 mm diameter at breast height (dbh) were identified, tagged, and mapped. Today, more than 40 such forest dynamics plot exist around the world and all follow the same protocols (Condit 1998). The plot network is managed by the Center for Tropical Forest Science at the Smithsonian Tropical Research Institute.

Barro Colorado Island has a moist, lowland, tropical climate, with 2500 mm rain per year, a strong 3.5-month dry season, and a year-round mean daily temperature of 27°C (Leigh 1999). The island is covered by a tall, high-biomass forest. The BCI plots exhibit a topographic habitat structuring (Harms et al. 2001). Slopes are wetter than plateaus (Daws et al. 2002; Leigh et al. 2004) and experience a shorter duration of drought during the annual 4-month dry season (Daws et al. 2002). The high plateau is the driest area, having lower dry season soil water availability than the low plateau or the slopes (Comita et al. 2007). Additionally, a seasonal swamp area is inundated during the 8-month wet season (Hubbell and Foster 1986; Daws et al. 2002). A prolonged drought occurred at the BCI plot through the first half of the study period with severe El Niño droughts in 1982–83 and 1997–98. The plot is described in detail by Hubbell and Foster (1983). Details on census methods can be found in Condit (1998), and further information is available at http://ctfs.arnarb.harvard.edu/webatlas/datasets/bci/

In general, it is desirable to use environmental covariates that are believed to be the causal, driving forces for the distribution and abundance of the organisms of interest, or at least correlated to them. To improve the interpretation of habitat models it is recommended that *variable reduction* be performed prior to the estimation of the intensity function. This is done to remove environmental covariates that do not show a significant association with the point pattern (here the tests presented in the three previous sections can be used) and to remove highly correlated covariates. To avoid spatial dependence in the analysis, one may stratify the observational data to remove spatial autocorrelation (Kanagaraj et al. 2011a) or add a spatial autocorrelation term to the model (Dormann et al. 2007). For data stratification, a grid may be placed over the observation window and only one randomly

TABLE 2.4

Results of the Habitat Analysis of the Species *Ocotea whitei*

	Estimate	Std. Error	*z* Value	Pr(>|*z*|)
(a) RSPF				
(Intercept)	−2.683*	1.248	−2.150	0.032
TWI	−0.080	0.108	−0.743	0.457
VDist.Chann	−0.784	0.461	−1.701	0.089
Slope	0.634***	0.071	8.885	≪2e−16
Aspect	−0.005***	0.001	−3.722	0.0002
(b) LLM	Estimate	Std. error	CI95.lo	CI95.hi
(Intercept)	−4.610	1.273	−7.106	−2.115
TWI	−0.15**	0.054	−0.255	−0.044
VDist.Chann	−0.42**	0.133	−0.681	−0.158
Slope	0.123***	0.013	0.097	0.15
Aspect	−0.003***	0.001	−0.004	−0.001
Elevation	−0.011	0.007	−0.024	0.003

Note: (a) Using the logistic resource selection probability function based on the approach of Lele (2009). The Hosmer and Lemeshow goodness of fit (GOF) test for the model yields a *p*-value of 0.0116 and we obtain an AUC of 0.788. (b) Using the log-linear model.
*$P < 0.05$; **$P < 0.01$; ***$P < 0.01$.

selected point within each grid cell is then retained. Kanagaraj et al. (2011a), for example, used the typical size of the home range of a female tiger as the size of a grid cell. Alternatively, cluster detection algorithms (see Section 3.4.1 "Cluster detection algorithm") may be used to remove data points in a way that only one representative individual within each cluster is retained for further analysis. However, it must be emphasized that the application of parametric estimates of the intensity function may be limited by two problems. First, if environmental dependency exists, but the critical covariates are not available, the intensity estimate may not do a good job of capturing the "real" heterogeneity. Second, the pattern may be heterogeneous, not because of environmental dependency, but because of internal mechanisms of population dynamics, such as dispersal limitation, invasion, or range contraction. In Section 4.1.2.3, we provide a more detailed treatment of these issues.

Techniques of statistical habitat modeling can also be applied as a formal test for heterogeneity in the intensity function. Three extreme cases of outcomes of such a habitat analysis may be expected:

1. No significant habitat model can be found. In this case, the pattern may indeed be homogeneous, the environmental variables included in the analysis do not drive a potentially existing heterogeneity, or the species pattern is heterogeneous, not because of environmental dependency, but because the species is constrained by dispersal limitation or is undergoing expansion or contraction, resulting in a

transient pattern. Thus, the available environmental covariates do not support the hypothesis of heterogeneity in the intensity function.

2. A significant habitat model can be found and the habitat model indicates a smoothly varying intensity function.

3. Finally, a significant habitat model can be found and the habitat model shows abrupt changes in suitability, with some areas having a high suitability and others a low suitability (Figure 2.27c). In this case, the habitat model provides reliable criteria for dividing the original observation window into two or more subwindows. Note that in this case the delineation of subareas is based on additional information regarding the relationship between environmental conditions and the distribution of the species and not purely on statistical grounds, as described in Sections 2.6.2.4 and 2.6.2.5.

★★ **2.6.3 Approaches for Dealing with Heterogeneity**

In the following, we will present relatively simple approaches that allow analyses of heterogeneous point patterns. Most of the methods are closely related to the methods used in detecting heterogeneity as presented above. Basically, we have four possibilities in accounting for heterogeneity: (1) we can ignore it (which is not necessarily a good idea); (2) we can avoid dealing explicitly with heterogeneity by selecting observation windows that omit heterogeneities; (3) we may adapt the null model to explicitly model the heterogeneity; and (4) we may factor out the effect of the heterogeneity by modifying the summary statistics.

2.6.3.1 Ignore Heterogeneity

Naively one might expect that large-scale heterogeneity in the intensity function would influence the values of point pattern summary statistics only at broad scales, but leave the values at small-scales untouched. Unfortunately, this is not the case for many important summary statistics in ecological applications. Large-scale heterogeneity imposes a signature at all spatial scales, especially if the cumulative K- or L-functions are used. For this reason, the CSR null model or other homogeneous point processes are generally not suitable benchmarks for point patterns, if the intensity of a pattern is not constant. Unfortunately, the ecological literature is riddled with many examples in which "clustering at all scales" is the diagnosis of pattern analyses, when using the K- or L-function (e.g., Figure 2.28c). The results are interpreted without considering the spurious effect of "virtual aggregation" caused by first-order heterogeneity (Wiegand and Moloney 2004).

The problem of "virtual aggregation" can be understood intuitively. Imagine a point pattern consisting of a single, internally homogeneous cluster of points in the center of the observation window (e.g., Figure 2.28a). The

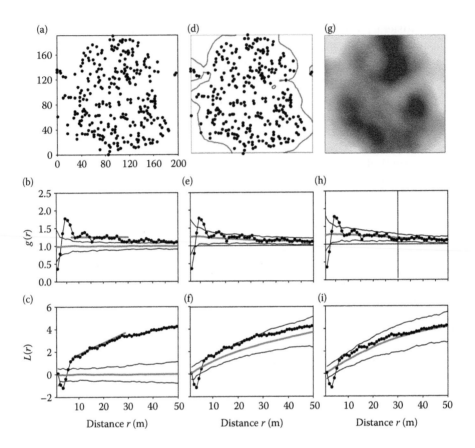

FIGURE 2.28

Comparing methods to account for heterogeneity. (a) Map of observed pattern of *Syagrus yatay* palms in the National Park El Palmar, Argentina. (b) Pair-correlation values for the pattern in panel (a), with simulation envelopes produced using a CSR model. (c) *L*-function values for the pattern in panel (a), with simulation envelopes produced using a CSR model. (d) Homogeneous area for the pattern delineated as shown in Figure 2.26. (e) Pair-correlation values with simulation envelopes produced by a CSR applied within the area of homogeneity. (f) *L*-values with envelopes produced as in panel (e). (g) Nonparametric intensity estimate of the pattern in panel (a), using an Epanechnikov kernel with a bandwidth of $R = 30$ m. (h) Pair-correlation values with simulation envelopes produced by a heterogeneous Poisson process based on the intensity surface in panel (g). (i) *L*-values with envelopes produced as in panel (h).

local density of points inside the cluster will be higher than the overall density of points in the entire observation window. As a consequence, the neighborhood density inside the cluster is greater than the intensity of the pattern (Figure 2.28b; see also Section 2.6.2.2). Thus, we have $O(r) > \lambda$ and the *K*- or *L*-function will indicate aggregation at smaller scales, even if the pattern is random inside the cluster. Therefore, if there is marked heterogeneity in the intensity of the pattern it is a mistake to not take this into account when conducting an analysis.

2.6.3.2 Analyze Homogeneous Parts of Heterogeneous Patterns

The simplest approach allowing the methods developed for homogeneous patterns to be retained in the analysis of a heterogeneous pattern is to select subareas of the pattern with a homogeneous appearance. The properties of the patterns within these subareas are then analyzed independently. This approach is still informative, if the aim is to characterize the small-scale correlation structure of the point configurations (Law et al. 2009), and useful in simple cases of heterogeneity, which show abrupt changes in point density rather than smooth gradients.

Clearly, the problem in applying this approach is to find rigorous methods for delineating homogeneous subareas. In the longleaf pine example (Figure 2.25), we visually judged where to place natural breaks in the pattern. In cases of simple heterogeneities, the approach of Pélissier and Goreaud (2001), described above (Section 2.6.2.5), may also be used to delineate the homogeneous subareas. Application of this method (Figure 2.26) to the pattern shown in Figure 2.28d removes a substantial part of the heterogeneity in the pattern (cf. Figure 2.28b and e), but it seems that the local neighborhood density is still somewhat too high, which causes a weak signal of heterogeneity that is indicated by the L-function (Figure 2.28f). The reason for this is that the separation criterion $n_i(r) = 5$ (see Figure 2.26b) for the two subareas was probably somewhat too conservative, as indicated by the mostly empty "buffer" around the points shown in Figure 2.28d.

In many cases, however, there is additional information available that may characterize underlying causes of heterogeneity (e.g., topography, changes in soil type, or land use patterns) and provide a mechanism for delineating, with some degree of objectivity, homogeneous subareas in the observation window. Under these circumstances, the most satisfying approach is to conduct a formal test of environmental dependence, such as through the *study of habitat suitability* for the species of interest, where the statistical techniques of habitat modeling can be employed (see "Habitat models" above). Ideally, a habitat model based on the appropriate information yields an approximate binary response by the species of interest to environmental conditions (i.e., suitable habitat vs. unsuitable habitat; Figure 2.27c). This would provide a potential basis for dividing the original observation window into smaller subwindows.

The analysis of the species *Ocotea whitei* in Figure 2.27 provides such a case. We calculated the intensity function for this species based on environmental variables using the logistic resource selection probability function (Lele 2009). The probability of occurrence of this species is shown in Figure 2.27a. To distinguish suitable from unsuitable areas we determined the threshold in the resource selection probability function for which 90% of all individuals were located in areas above the threshold (Figure 2.27c). The white area in Figure 2.27c is defined as the irregularly shaped study area and the 10% of individuals located in unsuitable habitat were excluded from the following analysis. We then fit a Thomas process (Box 2.5) to the pattern of the individuals of *Ocotea whitei* that were located in the suitable habitat and estimated the summary statistics using techniques for irregularly shaped observation windows (see Section 3.4.2). Figure 2.29a gives the pair-correlation function of the data (closed disks) and that of the fitted Thomas process (gray line). The corresponding *L*-functions are shown in Figure 2.29b. The fit yields a cluster radius of approximately 12.8 m and is satisfying in characterizing the small-scale (*r* < 25 m) pattern of the species *Ocotea whitei*. However, at distances greater than 25 m the pair-correlation function does not drop to values of 1, but remains slightly above the simulation envelopes. This indicates that the habitat model did not capture the full heterogeneity shown by the data. Indeed, Figure 2.27c indicates that some larger blocks of the suitable areas are empty. This is indicated by the pair-correlation function (Figure 2.29a) and the *L*-function (Figure 2.29b). However, the potentially suitable, but unoccupied, areas may also be the result of an additional amount of large-scale clustering shown by the species. The species may be affected by recruitment limitation and is not able to reach all suitable areas, or it is in the process of dynamic expansion or contraction. Further analyses and a consideration of the biological properties of the species are, therefore, required to explain this finding.

To test the fit with the Thomas process, we also analyzed the distribution function of the distances to the nearest neighbor (Figure 2.29c) and the spherical contact distribution $H_s(r)$ (Figure 2.29d), which were not used for model fitting. The $H_s(r)$ of the data shows good agreement with that of the Thomas

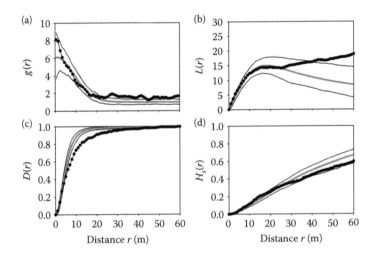

FIGURE 2.29
Pattern analysis of *Ocotea whitei* restricted to a consideration of suitable habitat. (a) Pair-correlation function. (b) The *L*-function. (c) The distribution function $D(r)$ of distances to the nearest neighbor. (d) The spherical contact distribution $H_s(r)$. Analyses were conducted on the pattern shown in Figure 2.27c, analyzing the pattern only in the suitable area (white in Figure 2.27c). See example in Section 2.6.2.9 for complete details.

process. There is a also good fit in the $D(r)$ at small distances ($r < 5$ m), but at larger distances we find a departure from the null model (Figure 2.29c). The latter indicates that the pattern consists of a certain proportion of points (say 15%) that have their nearest neighbor further away than expected by the Thomas process (the expectation at distance $r = 8$ is 0.15 units higher than the observation). One explanation for the occurrence of the "isolated" points is that we analyzed the pattern of all trees together and one can expect that larger trees would be more isolated.

Another approach to analyzing homogeneous subsets of the data would be to divide the observation window into a grid of small nonoverlapping subplots (e.g., subwindows of 20×20 m for a 1000×500 m plot) and use *regression tree analysis* (De'ath 2002; Kanagaraj et al. 2011b; Punchi-Manage et al. 2013) to characterize the relationship between environmental conditions and the local intensity of the point process. Regression tree analysis is an iterative algorithm that attempts to partition the plots into groups with similar intensity, based on threshold conditions in the environmental variables. In the first step of this process, the subplots are partitioned into two groups based on all possible threshold values of the different environmental variables. The threshold value of the partition that yields two groups with maximized within-group similarity is selected. The same procedure is then repeated for each subgroup until a stopping criterion is met.

When analyzing the *Ocotea whitei* data with regression trees (Figure 2.30a), we find that the only environmental variable that enters as a single variable is the slope s, with a threshold of $s = 6.46°$. The suitable habitat occurs in areas with a slope greater than 6.46° (Figure 2.30a, white area). The results of the analysis using the Thomas process within the irregularly shaped study area are similar to those shown in Figure 2.29, but the regression tree analysis appears to have a slightly better ability at discrimination (c.f. Figure 2.27c and 2.30a). This is also indicated by a slightly better fit of the pair-correlation and the L-function (Figure 2.30b,c) using this approach. However, there is still some larger-scale heterogeneity unaccounted for by the analysis.

2.6.3.3 Heterogeneous Poisson Process with Nonparametric Intensity Estimate

Two approaches are frequently used in estimating the intensity function without the additional use of information provided by covariates. The first approach is to *fit trend surfaces to the data*, for example, "log-cubic polynomial trends" (Baddeley and Turner 2000) or log-harmonic polynomial trends (Law et al. 2009). The other option, which we will present in more detail, is to *use kernel-smoothing approaches* (see Section 2.6.2.1).

The heterogeneous Poisson process based on a nonparametric kernel estimate of the intensity function is the simplest approach in which the null model explicitly characterizes the (first-order) heterogeneity. It is especially suited for cases with larger-scale heterogeneities. The idea behind this null

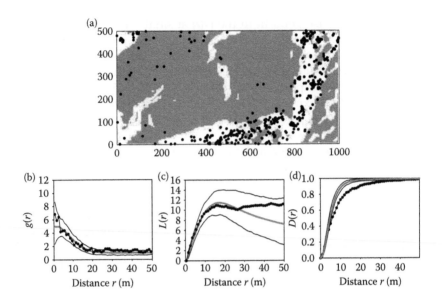

FIGURE 2.30
Application of regression tree analysis to account for heterogeneity. (a) Map of the distribution of *Ocotea whitei*, with the observation window categorized into suitable (white) and unsuitable (gray) habitat using regression tree analysis (see end of Section 2.6.3.2 for details). (b) Pair-correlation function. (c) *L*-function. (d) The distribution function $D(r)$ of distances to the nearest neighbor. In all cases, simulation envelopes were produced by a point-process model based on a Thomas process fit to the data within the suitable subarea shown in (a).

model is to condition on the observed larger-scale structure of the pattern and to selectively randomize the pattern at smaller scales. This can be done by randomly displacing each point of the pattern within a predefined neighborhood R around the point. We can calculate the spatially varying intensity $\lambda(x)$ of the pattern expected under this null model at any location x of the study area, by counting all individuals that are located within the neighborhood R around location x. Clearly, the nearer the individuals are to location x, the higher the chance that the random displacements of the null model will relocate them inside this neighborhood. In technical terms, this method of estimating $\lambda(x)$ corresponds to a nonparametric kernel estimate with bandwidth R (Box 2.4). This null model can be implemented as a heterogeneous Poisson process that uses a nonparametric intensity estimate with bandwidth R. Thus, the heterogeneous Poisson process "freezes" the larger-sale spatial structure, but randomizes the pattern at small scales. This allows the selective detection of small-scale effects conditioned on the observed large-scale pattern. Given this approach, we can only detect small-scale effects at distances $r < R$.

One important application of the heterogeneous Poisson process based on nonparametric kernel intensity estimates (Box 2.4) is in cases impacted by *"virtual aggregation"* (Wiegand and Moloney 2004). An example of this (Figure 2.28a; data provided by W. Batista and M. Lunazzi) is provided by the spatial distribution of *Syagrus yatay* palm trees in the National Park El Palmar in the Argentinean province of Entre Ríos (58°17′ Long. W; 31°50′ Lat. S). In this case, a large-scale pattern (e.g., caused by environmental variation) masks a small-scale pattern (e.g., interactions among points). Vegetation in El Palmar is dominated by a temperate savanna ecosystem. The pattern shows basically one large, apparently homogeneous patch with palm trees in the center of the observation window and void areas at the edges (Figure 2.28a). Nonparametric estimation of the intensity function with a bandwidth of $R = 30$ m confirms this diagnosis and additionally reveals some areas of somewhat higher point density (Figure 2.28g). As a consequence, the pair-correlation function and the *L*-function for distances larger than 10 m exhibit the typical shape of patch-type (or gradient) heterogeneity (c.f. Figures 2.22 and 2.28b,c). When applying the CSR null models, the typical result for virtual aggregation emerges: in particular, the *L*-function shows aggregation at almost all distances r (Figure 2.28c).

Application of the heterogeneous Poisson process, with a nonparametric intensity estimate, should remove most of the spurious virtual aggregation observed in the application of simple CSR techniques to the palm data (Figure 2.28b,c). To demonstrate this, we selected a relatively large bandwidth of $R = 30$ m to estimate the intensity function as shown in Figure 2.28g. Other than a significant effect at small distances ($r < 6$ m), we find that the *L*-function is now located well inside the simulation envelopes and approaches the expectations of the null model for distances greater than 30 m (Figure 2.28i). The same analysis based on the pair-correlation function reveals small-scale structure in the pattern, with repulsion at small distances (1–2 m) and aggregation at 4 m (Figure 2.28h). The species *S. yatay* reproduces exclusively by sexually produced seeds, which are then dispersed by gravity or animals. The observed pattern may be caused by superposition of competition and seed dispersal operating at different spatial scales. Seed dispersal by gravity may cause a distribution of seeds that is inversely related to the distance from the stem. However, strong competition in the neighborhood of a tree counteracts this effect and may cause overall repulsion at these scales. Therefore, the critical scale at 4 m may arise as competition becomes relatively weak for distances $r \geq 4$ m, but seeds are still aggregated at this scale (Wiegand and Moloney 2004).

An interesting aspect of this analysis is that we can confirm that a *separation of scales* occurs. A departure from CSR is observed at all distances less than the maximal distance of 50 m (Figure 2.28b,c). In the estimation of the intensity function, we used a bandwidth of $R = 30$ m. This means that we maintained spatial structure at scales greater than 30 m, but randomized the pattern at scales below 30 m. Thus, under separation of scales, significant effects should disappear at distances well below 30 m, but if separation of scales does not occur, significant effects disappear only at distances greater than 30 m. Indeed, analysis with the pair-correlation function and the *L*-function indicates that significant effects already disappear at distances greater than 7 m, when the heterogeneous Poisson process was used as the null model (Figure 2.28h,i).

Thus, conditioning the large-scale effects on a scale above 30 m removed most of the small-scale effects. This can be interpreted as indicating that the palm pattern responds to environmental heterogeneity at scales of 30 m or greater and small-scale interactions occur below 7 m. Indeed, as shown in Figure 2.28 the heterogeneity is a patch-type heterogeneity produced by a large patch, with a relatively homogeneous distribution of palms, surrounded by an area void of palms. By restricting the observation window to the area that contains palms (Figure 2.28d), we basically obtain the same results for the pair-correlation function and the L-function as using the heterogeneous Poisson process (cf. Figure 2.28e and h, and Figure 2.28f and i).

In the palm example, the selection of the bandwidth R was not motivated by additional information on the potential interaction range among palm trees. If such information is missing, several bandwidths should be analyzed. For example, in Wiegand and Moloney (2004) we used a bandwidth of $R = 15$ m for the same pattern and obtained the same results. This strengthens the diagnosis of separation of scales. While the above analysis can (and should) be conducted for a range of bandwidths R, it is desirable to use a bandwidth that is most likely to separate biological effects (Wang et al. 2010; Wiegand et al. 2012). Therefore, determination of an appropriate bandwidth R should be based on additional biological information.

In general, it is expected that direct interactions among points occur only over a limited range of spatial scales. The maximal range of expected interactions could be used as the bandwidth R in an analysis (e.g., Wiegand et al. 2007a). For example, it is expected that direct interactions among larger trees only occur within a limited spatial range (e.g., <30 m). Hubbell et al. (2001) found that the neighborhood effects of conspecific density on survival disappeared within approximately 12–15 m of the focal plant. Several other studies using individual-based analyses of local neighborhood effects on tree growth and survival have confirmed this result (e.g., Uriarte et al. 2004, 2005; Stoll and Newbery 2005), suggesting that direct plant–plant interactions in forests may fade away at larger scales. However, as shown earlier, one may also repeat the analysis with several distinct bandwidths to detect critical scales to determine where separation of scales occurs. Note also that even if a clear separation of scales is detected the effects may be caused by small-scale heterogeneities such as safe sites, local soil heterogeneities, and so on. Therefore, the results of the analysis need to be carefully interpreted in light of the known biology of the species analyzed. Ideally, the detection of small-scale effects by means of point-pattern analysis should be followed by field investigations to identify the driver behind the observed pattern. Point-pattern analysis can reveal spatial structures precisely and suggest underlying mechanisms, but cannot replace field investigations in confirming or rejecting such hypotheses.

The heterogeneous Poisson process generally works well as an extension of CSR, but if the pattern shows strong small-scale clustering, superposed on a large-scale heterogeneity, care is required in interpreting the results. In this case, the empirical pair-correlation function may produce values slightly below the simulation envelopes, indicating hyperdispersion at larger scales. This "regularity" is likely to be caused by the properties of the heterogeneous Poisson process. To understand this effect, imagine a pattern with dense clusters of points with a radius of say 10 m. The point density at a distance 30 m away from a point at the center of a cluster will be low. The heterogeneous Poisson process, which basically smooths the point density with a 30 m moving window, will "smear" the dense cluster and produce elevated point densities up to $10 + 30 = 40$ m. As a consequence, the pair-correlation function of the data will be slightly below the pair-correlation function of the heterogeneous Poisson process at scales around 30 m.

2.6.3.4 Heterogeneous Poisson Process with Parametric Intensity Estimate

The heterogeneous Poisson process can also be characterized using parametrically estimated intensity functions. This is theoretically the most satisfying approach for revealing second-order effects, because it incorporates additional information on environmental variables in reconstructing the first-order heterogeneity in the null model. However, this approach faces two problems in practice. First, if there are unmeasured environmental variables that critically influence the probability that a point is placed at a certain location, the resulting parametric intensity estimate may provide only an approximation of the intensity function. In this case, the heterogeneous Poisson process may not be able to fully factor out the first-order effects and we will observe departures of the pair-correlation function from the null model at larger distances r. Second, the pattern may exhibit heterogeneity that is not driven by environmental dependency. This might mean that there are larger areas of the observation window that are suitable habitat where the species of interest is currently not present. For example, tree species in tropical forests often show dispersal limitation, which prevents propagules from reaching all suitable areas. As a consequence, using a parametric estimate of the intensity function only based on environmental covariates may miss out on other sources of heterogeneity. In this case, the heterogeneous Poisson null model will not allow the small-scale effects of point interactions to be revealed (because the heterogeneity is not fully removed). In Section 4.1.2.3 we provide a more detailed treatment of these issues and present an approach that combines parametric and nonparametric estimation of the intensity function.

★★ 2.6.3.5 Inhomogeneous Summary Statistics

The approach presented in the previous section deals with first-order heterogeneous patterns by adapting the null model to the heterogeneity and using summary statistics developed for homogeneous patterns. An alternative method of dealing with first-order heterogeneous patterns is to adapt the summary statistics to the heterogeneity. Baddeley et al. (2000) developed summary statistics for first-order, heterogeneous patterns (cases B and D in Table 2.3) that behave in the same way as the classical counterparts for homogeneous patterns. With this approach, estimation of *inhomogeneous K-and g-functions* relies on a technical assumption that the observed point pattern results from a two-step process. In the first step, a homogeneous point process (which is unknown and a theoretical construct) is produced. The resulting pattern is then independently thinned, using the thinning function $p(x) = \lambda(x)/\lambda^*$, where $p(x)$ is the probability of accepting a tentative point at location x, $\lambda(x)$ is the intensity of the pattern at that location, and λ^* is the maximum value of $\lambda(x)$ within the observation window, which we can associate with the intensity λ of the homogeneous pattern. Thus, $p(x)$ ranges between 0, where a point at x is always deleted, and 1.0, where a point at x is always retained. Statisticians call such patterns *"second-order, intensity-reweighted stationary"* (Baddely et al. 2000). All heterogeneous Poisson processes and, in fact, all nonstationary point processes, which result from the thinning of a stationary point process, are second-order, intensity-reweighted stationary.

The underlying idea of the inhomogeneous *K*- and *g*-functions is simple. The probability that a point of the homogeneous pre-thinned pattern is located within an infinitesimal disk with area dx around locations x is λdx. In contrast, the probability that a point of the inhomogeneous, post-thinning pattern is located within an infinitesimal disk with area dx around locations x is $\lambda p(x)$ dx. Thus, in estimating the inhomogeneous g- and *K*-functions, each point at location x is weighted by the inverse of $p(x)$, since location x would, on average, have contained $1/p(x)$ points prior to thinning, given that a point currently lies on that location. Statisticians call this "re-weighting" (Baddeley et al. 2000). Thus, if the underlying intensity function $\lambda(x)$ has been correctly identified, the inhomogeneous pair-correlation and *K*-functions reveal the pure second-order properties of the pattern and have properties analogous to the equivalent homogeneous functions. If the pattern is homogeneous, all points initially placed are retained within the observation window, since $p(x) = 1$ [and $\lambda(x) = \lambda$].

Although we provide technical details on the estimation of second-order summary statistics in Section 3.1.2, we need some basic information on the estimators to understand the way inhomogeneous *K*- and *g*-functions work. Remember that the quantity $\lambda K(r)$ yields the expected number of points within distance r of the typical point of the pattern. Thus, the estimator of the *K*-function visits all points i of the pattern in turn, counts all further points j within distance r of point i, and averages these counts over the n focal points i, and divides by λ. More technically, the estimator of the *K*-function for a homogeneous pattern is given by

$$\hat{K}(r) = \frac{1}{\lambda}\frac{1}{n}\sum_{i=1}^{n}\sum_{j=1}^{n,\neq}\mathbf{1}(\|x_i - x_j\|,r)w_{ij}(r)$$

$$= \frac{1}{A}\sum_{i=1}^{n}\sum_{j=1}^{n,\neq}\mathbf{1}(\|x_i - x_j\|,r)\frac{w_{ij}(r)}{\lambda\lambda} \qquad (2.19)$$

where $\mathbf{1}(d, r)$ yields a value of 1 if $d \le r$ and zero otherwise (thus counting all further points j within distance r of point i), $w_{ij}(r)$ accounts for edge correction (to compensate for cases where the circle with radius r around points i is not fully located within the observation window), n is the number of points of the pattern, and A is the area of the observation window.

In estimating the inhomogeneous K-functions, each point at location x_i is weighted by the inverse of $p(x_i)$:

$$\hat{K}^{\text{BMW}}_{\text{inhom}}(r, \lambda(x)) = \frac{1}{A} \sum_{i=1}^{n} \sum_{j=1}^{n,\neq} \mathbf{1}(\|x_i - x_j\|, r) \frac{w_{ij}(r)}{\lambda(x_i)\lambda(x_j)}. \tag{2.20}$$

The superscript BMW refers to the three authors Baddeley, Møller, and Waagepetersen of a seminal 2000 article that introduced the second-order inhomogeneous summary statistics. If the pattern is homogeneous, we have $p(x) = 1$ and the homogeneous estimator (Equation 2.19) is recovered. The different estimators of inhomogeneous second-order statistics are presented in detail in Sections 3.1.2.5 through 3.1.2.7. It should be noted that this estimator contains edge correction and reweighting in one term [that is, $w_{ij}(r)/(\lambda(x_i)\lambda(x_j))$] associated with each point pair i–j. This is problematic because the weights can become large if values of the thinning surface at points i or j become small.

One problem with the reweighting approach is that it is generally not clear *a priori* if patterns exhibiting a nonrandom, second-order structure are indeed second-order, intensity-reweighted stationary, since their second-order structure may depend on location. This is not much different from patterns that have an intensity that is approximately constant, but also contain potential heterogeneities in the local point configurations. Therefore, the assumption of a constant second-order structure in the pattern should be tested before applying inhomogeneous summary statistics. For example, in Figure 2.25, we demonstrated that the longleaf pine data did not fulfill this assumption, since the second-order structure differed among regions within the observation window. There are, in fact, many possibilities as to how an intensity function may interact with the properties of a cluster processes. For example, we can easily imagine a biologically plausible cluster process where the probability of a cluster occurring and the number of points in a cluster both depend on an environmental covariate.

Second-order, intensity-reweighted stationarity can be tested in a relatively simple way by selecting several subwindows $W_1, \ldots, W_k$ within which the point distribution appears to be approximately homogeneous (e.g., Figure 2.24). For second-order, intensity-reweighted stationary patterns, the estimates of the pair-correlation functions within the different subwindows should be essentially the same and only vary due to stochastic fluctuations (Illian et al. 2008, p. 282). In addition, differences should not be due to systematic differences in second-order characteristics. This can be understood as follows: The neighborhood density functions within the (approximately internally homogeneous) subwindows W_i yield $O_i(r) = \lambda_i\, g_i(r)$, where λ_i is the local intensity in subwindow W_i. Thus, if the second-order structure of the patterns is approximately the same in all subwindows i we have $g_1(r) \approx g_2(r) \approx g_i(r)$. For example, the pattern of longleaf pine data is not second-order, intensity-reweighted stationary, because this condition does not hold (Figure 2.24d).

Although theoretically appealing, the practical problem with inhomogeneous functions is that estimation of the intensity function requires additional information. This is due to the fact that determining the intensity function by using nonparametric kernel estimates (Box 2.4) only based on the point-pattern data is potentially problematic, since the data of a single realization of the underlying point process are used for estimating both the intensity function $\lambda(x)$ (a first-order statistic) and the inhomogeneous K-function (a second-order statistic) nonparametrically (Diggle et al. 2007). Baddeley et al. (2000) found that the estimator of the inhomogeneous K-function that used the kernel estimate of the intensity function was biased downwards. This appeared to be due to a positive bias in the kernel estimate when it was evaluated at the data points. Additional problems may arise if the pattern exhibits a strong tendency toward small-scale clustering. Small clusters that are not caused by locally elevated habitat suitability, but are due to inherent clustering mechanisms, must either be accounted for or removed before the intensity function is estimated (i.e., de-clustering; Section 3.4.1). From an ecological point of view, the most appropriate method for estimating the intensity function would then be to use the statistical techniques of habitat or species distribution modeling (see Section 2.6.2.9), but this only works if the parametric estimate does a good job in characterizing the heterogeneity (see Section 2.6.3.4).

The best way to appreciate the complications involved, and to develop an intuitive feel for how to proceed, is by examining some examples of the approach, which we provide in the following. The first example presents a heterogeneous Poisson process (Box 2.3) with an underlying linear gradient in the intensity function (Figure 2.7). The pattern (Figure 2.31a) was created by thinning a homogeneous Poisson process by proposing potential points and accepting them based on the local value of the thinning surface $p(x) = \lambda(x)/\lambda^*$, as explained above. In this way, 491 points were distributed within a 100×100 m observation window (Figure 2.31a). In exploring the pattern, we first use the known intensity function to apply weights in estimating the inhomogeneous pair-correlation and L-functions, as described in Equation 2.20. We can see that the resulting inhomogeneous pair-correlation function approximates the expected value of $g_i(r) = 1.0$ reasonably well (Figure 2.31b). It falls well within the 95% simulation envelopes produced by a Monte Carlo simulation of the CSR null model, using the same number of points as in the "real" pattern. In comparison, the conventional

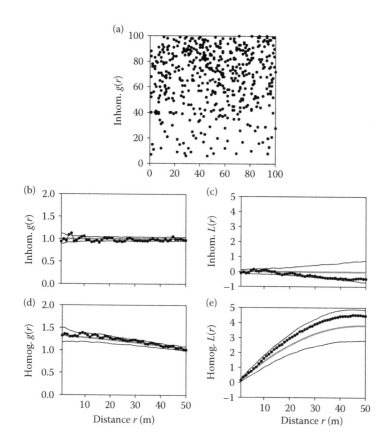

FIGURE 2.31
Homogeneous versus inhomogeneous summary statistics. (a) Heterogeneous pattern with a gradient heterogeneity. (b) Inhomogeneous pair-correlation function for the pattern shown in panel (a), based on the "exact" gradient intensity function. (c) Inhomogeneous L-function, based on the "exact" gradient intensity function. (d) Homogeneous pair-correlation function. (e) Homogeneous L-function.

"homogeneous" pair-correlation function exhibits substantial differences from the inhomogeneous function (Figure 2.31d). Nonetheless, this function falls within the 95% simulation envelopes of simulations of the heterogeneous Poisson process based on the linear intensity function. We see similar results when we examine the L-function. The inhomogeneous L-function produces values very close to the expected value of $L = 0$ for a CSR pattern (Figure 2.31c), whereas the "homogeneous" L-function differs substantially from the inhomogeneous function, but is located well within the simulation envelopes of the heterogeneous Poisson process (Figure 2.31e). This example clearly illustrates that, if the intensity function is known, the inhomogeneous functions can be successfully used to factor out the effect of smoothly varying point intensities. There is also a *significant advantage of using the inhomogeneous summary statistics* as no detailed analysis of the second-order properties of a pattern can be conducted if it is analyzed using the homogeneous summary statistics and the heterogeneous Poisson process is employed as a null model. In contrast, applying the inhomogeneous summary statistics (that reveal the true second-order effects) allows us to incorporate models of the specific point processes into the assessment of inhomogeneous summary statistics.

In a second example, we explore the application of inhomogeneous summary statistics in the analysis of a more complex cluster process. Figure 2.32a shows a stochastic realization of a Thomas process (Box 2.5) with 25 clusters and a cluster size of $r_C = (2 * \sigma) = 30$ m, where each cluster contains on average $\mu = 40$ points. Because of the stochastic effects, we cannot expect that a fit of the simulated data with a Thomas process will recover exactly the same parameters used for simulation, but the estimate of 28 clusters and a cluster size of $r_C = 28$ are satisfyingly close (Figure 2.32d). Note that points of simulated clusters that fall outside the study area reappear at the opposite edge, since the simulation is conducted on a torus (Figure 2.32a). If the clusters are relatively large compared to the dimensions of the observation window, this has the effect of producing an estimate of the number of clusters that may be slightly higher than the true number. Estimates of the size of the clusters are also slightly smaller than expected, given the original parameters.

Two inhomogeneous patterns generated by thinning the homogeneous pattern (Figure 2.32a) are shown in Figure 2.32b,c. For creation of the pattern shown in Figure 2.32b we retained a point of the original pattern with probability $p(x) = x/500$, that is, the larger the value of the x-coordinate the lower the probability that a point would be deleted. The pattern shown in Figure 2.32c was created in the same way, but with a probability of $p(y) = y/500$.

Plugging in the original intensity function used to create the pattern shown in Figure 2.32b, as a weight into the inhomogeneous pair-correlation function, allows us to recover the homogeneous pair-correlation function and parameters of the original process in good approximation: the fit yields the correct cluster size of $r_C = 30$ and the correct number of clusters, that is 23. The inhomogeneous pair-correlation function shows some stochastic variation and is slightly larger at smaller scales than the pair-correlation function of the original prethinned data (Figure 2.32e). The latter result is understandable because thinning also produces loss of information: about half of the points were deleted. This is also illustrated in the examples shown in Figure 2.32c,f, and i, which used the same original pattern of data, but a different thinning surface. Thus, the inhomogeneous functions are able to recover information on clustering, if the real intensity function is known and thinning does not cause too large of a loss of information regarding the spatial structure.

What happens if we ignore the heterogeneity of the pattern and use homogeneous summary statistics? The shape of the (homogeneous) pair-correlation function of the pattern exhibits some indication of heterogeneity,

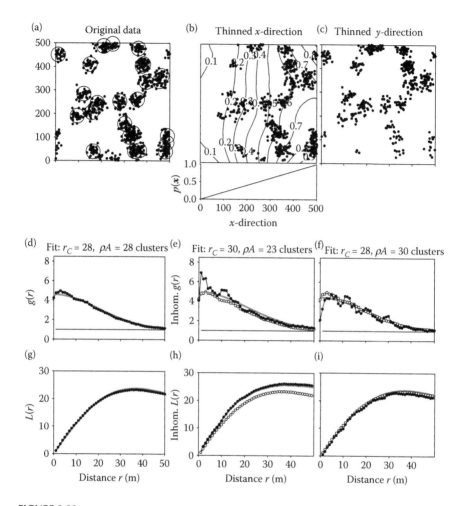

FIGURE 2.32
Inhomogeneous summary statistics. (a) Map of a stochastic realization of a Thomas process with 25 clusters and a cluster size of $r_C = (2 * \sigma) = 30$. (b) Map of the distribution in panel (a) thinned in the x-direction by a thinning surface of $p(x, y) = 1 - x/500$, with contours based on a nonparametric intensity estimate with bandwidth of $R = 250$ m. (c) Map of the distribution in panel (a) thinned in the y-direction by a thinning surface of $p(x, y) = 1 - y/500$. (d) Homogeneous pair-correlation function of the pattern in panel (a), fit with a Thomas process (solid gray line). (e) Inhomogeneous pair-correlation function of the pattern in panel (b), based on the original thinning surface (closed disks), fit with the inhomogeneous Thomas process (gray line) and the pair-correlation function of the original homogeneous pattern (open disks). (f) Inhomogeneous pair-correlation function of the pattern in panel (c) with graphic elements as in panel (e). (g)–(i) L-functions in the same sequence as the pair-correlation functions in panels (d)–(f).

since its value does not drop to the expected value of 1 at scales larger than 50 (compare Figure 2.33a with the pair-correlation function of the original homogeneous pattern shown in Figure 2.32d). The fit with homogeneous summary statistics yields an estimate of a cluster size of $r_C = 32$, which is surprisingly close to the parameter used to generate the pattern (Figure 2.32). (We used both the L- and the g-function together with minimum contrast methods in fitting the process; see Section 2.5.2.1, Wiegand et al. 2007c, 2009.) However, the fit yielded 16 clusters, which is substantially below the 25 clusters used to simulate the pattern. The reason for the surprisingly good estimate of the cluster size is that the (smooth) thinning did not destroy the internal structure of the clusters but, in a first approximation, randomly removed points from each cluster. Therefore, the information on cluster size was still present in the thinned pattern. The lower estimate of the number of clusters is also understandable. Comparison of Figure 2.32b with Figure 2.32a shows that thinning removed or depopulated several clusters; that is, clusters positioned with x-values below 250 retained only a few, if any, points. The clusters containing very few points were then statistically "overpowered" in the fit by the clusters that retained many points. Thus, the fit of the heterogeneous pattern with homogeneous summary statistics still revealed, in a way, the "average" structure of the pattern.

In summary, identifying cluster size seems to be quite robust when ignoring the underlying heterogeneity of the pattern, but identification of the number of clusters is more sensitive. This means that minor heterogeneities, which are not very obvious from visualizing the pattern (the heterogeneity in Figure 2.32b is e.g., is already quite obvious visually), may not be a significant problem when applying homogeneous summary statistics. Clearly, the gradient we used in the example for thinning is a relatively weak form of heterogeneity, which does not introduce strong spatial structures compared to those produced by clustering. However, the error when using homogeneous summary statistics

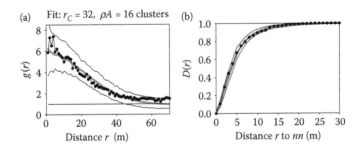

FIGURE 2.33
Analysis of an inhomogeneous pattern with homogeneous methods. (a) Pair-correlation function applied to the pattern shown in Figure 2.32b, fit with a homogeneous Thomas process and simulation envelopes of the fitted point process. (b) Distribution function of distances to the nearest neighbor applied to the same pattern as panel (a).

for assessment of heterogeneous patterns may be larger in cases where the structure of the intensity function is similar in scale to the cluster structure.

At this point, it is not clear how sensitive inhomogeneous summary statistics are to the form of the thinning surface. To explore this issue more fully, we used the original thinning surface shown in Figure 2.34a and generated two realizations of the associated heterogeneous Poisson process (e.g., as in Figure 2.32b). One realization was used as the observed pattern (see Figure 2.34g) and the second realization was used to generate a series of nonparametric kernel estimates (Box 2.4) of the intensity of the heterogeneous Poisson process with different bandwidths R. We thus used two different realizations of the underlying point process for estimation of the intensity function and for estimation of the second-order properties. The resulting thinning surfaces shown in Figure 2.34b–f were generated with the Epanechnikov kernel using bandwidths of $R = 60, 50, 40, 30, 20$, respectively. Although they approximate the original intensity function (shown in Figure 2.34a), the kernel estimation approach introduces additional, systematic spatial structure. The question is whether this spatial structure influences the estimators of the inhomogeneous summary statistics in the attempt to recover the original spatial structure of the heterogeneous cluster pattern and, if it does, how this changes our interpretation of the analysis.

Figure 2.34h demonstrates that moderate structure in the intensity function $\lambda(x, y)$ does not affect the shape of the inhomogeneous L-function severely (i.e., thinning surfaces shown in Figure 2.34b–d). As before, the estimates of cluster size do a good job of approximating the original values, but the estimate of the number of clusters declines with increasingly finer spatial structure in the thinning surface (Table 2.5). For example, the intensity function based on Figure 2.34d yields an estimate of only 17.5 clusters instead of 25. The reason for the lower estimate is that the spatial structure in the intensity function weights some clusters more than others. If a cluster is accidently located at a local minimum in the intensity function $\lambda(x, y)$ the points in the cluster will be weighted much more than clusters located at a local maxima of the intensity function. As a consequence, the inhomogeneous functions produce a spurious appearance of a cluster process involving fewer clusters.

The latter effect becomes aggravated if the spatial structure in the intensity function has local minima, with almost zero intensity, at locations where the original thinning surface is notably greater than zero (as in thinning surfaces E and F). Imagine that the thinning surface $p(x, y)$ has a value of 0.01 but that the original thinning surface has a value of 0.3. In this case, two nearby points in a cluster, which is accidently located at this local minimum, receive a weight in the estimator of the inhomogeneous functions proportional to $1/(0.01)^2 = 10000$. However, these points should only have a weight proportional to $1/(0.3)^2 = 11.1$. Thus, the weight in the estimator of the inhomogeneous function is inflated by factor of $10000/11.1 = 900.1$, which means that it basically depicts 30 times more points in this cluster than it should! As a consequence, the inhomogeneous L-functions based on the thinning surface

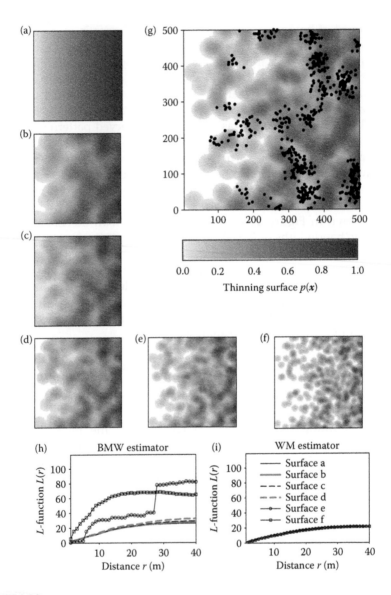

FIGURE 2.34
Impact of the approximation of the intensity function $\lambda(x)$ on the estimate of the inhomogeneous
L-function. (a) Original thinning surface $p(x) = \lambda(x)/\lambda^*$ used to generate the pattern (black dots)
shown in panel (g). (b)–(f) Thinning surfaces calculated from a different realization of the point
process that produced the pattern shown in (g), based on a nonparametric kernel estimate with
bandwidths of $R = 60, 50, 40, 30, 20$, respectively. (g) Heterogeneous pattern used to estimate the
inhomogeneous *L*-functions, shown with the thinning surface in panel (f). (h) Estimates of the
inhomogeneous *L*-function based on thinning surfaces (a–f) using the BMW estimator. (i) Same
as panel (h), but with estimates based the corresponding WM estimator.

TABLE 2.5

Exploration of the Impact of (Spurious) Spatial Structure in the Intensity Functions on the Estimation of the Inhomogeneous Summary Statistics

Bandwidth R for Estimation of Thinning Surface	Figure	BWM Estimator		WM Estimator	
		Cluster size $r_L = 2\sigma$	Number of Clusters	Cluster size $r_L = 2\sigma$	Number of Clusters
Original	(a)	27.4	25.4	27.6	31.9
60	(b)	29.4	21.9	27.2	32.7
50	(c)	30.2	20.3	27.6	32.4
40	(d)	31.2	17.5	28.1	32.9
30	(e)	43.4	4.0	28.4	32.2
20	(f)	14.2	8.8	28.9	31.2
Unthinned pattern		27.2	29.6	27.6	28.3

Note: Shown above are the results of the fit with the inhomogeneous Thomas process based on the different thinning surfaces and the original parameters of the unthinned pattern are shown above. We conducted this test with the BMW estimator (Equation 2.20) and the alternative WM estimator detailed in Section 3.1.2.6.

Figure 2.34e,f are very different from their expected values (Figure 2.34h). The pair-correlation functions are even more erratic in behavior (not shown).

In summary, an important point of this example is that the inhomogeneous pair-correlation and L-functions based on the BMW estimator are relatively robust against small uncertainties in the real underlying intensity function, as long as the thinning surfaces $p(x)$ used are sufficiently smooth and do not drop to low values (as stated earlier, the thinning surface is normalized between $p(x) = 0$, where a point is always removed, to $p(x) = 1$, where a point is always retained). As a consequence, before a thinning surface is used in estimating an inhomogeneous summary statistic, it should be critically inspected. If the thinning surface yields low values (say $p(x) < 0.1$) for some points of the pattern, it is quite likely that the estimate of the inhomogeneous summary statistics will be biased by the points which are accidently located in those areas of low-point density.

Later in the book we present alternative estimators of the inhomogeneous pair-correlation function (Section 3.1.2.6) and the K-function (Section 3.1.2.7), which are not affected by these problems. These estimator are not based on individual weighting of each point pair i–j with the term $w_{ij}(r)/(\lambda(x_i)\lambda(x_j))$ as is done with the BMW estimator (Equation 2.20). Instead, they are based on the philosophy of the Ohser edge correction (see Section 2.3.2.5) in which the weighting factor does not depend on the individual point pairs, but is a global factor within the observation window that depends only on distance r. Figure 2.34i shows that estimators of this form, such as the WM estimator, are indeed not affected by the problems associated with the BMW estimator of the inhomogeneous second-order statistics.

3

★★ *Estimators and Toolbox*

This chapter broadens the discussion on point-pattern analysis by introducing various technical aspects concerning the estimation of summary statistics. In Section 3.1, we present *methods for estimating the various summary statistics* introduced earlier in the book. Estimation, in particular, must explicitly account for the problem of edge correction, a topic briefly touched upon in Section 2.3.2.5. Here, we provide the theoretical foundation of the currently available methods and compile the different approaches into a common framework. In Section 3.2, we present methods for analyzing replicate patterns by *combining results into one "master" result*. This task is closely associated with the initial development of estimators, since combining the results of replicate patterns comes down to defining rules of how to estimate master summary statistics for replicate plots. Section 3.3 deals with the *independent superposition of point processes*. If the summary statistics of two-point processes are known, we can calculate, under certain conditions, the summary statistics that result from the independent superposition of the two-point processes. This operation considerably expands the range of point-process models we can use and allows us to describe observed patterns in a more realistic way.

Section 3.4 consists of a toolbox presenting different techniques and methods that are especially useful in a variety of applications, such as an algorithm to *detect clusters* (Section 3.4.1), methods to conduct point-pattern analysis in *observation windows of irregular shape* (Section 3.4.2), and methods of pattern reconstruction (Section 3.4.3), a nonparametric approach to *generating patterns with predefined statistical properties*. In most applications, the predefined statistical properties are the summary statistics of an observed pattern. The underlying simulated annealing algorithm generates patterns that match the (predefined) summary statistics of the observed pattern very closely. Or, in other words, it allows us to generate stochastic, replicate patterns consisting of the same, underlying statistical properties. This algorithm has two important applications: the reconstructed patterns can serve as realizations of the fundamental division for bivariate patterns (i.e., the independence null model we briefly touched upon in Section 2.2.1.2), and it allows us to determine how many (and which) different summary statistics are needed to accurately describe the statistical properties of point patterns.

★★★ 3.1 Estimators of Summary Statistics

In this chapter, we present the more technical aspects of calculating summary statistics for the different data types by presenting their estimators, which generally contain some form of edge correction. In general, *edge correction* is used to remove a bias in the estimation of a summary statistic that is caused by the data being confined to a finite observation window W. Corrections are required because points near the border of W will have fewer neighbors in some directions than others. This is due to the fact that a circle or ring with radius r centered on a point near the edge is often not completely located within the observation window W (i.e., disk C2 in Figure 3.1). For example, when estimating the K-function of a univariate pattern, we visit each point i of the pattern and count the number of additional points $n_i^c(r)$ lying within distance r of that point (the superscript c refers to "circle") and take the average over all points. This yields an estimate of $\lambda K(r)$, the number of additional points lying within distance r of the typical point (see Section 2.3.2.5). A naive estimator of $\lambda K(r)$ would, therefore, calculate the average of $n_i^c(r)$ over all points i, that is,

$$\bar{n}^c(r) = \frac{1}{n} \sum_{i=1}^{n} n_i^c(r) \tag{3.1}$$

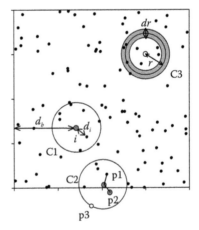

FIGURE 3.1
Need for edge correction in estimation of second-order and nearest neighbor summary statistics. As illustrated by circle C3, the O-ring statistic is determined by calculating the density of points lying within the gray ring of radius r and width dr centered on a typical focal point (open circle) of the process. No edge correction is required when the ring lies fully within the observation window, as shown by circle C1. In this example, the distance d_i to the nearest neighbor can also be correctly estimated. However, for circle C2 many points within distance r of the focal point are potentially located outside the observation window W, as is the nearest neighbor (p2).

This approach is fine if the focal point i is located further than distance r from the nearest border (e.g., the gray focal point in circle C1 in Figure 3.1). The corresponding estimate of $n_i^c(r)$ will be correct. However, a problem arises if the circle of radius r does not completely overlap W (e.g., circle C2 in Figure 3.1). Under these circumstances some points within distance r of the focal point i may be located outside W and will not be counted, resulting in an estimate of $n_i^c(r)$ that will be too low. If this effect is not accounted for correctly, the estimator $\bar{n}^c(r)$ given in Equation 3.1 will be biased. Clearly, this bias will be aggravated for larger radii r because the "mean overlap" of circles around the points of the pattern with W will be correspondingly low.

A similar problem arises for the distribution function $D(r)$ of the distances r to the nearest neighbor. Again, we visit each point i of the pattern in turn and estimate the distance d_i to the nearest neighbor. The naive estimator of $D(r)$ enumerates, for each distance r, the number of cases where $d_i < r$, that is,

$$\hat{D}(r) = \frac{1}{n} \sum_{i=1}^{n} \mathbf{1}(0 < d_i \le r) \qquad (3.2)$$

The indicator function $\mathbf{1}(0 < d_i \le r)$ yields a value of 1 if $d_i < r$ and 0 otherwise. If the nearest-neighbor distance d_i of point i within W is less than the shortest distance d_b to the border (e.g., the gray focal point in circle C1 in Figure 3.1), the corresponding estimate of d_i will be correct. However, if the distance d_i to the nearest neighbor within W is larger than the shortest distance d_b to the border (as with circle C2 in Figure 3.1), a problem arises in that a point outside W (point p2 in Figure 3.1) may, in fact, be the nearest neighbor. As a consequence, the estimated value of d_i may be too large, and if this effect is not corrected, the estimate $\hat{D}(r)$ of $D(r)$ given in Equation 3.2 may contain a bias.

Note that the effect of not using edge correction in calculating $D(r)$ is less severe than it is for second-order summary statistics. This can be easily understood. If the circle around a point is only partly located within W, the count $n_i^c(r)$ in Equation 3.1 will be, in most cases, too low. In the worst case, where a point is located at the corner of W, the count of $n_i^c(r)$ will yield only 1/4 of its expectation (because only 1/4 of the area of the circle overlaps W). However, for nearest-neighbor statistics there is, even in this worst-case scenario, still a 1/4 probability that the nearest neighbor is indeed located within W and can be correctly determined. Even if this is not the case, there is a high probability that the second, third, or kth nearest neighbor is located within W, and will provide a reasonable approximation of the distance to the nearest neighbor. However, this is not guaranteed if the number of points of the pattern is low. We can, therefore, expect larger differences among estimators if the number of points of the pattern is low.

While estimators of summary statistics for unmarked patterns usually require edge correction, this is not necessary for certain summary statistics of marked point patterns, such as mark-connection and mark-correlation functions. This is due to the conditional nature of these summary statistics, which are estimated as the quotient of two functions that are subject to the same bias. The biases simply cancel out in the quotient (see Sections 3.1.6 and 3.1.7).

3.1.1 General Edge Correction Strategies

Various solutions have been proposed to deal with the problem of edge correction for univariate patterns. In the simplest case, we may *ignore the problem of edge correction*. Indeed, if the observation window is large compared to the range of distances *r* considered, only a few focal points *i* would be located close to the border. In this case, even summary statistics of unmarked patterns could be estimated, in a first approximation, without edge correction, using estimators analogous to those given in Equations 3.1 and 3.2 (Figure 2.13a).

Besides ignoring the edge correction problem, there are basically four different strategies to correct for the biases produced through edge effects. First, in *minus sampling* only the data in W that contain no bias are used in the analysis (Illian et al. 2008, p. 185f). In this method, only focal points *i* that are located at least distance *r* from the border of W are used in estimating the summary statistics (Figure 2.13b). This produces a buffer zone between the outer edge of W and the data employed in the analysis. In this case, the full circle or ring around point *i* is located within W and the summary statistics can be estimated without bias. Although this is an exact method that completely avoids bias, it has the disadvantage of not fully using the data that are available. Using this approach, the estimators given in Equations 3.1 and 3.2 are modified as follows:

$$\bar{n}^c_{\ominus r}(r) = \frac{1}{n_{\ominus r}} \sum_{i=1}^{n_{\ominus r}} n^c_i(r) \tag{3.3}$$

$$\hat{D}(r) = \frac{1}{n_{\ominus r}} \sum_{i=1}^{n_{\ominus r}} \mathbf{1}(0 < d_i \le r) \tag{3.4}$$

where $n_{\ominus r}$ is the number of points in the reduced observation window $W^{\cdot}_{\ominus r}$ (Stoyan et al. 2001).

Plus sampling is conceptually similar to minus sampling, but in this case missing information outside W is collected (or reconstructed). Note that this is only different from minus sampling for summary statistics based on nearest-neighbor distances, since second-order summary statistics require that all points be mapped, not only the nearest neighbors. If additional information outside W cannot be collected, or if the summary statistics require knowledge of all points within distance *r*, one may also reconstruct the missing points outside W to ensure that all focal points in W have a full neighborhood. A variety of methods exist for plus sampling. For example, *periodic edge correction* reconstructs the points outside W by enlarging the point pattern in W by periodic continuation outside W. (Periodic edge correction is sometimes referred to as toroidal edge correction, since, by attaching a copy of W to each edge of W, *ad infinitum*, we effectively turn the plane defined by W into a torus; Figure 2.13c.) In this case, the estimators given in Equations 3.1 and 3.2 can be used after redefining the distances between points, utilizing toroidal geometry. Conceptually, similar to this approach is the *reflection method*, where the points outside W are not reconstructed by periodic continuation outside W, but by reflection at the border. However, these latter two methods are somewhat questionable, since points outside W are not actually sampled, but are inferred. Plus sampling can be used, but only as a crude approximation for analyzing point patterns across a broader range of scales. The interested reader can refer to Illian et al. (2008, p. 183ff) for more details regarding the different plus-sampling methods.

A third strategy of edge correction does not attempt to reconstruct the pattern outside the observation window, but is based on individual assessment of each point or point pair. In the case of nearest-neighbor statistics, such an estimator considers only points i for which the nearest neighbor is located within W (i.e., $d_i \leq d_b$; Figure 3.1) and uses a correction to compensate for the skipped points. This strategy for estimating the nearest-neighbor distribution functions goes back to Hanisch (1984) and will be called in the following *Hanisch edge correction* (Stoyan et al. 2001). For second-order statistics, these more sophisticated methods basically extrapolate the number of points contained within incomplete disks or rings to account for what is expected for complete disks or rings. In practice, this is accomplished by weighting each point pair with an appropriate correction factor. Differences among methods arise through the use of different weights. One example for this approach is a weight based on dividing the circumference of the full disk by the circumference of the disk lying inside W (i.e., *Ripley edge correction*). In this case, if one-third of the disk lies within W, the weight would be 3.

Finally, in a fourth strategy, *the estimator itself is adapted to compensate for the bias* by multiplying the naive estimator given in Equation 3.1 with a correction factor that depends only on the distance r, but not on a given point pair. In this case, the geometry of W is used to calculate the expected bias. This strategy for estimating second-order statistics can be traced to Ohser and Mücklich (2000) and will be called in the following *Ohser edge correction*. The first two strategies (i.e., minus and plus sampling) are thoroughly described in the literature (e.g., Haase 1995; Yamada and Rogerson 2003; Illian et al. 2008, pp. 183–186), but are only rarely used in modern applications of point-pattern analysis. We therefore focus here on the more sophisticated estimators in which edge correction weights and adapted estimators are used.

The selection of the appropriate edge correction method depends critically on the purpose of the analysis. If the absolute value of the summary statistics matters, for example, when fitting a specific point-process model to the data, an edge correction method is required that minimizes the bias. However, if the purpose is to detect spatial patterns in the data, for example, by comparing the observed summary statistics with those of null models, the bias in the summary statistic may cancel out because it appears in both the summary statistics of the observed and simulated data.

3.1.2 Second-Order Summary Statistics for Univariate Patterns (Data Type 1)

3.1.2.1 Different Strategies for Deriving Edge Correction Weights

Four different strategies for deriving edge correction weights $w_{i,j}$ for second-order summary statistics are useful in most practical applications. They all attempt to compensate for the missing area of rings, which are located outside the observation window. Differences among the second-order estimators discussed in this book arise only due to different strategies for deriving the weights $w_{i,j}$ that accomplish edge correction.

The first strategy for deriving individual weighting factors w_{ij} goes back to Ripley (1976). *Ripley edge correction* has an appealingly intuitive geometric interpretation. The Ripley weighting factor $w_{i,j}^R$ basically scales the area of an incomplete circle, with an origin at point i that passes through point j, by the area of the complete circle. For example, if only half of a circle is located within W, the weight yields $w_{i,j}^R = 2$. Thus, the estimator evaluates all pairs of points (with focal point i and a second point j) that are located approximately at distance r apart and weights each observation by $w_{i,j}^R$. If a circle with radius r around a focal point i is completely inside the observation window, there is no need for edge correction, that is, $w_{i,j}^R = 1$. However, if the circle is located only partly inside the observation window, as with circle C2 in Figure 3.1, edge correction is necessary to compensate for points potentially located on the circle outside the observation window W (such as point p3 in Figure 3.1) and weights of $w_{i,j}^R > 1$ are applied.

Ripley edge correction works by first identifying all the points of intersection between a circle of radius r centered on focal point i and the observation window W (i.e., the small open circles in Figure 3.2). Next, the angles of the "pieces of cake" that lie partly outside of the observation window are determined (e.g., φ_1 and φ_2 in Figure 3.2a and φ_1 in Figure 3.2b). The arcs of these angles, which represent the arc length of the circle that is located outside the observation window (fine lines in Figure 3.2), are then summed up. The Ripley weights are then obtained by dividing the arc length of the full circle (i.e., $2\pi r$) by the arc length of the circle located within W ($\varphi_{ij}r$; bold lines in Figure 3.2). For example $\varphi_{ij} = 2\pi - \varphi_1 - \varphi_2$ in Figure 3.2a. The resulting weight is

$$w_{i,j}^R = 2\pi / \varphi_{ij} \qquad (3.5)$$

(a)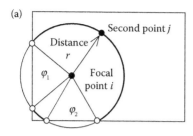

(b)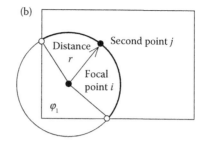

FIGURE 3.2
Edge correction based on the Ripley method. Ripley weights w_{ij} are obtained by taking the length of the full circle (i.e., $2\pi r$), centered on focal point i passing through point j, and dividing by the total length of the arcs lying within the observation window W (bold part of the circle in panels (a) and (b). (a) $w_{ij} = 2\pi/(2\pi - \varphi_1 - \varphi_2)$. (b) $w_{ij} = 2\pi/(2\pi - \varphi_1)$.

where φ_{ij} is the sum of all the central angles of the arcs lying entirely within W of the circle with center x_i and radius $r = \|x_i - x_j\|$. If the circle is completely inside the rectangular observation window W, we find that $w_{ij}^R = 1$. In the most extreme case, where point i is located on one of the four corners of the rectangle W, we find that $w_{i,j}^R = 4$.

The Ripley edge correction method is somewhat sensitive to clusters close to the corners (or boundary) of W, because the points of these clusters receive high weights when they are treated as focal points i. This may lead to overestimation of the degree of clustering in the pattern. Calculation of the weights $w_{i,j}^R$ involves basic trigonometric calculations, but is somewhat complicated by the various cases in which a circle can intersect W. Appendices B.4 and B.5 of Illian et al. (2008, p. 486f) provide elegant computer code for calculating the weights for $w_{i,j}^R$. Goreaud and Pelissier (1999) also provide explicit formulas, even for irregularly shaped observation windows.

An alternative strategy for deriving edge correction weights $w_{i,j}$ was proposed by Wiegand and Moloney (2004). Instead of assigning each point pair i–j an individual weight, as in the Ripley method, the WM *edge correction* addresses bias due to edge effects globally by using the same weight for all point pairs separated by a given distance r. This is done by dividing the area of a complete ring with radius r and width dr by the mean area $\bar{v}(r)$ of all the rings (inside W) with radius r (and width dr) centered on the points of the pattern, that is, $w_{i,j}^{WM} = 2\pi r dr/\bar{v}(r)$ (Figure 3.1). (A more formal approach to defining $\bar{v}(r)$ will be provided in Section 3.1.2.2 below.) Note that $w_{i,j}^{WM}$ belongs to the "Ohser" group of estimators that do not correct for the contribution of individual point pairs i and j to the estimator, but correct for the bias in the naive estimator with a factor that only depends on the distance r between points and the locations of all points of the pattern.

The *Ohser edge correction* goes back to Ohser and Mucklich (2000) and is related to the WM edge correction. As with WM edge correction, the Ohser weight corrects only the expected final bias. This is accomplished by using a weighting factor $w_{i,j}^O$ which is only a function of distance r, not of the specific locations of points. Recall that the denominator of the WM weight is the mean area of all the rings (inside W) with radius r (and width dr) centered on the points of the pattern. By dividing with the ring width dr we obtain the mean length of the circumferences of circles with radius r around the points i of the pattern that lie inside W. The Ohser weight is based on an idea similar to the WM weights, but instead of using the actual points of the pattern, it calculates the expected length of the circumference of a randomly placed circle with radius r lying within W. This is accomplished with the weight:

$$w^O(r) = A/\bar{\gamma}_W(r) \tag{3.6}$$

where $\bar{\gamma}_W(r)$ is the so-called isotropized set covariance, and A the area of the observation window.

The *isotropized set covariance* $\bar{\gamma}_W(r)$ is a quantity developed in set theoretical analysis (e.g., Stoyan and Stoyan 1994; Chapter 8). It is based on the set covariance $\gamma_W(\mathbf{r})$ for a rectangular observation window W with sides of length a_x and b_y (and area $A = a_x b_y$). Essentially, $\gamma_W(\mathbf{r})$ is the area of the intersection of W with W_r (the hatched area in Figure 3.3), where W_r is the rectangle W shifted in space by a vector $\mathbf{r} = (r_x, r_y)$ and $\gamma_W(\mathbf{r}) = |(a_x - r_x)(b_y - r_y)|$. Thus, $\gamma_W(\mathbf{r})$ is the area of overlap between the original rectangle and the rectangle shifted by $\mathbf{r}$. If $\mathbf{r} = 0$, then $\gamma_W(\mathbf{r}) = A$. Using polar coordinates, the vector $\mathbf{r}$ can be rewritten as $\mathbf{r} = (r, \varphi)$. The isotropized set covariance function $\bar{\gamma}_W(r)$ for rectangle W is then the average set covariance $\gamma_W(\mathbf{r})$ taken over all angles φ (Figure 3.4).

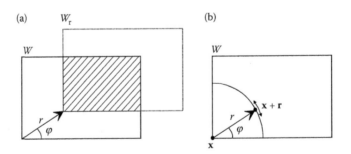

FIGURE 3.3
Isotropized set covariance. (a) *Definition of the set covariance.* The set covariance is the area of the intersection of window W and the translated window W_r where $\mathbf{r} = (r, \varphi)$ is a vector. (b) *Calculation of the isotropized set covariance.* A point x on W is shifted by vector r and counted if it is located within W. The result is that all rotations φ of r located in W are counted, which produces an estimate of the boundary length of the portion of the circle with center x and radius r lying in W. This is repeated for all x in W, thereby yielding the isotropized set covariance.

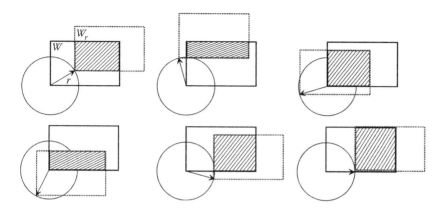

FIGURE 3.4

Calculation of the isotropized set covariance function $\bar{\gamma}_W(r)$. The window W is translated by a vector $\mathbf{r} = (r, \varphi)$ with length r and angle φ and the mean area of overlap (hatched area) that results from all angles φ and a constant value of r, is calculated. The different figures show the resulting area of overlap for different angles φ with a fixed value of r.

It is, therefore, the expected intersection of W and the translated window $W_\mathbf{r}$ without specifying the angle.

For a rectangle with sides $a_x \le b_y$, the isotropized set covariance can be calculated analytically (Stoyan and Stoyan 1994, p. 123). For the range $r \le a_x$, which is relevant for point-pattern analysis, this yields

$$\bar{\gamma}_W(r) = A - \frac{2r}{\pi}(a_x + b_y) + \frac{r^2}{\pi} \tag{3.7}$$

Thus, $\bar{\gamma}_W(r)$ decreases almost linearly with distance r.

Initially, it might be surprising that calculating the overlap between the original rectangle W and the set of shifted and rotated rectangles $W_\mathbf{r}$ might be an appropriate weight for edge correction, that is, $w^O(r) = A/\bar{\gamma}_W(r)$. However, there is a relatively simple geometric explanation. As shown in Figure 3.3, calculation of the set covariance $\gamma_W(\mathbf{r})$ involves shifting each point $\mathbf{x}$ within W by vector $\mathbf{r}$ and integrating over all shifted points $\mathbf{x} + \mathbf{r}$ that are still located within W (Figure 3.3a). The isotropized set covariance $\bar{\gamma}_W(r)$ is then calculated by rotating $W_\mathbf{r}$ over all angles φ and integrating the resulting values of $\gamma_W(\mathbf{r})$ for all rotations φ (Figure 3.4). If we switch the order of integration, we can obtain a quantity related to the Ohser weight as follows: First, we integrate over all points lying inside W that result from rotating the vector $\mathbf{r} = (r, \varphi)$ for $\varphi \in (0, 2\pi)$ around a point $\mathbf{x}$ that also lies in W. This yields the length of the circumference of the circle with radius r around point $\mathbf{x}$ that lies within W (Figure 3.3b). In the second step, we integrate the step over all points $\mathbf{x}$ that lie within W. The result is then divided by A, the area of W. This gives us the expected length of the circumference of a randomly placed circle, with radius r, lying inside W, that is, $2\pi r \bar{\gamma}_W(r)/A$. Taking the inverse and multiplying by $2\pi r$ yields the Ohser weight.

For a random pattern, the WM weight approximates the Ohser weight, as can be seen in Figure 3.5b. The expected area of incomplete rings with radius r and random location within W captured by the Ohser weight (solid line in Figure 3.5b) is closely approximated by the actual area of incomplete rings with radius r, centered on the points of the pattern that are used in calculating the WM weight (closed circles in Figure 3.5b). Interestingly, there are two small systematic departures between the WM and Ohser weights. These arise due to the fact that the WM weights are based on a single realization of a random process. The mean area of the "real" rings is somewhat less than expected up to a radius of 2 m. This indicates that a few more points than expected are located close to the border of W. In contrast, the area of the real rings is slightly greater than expected at distances of 6–8 m. Thus, the WM weights are "adapted" to the small irregularities of the pattern and may provide slightly better results than the Ohser weights. However, the Ohser weights do not depend on the actual locations of the points of the pattern and, therefore, have the advantage of allowing rapid computation. In a later section we will see that this difference is related to the so-called "adapted" estimator of the intensity λ used for estimation of the pair-correlation function (see Section 3.1.2.4).

A fourth form of edge correction, the *Stoyan edge correction*, is the one favored by Stoyan and Stoyan (1994) and Illian et al. (2008). It is based on the set covariance $\gamma_W(\mathbf{r})$, which for a rectangle with sides of lengths a_x and b_y yields

$$w_{i,j}^S(r) = A/\gamma_W(x_i - x_j) \qquad (3.8)$$

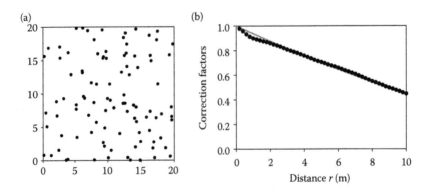

FIGURE 3.5
Ohser correction factor. (a) Random pattern of 100 points within a 20 × 20 m² observation window. (b) The inverse of the Ohser edge weight $\bar{\gamma}_W(r)/A$ (solid gray line) and inverse of the WM edge correction weight (closed disks) for the pattern in panel (a).

where $\gamma_W(\mathbf{r}) = |(a_x - r_x)(b_y - r_y)|$, as before. Thus, the Stoyan weight follows the first step of the calculation of the Ohser weights. It shifts each point $\mathbf{x}$ within W by vector $\mathbf{r}$ and integrates over all shifted points $\mathbf{x} + \mathbf{r}$ that are still located within W, resulting in a calculation of the set covariance $\gamma_W(\mathbf{r})$. It then uses the set covariance $\gamma_W(\mathbf{r})$ as a weight, thereby summing up and weighting all point pairs in W that have distance r separating them. Because the set covariance yields only a value of A (i.e., the area of W) when $r = 0$, all pairs of points separated by a distance $r > 0$ receive a weight greater than 1. Point pairs with small separation distances r receive, in general, small weights, whereas pairs of points separated by larger distances r receive larger weights. In contrast, the Ripley edge correction method weights only point pairs defining incomplete circles within W. In the latter case, only points close to the border are given weights greater than 1. The Stoyan weights, therefore, also consider the explicit locations of the points and correct for smaller amounts of stochastic irregularity in the pattern.

The formulas for the four edge correction weights are given below in Equation 3.9. The Ripley and Stoyan weights are applied as individual weights to each i–j point pair, whereas the Ohser weights are completely independent of the pattern being analyzed and depend only on the geometry of the observation window W and distance r. The WM weights are intermediate to these two extremes as they are specific to the given pattern analyzed, but are not applied to the specific point pairs i–j, depending only on the geometry of W and distance r, like the Ohser weights.

$$w^{WM}(r) = \frac{2\pi r dr}{\bar{v}(r)} \qquad \text{WM}$$

$$w^R_{i,j}(r) = \frac{2\pi}{\varphi_{ij}} \qquad \text{Ripley}$$

$$w^S_{i,j}(r) = \frac{A}{\gamma_W(x_i - x_j)} \qquad \text{Stoyan} \tag{3.9}$$

$$w^O(r) = \frac{A}{\gamma_W(r)} \qquad \text{Ohser}$$

In the next sections, we will combine the edge correction weights with the estimators for the different second-order statistics and explore the implications of using the different weighting schemes.

3.1.2.2 O-Ring Statistic

Now that we have introduced several approaches to edge correction, we are in a position to present the methods for estimating the various second-order point-pattern statistics. The first estimator we consider is the O-ring statistic

$O(r)$. $O(r)$ can be defined as the density of points within a ring of radius r and width dr around the typical point of the pattern (gray ring in Figure 3.1). Note that $O(r)$ is a conditional summary statistic; meaning that it yields the expected density of points within a ring, given the condition that there is a point at the center of the ring. This characteristic also means that the O-ring statistic is a neighborhood density function and, in some sense, a point-related analog to the location-related intensity function $\lambda(x)$. The naive estimator of $O(r)$ is also analogous to the estimator of $\lambda K(r)$ given in Equation 3.1 and is defined as follows:

$$\hat{O}^n(r) = \frac{\bar{n}^R(r)}{2\pi r dr} = \frac{1/n \sum_{i=1}^{n} n_i^R(r)}{2\pi r dr} \tag{3.10}$$

where $n_i^R(r)$ is the number of points located in a ring with radius r and width dr around point i (Figure 3.1). The superscript R refers to "ring." This is summed over all n points i in the pattern and divided by the area $2\pi r dr$ of a ring of radius r to obtain a density. Clearly, this estimator does not correct for edge effects.

A simple way to modify Equation 3.10, when some rings are not completely located within W, was proposed by Wiegand and Moloney (2004). This results in the WM edge correction factor described above. The basic idea is to estimate the mean number of points within complete and incomplete rings around points i, that is, $\sum n_i^R(r)/n$, and divide by the mean area of these rings. More technically, the neighborhood density is estimated as a point-related analog to the estimation of the location-related intensity λ, based on several plots (see Section 3.3.1). In essence, this means that the number of points n_i is counted within m subwindows W_i with area v_i located within a larger area W (Figure 3.6a). Following Illian et al. (2008, p. 261), the intensity of points within W can be estimated based on the counts within the m subwindows W_i:

$$\hat{\lambda} = \sum_{i=1}^{m} \hat{\lambda}_i \frac{v_i}{v} \tag{3.11}$$

where $\hat{\lambda}_i$ is the estimated intensity of points within the ith subwindow W_i ($=n_i/v_i$), and $v = \sum_{i=1}^{m} v_i$. We can rearrange Equation 3.11 to produce

$$\hat{\lambda} = \frac{\sum_{i=1}^{m} \hat{\lambda}_i v_i}{\sum_{i=1}^{m} v_i} = \frac{1/m \sum_{i=1}^{m} n_i}{1/m \sum_{i=1}^{m} v_i} \tag{3.12}$$

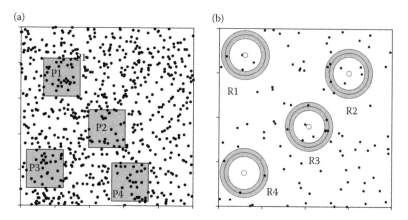

FIGURE 3.6
Estimation of the intensity and neighborhood density function $O(r)$ based on subplots. (a) *Intensity*. Points within several subplots P1–P4 obtained from a larger area are used to estimate the over all intensity of the observation window. (b) *Neighborhood density function*. Ring-shaped "subplots" R1, R2, R3,..., Rn within observation window W are constructed such that each subplot is centered on a point of the pattern (open disks). For clarity, we show here only the ring-shaped subplots for four focal points, the other points are indicated by closed disks. The O-ring statistics are the mean density of points within the n subplots.

which shows that this indeed yields the mean number of points within the m subwindows W_i divided by the mean area of these subwindows. This idea can be translated into an approach for estimating the neighborhood density function $O(r)$. In this case, the different subwindows are the rings R_i around the points i of the pattern (Figure 3.6b) and the resulting estimator is given by

$$\hat{O}^{WM}(r) = \frac{1/n \sum_{i=1}^{n} n_i^R(r)}{1/n \sum_{i=1}^{n} v_i} = \frac{\bar{n}^R(r)}{\bar{v}(r)}, \tag{3.13}$$

where v_i is the area of a potentially incomplete ring in W around point i and $n_i^R(r)$ is the number of points within that ring. Thus, the neighborhood density is estimated by dividing the average number of points in the rings surrounding the points of the pattern by the average area within each ring that lies in W.

Comparison of the naive estimator of the O-ring statistic given in Equation 3.10 with the method proposed in Wiegand and Moloney (2004) reveals that the area $2\pi r dr$ of the complete ring is replaced by the mean area of the rings within W for the given point pattern. Basically, Equation 3.13 is obtained by multiplying Equation 3.10 by the WM edge correction factor $w_{i,j}^{WM} = 2\pi r dr / \bar{v}(r)$.

In essence, this estimator is "adapted" to the point pattern because the same estimation rule is used in the numerator and denominator; in both cases, all points i are visited in turn and some property of the ring around point i is evaluated. It can, therefore, be expected that this estimator will compensate for stochastic irregularities in the distribution of the points and will be robust to small departures from heterogeneity. Note also that this estimator belongs to the fourth strategy of edge correction methods presented above (i.e., Ohser edge correction), which corrects for edge effects by multiplying the naive estimator (Equation 3.10) by a correction factor, yielding in this case $2\pi r dr / \bar{v}(r)$. The correction factor depends only on the distance r (and the locations of all points of the pattern), but not on a given point pair. Figure 3.5b shows the bias in $\bar{v}(r)/2\pi r dr$ (i.e., the mean area of the incomplete rings relative to that of a complete ring) resulting from the pattern shown in Figure 3.5a. This pattern comprises 100 points within a 20×20 m^2 plot. Interestingly, the bias increases almost linearly with distance r and reaches a value of 0.45 for a distance that corresponds to half of the width of the plot (i.e., $r = 10$ m). This shows that the bias cannot be ignored in most practical applications of the neighborhood density function and that edge correction is required.

Although the approach to estimation discussed above is geometrically intuitive, we need to introduce a more general approach for incorporating edge correction into our estimators. With this in mind, a modified approach based on the notation provided by Illian et al. (2008) will be first used to redefine the estimator presented in Equation 3.13. To do this, we need to introduce a general class of functions known as the *kernel function k(x)*. A kernel function essentially evaluates the distance between two points and returns a value based on that distance. Using this, we rearrange the numerator of Equation 3.13 by incorporating a kernel function in its definition. This is used to determine the number of points falling at distance r from a focal point i, which is given by $n_i^R(r)$ as follows:

$$n_i^R(r) = \sum_{j=1, j \neq i}^{n} k(\|x_i - x_j\| - r) \qquad (3.14)$$

The sum on the right-hand side of Equation 3.14, essentially compares all points of the pattern j to point i, where x_i and x_j are the coordinates of points i and j, respectively, and $d_{ij} = \|x_i - x_j\|$ is the distance separating the two points. The kernel function evaluates to 1 if the distance between i and j is close to the distance r, otherwise it evaluates to 0. Thus, what we described more loosely as a ring (e.g., R1–R4 in Figure 3.6b) can be described more formally with a *kernel function k(x)* which determines whether or not two points i and j are located distance r apart. The kernel function introduces a small "tolerance" interval $(r - dr/2, r + dr/2)$ for the distance r (called the bandwidth $h = dr/2$) within which two points are regarded as being located distance r apart. In this way, the kernel function defines rings with width dr and radius r around the focal point i (Figure 3.6b). Any point falling within the ring causes the kernel function to evaluate to a positive value, otherwise it returns zero.

Note that the bandwidth should not be too small, otherwise the number of points $n_i^R(r)$ within the ring will be low and cause spurious stochastic fluctuations in the estimator. However, bandwidths that are too large will produce smooth estimates, but important details regarding the pattern may be lost. Illian et al. (2008, p. 236) recommend, as a rule of thumb, an initial value of $h \approx 0.1/\lambda^{0.5}$. Thus, a pattern with 100 points within a 100×100 m^2 observation window will require a bandwidth of $h \approx 1$ m, and 100 points within a 500×500 m^2 observation window will require a bandwidth of $h \approx 5$ m. If the pattern is clustered, a somewhat smaller bandwidth could be selected, since the neighborhood density at small distances r is larger and more points are located within the rings.

Figure 3.7 shows two popular kernel functions. The *box kernel* (Figure 3.7a) is the simplest kernel function and evaluates as

$$k_h^B(d_{ij} - r) = \begin{cases} \dfrac{1}{2h} & |d_{ij} - r| \leq h \\ 0 & \text{otherwise} \end{cases} \tag{3.15}$$

where h is the bandwidth, that is, $dr/2$, and d_{ij} is the distance between points i and j (Figure 3.1). Illian et al. (2008) recommend use of this kernel function in the estimation of the product density. Note that the kernel function requires a small correction if $r - h/2 < 0$. Another popular kernel function is the *Epanechnikov kernel*, which was already presented in Box 2.4 in the context of nonparametric intensity estimators (Figure 3.7b):

$$k_h^E(d_{ij} - r) = \begin{cases} \dfrac{3}{4h}\left(1 - \dfrac{(d_{ij} - r)^2}{h^2}\right) & |d_{ij} - r| \leq h/2 \\ 0 & \text{otherwise} \end{cases} \tag{3.16}$$

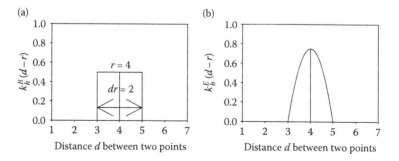

FIGURE 3.7
Kernel functions for selecting points separated by a distance of approximately 4 units (i.e., $r = 4$), using a bandwidth h of one unit. (a) The box kernel. (b) The Epanechnikov kernel.

Both kernel functions yield values of zero if $d_{ij} < r - h$ or $d_{ij} > r + h$ (i.e., they consider only points j within the ring with radius r and width $2h$ around the focal point i), but the Epanechnikov kernel weights point pairs with a separation distance d closer to r higher than those that are less close to r (Figure 3.7b).

A second modification, which formalizes the notation of the denominator of Equation 3.13, is more complicated. First, a disk with radius r and center x_i for point i is written as "$b(x_i,r)$" and the circumference of $b(x_i,r)$ is written as "$\partial b(x_i,r)$." Next, the symbol $\cap$ is used to define the intersection of two geometric elements, which in our case is the intersection of the circumference of a disk centered on x_i with the observation window W, that is, $W \cap \partial b(x_i,r)$. To express the length of the circumference located in W, the function $v_{d-1}()$ is used which gives the length of a one-dimensional geometric object (i.e., a line; $d = 2$), thus $v_{d-1}(W \cap \partial b(x_i,r))$ in formal notation. With these definitions the translation of the estimator given in Equation 3.13 into formal language yields:

$$\hat{O}^{WM}(r) = \frac{1/n \sum_{i=1}^{n} \left(\sum_{j=1}^{n,\neq} k_h^B(\, \|x_i - x_j\| - r) \right)}{1/n \sum_{i=1}^{n} v_{d-1}(W \cap \partial b(x_i,r))} \qquad (3.17)$$

A natural estimator of the pair-correlation function $g(r)$ based on 3.17 is to divide 3.17 by an estimate of the intensity, since $O(r) = \lambda\, g(r)$. The simplest way to accomplish this is to use the "natural" intensity estimator for this purpose, which is given by the number of points within W divided by the area of W; that is, $\hat{\lambda}_n(r) = n/v_d(W) = n/A$. In later sections we discuss other options for estimating the intensity using "adapted intensity estimators" (see Section 3.1.2.4).

3.1.2.3 The Product Density

A more formal method of estimating second-order summary statistics is based on the so-called "product density" $\rho^{(2)}(x_1, x_2)$ briefly introduced in Sections 2.3.1.1 and 2.3.2 (see also Figure 2.6b). This method is used in Illian et al. (2008). For greater clarity in notation we will omit the superscript (2) in the following sections.

For homogeneous patterns, the product density $\rho(r)$ is the conditional probability that one point of a given point process occurs within a disk with area dx_2 located at x_2 given that another point is located at distance r within a disk with area dx_1 located at x_1. This is, in fact, related to the neighborhood density function $O(r)$ as follows:

$$\rho(r) = \lambda\, O(r) \tag{3.18}$$

The general form of the estimators of the product density for univariate patterns is given by

$$
\begin{aligned}
\hat{\rho}(r) &= \frac{1}{2\pi r}\frac{1}{A}\sum_{i=1}^{n}\sum_{j=1}^{n,\neq} k(\|\mathbf{x}_i - \mathbf{x}_j\| - r)w_{i,j} \\
&= \hat{\lambda}_n\, \frac{1/n\sum_{i=1}^{n}\sum_{j=1}^{n,\neq} k(\|\mathbf{x}_i - \mathbf{x}_j\| - r)w_{i,j}dr}{2\pi r\, dr}
\end{aligned} \tag{3.19}
$$

The second expression for the function shown in Equation 3.19, incorporating the factor $\hat{\lambda}_n$ follows simply from the first expression, if we multiply the numerator and denominator by n and dr. The modified double sum can now be interpreted as the mean number of points located at a distance interval $(r - dr/2, r + dr/2)$ from the observed points of the pattern, with n/A yielding the natural estimator $\hat{\lambda}_n$ of the intensity. Written in this form, the equation can be interpreted as the following: the intensity $\hat{\lambda}_n$ represents the probability of having a point at a given focal location, and $1/n\sum_{i=1}^{n}\sum_{i=1}^{n,\neq} k(\|\mathbf{x}_i - \mathbf{x}_j\| - r)w_{i,j}\, dr$ is an estimate of the expected number of points located at a distance interval $(r - dr/2, r + dr/2)$ away from the focal location. Note that each point pair in the double sum is adjusted by an *edge correction weight* $w_{i,j}$, which compensates for rings that are partly located outside the observation window (Figure 3.1). Finally, the denominator in the second version of Equation 3.19 yields the area of a ring with radius r and width dr. Thus, we divide the expected number of points located within the ring $(r - dr/2, r + dr/2)$ centered on the focal location by the area of the ring, yielding the probability that there is an additional point at any location distance r away from the focal point.

Slight rearrangement of Equation 3.19 shows that this estimator basically multiplies the natural estimate $\hat{\lambda}_n = n/A$ of the intensity by the estimated mean density of points within rings with radius r around the points i of the pattern:

$$
\hat{\rho}(r) = \underbrace{\frac{n}{A}}_{\hat{\lambda}_n}\frac{1}{n}\sum_{i=1}^{n}\underbrace{\left(\frac{\sum_{j=1}^{n,\neq} k(\|\mathbf{x}_i - \mathbf{x}_j\| - r)w_{i,j}}{2\pi r}\right)}_{\text{Density of points in ring around point } i} \tag{3.20}
$$

This is reasonable since $\rho(r) = \lambda\, O(r)$ (Equation 3.18).

We can incorporate the four different weights adjusting for edge effects into the univariate product density estimator as follows:

$$\hat{\rho}^R(r) = \frac{1}{2\pi r A} \sum_{i=1}^{n} \sum_{j=1}^{n,\neq} k(\|x_i - x_j\| - r) \left[\frac{2\pi}{\varphi_{ij}} \right] \qquad \text{Ripley}$$

$$\hat{\rho}^S(r) = \frac{1}{2\pi r A} \sum_{i=1}^{n} \sum_{j=1}^{n,\neq} k(\|x_i - x_j\| - r) \left[\frac{A}{\gamma_W(x_i - x_j)} \right] \qquad \text{Stoyan}$$

$$\hat{\rho}^O(r) = \frac{1}{2\pi r A} \sum_{i=1}^{n} \sum_{j=1}^{n,\neq} k(\|x_i - x_j\| - r) \left[\frac{A}{\overline{\gamma}_W(r)} \right] \qquad \text{Ohser}$$

$$\hat{\rho}^{WM}(r) = \frac{1}{2\pi r A} \sum_{i=1}^{n} \sum_{j=1}^{n,\neq} k(\|x_i - x_j\| - r) \left[\frac{2\pi r}{(1/n) \sum_{i=1}^{n} v_{d-1}(W \cap \partial b(x_i, r))} \right] \qquad \text{WM}$$

$$(3.21)$$

A direct comparison of the weights resulting from the different edge correction methods for a random pattern is illustrated in Figure 3.8. Since the Ohser and WM weights depend only on the distance between the two points i and j, not on their location, they are approximately equal in value and cannot be distinguished in Figure 3.8. Interestingly, there are in general only small differences between the weights calculated by the Ohser and the Stoyan methods. In contrast, the Ripley weights may be much larger than the other weights, if the focal point is located close to the boundary of W. On average, the Ripley weights correspond, in good approximation, to the other weights, as can be seen by the Ripley regression lines shown in Figure 3.8. However, in the example shown in Figure 3.8, there is some departure for estimates at a distance of 150 m (Figure 3.8c). Thus, in general, we can expect that the four alternative estimators of the product density will yield similar results, since the noise of the Ripley and Stoyan weights will average out, even though the noise in the Ripley weights is somewhat greater.

Estimates of the product density utilizing the different edge correction techniques for a CSR pattern of 500 points (Figure 3.9a) within a 100×100 m² observation window are given in Figure 3.9b. We find that the estimates of the product density that consider edge correction for a random pattern are essentially the same for the four edge correction methods but that the naive estimator yields a considerable bias (Figure 3.9b). In contrast, Figure 3.9d explores how the estimators behave in the case of a strongly clustered pattern shown in Figure 3.9c, where some clusters are accidently located at the corners of the observation window W. The Ohser and Stoyan estimators cannot be easily distinguished (their curves overlap), but the Ripley estimator,

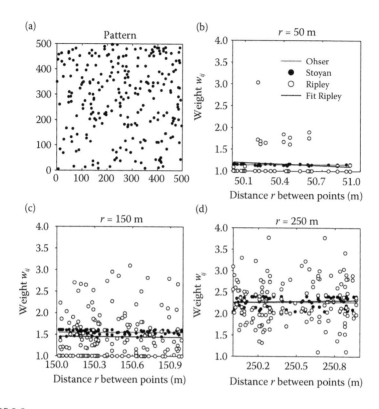

FIGURE 3.8
Comparison of edge correction weights w_{ij}. (a) Random pattern with 250 points within a 500×500 m² observational window, used in determining weights shown in panels (b)–(d). (b) Weights for pairs of points separated by distances of 50–51 m. (c) Weights for pairs of points separated by distances of 150–151 m. (d) Weights for pairs of points separated by distances of 250–251 m. In addition to the weights, we show a linear regression (black line) fit to the points of the Ripley weights, as a comparison to the Ohser weights.

as expected, indicates somewhat larger clustering at smaller distances (i.e., $r < 15$) (Figure 3.9d). This is consistent with the scale of the clusters in the lower and top left corners and occurs since the points located in the clusters receive greater weight, due to their proximity to the border. The result is a slightly higher estimate of the product density over a range of scales approximating the cluster size. The WM estimator ranges somewhat between the Ohser and the Ripley estimator and exhibits smaller fluctuations at larger scales (i.e., $r > 15$). However, note that the difference among estimators will appear smaller when the product density is plotted on a linear scale.

3.1.2.4 Estimators of the Pair-Correlation Function

A simple estimator of the pair-correlation function $\hat{g}(r)$, based on the different estimators of the product density shown in Equation 3.21, would be

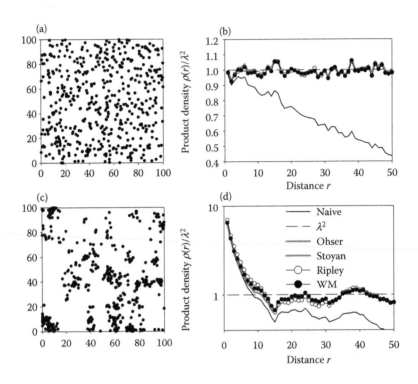

FIGURE 3.9
Comparison of estimators of the product density. (a) Random pattern. (b) Estimators for the random pattern in panel (a). (c) Clustered pattern. (d) Estimators for the clustered pattern in panel (c). Note the logarithmic scale used in panel (d).

to use the "natural" estimator of $\hat{\lambda}_n^2 = n(n-1)/A^2$, that is, $\hat{g}(r) = \hat{\rho}(r)/\hat{\lambda}_n^2$, since $\rho(r) = \lambda^2 g(r)$. Note that $(n-1)$ is required in the numerator of the natural estimator, since the double sum of the estimator in Equation 3.21 considers only points $i \neq j$. Although these estimators were used in earlier studies (e.g., Stoyan and Stoyan 1994), Stoyan and Stoyan (2000) and Illian et al. (2008) now recommend use of an estimator of λ that is especially adapted to the estimator of the product density. The objective of this is to reduce variability in the estimator.

The need to use a correction in estimating the intensity λ for the pair-correlation function can be understood from Figure 3.5b. The figure compares the mean area of the partly incomplete rings around the points of the pattern (dots) with the expectation for randomly distributed points provided by the isotropized set covariance (line). Small irregularities may cause slightly more (or fewer) points than expected being located close to the border. This decreases (increases) the mean area of the rings (as shown in Figure 3.5b), whereas the expectation of the isotropized set covariance produces no such irregularities. To compensate for stochastic variation in estimating the pair-correlation function, Stoyan (2006a; his Equation 19) proposed an adapted estimator of the intensity λ given by

$$\hat{\lambda}_S(r) = \frac{\sum_i^n v_{d-1}(W \cap \partial b(x_i, r))}{2\pi r \bar{\gamma}_W(r)} \tag{3.22}$$

where $v_{d-1}(W \cap \partial b(x_i, r))$ is the length of the circumference of a circle of radius r around point i in W. This formal notation was introduced in Section 3.1.2.2. The length of the circumference of a circle of radius r around point i in W yields $2\pi r$, if the circle is fully located within W, but it may be smaller if the circle only partly overlaps W. By factoring out the natural estimator $\hat{\lambda}_n(r) = n/A$ of lambda and slightly rearranging Equation 3.22 we find that

$$\hat{\lambda}_S(r) = \frac{\sum_i^n v_{d-1}(W \cap \partial b(x_i, r))}{2\pi r \bar{\gamma}_W(r)}$$

$$= \hat{\lambda}_n \frac{(1/n)\sum_i^n v_{d-1}(W \cap \partial b(x_i, r))}{2\pi r} \frac{A}{\bar{\gamma}_W(r)},$$

$$= \hat{\lambda}_n \frac{w_{ij}^O(r)}{w_{ij}^{WM}(r)} = \hat{\lambda}_n \times \text{corr}(r) \tag{3.23}$$

Thus, the adapted estimator of lambda $\hat{\lambda}_S(r)$ is basically the natural estimator $\hat{\lambda}_n(r)$ multiplied by the ratio of the Ohser and the WM edge correction weights, that is, corr(r). The correction term corr(r), itself, is basically the ratio of the average length of the circumference of circles centered on points i lying inside of W to the circumference expected for random points in W.

Further consideration also shows that the WM estimator of the neighborhood density function $O(r)$ given in Equation 3.17 is identical with the estimator that results from combining the estimator of the product density based on Ohser edge correction (Equation 3.21) with the adapted intensity estimator in Equation 3.22, that is, $\hat{O}^{WM}(r) = \hat{\rho}^O(r)/\hat{\lambda}_S(r)$. This integrates the estimator of the O-ring statistic suggested in Wiegand and Moloney (2004) into the more general framework of estimators presented in Illian et al. (2008).

The main effect of using $\hat{\lambda}_S(r)$ to calculate the pair-correlation function is that the mean-squared error (mse) of the estimate is smaller than it would be if we used the natural estimator of λ (Illian et al. 2008, p. 232). Figure 3.10 illustrates the correction factor applied to different patterns. For the random

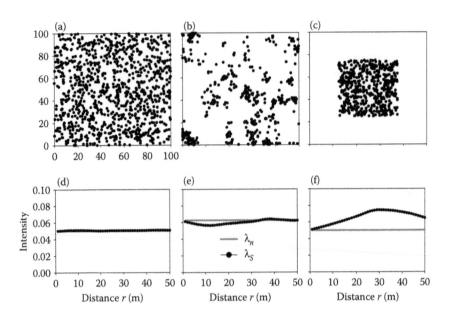

FIGURE 3.10
The adapted intensity estimate for three contrasting patterns. (a) Random pattern. (b) Clustered pattern. (c) Heterogeneous pattern. (d)–(f) The naive λ_n and adapted λ_S (Equation 3.22) intensity estimators, corresponding to the patterns lying directly above the respective panels.

pattern shown in Figure 3.10a, the adapted estimator departs somewhat from the expectation, but only a small correction is required. The points of the pattern exhibit only a slight tendency to be located away from the border (i.e., the area of the corresponding rings is slightly large than expected) yielding a correction > 1 at smaller distances (Figure 3.10d). However, in the case of the clustered pattern, where some clusters with many points are located close to the (lower and upper left) corners (Figure 3.10b), we find that the rings around the points of the pattern have, on average, a smaller area than expected inside W. Consequently, the Ohser weight yields an underestimation that must be corrected for by a value of $\hat{\lambda}_S(r) < \hat{\lambda}_n$ (Figure 3.10e). For illustrative purposes, we also show the intensity estimates for an extreme pattern, as shown in Figure 3.10c that has no points close to the border (Figure 3.10f). In this case, the rings around the points of the pattern have, on average, a larger area than expected inside W (because they are mostly inside W) leading to a correction with $\hat{\lambda}_S(r) > \hat{\lambda}_n$ (Figure 3.10f).

Following the suggestion by Stoyan (2006a) and Illian et al. (2008), we therefore obtain three alternative estimators for the pair-correlation function based on the different edge correction methods (i.e., the Ripley, Stoyan, and Ohser weights shown in Equation 3.21) that divide the estimators for the product by the square of the adapted intensity estimator (Equation 3.22):

$$\hat{g}^R(r) = \frac{\hat{\rho}^R(r)}{(\hat{\lambda}_S(r))^2} \qquad \text{Ripley}$$

$$\hat{g}^S(r) = \frac{\hat{\rho}^S(r)}{(\hat{\lambda}_S(r))^2} \qquad \text{Stoyan}$$

$$\hat{g}^O(r) = \frac{\hat{\rho}^O(r)}{(\hat{\lambda}_S(r))^2} \qquad \text{Ohser}$$

$$\hat{g}^{WM}(r) = \frac{\hat{\rho}^{WM}(r)}{\hat{\lambda}_n^2} = \frac{\hat{\rho}^O(r)}{\hat{\lambda}_S(r)\hat{\lambda}_n} \qquad \text{WM}$$

(3.24)

For completeness, we also show the estimator $\hat{g}^{WM}(r)$ of the pair-correlation function presented in Wiegand and Moloney (2004), which divides the estimator of the O-ring statistic by the natural estimator of the intensity (i.e., n/A). Equation 3.20, however, shows that the estimator of the product density contains the naive estimator of the intensity. This suggests that it may be more appropriate to divide the estimator of the product density (Equation 3.21), one time, by the naive estimator of the intensity and the second time by the adapted estimator, as is done with the WM estimator.

We apply the different estimators to an extreme case of a heterogeneous pattern to better understand their behavior. We do this using a random pattern within a 100×100 m² area (denoted W_i) with area A_i, which is enlarged by an empty buffer of 50 m, thus yielding an observation window W of 200×200 m² (Figure 3.11). The expected average neighborhood density of this pattern can be calculated analytically yielding $O_{exp}(r) = (n/A_i)(\bar{\gamma}_{W_i}(r)/A_i)$ (Figure 3.11c). The expected average neighborhood density of the points of this pattern is always smaller than n/A_i because rings close to the border of the patch overlap the patch only partly. The product densities calculated with Equation 3.21 differ among estimators. The WM, Ohser, and Stoyan estimators cannot be distinguished, but differ quite a bit from the Ripley estimator (Figure 3.11b). This is because the weights for the Ripley edge correction always yield $w_{ij}^R = 1$ (since all the rings around the points of the pattern are fully within W), but the Ohser and Stoyan weights are always greater than 1. When calculating the neighborhood density $O(r)$ we find that the WM estimator (Equation 3.17) and the Ohser and Stoyan estimators (based on Equations 3.21 and the adapted intensity estimate Equation 3.22) agree with the expected neighborhood density, but the estimator based on Equations 3.21 and 3.22, for the Ripley edge correction method, underestimate the neighborhood density at greater distances (Figure 3.11c).

This result illustrates quite well the problem of Ripley edge correction. In particular, if more points (than expected) are located close to the border, Ripley edge correction will overestimate the neighborhood density. In contrast, if fewer points are, by accident, located close to the border, the neighborhood density will be underestimated. The Ohser and Stoyan estimators are

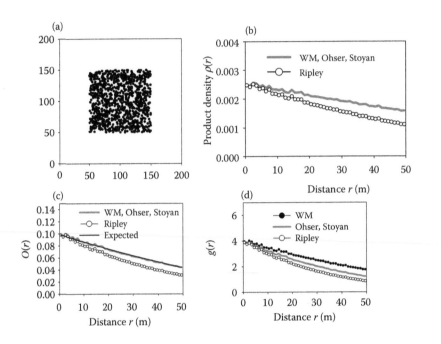

FIGURE 3.11
Comparison of estimators for a heterogeneous pattern. (a) Heterogeneous pattern used for comparing estimators. (b) Product density. (c) Neighborhood density. (d) Pair-correlation function.

robust against such effects and are therefore preferable. Finally, if we calculate the pair-correlation function with the estimators given in Equation 3.24, the difference between the Ohser and Ripley methods persists, with lower estimates given by the Ripley method (Figure 3.11d). At larger distances, the WM estimator, which uses the adapted intensity estimator (3.23) only once in the calculation of the pair-correlation function, produces somewhat higher estimates than the Ohser estimator (Figure 3.11d).

The example shown in Figure 3.11, which contains an extreme case of heterogeneity, was only used to develop a better understanding of the behavior of the different estimators. We see that, even in this extreme case, differences among the four estimators of the pair-correlation function provided in Equation 3.24 are relatively small. Thus, we can expect that all four estimators should produce robust results, with only small differences among them, especially for patterns that only depart slightly from homogeneity. However, an important lesson is that the neighborhood density $O(r)$ provides, even in the case of an inhomogeneous pattern, the correct result, if the Ohser or Stoyan estimator is used. However, it produces a notable bias if the Ripley estimator is used. Clearly, in the case of a heterogeneous pattern, we will need more precise methods to factor out the effects of heterogeneity. These will be discussed in the next section.

3.1.2.5 Inhomogeneous Pair-Correlation Function Using Reweighting

For a homogeneous pattern (i.e., one that has the same statistical properties throughout the observation window), the estimators of the product density, the neighborhood density, and the pair-correlation function have an intuitive interpretation, as they represent the (neighborhood) properties of the typical point of the pattern. However, each point may have somewhat different neighborhood properties, meaning that the pattern is not homogeneous. In this case, we can still apply the estimators for homogeneous patterns (Illian et al. 2008, p. 280), but the interpretation is more complicated, because no typical point exists. Under these circumstances we can only estimate "average" neighborhood properties. It would, therefore, be desirable to derive estimators for nonstationary patterns that would be able to factor out the effects of the heterogeneity and reveal the "pure" underlying neighborhood (second-order) properties. Indeed, this is possible for certain types of heterogeneous patterns.

Baddeley et al. (2000) generalized the product density estimator (Equation 3.21) for certain types of heterogeneous patterns using the Ripley or Stoyan edge correction method. The basic assumption of their approach was that the observed pattern φ_o is only heterogeneous with respect to the intensity function $\lambda(x)$, but that the other properties of the pattern are not dependent on location x (i.e., homogeneous point interactions are "modulated" by the intensity function). A pattern like this can be conceptualized as resulting from a two-step process: Initially a homogeneous point pattern φ_h is independently thinned using a thinning function $p(x) = \lambda(x)/\lambda^*$, where λ^* is the maximal value of the intensity function $\lambda(x)$ within the observation window. In the thinning operation, all points of the initially homogeneous pattern φ_h are visited in turn and are retained in the pattern with probability $p(x)$. This results in a heterogeneous pattern with intensity function $\lambda(x) = \lambda\, p(x)$.

The idea of the estimator introduced by Baddeley et al. (2000) is to reconstruct the pair-correlation function of the initial homogeneous pattern φ_h, based on the information provided by the points of the observed heterogeneous pattern φ_o and an estimate of the intensity function. To do this, we first look at the probability that a point of φ_h is located within an infinitesimal disk of area dx centered on location x. This probability yields λdx. Consequently, the probability that a point of φ_o is located within an infinitesimal disk with area dx centered on location x yields $\lambda p(x)dx = \lambda(x)dx$. Thus, to recover the properties of φ_h, an estimator of the "inhomogeneous" pair-correlation function must weight a point at location x with the inverse of $p(x)$. The reason for this is that the location contained a point with a probability elevated, on average, by a factor of $1/p(x)$ before thinning was applied. Statisticians call this "re-weighting" (Baddeley et al. 2000). The corresponding estimator of the pair-correlation function therefore yields:

$$\hat{g}^{BMW}(r, \lambda(x)) = \frac{1}{2\pi r A} \sum_{i=1}^{n} \left[\sum_{j=1}^{n,\neq} k(\|x_i - x_j\| - r) \frac{w_{ij}(r)}{\lambda(x_i)\lambda(x_j)} \right] \tag{3.25}$$

The superscript BMW stands for the three authors Baddeley, Møller, and Waagepetersen of the seminal 2000 article describing this approach.

The inhomogeneous pair-correlation function reveals the pure second-order properties of the pattern and has properties analogous to a homogeneous pair-correlation function, but only if the underlying intensity function $\lambda(x)$ has been correctly identified and fulfills certain conditions (i.e., the observed pattern φ_o can indeed be obtained by thinning of a homogeneous pattern φ_h, and the thinning function $p(x)$ is a smooth function that does not contain very low values). Note that for each individual point pair i–j, a common weighting term $w_{ij}(r)/[\lambda(x_i)\lambda(x_j)]$ in the estimator is used for both edge correction and compensating for the heterogeneity. Therefore, it makes sense to apply this together with the Ripley and Stoyan edge correction methods, which weight point pairs individually (Equation 3.21).

Although it has some useful properties, the BMW estimator 3.25 has two problems. First, it cannot be applied for all intensity functions $\lambda(x)$ because $\lambda(x) > 0$ is required for all x in W. Second, the estimator is highly unstable if the value of $p(x_i)$ and/or $p(x_j)$ is small in some locations, because the weights can become very large (see Section 2.6.3.5). Because of these problems, estimator 3.25 cannot be applied for a wide range of situations that are however highly relevant in practical applications. We therefore present, in the following, an alternative estimator of the inhomogeneous pair-correlation function that is not subject to these restrictions. This estimator generalizes the WM and Ohser edge correction methods.

3.1.2.6 Alternative Estimators of Inhomogeneous Pair-Correlation Functions

A naive estimator of the mean number of points at distance r from the points of the pattern is given by

$$\hat{Z}_n^g(r) = \frac{1}{n} \sum_{i=1}^{n} \underbrace{\left(\sum_{j=1}^{n,\neq} k(\|x_i - x_j\| - r) \right) dr}_{\substack{n_i^R(r) = \text{\# points } j \text{ within a ring} \\ \text{around point } i}}$$

(3.26)

and a naive estimator of the expected number of points within a ring with radius r and width dr is given by

$$\hat{N}_n^g(r) = 2\pi r \, dr \, \lambda.$$

(3.27)

Thus, a naive estimator of the pair-correlation function yields $\hat{g}_n(r) = \hat{Z}_n^g(r)/\hat{N}_n^g(r)$. The bias due to edge effects can be corrected in the numerator or the denominator. The Ripley and Stoyan edge correction methods correct the numerator, leaving the denominator proportional to $2\pi r \lambda$.

Consequently, the corresponding estimator of the inhomogeneous pair-correlation function yields the BMW estimator (Equation 3.25), which is based on individual reweighting of point pairs.

Instead of altering the numerator (Equation 3.26), estimation of the inhomogeneous pair-correlation function using the Ohser and WM methods of edge correction modifies the denominator $\hat{N}_n^g(r)$, which is the expected number of points in rings with radius r and width dr. For homogeneous patterns, the expectation is given by the homogeneous Poisson process. However, for heterogeneous patterns, where φ_o results from independent thinning of a homogeneous pattern φ_h, the expectation is given by a heterogeneous Poisson process based on the intensity function $\lambda(x)$ used in thinning. In contrast to the BMW estimator, no further assumptions need to be made on the intensity function $\lambda(x)$.

Modification of the denominator can be accomplished in two ways, depending on whether or not we use an adapted estimator. The nonadapted estimator of the expected number of points based on the Ohser edge correction method yields:

$$\hat{N}_O^g(r) = \frac{1}{n} \int\limits_W \lambda(x) \underbrace{\left[\int\limits_W \lambda(a)k(\|x - a\| - r)da \right]}_{\text{Expected \# points in } W \text{ located in rings with radius } r \text{ around points } x} dx \tag{3.28}$$

To understand this, consider that under the heterogeneous Poisson process, the probability that a point is located within a small disk with area da centered at location a yields $\lambda(a)\,da$. Thus, to determine the number of points $w(x, r)$ expected to be located within W in a ring of radius r centered on x, we need to integrate $\lambda(a)\,da$ over all locations a in W that are located within the ring, that is, $w(x,r) = \int_W \lambda(a)k(\|x - a\| - r)da$. Note that the kernel function $k(\cdot)$ yields a value of 1 if a point is located within the ring and 0 otherwise. Finally, we need to average over all potential point locations x. This is done by multiplying $w(x, r)$ by $\lambda(x)\,dx$, the probability that there is a point within disk dx centered at x, and integrate over all locations in W to end up with Equation 3.28. The corresponding estimator of the pair-correlation function thus yields

$$\hat{g}_{\text{inhom}}^O(r, \lambda(x)) = \frac{(1/n)\sum\limits_{i=1}^{n}\left(\sum\limits_{j=1}^{n,\neq} k(\|x_i - x_j\| - r) \right)}{(1/n)\int\limits_W \lambda(x)\left[\int\limits_W \lambda(a)k(\|x - a\| - r)da \right]dx}$$

$$= \frac{1}{\hat{\lambda}_n^2} \frac{1}{2\pi r A} \sum\limits_{i=1}^{n}\left(\sum\limits_{j=1}^{n,\neq} k(\|x_i - x_j\| - r) \right) \frac{A}{\gamma_W(r, \lambda(x))} \tag{3.29}$$

where $\hat{\lambda}_n = n/A$ is the natural estimator of the intensity. The denominator contains a generalized, isotropized set covariance

$$\bar{\gamma}_W(r, \lambda(x)) = \frac{1}{2\pi r} \frac{1}{\hat{\lambda}_n^2} \int_W \lambda(x) \left[\int_W \lambda(a)k(\|x - a\| - r)da \right] dx \qquad (3.30)$$

which needs to be estimated numerically.

The analogous adapted estimator that modifies the estimator $\hat{N}_n^g(r)$ in Equation 3.27 does not average over all focal points x within W, but only over the points x_i of the pattern

$$\hat{g}_{\text{inhom}}^{WM}(r, \lambda(x))$$

$$= \frac{(1/n)\sum_{i=1}^{n}\left(\sum_{j=1}^{n,\neq}k(\|x_i - x_j\| - r)\right)}{(1/n)\sum_{i=1}^{n}\left[\int_W \lambda(a)k(\|x_i - a\| - r)da\right]} \qquad \text{contains intensity function } \lambda(x)$$

$$= \frac{m}{n-1} \frac{\sum_{i=1}^{n}\left(\sum_{j=1}^{n,\neq}k(\|x_i - x_j\| - r)\right)}{\sum_{i=1}^{n}\left(\sum_{j=1}^{n,\neq}k(\|x_i - a_j\| - r)\right)} \qquad \begin{array}{l}\text{approximates } \lambda(x) \text{ by points } a_i \\ \text{of an auxilary pattern}\end{array}$$

$$(3.31)$$

Note that this estimator does not need to weight the points x_i with the intensity function $\lambda(x_i)$ as required in the estimator using the generalized Ohser edge correction (i.e., 3.29 and 3.30). This is because the points x_i of the pattern are assumed to be based on the intensity function $\lambda(x)$. This becomes clear when comparing the denominators of Equations 3.29 and 3.31 where the integration "$(1/n)\int_W \lambda(x)$" over the entire observation window W is replaced in Equation 3.31 by the sum "$(1/n)\sum_{i=1}^{n}$" over the points x_i of the observed pattern in Equation 3.29.

In practice, the denominator of Equation 3.31 can be approximated numerically by generating m points a_j based on an auxiliary pattern that follows a heterogeneous Poisson process with intensity function $\lambda(x)$. If m is large (e.g., $m = 20,000$) it will approximate the continuous integral, but may somewhat slow down the computation. Because estimator 3.31 is composed of the ratio of two quantities that are estimated in exactly the same way around the points x_i of the pattern (i.e., it is an adapted estimator) we expect that effects of small irregularities in the pattern will cancel out.

Estimators 3.29 and 3.31 of the inhomogeneous pair-correlation function have three principal advantages over the BMW estimator 3.25. First, they allow estimation of the inhomogeneous pair-correlation function for all intensity functions $\lambda(x)$, including those with zero values at some locations x.

Second, because they do not weight each point pair *i*–*j* individually, they do not produce very large weights if the intensity function contains very low values. Thus, it can be expected that these estimators yield robust estimates of the pair-correlation function, even for rugged intensity functions that include zero values. These properties open up many possibilities for practical applications. Third, estimators 3.29 and 3.31 reconcile edge correction and heterogeneity and allow for a simple approach to edge correction for irregular observation windows (see Section 3.4.2.2). For example, the pair-correlation function of a homogeneous pattern within an irregular observation window can be estimated with Equations 3.29 and 3.31, if the intensity function is defined as being zero outside the observation window and λ inside. The generalized isotropized set covariance in Equation 3.30 will then collapse to the "standard" isotropized set covariance for the irregularly shaped observation window W. However, these estimators have the disadvantage that they must be approximated numerically.

A simple example illustrating how the inhomogeneous pair-correlation function works is given in Figure 3.12. The pattern arises from a random

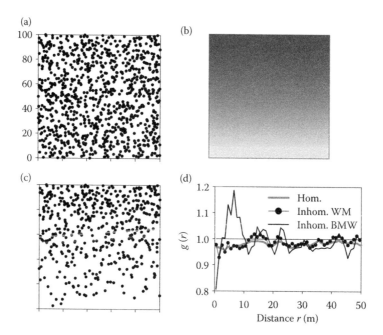

FIGURE 3.12

Application of inhomogeneous pair-correlation functions to a pattern with a gradient heterogeneity. (a) Original homogeneous random pattern φ_h, with 961 points within a 100×100 m² observation window. (b) Thinning function $p(x)$ used to generate the heterogeneous pattern φ_o shown in panel (c), darkness of shading is proportional to intensity. (c) Thinned pattern. (d) Analysis of the inhomogeneous pattern φ_o in panel (c). The expected homogeneous pair-correlation function for the original pattern in panel (a) is given as a gray line.

pattern φ_h (Figure 3.12a) that was independently thinned with a gradient intensity function $\lambda(x)$ (Figure 3.12b), yielding an inhomogeneous pattern φ_o (Figure 3.12c). Figure 3.12d shows the homogeneous pair-correlation function of the original homogeneous pattern φ_h (gray line), and the inhomogeneous pair-correlation function estimated with the WM method (Equation 3.31) from the inhomogeneous pattern φ_o and the intensity function $\lambda(x)$ (dots). The two estimates are in excellent agreement. Figure 3.12d also shows that the corresponding inhomogeneous pair-correlation function estimated with the BMW method yields essentially the same results. However, it also includes some erratic fluctuations, which are caused by the two points located in the bottom left at locations with low values of the intensity function (Figure 3.12c).

The inhomogeneous pair-correlation function also works very well for more complex patterns produced by complex intensity functions, as we show in Figure 3.13. This provides an example of a homogeneous pattern generated by a double-cluster Thomas process, where 468 small clusters with an approximate radius of 8.8 m and an average of 27 points were nested within 183 large clusters with approximate radii of 36 m (Figure 3.13a). This rather complex pattern was then independently thinned with the thinning function shown in Figure 3.13b. Most areas in the pattern have very low or zero values (light gray) for the intensity. However, there are also high values of the intensity function in a continuous band extending from the lower part of the plot toward the eastern border (dark gray). Independent thinning with this intensity function removed almost 90% of the points of the original pattern yielding the pattern shown in Figure 3.13c. As a result, most of the very dense small clusters were removed and only the pattern at the lower middle of the plot was retained. Thus, much of the structure of the original homogeneous pattern was destroyed and the question is whether or not the inhomogeneous pair-correlation function is able to recover the original second-order structure.

Indeed, Figure 3.13d demonstrates that the inhomogeneous pair-correlation function estimated with the WM method recovers the pair-correlation function of the original homogeneous pattern very well (cf. dark gray line with dotted line). Figure 3.13d shows that the corresponding inhomogeneous pair-correlation function estimated with the BMW method yields essentially the same results, but shows some erratic fluctuations, which are caused by points located at locations with low values of the intensity function. Thus, the two examples of Figures 3.12 and 3.13 show that the inhomogeneous pair-correlation function is able to recover the "pure" second-order structure of an inhomogeneous pattern that resulted from thinning of a homogeneous pattern, if the underlying intensity function is known. The examples also suggest that the BMW estimation method is, as expected, somewhat sensitive to intensity functions with near-zero values, whereas this problem does not arise with the WM estimation method.

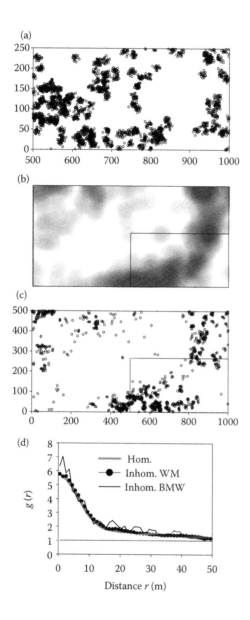

FIGURE 3.13

Application of inhomogeneous pair-correlation functions to a pattern with complex heterogeneity. (a) Original homogeneous pattern φ_h consisting of 11,519 points within a 1000×500 m² observation window. The pattern was generated by a double-cluster Thomas process containing 468 small clusters, with an approximate radius of 8.8 m (and an average of 27 points), nested within 183 large clusters with approximate radii of 36 m. (b) Thinning function $p(x)$ used to generate the pattern shown in panel (c). The gray scale is proportional to intensity, ranging from black, with $p(x) = 1$, to white, with $p(x) = 0$. (c) Pattern resulting from thinning. (d) Analysis of the pattern in panel (c). The expected homogeneous pair-correlation function for the original pattern in panel (a) is given as a gray line.

3.1.2.7 Ripley's K-Function

The estimators for the K-function are fully analogous to those of the pair-cor-relation function. As a consequence, the same principles for edge correction and generalization for inhomogeneous patterns can be applied. For homoge-neous patterns, the quantity $\lambda K(r)$ can be interpreted as the number of points lying within distance r of the typical point of the pattern. Thus, $\lambda K(r)$ counts the number of further points that are expected within a neighborhood r of a given point. Following Illian et al. (2008), the estimators of the K-function are based on the quantity $\kappa = \lambda^2 K(r)$ which is analogous to the product den-sity $\rho(r) = \lambda^2 g(r)$. We estimate the quantity κ, with an estimator analogous to Equation 3.19, based on pairs of points lying within distances d that are less than r:

$$
\hat{\kappa}(r) = \frac{1}{A} \sum_{i=1}^{n} \sum_{j=1}^{n,\neq} \underbrace{1(\|x_i - x_j, r\|)}_{\text{Count function}} \times \underbrace{w_{i,j}}_{\substack{\text{Edge} \\ \text{correction}}}
$$

$$
= \underbrace{\frac{n}{A}}_{\lambda_n} \left(\underbrace{\frac{1}{n} \sum_{i=1}^{n} \sum_{j=1}^{n,\neq} 1(\|x_i - x_j, r\|) \times w_{i,j}}_{\substack{\text{Mean number of points within distance } r \\ \text{of points } i}} \right) \tag{3.32}
$$

The *indicator function* $1(d, r)$ plays the role of the kernel function used in estimating the product density and has a value of 1 if point j is located within distance r or less of point i and zero otherwise. By slight rearrangement of the terms, we see that indeed $\kappa = \lambda_n[\lambda K(r)]$ (Equation 3.32).

The *edge correction terms* w_{ij} in Equation 3.32 are the same as in case of the product density (i.e., Equation 3.9, 3.21), although the Ohser edge correction term $w_{ij}^{O} = A/\bar{\gamma}_W(r)$ now depends on the distance d between the points i and j, because the isotropisized set covariance depends on distance r. As a con-sequence, it cannot be factored out in the corresponding estimator. To yield an edge correction weight w_{ij} that is independent of the locations of the two points i and j, and only depends on their distance $d = \|x_i - x_j\|$, Wiegand and Moloney (2004) used the edge correction weight

$$
w_{ij}^{WM}\left(\|x_i - x_j\|, r\right) = \frac{\pi r^2}{(1/A)2\pi \int_0^r t\bar{\gamma}_W(t)dt} \quad \text{if } \|x_i - x_j\| \le r. \tag{3.33}
$$

This weighting factor is composed of the area of the full circle (numerator) divided by the average area (within W) of disks with radius r located in W

(denominator). The WM weight should provide a good approximation of the Ohser weights for shorter distances of r. Ward and Ferrandino (1999) derived basically the same edge correction term as Equation 3.33. When inserting the formula for $\bar{\gamma}_W(t)$ given in Equation 3.7 into Equation 3.33 we can derive an expression analogous to Equation 3.7 by integration of

$$\int_0^r \bar{\gamma}_W(t)2\pi t\,dt = \left(A - \frac{4r}{3\pi}(a_x + b_y) + \frac{r^2}{2\pi} \right)(\pi r^2).$$

Given the preceding considerations, the four estimators of the univariate quantity $\hat{\kappa}(r)$ yield

$$\hat{\kappa}^R(r) = \frac{1}{A}\sum_{i=1}^{n}\sum_{j=1}^{n,\neq} 1(\|x_i - x_j\|, r)\frac{2\pi}{\varphi_{ij}} \qquad \text{Ripley}$$

$$\hat{\kappa}^S(r) = \frac{1}{A}\sum_{i=1}^{n}\sum_{j=1}^{n,\neq} 1(\|x_i - x_j\|, r)\frac{A}{\gamma_W(x_i - x_j)} \qquad \text{Stoyan}$$

$$\hat{\kappa}^O(r) = \frac{1}{A}\sum_{i=1}^{n}\sum_{j=1}^{n,\neq} 1(\|x_i - x_j\|, r)\frac{A}{\bar{\gamma}_W(\|x_i - x_j\|)} \qquad \text{Ohser}$$

$$\hat{\kappa}^{WM}(r) = \frac{1}{A}\sum_{i=1}^{n}\sum_{j=1}^{n,\neq} 1(\|x_i - x_j\|, r)\frac{A\pi r^2}{2\pi\int_0^r t\bar{\gamma}_W(t)dt} \qquad \text{WM}$$

(3.34)

Because the expectation of $\hat{\kappa}(r)$ increases at a rate of πr^2, we compare the different estimators based on the quantity $\hat{\kappa}(r)/\pi r^2$. Figure 3.14a shows the values of the four estimators of $\hat{\kappa}(r)/\pi r^2$ for a random pattern with 1000 points within a 100×100 m^2 observation area (Figure 3.14a). In contrast, Figure 3.14b shows the estimators for a clustered pattern shown in Figure 3.10b. The WM, Ohser, and Stoyan estimators yield virtually identical results, whereas the Ripley estimator yields slightly lower values for the random pattern and somewhat larger values for the clustered pattern (Figure 3.14a,b). This result parallels the results for the product density function (e.g., Figure 3.11b).

Analogous to the estimation of the pair-correlation function, where the adapted intensity estimator $\hat{\lambda}_S(r)$ was used (Equations 3.22 and 3.24), one can also define $\hat{\lambda}_V(r)$, an estimator of the intensity that is adapted to the K-function (Illian et al. 2008, p. 194). While the adapted intensity estimator $\hat{\lambda}_S(r)$ used for the pair-correlation function considered the length of the

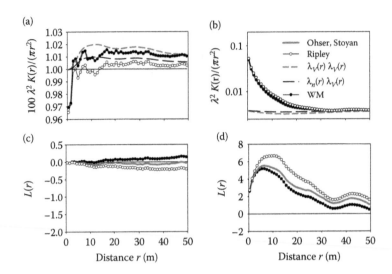

FIGURE 3.14
Estimators of the K-function. (a) Estimators of $\lambda^2 K(r)$ for a random pattern of 1000 points within a 100×100 m² area. (b) Estimators of $\lambda^2 K(r)$ for the clustered pattern in Figure 3.10b, consisting of 626 points within a 100×100 m² area. (c) Estimates of the L-function for a random pattern as in panel (a). (d) Estimates for the L-function of the clustered pattern in Figure 3.10b. Panels (a) and (b) show values for $\lambda_V \lambda_n$ and $\lambda_V \lambda_V$ in addition to K-function estimates.

circumference of the circles of radius r around the points i of the pattern that lay inside the observation window W, the intensity estimator adapted for the K-function considers the areas of disks with radius r around the points i of the pattern that lie inside W. As a consequence, there are two small modifications required in the intensity estimator. First, the length $v_{d-1}(W \cap \partial b(x_i, r))$ of the circumference of a circle of radius r around point i in W has to be replaced by the corresponding area $v(W \cap b(x_i, r))$ of a circle of radius r around point i which overlaps W. Second, the term $r \, \bar{\gamma}_W(r)$ has to be replaced by the integral $\int_0^r t\bar{\gamma}_W(t)dt$ to yield the expected area of a disk with radius r within W:

$$\hat{\lambda}_S(r) = \frac{\sum_i^n v_{d-1}(W \cap \partial b(x_i, r))}{2\pi r \bar{\gamma}_W(r)}$$

$$\hat{\lambda}_V(r) = \frac{\sum_i^n v(W \cap b(x_i, r))}{2\pi \int_0^r t\bar{\gamma}_W(t)dt} = \hat{\lambda}_n \left[\frac{(1/n)\sum_i^n v(W \cap b(x_i, r))}{2\pi \int_0^r t\bar{\gamma}_W(t)dt/A} \right] \qquad (3.35)$$

When factoring out $\hat{\lambda}_n = n/A$ in Equation 3.35, we find that $\hat{\lambda}_V(r)$, the estimator of lambda, is basically given by the naive estimator of the intensity multiplied by a correction factor (i.e., the expression within parentheses in the numerator of Equation 3.35). The correction term adjusting the value of the naive estimator is composed of a numerator, which is the average area of disks $b(x_i, r)$ overlapping W calculated from the points of the pattern, and denominator, which is the expected area overlapping W of disks centered in W. This is analogous to the case of the intensity estimator adapted to the pair-correlation function (Equation 3.23).

Continuing the example shown in Figure 3.14, we calculated the squared adapted intensity estimate $\hat{\lambda}_V^2(r)$, which is shown as dashed dark gray lines in Figure 3.14a and b. These figures confirm that the correction factor indeed captures some of the small bias in $\hat{\kappa}(r)/\pi r^2$ as also shown for the product density (Figure 3.10). The $\hat{\lambda}_V(r)$ will be smaller/larger than $\hat{\lambda}_n = n/A$ if many/few points are accidently located close to the border.

Wiegand and Moloney (2004) proposed an estimator of $\lambda K(r)$ based on a grid approximation to facilitate calculation of the K-function for irregularly shaped study areas. The quantity $\lambda K(r)$ yields the expected number of points of the pattern within distance r of the typical point of the pattern. The quantity $\lambda K(r)$ is thus a conditional summary statistic, similar to $O(r) = \lambda g(r)$; that is, it yields the expected number of points within a disk, given that there is a point at the center of the disk. With this definition, the WM estimator first counts, for each point i of the pattern, the number of points j located within distance r of i that lie inside W (i.e., $\sum_{j=1}^{n,\neq} \mathbf{1}(\|x_i - x_j\|, r)$) and then averages the number of points meeting this criterion for all of the points i of the pattern: $(1/n)\sum_{i=1}^{n}\left(\sum_{j=1}^{n,\neq} \mathbf{1}(\|x_i - x_j\|, r)\right)$. The estimate of $\lambda K(r)$ is then obtained by dividing the average number of points in disks with radius r by the average area of the disks that lie inside the observation window W (i.e., $(1/n)\sum_{i=1}^{n} v(W \cap b(x_i, r))$):

$$\widehat{\lambda K^{WM}}(r) = \pi r^2 \frac{(1/n)\sum_{i=1}^{n}\left(\sum_{j=1}^{n,\neq} \mathbf{1}(\|x_i - x_j\|, r)\right)}{(1/n)\sum_{i=1}^{n} v(W \cap b(x_i, r))} = \frac{\hat{\kappa}^{WM}(r)}{\hat{\lambda}_V(r)} \qquad (3.36)$$

It can be easily verified that dividing $\hat{\kappa}^{WM}(r)$ of Equation 3.34 by the adapted intensity estimator $\hat{\lambda}_V(r)$ of Equation 3.35 yields the estimator of $\lambda K(r)$ given in Equation 3.36.

Following the suggestion by Stoyan (2006a) and Illian et al. (2008), we obtain three alternative estimators for the K-function, based on the different edge correction methods (i.e., the Ripley, Stoyan, and Ohser weights shown in Equation 3.34). These divide $\hat{\kappa}(r)$ by the square of the adapted intensity estimator (Equation 3.35):

$$\hat{K}^{WM}(r) = \frac{\hat{\kappa}^{WM}(r)}{\hat{\lambda}_V(r)\hat{\lambda}_n} \qquad \text{WM}$$

$$\hat{K}^O(r) = \frac{\hat{\kappa}^O(r)}{\hat{\lambda}_V(r)^2} \approx \frac{\hat{\kappa}^{WM}(r)}{\hat{\lambda}_V(r)^2} \qquad \text{Ohser}$$

$$\hat{K}^S(r) = \frac{\hat{\kappa}^S(r)}{\hat{\lambda}_V(r)^2} \qquad \text{Stoyan} \tag{3.37}$$

$$\hat{K}^R(r) = \frac{\hat{\kappa}^R(r)}{\hat{\lambda}_V(r)^2} \qquad \text{Ripley}$$

For completeness, we also show the estimator $\hat{K}^{WM}(r)$ presented in Wiegand and Moloney (2004), which divides the estimator of $\lambda K(r)$ by the natural estimator of the intensity (i.e., n/A).

Continuing our example shown in Figure 3.14, we calculated the four K-functions for the two patterns and show the corresponding L-functions, where $L_1(r) = \sqrt{\pi^{-1}K(r)} - r$. The resulting differences among the estimators for the L-function are relatively small. In general, the L-function estimated with the Ohser and the Stoyan estimators (which are almost identical) falls somewhere in-between the estimates obtained with the Ripley and the Wiegand–Moloney methods (Figure 3.14c,d). For clustered patterns, with a large cluster in one corner of the observation window (Figure 3.10b), the Ripley method overestimates the clustering (Figure 3.14d) and the Wiegand–Moloney method tends to indicate somewhat reduced clustering.

The approach to developing estimators for the inhomogeneous K-function is fully analogous to the approach used for the pair-correlation function. The corresponding BMW estimator of the inhomogeneous pair-correlation function corresponding to Equation 3.25, therefore, yields

$$\hat{K}^{BMW}(r, \lambda(x)) = \frac{1}{A} \sum_{i=1}^{n} \sum_{j=1}^{n,\neq} \mathbf{1}(\|x_i - x_j\|, r) \frac{w_{ij}(r)}{\lambda(x_i)\lambda(x_j)}. \tag{3.38}$$

For each individual point pair i–j lying closer than distance r apart, the estimator applies the common weighting term $w_{ij}(r)/[\lambda(x_i)\lambda(x_j)]$ for edge correction and adjustment for the heterogeneity. The Ripley and Stoyan edge corrections methods are therefore applied, since they weight point pairs individually (Equation 3.9).

The alternative WM method for estimation of the inhomogeneous K-function is based on the ratio of the mean number of points $\hat{Z}^K_{WM}(r)$ within distance r of the points of the pattern, relative to the corresponding expected number of points $\hat{N}^K_{WM}(r)$:

$$\hat{K}^{WM}(r) = \pi r^2 \frac{\hat{Z}^K_{WM}(r)}{\hat{N}^K_{WM}(r)} \tag{3.39}$$

The mean number of points within distance r of the points of the pattern yields, as before, $\hat{Z}^K_{WM}(r) = (1/n)(\sum_{i=1}^{n} \sum_{j=1}^{n,\neq} \mathbf{1}(\|x_i - x_j\|, r))$, cf. Equation 3.36. It then follows that the expectation under the heterogeneous Poisson process yields an equation analogous to Equation 3.28 for the nonadapted case:

$$\hat{N}^K_O = \frac{1}{n} \int_W \lambda(x) \underbrace{\left[\int_W \lambda(a)\mathbf{1}(\|x - a\|, r)da \right]}_{\substack{\text{Expected \# points in } W \text{ located in} \\ \text{disk with radius } r \text{ around point } x}} dx \tag{3.40}$$

The corresponding estimator of the K-function becomes

$$K^O_{inhom}(r, \lambda(x)) = \pi r^2 \frac{(1/n)\sum_{i=1}^{n} \left(\sum_{j=1}^{n,\neq} \mathbf{1}(\|x_i - x_j\|, r) \right)}{(1/n)\int_W \lambda(x) \left[\int_W \lambda(a)\mathbf{1}(\|x - a\|, r)da \right] dx} \tag{3.41}$$

The analogous adapted estimator that modifies the estimator $\hat{N}_n(r)$ does not average over all focal points x within W, but only over the points x_i of the pattern:

$$\hat{K}^{WM}_{inhom}(r, \lambda(x))$$

$$= \pi r^2 \frac{\sum_{i=1}^{n} \left(\sum_{j=1}^{n,\neq} \mathbf{1}(\|x_i - x_j\|, r) \right)}{\sum_{i=1}^{n} \left[\int_W \lambda(a)\mathbf{1}(\|x_i - a\|, r)da \right]} \qquad \text{contains intensity function } \lambda(x)$$

$$= \pi r^2 \frac{m}{n-1} \frac{\sum_{i=1}^{n} \left(\sum_{j=1}^{n,\neq} \mathbf{1}(\|x_i - x_j\|, r) \right)}{\sum_{i=1}^{n} \left(\sum_{j=1}^{n} \mathbf{1}(\|x_j - a_j\|, r) \right)} \qquad \text{approximates } \lambda(x) \text{ by points } a_i \text{ of an auxiliary pattern}$$

$$\tag{3.42}$$

3.1.3 Nearest-Neighbor Summary Statistics for Univariate Patterns (Data Type 1)

3.1.3.1 Distribution Functions for Nearest-Neighbor Distances

Summary statistics based on distances to the kth neighbor provide an alternative way of characterizing the small-scale correlation structure of point patterns and complement the use of the second-order summary statistics. For homogeneous patterns, the function $D^k(r)$ represents the (cumulative) distribution of the distances r to the kth nearest neighbor as measured from the typical point of the pattern. $D(r)$ is used to refer to the distribution function of distances to the nearest neighbor [i.e., $k = 1$, $D(r) = D^1(r)$]. Nearest-neighbor statistics are "short-sighted" and sense only the immediate neighborhood of the typical point. They are, therefore, particularly sensitive to local cluster structures and can differentiate among dissimilar cluster processes that have identical second-order summary statistics (e.g., Wiegand et al. 2007c, 2009).

In estimating $D^k(r)$, we first need to determine the distance d_{ik} to the kth neighbor for each point i in W. The *naive estimator of $D^k(r)$* without edge correction is given by

$$\hat{D}^k(r) = \frac{1}{n} \sum_{i=1}^{n} 1(0 < d_{ik} \leq r) \tag{3.43}$$

where $1(0 < d_{ik} \leq r)$ is an indicator function, which yields a value of 1, if the kth neighbor is within distance r (i.e., $0 < d_{ik} < r$), and zero otherwise. However, the "real" d_{ik} for some of the points may be less than the estimated d_{ik}, since only information on points within W is available. This error occurs when an unobserved point lying outside W is the true kth neighbor for a focal point, that is, d_{ik} is greater than the distance β_i from point i to the nearest border of W (Figure 3.15a). This edge effect produces a small bias in the estimation of $D^k(r)$.

The *buffer method*, which was introduced by Ripley (1977), is able to avoid this bias. It uses only focal points i located within an "inner window" surrounded by a buffer of width b in calculating $D^k(r)$ (Figure 3.15b). If we restrict the calculation of $D^k(r)$ to distances $r \leq b$, $D^k(r)$ can be correctly estimated based on the points of the pattern within W. The estimator that uses the buffer method is similar to the estimator without edge correction (Equation 3.43), but only includes focal points i that are located within the inner window:

$$\hat{D}^{k,b}(r) = \frac{\sum_{i=1}^{n} 1(0 < d_{ik} \leq r) \times 1(b < \beta_i)}{\sum_{i=1}^{n} 1(b < \beta_i)}, \quad r \leq b \tag{3.44}$$

The indicator function $1(b < \beta_i)$ yields a value of 1 if $b < \beta_i$ and zero otherwise. Although Equation 3.44 removes the bias in the estimate of $D^k(r)$, it has the disadvantage of only producing reliable results for a small range of distances

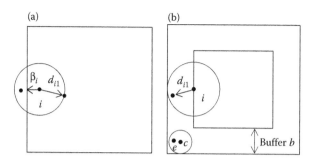

FIGURE 3.15

Edge correction for the estimation of the distribution function of the distances to the nearest neighbor. (a) A focal point i located distance β_i from nearest border, d_{ik} from the observed nearest neighbor, but whose "real" nearest neighbor lies outside the observation window. (b) Buffer edge correction method. Only focal points i, located further than b from the nearest border of the observation window are used as focal points. For these points, the nearest neighbor can be correctly determined. The buffer method disregards information for points in the buffer area for which the nearest neighbor can be correctly estimated (e.g., point c with nearest neighbor point e).

$r \leq b$. For larger values of the buffer b, the inner window becomes small compared to W and almost no focal points remain to be analyzed. An additional disadvantage of this estimator is that it disregards valuable information for points in the buffer for which the distance to the kth neighbor can be correctly evaluated. This can be seen in Figure 3.15b where the nearest neighbor for point c lies within W, that is, point e is closer to c than c is to the border.

Some of the issues related to the use of the buffer method can be illustrated by exploring some examples. Figure 3.16a,b show two patterns together with a buffer area of 100 m. The points lying outside the inner rectangle are not used as foci for estimation of $D^k(r)$, if we apply the buffer method. As a consequence, the buffer method excludes the information for many points that are located within the buffer area, but for which the kth nearest neighbor can be, in fact, correctly estimated. For example, the open circles are those points for which the border is closer than the first nearest neighbor, which means that the nearest neighbor (i.e., $k = 1$) cannot be accurately determined. It can be seen that there are only a few points for which this is the case, so most points can actually be included in the analysis without producing a bias. However, as the rank k of the neighbor increases, fewer focal points can be accurately assessed. Even so, the buffer method excludes a number of points that can be included in the analysis, since the kth nearest neighbor in W lies closer to the focal point than the distance between the focal point and the border. For example, the closed black disks in Figure 3.16a,b represent points that can be included in an analysis for $k = 25$. In essence, we may lose a substantial amount of information if we do not consider the points within the buffer area for which the distance to the kth neighbor can be correctly determined.

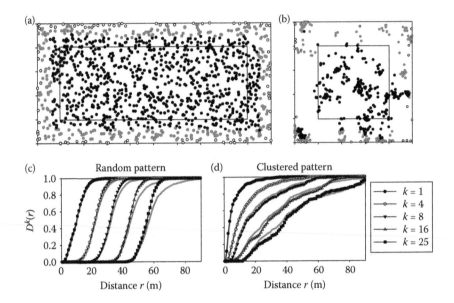

FIGURE 3.16

Comparison of estimators for the nearest-neighbor distribution function $D^k(r)$. (a) Random pattern with 1296 points within a 1000×500 m² observation window. (b) Clustered pattern with 626 points within a 500×500 m² observation window. (c) Nearest-neighbor distribution functions $D^k(r)$ for the pattern in panel (a). (d) Nearest-neighbor distribution functions $D^k(r)$ for the pattern in panel (b). The rectangles in panels (a) and (b) enclose the area of focal points used for the buffer method; open points are those excluded by the Hanisch estimator for $k = 1$, and gray points were those additionally excluded for $k = 25$. In panels (c) and (d), Hanisch estimators are given (symbols), and estimators without edge correction are shown as solid gray lines which often overlaps with the estimates of the Hanisch estimator.

Hanisch (1984) proposed an estimator of $D^k(r)$ that uses more information than allowed by Equation 3.44 and overcomes the disadvantages of the buffer estimator. It is similar to the buffer estimator, but instead of the buffer being fixed for all points, it is individually adjusted for each focal point i. In this way, the *Hanisch estimator* applies edge correction by using only focal points i for which the kth neighbor is definitely located within W. The kth neighbor of point i can be assessed if the distance d_{ik} to its kth neighbor is shorter than the distance β_i to the nearest border of W. For points not meeting this condition, the kth neighbor may potentially be located outside W and the point is therefore not considered in the analysis. Given these considerations, the Hanisch estimator can be defined as follows:

$$\hat{D}^{k,H}(r) = \frac{\sum_{i=1}^{n} \mathbf{1}(0 < d_{ik} \leq r) \times \mathbf{1}(d_{ik} < \beta_i) \times w(d_{ik})}{\sum_{i=1}^{n} \mathbf{1}(d_{ik} < \beta_i) \times w(d_{ik})} \tag{3.45}$$

The indicator function $\mathbf{1}(d_{ik} < \beta_i)$ tests for the condition that the distance d_{ik} to the kth nearest neighbor is shorter than the distance β_i to the nearest border and yields a value of 1 if $d_{ik} < \beta_a$ and zero otherwise. A weighting factor $w(d_{ik})$ is included in Equation 3.45 to compensate for excluded points. This can be defined by noting that the proportion of points in W located further from the border than distance d_{ik} is given for homogeneous patterns by $(l_x - d_{ik})(l_y - d_{ik})/l_x l_y$ for W defined by a rectangle with sides of length l_x and l_y. Thus, for $r < \min(l_x/2, l_y/2)$, an appropriate weighting factor $w(d_{ik})$ that compensates for the points for which $d_{ik} > \beta_i$ is given by

$$w(d_{ik}) = \frac{1}{(l_x - d_{ik})(l_y - d_{ik})} \tag{3.46}$$

However, for large values of r the Hanisch estimator becomes unreliable because few points will exist that are farther than distance r from the nearest border. As an upper limit, r is certainly constrained to be less than the maximal distance inside W from the nearest border. For a rectangle, this is half of the length of its smaller side.

Equation 3.47 outlines the similarity between the buffer method and the Hanisch method by presenting a slightly rewritten version of the buffer method (Equation 3.44):

$$\hat{D}^{k,H}(r) = \frac{\sum_{i=1}^{n} \mathbf{1}(0 < d_{ik} \leq r) \times \mathbf{1}(d_{ik} < \beta_i) \times w(d_{ik})}{\sum_{i=1}^{n} \mathbf{1}(d_{ik} < \beta_i) \times w(d_{ik})}$$

$$\hat{D}^{k,b}(r) = \frac{\sum_{i=1}^{n} \mathbf{1}(0 < d_{ik} \leq r) \times \mathbf{1}(b < \beta_i) \times w(b)}{\sum_{i=1}^{n} \mathbf{1}(b < \beta_i) \times w(b)}, \tag{3.47}$$

Basically, the distance d_{ik} to the kth neighbor has to be replaced by the constant buffer width b. Because the weight $w(b)$ of the buffer method is the same for all focal points i it cancels out in Equation 3.47, thus yielding Equation 3.44. In contrast, the Hanisch method determines for each individual point i whether or not it is included in the calculations. Note that the estimators given in Equation 3.47 are adapted estimators that use a rationale similar to that used for the adapted estimators of the second-order statistics presented in the previous sections. The numerator is an estimator of $\lambda D^k(r)$ and the denominator is an adapted estimator of the intensity λ (Stoyan 2006b).

To obtain an intuitive appreciation of the Hanisch method of edge correction, we show the points of the pattern used by the Hanisch method as focal points in estimating $D^k(r)$'s for a few chosen values of k and r (Figure 3.16a,b). For the nearest neighbor (i.e., $k = 1$), only a few points very close to the border are excluded from the calculation (open points in Figure 3.16a,b). In this case, the estimates using the Hanisch method versus not applying edge correction should be very close. However, for higher neighbor ranks ($k = 25$; Figure 3.16a,b) substantially more points are excluded as focal points by the Hanisch method in estimating $D^k(r)$ (gray points). All of the excluded points are suspected to have their 25th nearest neighbor lying outside of W. It is also evident from Figure 3.16 that the buffer method excludes substantially more points as compared to the Hanisch method.

Figure 3.16 also provides a comparison of the estimator with no edge correction and that of the Hanisch method. The estimates of the buffer and the Hanisch method coincide for the random pattern (not shown). For small neighborhood ranks of say $k \leq 8$, the estimates without edge correction are also very close to those with edge correction (Figure 3.16c). However, for larger neighborhood ranks we find differences between the method without edge correction (gray lines) and the two methods that use edge correction (Figure 3.16c). While the estimates for $D^k(r)$ still agree for smaller values of $D^k(r)$, the farthest distances to the kth neighbor [i.e., the approach of $D^k(r)$ to a value of 1] are overestimated when no edge correction is applied. This means that the proportion of points having a nearby kth neighbor (i.e., clustered points) is usually well estimated by the estimator that does not use edge correction, but the proportion of points having their kth neighbor farther away (i.e., more isolated points) shows a bias. This is expected because the estimator without edge correction may in many cases overestimate the distances to the kth neighbor for points close to the border. This effect will be more pronounced for more isolated points (i.e., points with larger distances to the kth neighbor) and for larger neighborhood ranks k. Keeping these results in mind is important for correctly interpreting results of $D^k(r)$ when edge correction is not applied.

Figure 3.16d shows the results of the two methods of edge correction applied to nearest-neighbor analysis for a clustered pattern (Figure 3.16b). Interestingly, differences among the three estimators are smaller than for the random pattern. For all distances r, the estimates obtained agree in close approximation. Because the pattern is highly clustered, the distribution functions to even the 25th nearest neighbor agree in all three estimators. In this case, the summary statistics $D^k(r)$ mostly measure the properties of the cluster of points surrounding the focal points or the points of the nearest cluster and problems with isolated points observed for the random pattern do not arise. Summarizing the two examples, a good approximation of $D^k(r)$ will arise even if edge correction in the estimation of the distribution function to the kth neighbor is not employed, unless the pattern comprises a substantial proportion of isolated points. In the latter case, bias is introduced if the neighborhood rank k is high, but only as $D^k(r)$ approaches a value of 1. However, it is also clear that we may expect a bias, especially for patterns with few points. More technical details and comparisons among estimators of the nearest neighbor distance distribution function can be found in Stoyan (2006b).

3.1.3.2 Mean Distance to kth Neighbor

The mean distance to the nearest-neighbor m_D is a measure that has been frequently used to characterize univariate point patterns, particularly in earlier studies. It is closely related to the classical Clark–Evans index CE (Clark and Evans 1954) that divides m_D by the corresponding expectation of m_D under CSR and has been used to determine whether a pattern is hyperdispersed, random, or clustered. The m_D metric can be generalized to characterize a pattern by means of the mean distances $nn(k)$ to the kth neighbor. This summary statistic has the convenient property of approximating a power law (Hubbell et al. 2008) and thus may have the potential to reveal scale independent features of a pattern.

For a random pattern, $nn(k)$ can be estimated analytically and approximated by a power law (Table 2.2):

$$nn(k) = \frac{1}{\sqrt{\lambda}} \frac{k(2k)!}{(2^k k!)^2} \approx \frac{1}{\sqrt{\pi\lambda}} k^{1/2} \tag{3.48}$$

Figure 3.17 shows the estimates for $nn(k)$ for three different patterns, employing the Hanisch edge correction method (disks) and estimates without edge correction (lines in Figure 3.17d). The estimates for the random pattern agree almost perfectly with those of the power law (Figure 3.17d). The slope fit to the observed values for the Hanisch estimator yield a value of 0.507 instead of the expected value of 0.5 and the observed and expected mean neighborhood distances are 8.5 and 8.9, respectively, which is in good agreement. Figure 3.17e shows that $nn(k)$ for clustered patterns also follows the power law very well, but the slope differs from 0.5. Clearly, the value at $k = 1$, the mean nearest-neighbor distance, depend on the intensity of the process and the local cluster structure. If the intensity and/or the clustering are greater, the mean nearest-neighbor distance is shorter and $nn(k = 1)$ is smaller. The slope depends on the clustering of the pattern; the more clustered in the pattern, the larger the distance to the kth neighbor becomes for large neighborhood ranks k. Additionally, the fewer the number of individuals there are, the greater the slope (Hubbell et al. 2008). Figure 3.17d shows the estimates of $nn(k)$ based on the Hanisch edge correction method (disks) and the estimates without edge correction as lines. For the first 64 neighborhood ranks the mean distance to the kth neighbor is approximated well, but at ranks greater than 64 the difference between the methods becomes visible.

3.1.3.3 Spherical Contact Distribution

Second-order and nearest-neighbor statistics describe the properties of a pattern from the viewpoint of the points of the pattern and, therefore, mostly characterize neighborhood properties of the typical point. However, they

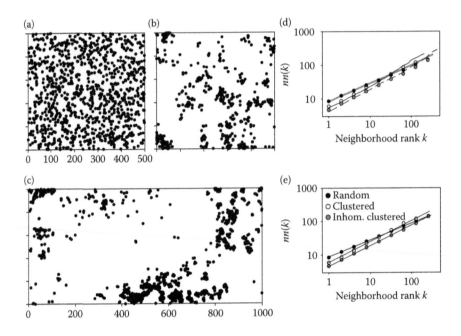

FIGURE 3.17
Summary statistic for the mean distance $nn(k)$ to the kth neighbor for three patterns (a) Random pattern with 1000 points in a 500×500 m^2 study area. (b) Clustered pattern with 626 points in a 500×500 m^2 study area. (c) Inhomogeneous pattern with 1296 points in a 1000×500 m^2 study area. (d) Summary statistic $nn(k)$ estimated for the three patterns (see legend in panel (e)) with continuous lines representing estimates without edge correction and dots estimates using the Hanisch method. (e) Same as panel (d), but lines show a fit with a power law to the estimates of mean nearest-neighbor distances using the Hanisch method.

are not well suited to provide a statistical description of the properties of the "holes" of the pattern. This gap is filled with the spherical contact distribution $H_s(r)$ that characterizes the pattern from the perspective of "test points" that are randomly or regularly distributed in the observation window. Because of this property, the $H_s(r)$ provides additional information for characterizing patterns that can be used, for example, when testing the fit of parametric point-process models.

The spherical contact distribution can be viewed as a bivariate nearest-neighbor distribution function in which the focal pattern is a "test pattern" and the second pattern is the observed pattern. Therefore, the spherical contact distribution $H_s(r)$ is the distribution of the distances r from the typical test location to the nearest point of the pattern. A random pattern can be used to generate the test locations or they can be based on the nodes of a grid. In general, the test locations should "sufficiently" cover the observation window to produce good resolution in the estimate of $H_s(r)$.

The spherical contact distribution can be estimated as a bivariate distribution function to the nearest neighbor using Hanisch edge correction:

$$\hat{H}_s(r) = \frac{\sum_{t=1}^{n} \mathbf{1}(0 < d_t \le r) \times \mathbf{1}(d_t < \beta_t) \times w(d_t)}{\sum_{t=1}^{n} \mathbf{1}(d_t < \beta_t) \times w(d_t)} \tag{3.49}$$

where d_t is the distance from test location t to the nearest point of the pattern and β_t is the distance from the test location to the nearest border of W. The weighting factor $w(d_t)$ is the same as in Equation 3.46.

An alternative way of estimating the spherical contact distribution is to take advantage of its interpretation as a morphological function (Mecke and Stoyan 2005). In this case, each point of the pattern is enlarged to be a disk of radius r. As a result, a pattern of potentially overlapping disks is obtained. The topology or morphology of this pattern is then analyzed for different values of r. On first view, the spherical contact distribution is, somewhat surprisingly, the fraction of the observation window covered by the union of all disks of radius r centered on the points of the pattern (Illian et al. 2008, p. 200). This quantity, referred to as $A_A(r)$, yields the probability that the first neighbor to an arbitrary "test point" in W is within distance r (Figure 2.14). Illian et al. (2008, p. 204) recommend the use of minus sampling edge correction with the estimator

$$\hat{H}_s(r) = \frac{v(X_r \cap W_{\ominus r})}{v(W_{\ominus r})} \tag{3.50}$$

where X_r is the union of all (possibly overlapping) disks of radius r centered on the points of a pattern (the gray shaded area in Figure 2.14), the symbol $v(X)$ denotes the area of X, and $W_{\ominus r}$ is the inner rectangle with area $v(W_{\ominus r}) = (a - r)$ $(b - r)$ of the rectangle W with sides of length a and b. The numerator estimates the area of X_r within the reduced window $W_{\ominus r}$. It is often estimated based on an underlying grid. Brodatzki and Mecke (2002) provide an algorithm that is able to estimate (3.50) without approximation.

3.1.4 Summary Statistics for Bivariate Patterns (Data Type 2)

3.1.4.1 Partial Product Densities

The estimation of the second-order statistics for bivariate patterns is completely analogous to the approach used for univariate patterns. We, therefore, do not present the estimators for bivariate patterns in much detail. The general form of the estimator of the product density given in Equation 3.19

for univariate patterns is maintained. However, a new indicator function $C_{lm}(x_i, x_j)$ is required that returns a value of 1 for point pairs when the first point x_i is of type l and the second point x_j is of type m, and a value of zero otherwise. The general form of the estimator of the "partial" product densities for a pattern of n points, which may be of different types (e.g., species), therefore yields

$$\hat{\rho}(r) = \frac{1}{2\pi r}\frac{1}{A}\sum_{i=1}^{n}\sum_{j=1}^{n,\neq} k(\|x_i - x_j\| - r)w_{i,j} \quad \text{univariate}$$

$$\hat{\rho}_{lm}(r) = \frac{1}{2\pi r}\frac{1}{A}\sum_{i=1}^{n}\sum_{j=1}^{n,\neq} \underbrace{C_{lm}(x_i, x_j)}_{\substack{\text{First point type } l, \\ \text{second point type } m}} \underbrace{k(\|x_i - x_j\| - r)}_{\text{Kernel function}} \underbrace{w_{i,j}}_{\substack{\text{Edge} \\ \text{correction}}} \quad \text{bivariate} \quad (3.51)$$

where $k()$ is the kernel function that selects point pairs that are located approximately at distance r apart, $C_{lm}(x_i, x_j)$ selects point pairs where the first point is of type l and the second of type m, and w_{ij} is the edge correction weight for the point pair i–j. All four edge correction weights (summarized in Equation 3.21) apply without change. Note that the estimator of the univariate "partial" pair-correlation function for type m points yields

$$\hat{\rho}_{mm}(r) = \frac{1}{2\pi r}\frac{1}{A}\sum_{i=1}^{n}\sum_{j=1}^{n,\neq} C_{mm}(x_i, x_j)k\left(\|x_i - x_j\| - r\right)w_{i,j}$$

$$= \frac{1}{2\pi r}\frac{1}{A}\sum_{i=1}^{n_m}\sum_{j=1}^{n_m,\neq} k\left(\|x_i^m - x_j^m\| - r\right)w_{i,j} \quad (3.52)$$

where n_m is the number of type m points, and the superscript m indicates that the points x_i^m and x_j^m of the partial pattern consist of type m points.

3.1.4.2 Partial Pair-Correlation Functions

For the bivariate case, the (partial) pair-correlation functions based on the product densities given in Equation 3.51 require two adapted estimators of the intensity functions, one for the partial pattern of type l points, and one for the partial pattern of type m points. For points of type m

$$\hat{\lambda}_S^m(r) = \frac{\sum_i^n C_m(x_i) \times v_{d-1}(W \cap \partial b(x_i, r))}{2\pi r \bar{\gamma}_W(r)} \quad (3.53)$$

Equation 3.53 includes the indicator function $C_m(x_i)$ that selects only points of type m from the pattern. Thus, the estimators for the partial pair-correlation functions yield

$$\hat{g}_{lm}^R(r) = \frac{\hat{\rho}_{lm}^R(r)}{\hat{\lambda}_S^l(r)\hat{\lambda}_S^m(r)} \qquad \text{Ripley}$$

$$\hat{g}_{lm}^S(r) = \frac{\hat{\rho}_{lm}^S(r)}{\hat{\lambda}_S^l(r)\hat{\lambda}_S^m(r)} \qquad \text{Stoyan}$$

$$\hat{g}_{lm}^O(r) = \frac{\hat{\rho}_{lm}^O(r)}{\hat{\lambda}_S^l(r)\hat{\lambda}_S^m(r)} \qquad \text{Ohser} \qquad (3.54)$$

$$\hat{g}_{lm}^{WM}(r) = \frac{\hat{\rho}_{lm}^{WM}(r)}{\hat{\lambda}_n^l \hat{\lambda}_n^m} = \frac{\hat{\rho}_{lm}^O(r)}{\hat{\lambda}_n^l(r)\hat{\lambda}_S^m(r)} \qquad \text{WM}$$

Note that the WM estimator uses the adapted intensity estimator for the second pattern (i.e., type m points), but the natural estimator for the focal pattern l. The reason for this is that the estimator of the bivariate product density in Equation 3.51 can be rewritten as $(1/2\pi r)(n_l/A)(1/n_l)$ $\sum_{i=1}^{n}\sum_{j=1}^{n,\neq} C_{lm}(x_i, x_j)k(\|x_i - x_j\| - r)w_{i,j}$, which includes the natural estimator of the intensity of pattern l.

Figure 3.18 compares the different estimators provided in Equation 3.54 for a bivariate pattern comprising a clustered focal pattern and an independent random pattern (Figure 3.18a). Because the two patterns are independent by construction, the expectation for this pattern yields $g_{12}(r) = 1$. We find that the WM, Ohser and Stoyan estimators yield almost identical estimates, whereas the Ripley estimator shows a slight positive bias (Figure 3.18b). This is because the focal pattern is a clustered pattern with several clusters close to the border, which yields weights that are too large, analogous to the univariate case for the focal pattern (Figure 3.14d). For the "real life" pattern shown in Figure 3.18d, in which there is no strong clustering in the focal pattern close to the border, we find only small differences among the various estimators (Figure 3.18e,f).

To put the small departures among estimators into perspective, we also show in Figure 3.18b the simulation envelopes of a null model, based on the WM estimator. In the simulations, the clustered focal pattern is unchanged, but pattern 2 is randomized based on CSR. The width of the simulation envelopes for the pair-correlation function, at any distance r, is greater than the differences among estimators. However, the Ripley estimator shows departure from the other estimators, approaching a magnitude of half of the width of the simulation envelopes, at largest distances r (Figure 3.18b). These examples outline that the differences among the three estimators for the bivariate pair-correlation function (WM, Ohser, and Stoyan) are negligible in practical applications, but that the Ripley estimator may in some cases produce biased results.

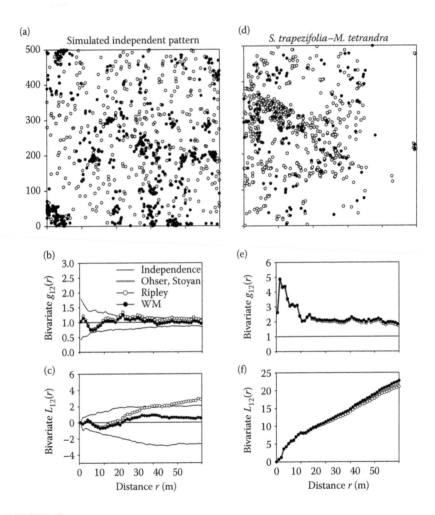

FIGURE 3.18
Estimators for bivariate second-order statistics. (a) Simulated bivariate pattern composed of a clustered focal pattern (closed disks) and a second, random pattern (open disks). (b) Estimators of the bivariate pair-correlation functions for the pattern shown in panel (a) together with simulation envelopes of the CSR null model based on the WM estimator. (c) Same as (b) but for the bivariate *L*-function. (d) Bivariate pattern of two tropical tree species. (e) Estimators of the bivariate pair-correlation functions for the pattern shown in panel (d). (f) Estimators of the bivariate *L*-function for the pattern shown in panel (d).

3.1.4.3 Partial Inhomogeneous Pair-Correlation Functions

The estimators of the inhomogeneous (cross) pair-correlation functions require two separate estimates of the intensity functions $\lambda_l(x)$ and $\lambda_m(x)$ for the type *l* and *m* patterns, respectively. Analogous to Equation 3.25 for the univariate, pair-correlation function, the BMW estimator of the inhomogeneous partial pair-correlation function yields

$$\hat{g}_{lm}^{\text{BMW}}(r, \lambda_l(x), \lambda_m(x))$$

$$= \frac{1}{2\pi r A} \sum_{i=1}^{n} \left[\sum_{j=1}^{n,\neq} C_{lm}(x_i, x_j) \times k\left(\|x_i - x_j\| - r\right) \frac{w_{ij}(r)}{\lambda_l(x_i)\lambda_m(x_j)} \right] \quad (3.55)$$

And, analogous to Equation 3.29, the estimator of the inhomogeneous partial pair-correlation function based on a generalization of the Ohser edge correction yields

$$\hat{g}_{lm}^{O}(r, \lambda_l(x)\lambda_m(x))$$

$$= \frac{1}{\hat{\lambda}_l \hat{\lambda}_m} \frac{1}{2\pi r A} \sum_{i=1}^{n} (\sum_{j=1}^{n,\neq} C_{lm}(x_i, x_j) k(\|x_i - x_j\| - r) \frac{A}{\bar{\gamma}_W(r, \lambda_l(x_i)\lambda_m(x_j))} \quad (3.56)$$

where the $\hat{\lambda}_l$ and $\hat{\lambda}_m$ are the natural estimators of the intensity of type l and type m points, respectively. In this case, the generalized, isotropic-set covariance yields

$$\bar{\gamma}_W(r, \lambda_l(x)\lambda_m(x)) = \frac{1}{\hat{\lambda}_l \hat{\lambda}_m} \frac{1}{2\pi r} \int_W \lambda_l(x) \left[\int_W \lambda_m(a) k(\|x - a\| - r) da \right] dx \quad (3.57)$$

Analogous to Equation 3.31, the estimator of the inhomogeneous, partial pair-correlation function based on a generalization of the WM edge correction yields

$$\hat{g}_{lm}^{\text{WM}}(r, \lambda_l(x), \lambda_m(x)) = \frac{(1/n_l) \sum_{i=1}^{n} \left(\sum_{j=1}^{n,\neq} C_{lm}(x_i, x_j) k\left(\|x_i - x_j\| - r\right) \right)}{(1/n_l) \sum_{i=1}^{n} C_l(x_i) \left[\int_W \lambda_m(a) k(\|x_i - a\| - r) da \right]}$$

$$\approx \frac{\sum_{i=1}^{n} \left(\sum_{j=1}^{n,\neq} C_{lm}(x_i, x_j) k\left(\|x_i - x_j\| - r\right) \right)}{(n_m - 1)/s \sum_{i=1}^{n} \left(\sum_{j=1}^{s} C_{lm}(x_i, a_j) k\left(\|x_i - a_j\| - r\right) \right)} \quad (3.58)$$

As can be seen in the lower expression, the integral in the denominator can be approximated by distributing s type m points (i.e., a_j) within the observation window, using a heterogeneous Poisson process with intensity function $\lambda_m(x)$. The n_l are the number of points of type l. Note that the intensity function of the focal pattern l does not enter into this estimator, because the type *l* points are used as focal points and are assumed to already be the result of thinning a homogeneous pattern with the intensity function $\lambda_l(x)$.

3.1.4.4 Partial K-Functions

The quantity $\lambda_l \lambda_m K_{lm}(r)$ is used for the estimation of the partial *K*-functions and is analogous to Equation 3.32:

$$\hat{K}_{lm}(r) = \frac{1}{A} \sum_{i=1}^{n} \sum_{j=1}^{n,\neq} \underbrace{C_{lm}(x_i, x_j)}_{\substack{\text{First point type } l, \\ \text{second point type } m}} \times \underbrace{1(\|x_i - x_j\|, r)}_{\text{Count function}} \times \underbrace{w_{i,j}}_{\substack{\text{Edge} \\ \text{correction}}} \tag{3.59}$$

The edge correction weights for the univariate case given in Equation 3.34 can be used. The adapted estimator of the intensity of type *l* points yields

$$\hat{\lambda}_V^l(r) = \frac{\sum_{i}^{n} C_l(x_i) \times v(W \cap b(x_i, r))}{2\pi \int_0^r t \bar{\gamma}_W(t) dt} \tag{3.60}$$

where the indicator function $C_l(x_i)$ selects all points *i* of the focal type *l*. Analogous to Equation 3.37, the various estimators of the *K*-function are

$$\hat{K}_{lm}^{WM}(r) = \frac{\hat{K}_{lm}^{WM}(r)}{\hat{\lambda}_n^l \hat{\lambda}_V^m(r)} \quad \text{WM}$$

$$\hat{K}_{lm}^{O}(r) = \frac{\hat{K}_{lm}^{O}(r)}{\hat{\lambda}_V^l(r) \hat{\lambda}_V^m(r)} \quad \text{Ohser}$$

$$\hat{K}_{lm}^{S}(r) = \frac{\hat{K}_{lm}^{S}(r)}{\hat{\lambda}_V^l(r) \hat{\lambda}_V^m(r)} \quad \text{Stoyan} \tag{3.61}$$

$$\hat{K}_{lm}^{R}(r) = \frac{\hat{K}_{lm}^{R}(r)}{\hat{\lambda}_V^l(r) \hat{\lambda}_V^m(r)} \quad \text{Ripley}$$

where the type of point used in estimating lambda is indicated as a superscript.

The estimators of the bivariate *L*-function derived from the *K*-function (Equation 3.61) mirror the results for the bivariate, pair-correlation function (Figure 3.18). The estimates based on the WM, Ohser and Stoyan estimators are practically identical, whereas the estimate of the *L*-function based on the Ripley edge correction shows a notable difference, with a positive bias in the example pattern (Figure 3.18c). When contrasting the *L*-function and the corresponding simulation envelopes for the WM estimator (Figure 3.19b) with

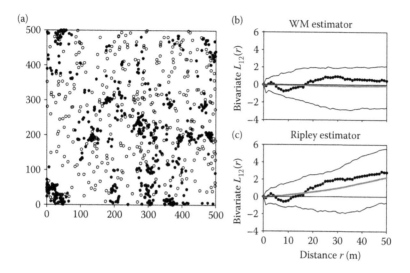

FIGURE 3.19
Simulation envelopes for the WM and Ripley estimators of the *L*-function. (a) Bivariate pattern where the random, "second" pattern (open circles) is independent of the clustered focal pattern (closed disks). (b) WM estimator used to estimate the *L*-function and simulation envelopes (black solid lines) of the CSR null model randomizing the "second" pattern. (c) Ripley estimator used to estimate the *L*-function with simulation envelopes as in panel (b).

those of the Ripley estimator (Figure 3.19c), we find that the Ripley estimator shows a considerable bias in the expectation (gray line) not shown by the WM estimate (and the Ohser and Stoyan estimator; not shown). This is because the Ripley estimator weights the large number of points located in the clusters close to the border of *W* with too high a value, resulting in an overestimation based on the number of nearby points in the second pattern. This creates increasingly positive values of the *L*-function with increasing distance *r* (Figure 3.19c). However, if the inference is based on simulation envelopes, this is not a problem because the estimates of the realizations of the null model are subject to the same bias and Figure 3.19c still indicates independence. However, the bias in the Ripley estimator may become a problem if a parametric fit is used.

Again, we see a situation paralleling the inhomogeneous case for the univariate estimators in developing estimators for the bivariate case. So briefly, the BMW estimator of the inhomogeneous cross *K*-function yields (cf. Equation 3.38)

$$\hat{K}^{BMW}_{lm}(r, \lambda_l(x), \lambda_m(x)) = \frac{1}{A} \sum_{i=1}^{n} \sum_{j=1}^{n, \neq} C_{lm}(x_i, x_j) \times \mathbf{1}(\|x_i - x_j\|, r) \frac{w_{ij}(r)}{\lambda_l(x_i)\lambda_m(x_j)}, \quad (3.62)$$

and the Ohser edge correction yields (cf. Equation 3.41)

$$\hat{K}_{lm}^{O,\text{inhom}}(r, \lambda_l(x), \lambda_m(x)) = \pi r^2 \frac{\sum_{i=1}^{n}\left(\sum_{j=1}^{n,\neq} C_{lm}(x_i, x_j) \times \mathbf{1}(\|x_i - x_j\|, r)\right)}{\int_W \lambda_l(x)\left[\int_W \lambda_m(a)\mathbf{1}(\|x - a\|, r)\mathrm{d}a\right]\mathrm{d}x} \tag{3.63}$$

and finally the WM edge correction yields (cf. Equation 3.42)

$$\hat{K}_{lm}^{\text{WM,inhom}}(r, \lambda_l(x), \lambda_m(x)) = \pi r^2 \frac{\sum_{i=1}^{n}\left(\sum_{j=1}^{n,\neq} C_{lm}(x_i, x_j) \times \mathbf{1}(\|x_i - x_j\|, r)\right)}{\sum_{i=1}^{n} C_l(x_i)\left[\int_W \lambda_m(a)\mathbf{1}(\|x_i - a\|, r)\mathrm{d}a\right]} \tag{3.64}$$

3.1.4.5 Partial Nearest-Neighbor Summary Statistics

The estimators of the nearest-neighbor distribution functions for bivariate patterns can be derived in a natural way from those for univariate patterns. If we have a homogeneous and bivariate pattern composed of type l and type m points, the summary statistic $D_{lm}^k(r)$ is the distribution function of the distances from a typical point of type l to the kth nearest point of type m. In this case, we can generalize the estimators of the univariate case by introducing an indicator function $C_{lm}(x_i, x_j)$ that selects from among all point pairs x_i and x_j only those where the first point is of type l (i.e., point x_i) and the second is of type m (i.e., point x_j). If, as before, we define d_{ik} as the distance from point x_i to the kth neighbor of type m (i.e., points x_j), the *naive estimator of $D_{lm}^k(r)$ without edge correction* is then analogous to Equation 3.43:

$$\hat{D}_{lm}^k(r) = \frac{\sum_{i=1}^{n} \mathbf{1}(0 < d_{ik} \leq r) \times C_{lm}(x_i, x_j)}{\sum_{i=1}^{n} C_l(x_i)} \tag{3.65}$$

where the indicator function $C_{lm}(x_i, x_j)$ selects pairs of points in which the first point i is of type l and the second point j is of type m. $C_l(x_i)$ selects all points i of the focal type l and the indicator function $\mathbf{1}(0 < d_{ik} \leq r)$ yields the value 1, if d_{ik} is smaller than or equal to r, and zero otherwise. Thus, the summation counts, for a given distance r, all focal points i which are of type l and have their k-nearest-neighbor type m within distance r. The denominator then normalizes by counting the number of focal points i which are of type l.

The estimator using the *buffer method* is analogous to the univariate estimator, but now we need to add the condition that the distance β_i of the focal points x_i to the border is greater than the buffer distance b. This condition is implemented as the indicator function $\mathbf{1}(b < \beta_i)$ which yields a value of 1 if distance β_i is smaller than b and zero otherwise:

$$\hat{D}^{k,b}_{lm}(r) = \frac{\sum_{i=1}^{n} \mathbf{1}(0 < d_{ik} \le r) \times [C_{lm}(x_i, x_j) \times \mathbf{1}(b < \beta_i)]}{\sum_{i=1}^{n} [C_l(x_i) \times \mathbf{1}(b < \beta_i)]}, \quad r \le b \quad (3.66)$$

The *Hanisch estimator* selects the focal points i not based on the buffer, but decides for each point i of type l, individually, if it is used as a focal point. To this end, focal points are selected only if the k-nearest-neighbor of type m can be correctly determined. The condition for this is that the distance d_{ik} to the k-nearest-neighbor of type m is smaller than the distance β_i to the border. This condition is implemented by using the function $\mathbf{1}(d_{ik} < \beta_i)$, which yields a value of 1 if the d_{ik} is smaller than β_i and zero otherwise:

$$\hat{D}^{k,H}_{lm}(r) = \frac{\sum_{i=1}^{n} \mathbf{1}(0 < d_{ik} \le r) \times [C_{lm}(x_i, x_j) \times \mathbf{1}(d_{ik} < \beta_i) \times w(d_{ik})]}{\sum_{i=1}^{n} [C_{lm}(x_i, x_j) \times \mathbf{1}(d_{ik} < \beta_i) \times w(d_{ik})]} \quad (3.67)$$

The weighing factor $w(d_{ik})$ compensates for the points of type l which are not used as focal points and, for a rectangular observation window with side length l_x and l_y, yields

$$w(d_{ik}) = \frac{1}{(l_x - d_{ik})(l_y - d_{ik})} \quad (3.68)$$

Comparing Equations 3.66 and 3.67 shows that the indicator function $\mathbf{1}(b < \beta_i)$, which defines the buffer in Equation 3.66, is replaced by the weighted indicator function $\mathbf{1}(d_{ik} < \beta_i) \times w(d_{ik})$, which defines the focal points of the Hanisch estimator, in Equation 3.67.

While the estimators for bivariate patterns are completely analogous to those for univariate patterns, there is one important difference with respect to *sensitivity of the estimators to departures from homogeneity*. To illustrate this potential problem, we use an extreme case of a heterogeneous bivariate pattern (Figure 3.20a) in which type 1 points (indicated as closed disks) are randomly distributed in the western half ($x < 250$) of a 500×500 m² observation window and type 2 points (open disks) are randomly distributed in the eastern half ($x > 250$). In this case, we observe a large difference between the naive estimator without edge correction and the Hanisch estimator (Figure 3.20c). Somewhat

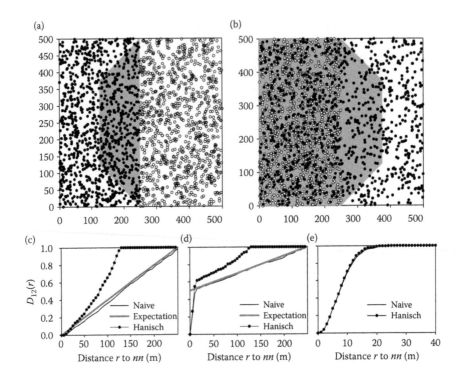

FIGURE 3.20
Potential problems for estimators of the nearest-neighbor distribution function applied to complex bivariate patterns. (a) Pattern where the focal points (type 1, black disks) are randomly distributed in the western half of the observation window W and type 2 points (open disks) are randomly distributed in the eastern half. (b) Pattern where the focal points (type 1, black disks) are randomly distributed throughout W and type 2 points are randomly distributed in the western half (open disks). Gray shaded areas in panels (a) and (b) indicate locations i for which the distance to the nearest type 2 neighbor is potentially less than the distance to the nearest border of W. (c) Analysis of the pattern in panel (a) with different estimators of the bivariate $D_{12}(r)$. (d) Analysis of the pattern shown in panel (b) with different estimators of the bivariate $D_{12}(r)$. (e) Analysis of the pattern in panel (b) reversing the roles of patterns one and two (i.e., pattern 1: open disks and pattern 2: closed disks).

surprisingly, the Hanisch estimator of $D_{12}(r)$ yields, on first examination, a value of 1 for distances greater than 125 m, whereas the naive estimator yields a value of approximately 1 only at distances greater than 250 m. The latter was expected because points with an x-coordinate close to zero will have their nearest neighbor of type 2 approximately at a distance of 250 m, and all points with an x-coordinate larger than zero will have their nearest type 2 neighbor at distances below 250 m. This suggests that, the naive estimator approximates the true $D_{12}(r)$ better than the Hanisch estimator, at least in this example employing an extremely heterogeneous pattern.

The difference between the Hanisch and naive estimators can be explained by the characteristics of the two in relation to the pattern. The Hanisch estimator uses only focal points i for which the distance d_{i1} to the nearest neighbor is smaller than its distance to the nearest border of the observation window. In our particular example, those focal points are located inside the gray shaded trapezoid as shown in Figure 3.20a. In the univariate case, there is no reason to assume that the (true) nearest-neighbor distances of points closer to the border (i.e., outside the trapezoid) would differ principally from those used for estimation (i.e., inside). However, this is not the case in our example. The distances to the nearest type 2 neighbor inside the trapezoid are always smaller than 125 m, but outside they may reach distances up to 250 m (for points with x-coordinates close to zero). In particular, the nearest-neighbor distance is greater than 125 m for all points with x-coordinates smaller than 125 m. For this reason, the Hanisch estimator yields $D_{12}(r = 125$ m$) \approx 1$. However, this means that the focal points selected by the Hanisch (and the buffer) estimator are not "representative" of the type 1 points (with respect to their type 2 neighbors) and this particular selection of focal points results in the bias observed in Figure 3.20c.

This extreme example illustrates that the buffer and the Hanisch estimator for bivariate patterns may not be robust against heterogeneities of a specific type, in which the selection of the focal points is not "representative" with respect to the nearest-neighbor distances. However, if type 2 points do overlap sufficiently well with the area occupied by type 1 points, this problem does not occur. This is illustrated in Figure 3.20e, which provides the results of a bivariate pattern analysis, with the focal type 1 points located only in the western part of the plot (as in Figure 3.20a) and the type 2 points randomly distributed within the entire plot (as for the pattern in Figure 3.20b, but with the open disks being pattern 1 and the closed points being pattern 2). In this case, the type 2 points overlap the area covered by type 1 points and the Hanisch and naive estimators agree in their results (Figure 3.20e). However, if we switch the role of type 1 and 2 (Figure 3.20b), the problem reappears as shown in Figure 3.20d. In this case, the type 1 focal points are randomly distributed within the entire observation window, but type 2 points overlap only half of this area. The resulting expectation of $D_{12}(r)$ is a straight line between points $(0, 0.5)$ and $(250, 1)$ (Figure 3.20d), which is captured by the naive estimator. The straight line actually arises because half of the focal points have their type 2 nearest neighbor quite close by (those in the western part of the plot), whereas those in the eastern part of the window have their nearest type 2 neighbor located at an x-coordinate of approximately 250.

The above examples suggest that it may be safer to use the naive estimator without edge correction for bivariate patterns that are suspected of heterogeneity, due to the focal points not being "representative" of the nearest-neighbor distances found within the entire observation window. The smaller bias introduced by not always having the true nearest neighbor inside the observation window may be smaller in this case than the bias arising through edge correction.

3.1.5 Summary Statistics for Multivariate Patterns (Data Type 3)

Although the modification of summary statistics to accommodate bivariate patterns is, in general, straightforward, generalization to truly multivariate summary statistics is not. The issues for accomplishing this are, in fact, not yet entirely resolved. As a result, truly multivariate summary statistics that would be able to summarize the spatial structure of multivariate point-pattern data, in a simple and ecologically meaningful way, are lacking. The few attempts to develop summary statistics for multivariate patterns represent logical extensions of the summary statistics for uni- or bivariate patterns. They can, in fact, be expressed as the sum of bivariate summary statistics. Each one generally characterizes only one specific aspect of the potentially complex structure of multivariate patterns and often represents a spatial analog to classical nonspatial diversity indices.

3.1.5.1 Spatially Explicit Simpson Index

One approach to analyzing multivariate patterns, which has been applied to multispecies communities, such as tropical forests, considers summary statistics that average the various possible pairwise, partial, summary statistics into a community average. For example, the *spatially explicit Simpson index* is basically the conditional probability $\beta(r)$ that two points separated by distance r, belong to different types (Plotkin et al. 2000; Shimatani 2001; Chave and Leigh 2002). The formula for this is given by

$$\beta(r) = 1 - \sum_{m=1}^{S} \frac{\lambda_m^2 g_{mm}(r)}{\lambda^2 g(r)} \tag{3.69}$$

where S is the number of species in the community, $g(r)$ is the pair-correlation function for the entire community (i.e., all individuals of all species are treated as one pattern), $g_{mm}(r)$ is the partial pair-correlation function of species m, λ_m is the intensity of type m points, and λ is the intensity of all points of the multivariate pattern. If all patterns are random patterns, the expectation of $\beta(r)$ is a constant and yields $\beta^* = 1 - \sum(\lambda_m/\lambda)^2$. This is the *classical Simpson index* (Simpson 1949) and yields the probability that two arbitrarily chosen individuals (without replacement) are different species (Shimatani and Kubota 2004). This index quantifies the degree of evenness in species composition. β^* is small if one (or a few) species dominate the community, while β^* reaches its maximum, $\beta^* = 1 - 1/S$, if the abundances n_m of all species are equal, that is, $n_m = n/S$, where n is the total number of individuals in the community and S the total number of species.

If the patterns of the individual species tend to be clumped, the spatially explicit Simpson index $\beta(r)$ will be below the expectation β^* for small distances r and, if the individual species tend toward hyperdispersed patterns, $\beta(r)$ will be above the expectation β^* for small distances r. In the case of conspecific clustering and heterospecific segregation, the index $\beta(r)$ would yield substantially lower values than β^* at small distances, because in this case most neighbors would be conspecifics.

Figure 3.21 provides an example of the spatially explicit Simpson index $\beta(r)$ calculated for recruits from the 2005 census at the BCI forest plot in Panama (Box 2.6). In this case, recruits are defined to be trees that were smaller than 1 cm diameter at breast height in 2000 and were larger than that in 2005. There were, in total, 23,290 recruits of 246 species (Figure 3.21a). Thus, the classical Simpson index β^* for an evenly distributed community would yield $\beta^* = 0.996$, however, we find $\beta^* = 0.968$. Thus, the recruits are clearly not uniformly distributed. Figure 3.21b shows the spatially explicit Simpson index $\beta(r)$. It indicates that there is a strong, spatial structure at shorter distances of r (<10 m), at which the recruits of different species

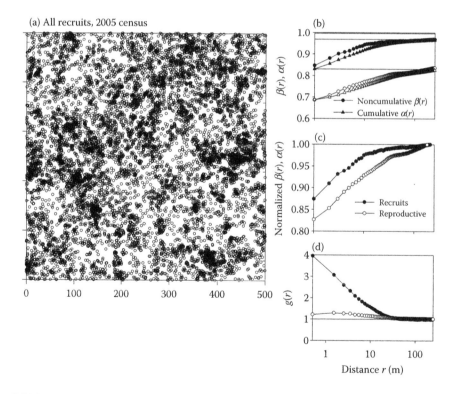

FIGURE 3.21

Spatially explicit Simpson index $\beta(r)$ for β diversity. (a) Spatial pattern of all recruits (i.e., individuals that grew to a dbh > 1 cm during a five-year period) for a 500 × 500 m² area of the BCI plot. (b) Simpson index $\beta(r)$ for the pattern of recruits (closed disks) and for all reproductive individuals (open disks), as well as the corresponding cumulative indices $\alpha(r)$ (triangles). The horizontal lines show the asymptotic expectation β^* for $\beta(r)$ at large scales. (c) Same as panel (b), but for the normalized indices $\beta(r)/\beta^*$ and $\alpha(r)/\beta^*$. (d) Pair-correlation function of all recruits (closed disks) and all reproductive individuals (open disks).

tend to be clustered. Two recruits located at distance $r = 1$ m are, with an 85% probability, not of the same species, but for the nonspatially explicit Simpson index $\beta^* = 0.968$ the expectation is that 97% of all randomly selected pairs of individuals will be heterospecific. Two recruits located at distance $r = 10$ m still have a 94.7% probability of being heterospecific. The pair-correlation function of all recruits shows strong clustering up to distances of say $r = 30$ m (Figure 3.21d), with recruits having a 4 times higher neighborhood density at 0.5 m than expected by their over all density [i.e., $g(r = 0.5) = 4$]. This indicates that the community of recruits exhibits clustering (see also Figure 3.21a), with some intraspecific clustering at smaller scales. However, the relatively high values of $\beta(r)$ at short distances [e.g., $\beta(r = 1$ m$) = 0.85$] indicate that the clusters are composed of a mixture of species.

Figure 3.21b also shows an analysis for all reproductive individuals (i.e., individuals with a diameter above a species-specific threshold; open symbols). We find in total 71,597 reproductive individuals from 272 species. Thus, the classical Simpson index β^* for an evenly distributed community would yield $\beta^* = 0.996$; however, we find $\beta^* = 0.831$. Clearly, the abundances of reproductive individuals are not uniformly distributed and are even more unevenly distributed than recruits. Figure 3.21b shows the spatially explicit Simpson index $\beta(r)$. We find a strong spatial structure at distances r of less than approximately <30 m, at which scales the different species tend to be clustered. Two reproductive trees located at distance $r = 1$ m are, with a 69% probability, heterospecifics, and two reproductive individuals at distance $r = 10$ m still have a 77% probability of being heterospecifics. The pair-correlation function of all reproductive individuals shows clustering up to distances of say $r = 30$ m (Figure 3.21d). This pattern, however, is much weaker than for recruits.

If the multivariate pattern does not show strong heterogeneity, the spatially explicit Simpson index $\beta(r)$ approaches the nonspatial Simpson index for large distances r. Because of this, it is somewhat complicated to compare the pure spatial effects in the decline of similarity in local species composition (i.e., Figure 3.21b). To compare the spatial structure between different communities, we may therefore normalize the spatially explicit Simpson index $\beta(r)$ with its nonspatial expectation β^* (Figure 3.21c). This analysis shows that reproductive individuals have a stronger spatial structure than recruits, where the species richness is less evenly distributed locally than that of the recruit community.

Note that the numerator and denominator of Equation 3.69 are product densities and that the partial product density of species m, divided by the full product density, yields a mark connection function, which is discussed in more detail in the next section. If we insert the product densities from Equation 3.51 into Equation 3.69, we obtain an estimator of the spatially explicit Simpson index $\beta(r)$:

$$\hat{\beta}(r) = 1 - \sum_{m=1}^{S} \left[\frac{\sum_{i=1}^{n} \sum_{j=1}^{n,\neq} C_{mm}(x_i, x_j) \times k(\|x_i - x_j\| - r)w_{i,j}}{\sum_{i=1}^{n} \sum_{j=1}^{n,\neq} k(\|x_i - x_j\| - r)w_{i,j}} \right]$$

$$= 1 - \frac{\sum_{i=1}^{n} \sum_{j=1}^{n,\neq} \mathbf{1}(m_i = m_j) \times k(\|x_i - x_j\| - r)w_{i,j}}{\sum_{i=1}^{n} \sum_{j=1}^{n,\neq} k(\|x_i - x_j\| - r)w_{i,j}} \qquad (3.70)$$

The index can be simplified by using the indicator function $\mathbf{1}(m_i = m_j)$, which yields one if the species identifier of point i (= m_i) is the same as the species identifier of point j (= m_j) and zero otherwise. If the Ohser weights are used, the edge correction factors w_{ij} of the numerator and denominator cancel out, because they are independent of the concrete point pair i–j. This suggests that this estimator, which is the ratio of two product densities, does not need edge correction. Note that the index $\beta(r)$ is conditional by nature, since dividing by $g(r)$ factors out the over all pattern of the entire community.

As we will see in Section 3.1.7.6, the spatially explicit Simpson index can be generalized to consider different levels of similarity between species. While Equation 3.70 treats all heterospecific individuals equally [i.e., $\mathbf{1}(m_i = m_j) = 1$ for $i \neq j$], a more nuanced view of the spatial structures in a community can be gained by considering alternative axes of biodiversity, such as phylogenetic and functional diversity. For example, closely related (or functionally similar) species might be found in similar environments, but due to competitive exclusion the local neighborhoods around species may be characterized by more distantly related species (e.g., Weiher and Keddy1995; Cavender-Bares et al. 2006; Swenson et al. 2006). Thus, consideration of species similarity in spatially explicit, summary statistics of multivariate patterns may allow us to better understand the mechanisms underlying observed biodiversity patterns. In Section 3.1.7.6 we will extend the spatially explicit Simpson index by including a continuous measure of species dissimilarity into Equation 3.70, instead of the indicator function $\mathbf{1}(m_i = m_j)$. Since this index works with quantitative marks provided by the continuous dissimilarity index, the resulting summary statistics are mark-correlation functions and are treated therefore in detail in Section 3.1.7.

The *cumulative spatially explicit Simpson index* $\alpha(r)$ is based on K-functions and is given by

$$\alpha(r) = 1 - \sum_{m=1}^{S} \frac{\lambda_m^2 K_{mm}(r)}{\lambda^2 K(r)} \qquad (3.71)$$

Equation 3.71 expresses the probability that two points with a separation distance of no more than r are of different types (Shimatani 2001). The index

$\alpha(r)$ is based on an estimation of the quantity $\lambda^2 K(r)$. Here, the WM edge correction weight (Equation 3.34), which approximates the Ohser weight, also cancels out. Figure 3.21b shows that this index behaves in a manner similar to $\beta(r)$, but because it is a cumulative index it shows scale effects less clearly.

3.1.5.2 Individual Species–Area Relationship ISAR

Another summary statistic that has been proposed for multivariate point patterns is the *spatial species richness* $d(r)$ (Shimatani and Kubota 2004). It is defined as the expected number of different species present within distance r from an arbitrarily chosen point in an observation window W. Thus, this summary statistic is the point-pattern analog to the species–area relationship (SAR), if distance r is converted to area πr^2. Usually, the SAR is estimated by dividing the observation window into quadrats, counting the number of species in every quadrat, and calculating the average (Shimatani and Kubota 2004). However, the spatial species richness $d(r)$ can be calculated by decomposing the index into the sum of the spherical contact distributions of all the univariate patterns:

$$d(r) = \sum_{m=1}^{S} H_s^m(r) \tag{3.72}$$

where $H_s^m(r)$ is the partial spherical contact distribution for species m. Decomposing the spatial species richness $d(r)$ into the "detectability" $H_s^m(r)$ of each species allows the role of each species in the resulting over all species diversity to be examined (Shimatani and Kubota 2004).

Figure 3.22 shows the spatial species richness $d(r)$ for the recruits and reproductive individuals of the 2005 BCI census (Box 2.6). The species at the BCI forest are not well mixed. At a distance of 10 m, only 4.4% of all recruit species can be found on average at a random location within the plot (Figure 3.22b). At sample plots with a 40 m radius we find on average 27.2% of all species and, at sample plots with a radius of 88 m, we find 50% of all species (Figure 3.22b). This is, in part, a consequence of the spatial structure, as revealed by the spatially explicit Simpson index, but is also due to many species with low abundances. Only 163 species (= 66% of the total of 246 species represented by recruits) have more than 10 recruits. A similar result was found for the reproductive community, in which 175 species had more than 10 individuals. A more uneven distribution of abundances of reproductive individuals (compared with recruits), together with the spatial structure revealed by the spatially explicit Simpson index, yields a $d(r)$ index below that of the recruits, even if total species richness is greater (272 vs. 246 species) (Figure 3.22a).

Another summary statistic closely related to $d(r)$ is the *individual species–area relationship* ISAR$_f(r)$ introduced by Wiegand et al. (2007b). This summary statistic is defined to be the expected number of species present within

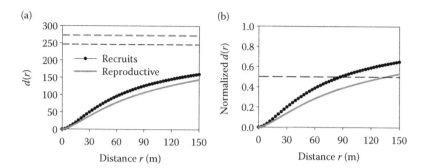

FIGURE 3.22

Spatial species richness $d(r)$: a point-pattern analog to the SAR. (a) Spatial species richness $d(r)$ for the community of recruits (black line) and reproductive trees (gray line) for the BCI forest (2005 census). Dashed lines represent the total number of species (246 for recruits and 272 for reproductive trees) (b) Same as panel (a), but normalized for total number of species. The intersections of the functions with the dashed line yield the distances r at which 50% of the species are encountered.

distance r of the typical individual of a focal species f. It determines the average neighborhood species richness around individuals of the focal species and is therefore the point-centered analog to $d(r)$. The ISAR can be calculated by decomposing the index into the sum of the cross, nearest neighbor, distribution functions between the focal species f and all other species m:

$$\text{ISAR}_f(r) = \sum_{m=1}^{S} D_{fm}(r) \tag{3.73}$$

where $D_{fm}(r)$ is the probability that an individual of the focal species f has its nearest neighbor of species m within distance r. Given the potential problems with the Hanisch estimator (Figure 3.20), it is safer to estimate the ISAR function without edge correction or use the buffer method (as in Wiegand et al. 2007b). To complete the analogy with the other summary statistics presented in this section, the ISAR can also be averaged over all species:

$$\overline{\text{ISAR}}(r) = \frac{1}{n}\sum_{f=1}^{S} n_f \text{ISAR}_f(r) = \frac{1}{n}\sum_{f=1}^{S} n_f \sum_{m=1}^{S,\neq} D_{fm}(r) \tag{3.74}$$

where n_f is the number of individuals of the focal species and n is the total number of individuals for the entire community. This summary statistic yields the average number of species within distance r of the typical individual of the community and is the point-centered analog to the location related, spatial species richness $d(r)$.

The special characteristic of the ISAR is that it views the structures in local species richness from the "plant's-eye" viewpoint of individuals of a given focal species (Wiegand et al. 2007b). Figure 3.23a shows an example of the community of all large trees (dbh > 10 cm) at the BCI plot (1995 census; Box 2.6) and sampling areas with a 10 m radius around the individuals of the focal species *Simarouba amara*. The ISAR function determines the number of species in each of these sampling areas and then their average value. If we vary the neighborhood radius r we obtain the full $ISAR_f(r)$ function that quantifies scale-dependent spatial structures in local species richness around a given focal species f (Figure 3.23c).

The ISAR function can be used in different ways to explore how the individuals of a focal species are located within the *"landscape" of local species richness* as shown in Figure 3.23a. For example, it allows us to quantify if a given species is predominately located in areas of below or above average species richness across a range of spatial scales. This can be tested by contrasting the observed ISAR curve to that of a null model that repeatedly

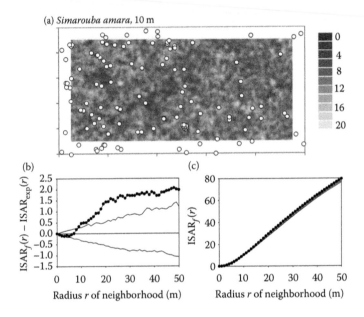

(a) *Simarouba amara*, 10 m

FIGURE 3.23
(See color insert.) ISAR for large trees (dbh > 10 cm) at the BCI plot, with *Simarouba amara* used as an example. (a) "Landscape of local species richness" for neighborhoods of 10 m, estimated as the number of species within 10 m of nodes on a regular grid. Open disks are the locations of the individuals of the species *S. amara*. (b) Observed values of $ISAR_f(r)$ adjusted by the expectation for the null model of spatial species richness for *S. amara*. (c) Observed values of $ISAR_f(r)$ for *S. amara*. In panels (b) and (c), closed disks represent the observed values, whereas solid lines represent the simulation envelopes and expectation. The null model (i.e., CSR) was implemented by randomly relocating *S. amara* within the observation window W and calculating the resulting $ISAR_f(r)$ values. *S. amara* is located, on average, in areas of higher local species richness, as indicated by the positive departure from the null model in panel (b). Because we subtracted the expectation of the null model (i.e., $ISAR_{exp}$) in (b) from the ISAR index, the expectation under the null model is zero at all scales.

relocates the individuals of the focal species to random locations within the entire plot (Wiegand et al. 2007b). This null model, which approximates the spatial species richness $d(r)$ (SAR; Shimantani and Kubota 2004), is the point-pattern analog of the common SAR. It estimates the mean number of species in neighborhoods of test points and links the ISAR concept with the SAR concept. The key question here is to find out whether the $ISAR_f(r)$ function of species f is significantly different from the point-pattern SAR, and identify at what spatial scales this occurs. Figure 3.23b shows that the species *S. amara* is predominantly located in areas of higher local species richness.

3.1.5.3 Phylogenetic Extension of the ISAR

In Section 3.1.5.1, we argued that measures of species diversity, when used alone, are relatively information poor, because they assign each species pair the same level of dissimilarity. Accordingly, we proposed a way of including continuous measures $d(f, m)$ of species dissimilarity (for species f and m) into summary statistics of multivariate patterns. The phylogenetic Simpson index, which generalizes the spatially explicit Simpson index, is based on the sum of bivariate (partial) pair-correlation functions (see Section 3.1.7.6). A similar approach is also possible for multivariate summary statistics derived from the sum of bivariate (partial) nearest-neighbor distribution functions. Thus, we can generalize the ISAR to yield a multivariate summary statistic that includes a continuous measure of species dissimilarity.

Recall that the individual species–area relationship $ISAR_f(r)$ for species f can be estimated as the sum of the probabilities $D_{fm}(r)$ over all species m that an individual of the focal species f has its nearest species m neighbor within distance r. We generalize this expression by weighting the different species m according to their phylogenetic (or functional) distance $d(f, m)$ from the focal species f:

$$\text{ISAR}_f(r) = \sum_{m=1}^{S,m \neq f} D_{fm}(r)$$

$$\text{PISAR}_f(r) = \sum_{m=1}^{S} d(f,m) D_{fm}(r)$$

(3.75)

where S is the total number of species in the observation window. The PISAR function estimates the average phylogenetic dissimilarity $d(f, s)$ (or phylogenetic distance) between all species m located within a neighborhood of radius r around the typical individual of the focal species f. Or, in other words, it is the mean phylogenetic distance (MPD) of the typical focal individual to all species that are located in its neighborhood.

The PISAR function given in Equation 3.75 contains a signal of both the multivariate spatial structure of the community (captured by the ISAR function) and a signal of phylogenetic spatial structure. If there is no phylogenetic spatial structure present, or if all heterospecific species show the same dissimilarity (i.e., a star shaped phylogeny), the phylogenetic PISAR function will basically collapse to the ISAR function. Thus, the difference between PISAR and ISAR is due to the phylogenetic spatial structure embedded in the multivariate pattern. To investigate the "pure" spatial phylogenetic signal, independent of species richness within the neighborhood, we normalize the PISAR$_f(r)$ with the ISAR$_f(r)$ to yield the rISAR function:

$$r\text{ISAR}_f(r) = \frac{\text{PISAR}_f(r)}{\text{ISAR}_f(r)} = \frac{\sum_{m=1}^{S} d(f,m)D_{fm}(r)}{\sum_{m=1}^{S,m\neq f} D_{fm}(r)} \qquad (3.76)$$

The rISAR function has a straightforward interpretation; it is the expected phylogenetic distance between the focal species and an arbitrarily chosen species from a neighborhood with radius r. We can calculate the expectation Δ_f^P of rISAR$_f(r)$ for very large neighborhoods r. If the neighborhoods approach the size of the plot, the $D_{fm}(r)$ has an approximate value of 1 (i.e., species m is present in the plot) and we find

$$\Delta_f^P = \frac{\sum_{m=1}^{S} d(f,m)}{S-1} \qquad (3.77)$$

Thus, rISAR$_f(r)$ will asymptotically approach the mean phylogenetic distance Δ_f^P of the focal species to all other species in the observation window. Note that the index Δ_f^P is analogous to the index Δ^P in Hardy and Senterre (2007) that measures phylogenetic distinctness based on species incidence within a given community. However, we restrict Δ_f^P to comparisons of the focal species f with all other species m present in the plot (i.e., we do not average over species f), similar to the manner in which we handled the ISAR. If the rISAR function is larger than Δ_f^P in small neighborhoods r, the focal species is locally surrounded by a subset of phylogenetically more dissimilar species (compared to all species present in the plot), and if the rISAR is less than Δ_f^P the focal species is surrounded by a subset of phylogenetically more similar species.

Analogous to the approach used with ISAR, we can define an underlying "landscape of phylogenetic neighborhood dissimilarity" for the rISAR function, which measures, at each location x of the observation window, the mean pairwise phylogenetic distance $\Delta_f^P(x,r)$ between the focal species f and all species located within a given distance r. Figure 3.24a shows an example of such a landscape for the species *Gustavia superba* in the BCI forest dynamics plot (Box 2.6). For some areas of the plot, we find that the phylogenetic neighborhoods are more dissimilar than expected (yellow to red areas), given the over all phylogenetic dissimilarity Δ_f^P of the focal species to all species

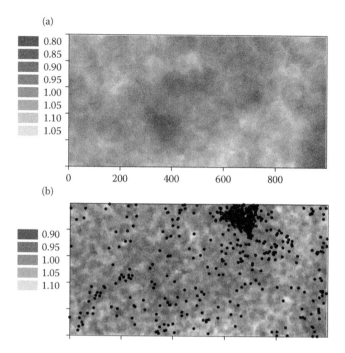

FIGURE 3.24

(**See color insert.**) Landscape maps of phylogenetic neighborhood dissimilarity for large individuals (i.e., dbh > 10 cm) in the 1000 × 500 m² BCI forest dynamics plot, using the species *Gustavia superba* as an example. (a) Map of phylogenetic dissimilarity for the focal species *G. superba*, defined at each location x as the mean pairwise phylogenetic distance $\Delta_f^P(x, r)$ between the focal species and all species present within 50 m of the focal locations. The map is normalized with the corresponding value Δ_f^P for the entire plot. (b) Same as panel (a), but for a neighborhood radius of $r = 10$ m, with the addition of the locations of all large focal individuals with dbh > 10 cm as closed disks. Phylogenies of the tree community in the BCI plot were constructed using APG (Angiosperm Phylogeny Group) III (http://www.phylodiversity.net).

present in the plot. However, there are also areas where the phylogenetic neighborhoods are more similar than expected (green to blue areas in Figure 3.24a). Figure 3.24b shows the same landscape, but now derived for neighborhoods with radius $r = 10$ m. The over all pattern of neighborhood dissimilarity is maintained, but additional small-scale structures appear. Note that the landscape of phylogenetic neighborhood dissimilarity will be different for each focal species.

The *r*ISAR function can be used in different ways to explore whether the location of an individual species is correlated with the landscape $\Delta_f^P(x, r)$ of phylogenetic neighborhood dissimilarity; that is, revealing spatial structures in the phylogenetic relatedness of focal species to the other species in the community. For example, we can use the homogeneous Poisson null model, which relocates the individuals of the focal species to random positions

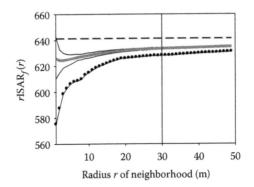

FIGURE 3.25
Observed *r*ISAR values for *Gustavia superba* from the map shown in Figure 3.24. *G. superba*
*r*ISAR$_f$(r) values are given as closed disks, simulation envelopes of the CSR null model as black
lines, expectations under the null model as bold gray lines and the expectation Δ_f^P for a large
neighborhood as a dashed horizontal line. *G. superba* is located in areas with more similar
phylogenetic neighborhoods than expected by the over all landscape of phylogenetic neighborhood dissimilarity.

within the plot, to reveal whether a focal species is surrounded locally by
species that are phylogenetically more similar or dissimilar, compared with
what is found on average within the entire plot. An even better null model
might additionally maintain the observed spatial autocorrelation structure
by using pattern reconstruction (see Section 4.3.3.3).

Plotting the locations of *G. superba* individuals within the landscape of
phylogenetic neighborhood dissimilarities [i.e., the $\Delta_f^P(x, r)$] indicates that
this species may show significant association with areas surrounded by
phylogenetically more similar species (Figure 3.24b). Application of the null
model shows that the observed *r*ISAR$_f$(r) is below the expectations of the
null model at all spatial scales, which indicates a highly significant effect.
This suggests that the distribution pattern of the species *G. superba* could be
influenced by habitat filtering of phylogenetically similar species (Figure
3.25). This species dominates ca. 2 ha of forest (4% of the total area) located
at the north-western part of the plot [centered on coordinate (700, 450)]. This
area was cleared during the nineteenth century and is occupied by species
that are phylogenetically similar to *G. superba. G. superba* is also surrounded
by more similar local species assemblages in other parts of the plot (Figure
3.24b).

3.1.6 Summary Statistics for Qualitatively Marked Point Patterns

Qualitatively marked patterns carry a qualitative (discrete or categorical) mark
that is descriptive of their state, such as surviving versus dead trees of a
given species. Importantly, the qualitative mark is produced by a process
acting *a posteriori* over a given univariate (unmarked) pattern. Examples of

qualitatively marked patterns are infected vs. noninfected plants or surviving vs. dead trees. This represents a fundamental difference to bivariate patterns in which the qualitative mark distinguishes among points where the marks were produced *a priori,* a good example of which is given by points whose marks represent different species (Goreaud and Pélissier 2004).

The basic interest in the analysis of qualitatively marked patterns is to determine whether the process that distributed the marks among points acted in a spatially uncorrelated way. To this end, several test statistics of qualitatively marked patterns are compared with those of the null model of *random labeling,* where the qualitative mark is randomly shuffled over the points of the univariate (unmarked) pattern.

In the following, we present summary statistics for two data types of qualitatively marked patterns that are relevant in ecological applications. The first data type is the standard situation, where a univariate pattern is qualitatively marked. The second data type also comprises a qualitatively marked pattern, but here the focus is on exploring the influence of an additional focal pattern on the marking. We have thus a bivariate pattern where the second pattern carries an additional qualitative mark. In this case, random labeling of the qualitatively marked subpattern provides the appropriate null model. Since this data type carries three types of points, it is referred to as *trivariate random labeling.* For these types of marked point pattern, we present *mark connection functions* as the appropriate adapted summary statistics.

3.1.6.1 Random Labeling (Data Type 4)

The simplest data structure with qualitative marks is a univariate pattern that carries a qualitative mark, for example, trees of a given species that are surviving versus dead. Because this data type consists of two types of points, similar in essence to a bivariate pattern, bivariate summary statistics, such as the pair-correlation function or Ripley's K-function, can be used as test statistics to characterize the spatial correlation structure of the marks. However, the bivariate summary statistics still contain the signal of the fixed spatial structure of the underlying univariate pattern (e.g., both dead and surviving trees), which is often not of primary interest. Mark connection functions $p_{lm}(r)$ are designed with this in mind, as they factor out the signature of the underlying univariate pattern.

Intuitively, the idea behind the mark connection function $p_{lm}(r)$ is that it represents the conditional probability that a pair of points, picked at random and separated by distance r, consists of a focal point of type l and a second point of type m. Thus, p_{lm} is symmetric and $p_{lm}(r) = p_{ml}(r)$. We can also define $p_{mm}(r)$ as the conditional probability that both points are of type m. It follows that the sum of all mark connection functions yields one. In the case of the two types l and m, we have $p_{ll}(r) + p_{lm}(r) + p_{ml}(r) + p_{mm}(r) = 1$.

A mark connection function can be expressed in terms of product densities or pair-correlation functions (Illian et al. 2008, p. 331):

$$p_{lm}(r) = \frac{\rho_{lm}(r)}{\rho(r)} = \frac{\lambda_l \lambda_m g_{lm}(r)}{\lambda^2 g(r)} = p_l p_m \frac{g_{lm}(r)}{g(r)}. \tag{3.78}$$

for $r > 0$ and $\rho(r) > 0$, with $p_l = \lambda_l/\lambda$ being the proportion of points having mark l. The denominators $\rho(r)$ and $g(r)$ are, respectively, the univariate product density and pair-correlation function applied to the over all pattern (i.e., points of both mark types). The numerators are the partial functions for points with mark l and m.

On the basis of Equation 3.78, the mark connection functions are best estimated using the ratio of the product densities (Equation 3.51):

$$\hat{p}_{lm}(r) = \frac{\hat{\rho}_{lm}(r)}{\hat{\rho}(r)} = \frac{\sum_{i=1}^{n} \sum_{j=1}^{n,\neq} C_{lm}(x_i, x_j) \times k(\|x_i - x_j\| - r) \times w_{i,j}}{\sum_{i=1}^{n} \sum_{j=1}^{n,\neq} k(\|x_i - x_j\| - r) \times w_{i,j}} \tag{3.79}$$

The indicator function $C_{lm}(x_i, x_j)$ yields one if the focal point i is of type l and the second point j of type m, and zero otherwise, and the w_{ij} are the edge correction weights for the different estimators derived in Section 3.1.2.1 and shown in Equation 3.21. Because the Ohser edge correction weights do not depend on the individual points i and j, they cancel out and the corresponding estimator

$$\hat{p}_{lm}^O(r) = \frac{\sum_{i=1}^{n} \sum_{j=1}^{n,\neq} C_{lm}(x_i, x_j) \times k(\|x_i - x_j\| - r)}{\sum_{i=1}^{n} \sum_{j=1}^{n,\neq} k(\|x_i - x_j\| - r)} \tag{3.80}$$

can be interpreted as the mean value of the "test function" $C_{lm}(x_i, x_j)$ over all pairs of points i and j, which are separated approximately by distance r [the latter is captured by the kernel function $k()$]. The test function $C_{lm}(x_i, x_j)$ yields a value of 1 if point i is of type l and point j of type m and zero otherwise. Thus, the estimator of $p_{lm}(r)$ shown in Equation 3.80 gives the proportion of point pairs separated by distance r where the first point is of type l and the second point is of type m. Equation 3.80 also shows that the estimator of the spatially explicit Simpson index (Equation 3.70) can be interpreted as a multivariate mark connection function, with a test function $1(m_i = m_j)$ that yields one if the species identifier of point i ($= m_i$) is the same as the species identifier of point j ($= m_j$) and zero otherwise.

Figure 3.26 provides an example for the application of mark connection functions. We used for this purpose the pattern of small surviving and dead

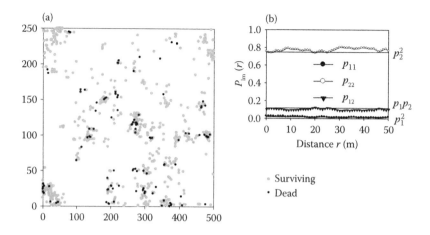

FIGURE 3.26

Mark connection functions. (a) Surviving and dead individuals with dbh < 10cm of the species *Shorea congestiflora*, a dominant species within a 25-ha plot in a rain forest at Sinharaja World Heritage Site (Sri Lanka). Living individuals are maked by a gray disk and dead individuals by a small black disk. (b) The three mark connection functions calculated for the pattern shown in (a). The horizontal lines represent the expectations of the mark connection functions for random mortality.

individuals (dbh < 10cm) of the species *Shorea congestiflora* at the Sinharaja World Heritage Site (Sri Lanka) (Figure 3.26a). Although the pattern of small trees is strongly clustered, the empirical mark connection functions are in good agreement with the expectations of the random mortality hypothesis (solid horizontal lines in Figure 3.26b). This results suggests that mortality of small trees of *S. congestiflora* is not subject to density dependent effects as for example shown in Figure 2.15.

3.1.6.2 Trivariate Random Labeling (Data Type 5)

Trivariate random labeling explores the effect of an antecedent focal pattern f on the process that distributes a qualitative mark (type l and type m) over a second pattern. Based on this data structure we can explore whether the qualitative mark of the second pattern depends on the distance from a point of the antecedent pattern f. For example, fire-induced mortality of a shrub species 2 may depend on proximity to individuals of a shrub species 1 that are readily killed by fire (Biganzoli et al. 2009). Killing its neighbor could be an evolved strategy of (focal) species 1 to escape from competition with nearby shrubs of species 2.

We thus have points of a focal pattern (subscript f), and type l and type m points of a qualitatively marked pattern. The appropriate summary statistic for this data structure selects pairs of points separated by distance r, where the first point is from the focal pattern (i.e., type f) and the second point is from the second qualitatively marked pattern (i.e., type l or m), and calculates the probability that the second point is of type l:

$$p_{fl}(r) = \frac{\lambda_l}{(\lambda_l + \lambda_m)} \frac{g_{fl}(r)}{g_{f,l+m}(r)} = \frac{\lambda_f \lambda_l}{\lambda_f(\lambda_l + \lambda_m)} \frac{g_{fl}(r)}{g_{f,l+m}(r)} \tag{3.81}$$

Here, $(\lambda_l + \lambda_m)$ and λ_l are the intensities of the points of the qualitatively marked pattern and its type l points, respectively, and $g_{f,l+m}(r)$ and $g_{fl}(r)$ are the corresponding bivariate, pair-correlation functions. Note that the intensity λ_f of the focal pattern cancels out; however, if we leave λ_f in Equation 3.81, we see that the summary statistic is analogous to a mark connection function:

$$p_{fl}(r) = \frac{\rho_{fl}(r)}{\rho_{f,l+m}(r)} \tag{3.82}$$

which can be calculated as the ratio of two bivariate product densities.

An estimator of the summary statistic for trivariate random labeling is given by

$$\hat{p}_{fl}(r) = \frac{\hat{\rho}_{fl}(r)}{\hat{\rho}_{f,l+m}(r)} = \frac{\sum_{i=1}^{n_f} \sum_{j=1}^{n_{l+m}} C_l(x_j) \times k(\|x_i - x_j\| - r) \times w_{i,j}}{\sum_{i=1}^{n_f} \sum_{j=1}^{n_{l+m}} k(\|x_i - x_j\| - r) \times w_{i,j}} \tag{3.83}$$

where n_f is the number of points of the focal pattern and n_{l+m} the number of points of the qualitatively marked pattern. The indicator function $C_l(x_j)$ yields a value of 1 if the second point j is type l from the qualitatively marked pattern. Given these definitions, we can see that the numerator counts all type f–l point pairs at distance r and the denominator counts all f–l and f–m point pairs at distance r. Thus, this summary statistic yields the proportion of type l points at distance r from type f points. In our example above, it therefore produces the proportion of burned shrubs of species 2 at distance r from shrubs of species 1.

Figure 3.27 provides an example of trivariate random labeling. We used a clustered pattern with 500 individuals and generated 200 dead individuals (pattern 1) and 300 surviving individuals (pattern 2) in such a way that the probability of mortality was dependent on the number of points of a third (focal) pattern within 40 m (shown in Figure 3.27a as open squares). Analysis with the mark connection functions shows some indication of nonrandom mortality with small-scale clustering of dead individuals, and some repulsion between surviving and dead individuals

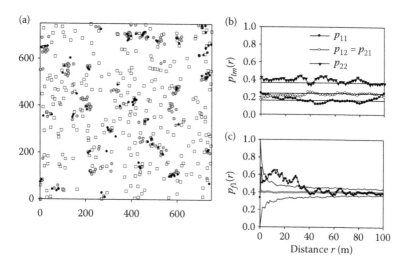

FIGURE 3.27

Trivariate random labeling. (a) Clustered pattern of 500 points with 200 dead (closed disks; pattern 1) and 300 surviving individuals (gray disks; pattern 2). Mortality of a given individual was dependent on the density of points of a third pattern (open squares; focal pattern) within 40 m. (b) Conventional analysis using mark connection functions of the dead (pattern 1) vs. surviving (pattern 2) individuals. The horizontal lines represent the expectations of the mark connection functions for random mortality. (c) Trivariate analysis with random labeling between surviving and dead individuals and using a summary statistic $p_{f1}(r)$ that yields the probability of mortality of an individual dependent on the distance r to the points of the focal pattern f.

(Figure 3.27b). However, the effects are difficult to interpret. Clearly, this is because mortality was not density dependent, but dependent on points of a focal pattern, which are not considered in this analysis (open squares in Figure 3.27a).

The dependence in the previous example can be revealed by employing trivariate random labeling (Figure 3.27c). We used the summary statistic $p_{f1}(r)$, which gives the probability of mortality $p_{f1}(r)$ for an individual, which is dependent on the distance r to points of the focal pattern. We then compare this summary statistic to simulation envelopes arising from random mortality (i.e., random permutation of labels 1 and 2). As expected, the probability of mortality, is substantially higher than expected at random, if a focal point is located at a distance less than 40 m from the type 1 or 2 individuals. The aggregation of dead individuals at small distances can be explained by collective mortality of a cluster, if one or two focal points were close by. Proximity or nonproximity of a focal point, therefore, produced dead or surviving clusters, respectively, and imprinted a signal of segregation between surviving and dead individuals on the over all pattern.

3.1.7 Summary Statistics for Quantitatively Marked Point Patterns

The marks carried by a univariate pattern can be quantitative, as well as qualitative. For example, the trees of a given species may be characterized by their diameter at breast height (dbh) or height. What is generally of interest in analyzing quantitative marks is to find out whether the marks show some sort of spatial correlation, conditional on the locations of the corresponding unmarked pattern. The summary statistics adapted to quantitative marks are "mark-correlation functions," a concept introduced by Stoyan (1984). Mark-correlation functions are only rarely used in the ecological literature and then mostly in their simplest form, where one quantitative mark is attached to a univariate pattern. However, a plethora of data types are possible for quantitative marks (see Section 2.2.3), since one or more quantitative marks may be attached to bivariate and qualitatively marked patterns (Figure 2.3, data types 6–9).

Here, we present mark-correlation functions for univariate, quantitatively marked patterns, as developed by Illian et al. (2008, their Section 5.3.3), and analogous functions for patterns with two marks. The latter includes cases where two quantitative marks are attached to a univariate pattern (data type 7 in Figure 2.3), cases in which the pattern carries one qualitative and one quantitative mark (data type 8 in Figure 2.3), and cases where a bivariate pattern carries one quantitative mark (data type 9 in Figure 2.3). An interesting new perspective in the application of mark-correlation functions is provided by analyses of spatial structures in traits or phylogenetic spatial structures in fully mapped communities. For example, a plant trait such as leaf area, wood density, or maximal height can be defined as a quantitative mark, and mark-correlation functions can then be used to reveal spatial structures in the trait distribution within communities. Similarly, the phylogenetic distance between two individuals in a community can be used as a test function and phylogenetic, mark-correlation functions can be used to calculate the mean phylogenetic distance between two individuals of the community that are separated by distance r (Section 3.1.7.6).

3.1.7.1 Univariate Quantitatively Marked Pattern (Data Type 6)

This is the simplest case of a quantitatively marked pattern, where one quantitative mark (e.g., size) is attached to a univariate pattern (e.g., trees of a given species). Analogous to the use of mark connection functions for qualitative marks, mark-correlation functions are used to analyze the spatial relationships among points containing quantitative marks. Point i at location x_i has mark m_i, where m_i is usually a positive, real number. The basic goal is to determine whether the joint properties of the marks of two points depend on the distance r separating them. Univariate mark-correlation functions are the summary statistics adapted to this data structure, and the empirical summary statistics are compared to those arising from *independent marking*, which represents the fundamental division related to the absence of any spatial structure in the marks.

The basic idea of mark-correlation functions is analogous to that of the interpretation of the mark connection function in Equation 3.79 as described above: they are the mean value of a test function of the marks of all point pairs i and j that are separated approximately by distance r (Illian et al. 2008, p. 341). Thus, the first step in developing mark-correlation functions is to identify appropriate test functions that characterize some relationship between the marks m_i and m_j of two points i and j. One example of a test function is simply to use the value of the mark of the second point $(= m_j)$. The associated mark-correlation function thus yields the mean mark of a point, given that this point has a neighbor at distance r. Note that this conditional mean may be quite different from the (unconditional) mean mark.

Intuitively, an estimator of the mark-correlation functions visits all pairs of points separated by distance r and calculates the mean value of the test function over these pairs. This is then repeated for different distances r. Thus, analogous to the estimator of the mark connection function (Equation 3.79), we find

$$\hat{c}_t(r) = \frac{\sum_{i=1}^{n} \sum_{j=1}^{n,\neq} t(m_i, m_j) \times k(\|x_i - x_j\| - r) \times w_{i,j}}{\sum_{i=1}^{n} \sum_{j=1}^{n,\neq} k(\|x_i - x_j\| - r) \times w_{i,j}} \tag{3.84}$$

where $t(m_i, m_j)$ is the test function that uses the mark m_i of point i and the mark m_j of point j, $k()$ is the kernel function that defines which points are located approximately at distance r and w_{ij} is the edge correction. This estimator is based on the analogous second-order product density for quantitatively marked patterns (Illian et al. 2008, p. 354 their Equation 5.3.5.3):

$$\hat{\rho}_t(r) = \frac{1}{2\pi r A} \sum_{i=1}^{n} \sum_{j=1}^{n,\neq} t(m_i, m_j) \times k(\|x_i - x_j\| - r) \times w_{i,j} \tag{3.85}$$

where the inner sum $\sum_{j=1}^{n} t(m_i, m_j) \times k(\|x_i - x_j\| - r) \times w_{i,j}$ sums up all points j located at distance r from the focal point i, and weights them with the test function $t(m_i, m_j)$. Thus, the second-order product density $\rho_t(r)$ for a pattern with quantitative marks contains information on both the spatial structure of the pattern and the correlation structure of the marks. As a parallel to the development of the mark connection functions (Equation 3.79), we factor out the conditional spatial structure of the points contained in the product density and define the *(nonnormalized) mark-correlation function* $c_t(r)$ based on test function t as

$$\hat{c}_t(r) = \frac{\hat{\rho}_t(r)}{\hat{\rho}(r)}, \tag{3.86}$$

where $\hat{\rho}_t(r)$ is the second-order product density for a pattern with quantitative marks and test function t and $\hat{\rho}(r)$ is the second-order product density for the corresponding univariate pattern. It can be easily verified that this definition yields the estimator given in Equation 3.84.

If there is no spatial correlation among the marks, the nonnormalized, mark-correlation function $c_t(r)$ yields the average c_t of the test function $t(m_i, m_j)$ over all pairs of points i and j of the pattern, that is, the nonspatial mean of the test function. It is, therefore, useful to normalize the $c_t(r)$ with its asymptotic value c_t to make *mark-correlation functions* $k_t(r)$ independent of the distribution and values of the marks:

$$\hat{k}_t(r) = \frac{1}{c_t} \frac{\hat{\rho}_t(r)}{\rho(r)}. \tag{3.87}$$

The subscript t in Equation 3.87 refers to the particular test function t used. Depending on the test function t, different mark-correlation functions $k_t(r)$ arise. Important examples of test functions used to construct (univariate) mark-correlation function, as pointed out in Illian et al. (2008, p. 343), are

$$k_{mm}(r): t_1(m_i, m_j) = m_i m_j$$

$$k_{m.}(r): t_2(m_i, m_j) = m_i$$

$$k_{.m}(r): t_3(m_i, m_j) = m_j$$

$$\gamma_m(r): t_4(m_i, m_j) = (m_i - m_j)^2/2 \tag{3.88}$$

$$I_m(r): t_5(m_i, m_j) = (m_i - \mu)(m_j - \mu)$$

$$I_m(r): t_6(r, m_i, m_j) = [m_i - \mu(r)][(m_j - \mu(r)]$$

where m_i and m_j are the marks of the two points i and j, μ is the mean mark over all points in the pattern, and $\mu(r)$ is the mean mark of points pairs separated by distance r. Note that test functions $t_1(m_i, m_j)$ to $t_5(m_i, m_j)$ are defined for any two points i and j, whereas, for reasons to be explained later, the test function $t_6(r, m_i, m_j)$ is only defined for points that are separated by distance r. The normalizing factors of the different test functions given in Equation 3.88 yield

$$c_1 = \mu^2$$

$$c_2 = \mu$$

$$c_3 = \mu \tag{3.89}$$

$$c_4 = \sigma_\mu^2$$

$$c_5 = \sigma_\mu^2$$

$$c_6 = \sigma_\mu^2$$

where μ is the mean mark over all points in the pattern and σ_μ^2 is the mark variance.

The simplest test functions, t_2 and t_3, result in mark-correlation functions that yield the mean mark of a point that has another point at distance r. These summary statistics are called "*r-mark-correlation functions*" $k_{m\cdot}(r)$ and $k_{\cdot m}(r)$, respectively. Note that the point in the subscripts indicate that the marks of the second point ($= m_j$) and the first point ($= m_i$) not used for estimation of the r-mark-correlation functions $k_{m\cdot}(r)$ and $k_{\cdot m}(r)$, respectively (see Equations 3.84 and 3.88). They describe a basic statistical property of the spatial correlation structure between the marks of two points separated by distance r, that is, their (conditional) mean value. The basic interests here are to find out how the conditional mean mark depends on distance r and if it departs significantly from the mean mark μ.

Figure 3.28b provides examples for the r-mark-correlation function $k_{\cdot m}(r)$. The constructed marks of the pattern shown in Figure 3.28a are the inverse of the number of neighbors within distance 10 m (including the focal point), thus isolated points have larger marks and points occurring in clusters have smaller marks. The marks in Figure 3.29a are the size of the trees of the species of *Shorea congestiflora*, a dominant species within a 25-ha plot in a rain forest at Sinharaja World Heritage Site (Sri Lanka). The r-mark-correlation function for both patterns shows a significantly negative departure from the null model at short distances r, indicating that nearby trees are smaller on average than trees selected at random. This was expected for the pattern shown in Figure 3.28a due to the way it was constructed. For the pattern of *S. congestiflora*, the significant effect was caused by small trees occurring together in clumps. However, Figure 3.29a shows that larger trees are often close to clusters of smaller trees. For this reason, the response of the r-mark-correlation function (Figure 3.29b) is weaker than for the constructed marks shown in Figure 3.28b.

Schlather et al. (2004) proposed the test function t_6 and its corresponding summary statistic to investigate how the marks of two points separated by distance r differ from their conditional mean value $\mu(r)$. This results in a *Moran's I type summary statistic* $I_m(r)$, which is a spatial variant of the classical Pearson correlation coefficient (Shimatani 2002). It characterizes the covariance, a second basic property of the spatial correlation structure between the marks of two points separated by distance r. The basic interest here is to determine whether the marks show a spatial correlation, for example, individuals may be smaller than average if they are located close to a neighbor

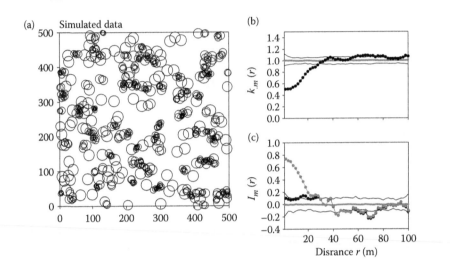

FIGURE 3.28
Mark-correlation functions together with the independent marking null model. (a) Simulated data composed of the independent superposition of 100 random points with a clustered pattern (with 500 points), where the mark attached to a point is the inverse of the number of neighbors within a distance of 10 m. Thus, isolated points are larger than points occurring in clusters. (b) The r-mark-correlation function of the pattern shown in panel (a). (c) The Schlather version of Moran's I mark-correlation function based on test function t_6 (black) and the common version based on test function t_5 (gray) for the pattern shown in panel (a).

and larger than average if they are located farther away from a neighbor. Note that the test function t_5, which is similar to t_6, adjusts for μ, the population mean, not the conditional mean $\mu(r)$, as required for a correlation coefficient. Figure 3.28c shows that the two variants of $I_m(r)$, based on t_5 and t_6, differ a great deal for the constructed marks. The reason for this is the strong effect of the r-mark-correlation function. Test function t_5 compares the marks m_1 and m_2 to the over all mean mark μ, whereas test function t_6 compares to the conditional-mean mark $\mu(r)$. Or, in other words, the mark-correlation function resulting from the Schlather test function t_6 looks only at pairs of points separated by distance r and compares the marks m_1 and m_2 for each point pair with the actual mean mark $\mu(r)$ resulting from all pairs of points separated by distance r. Clearly, test functions t_5 and t_6 will yield similar results, as shown in Figure 3.29e, if the conditional mean $\mu(r)$ agrees with the over all mean μ (this is tested by the r-mark-correlation function; Figure 3.29b). However, if $\mu(r)$ differs strongly at smaller distances r from the mean μ mark, as shown in Figure 3.28b, the resulting $I_m(r)$ summary statistics differ as well at these distances (Figure 3.28c). Thus, the Schlather test function t_6 must be selected if we want to obtain a mark-correlation function with an interpretation as a Moran's $I_m(r)$ summary statistic, whereas the test function t_5 may indicate spurious correlation caused by departures from the mean $\mu(r)$

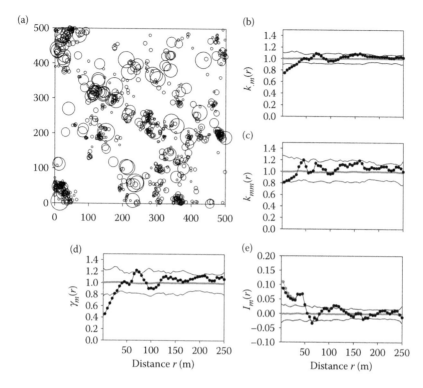

FIGURE 3.29

Application of mark-correlation functions and the independent marking null model to "real" data. (a) Map of the locations of individuals of the tree species *Shorea congestiflora*, a dominant species within a 25-ha plot of the rain forest at Sinharaja World Heritage Site, Sri Lanka. The marks represent diameter at breast height (dbh) and are drawn as proportional to $dbh^{0.5}$. (b) The r-mark-correlation function giving the mean size of trees located at distance r away from the typical tree, divided by the mean tree size. (c) The mark-correlation function giving the mean product of the sizes of trees, which are distance r apart, divided by the expectation without spatial structure in the mark. (d) The mark variogram. (e) The Schlather version of the Moran's I mark-correlation function based on test function t_6 (black) and the common version of test function t_5 (gray).

that are not due to correlations in the marks of points separated by distance r (i.e., Figure 3.28c).

Test function t_4 characterizes the difference between the marks of point pairs as a function of distance r and yields what is called the *mark variogram* $\gamma_m(r)$. Small values in the mark variogram indicate similarity in magnitude between points separated by distance r. Figure 3.29d shows that the sizes of two nearby *S. congestiflora* trees tend to be more similar than expected compared to two trees taken at random. Finally, test function t_1 yields the mean product of the two marks of points separated by distance r and the corresponding summary statistic is called the *mark-correlation function* $k_{mm}(r)$. It is interesting to note that the mark-correlation function $k_{mm}(r)$ does not show a significant effect for the species *S. congestiflora* (Figure 3.29c). Because the *r*-mark-correlation function shows a significant effect, this means that we may often have a large tree close to a small tree, which "neutralizes" the mark product. Thus, the mark-correlation function may miss out on important effects shown by the *r*-mark-correlation function and can only be properly interpreted in conjunction with the *r*-mark-correlation function.

The second-order product density for quantitatively marked patterns given in Equation 3.85 can also be interpreted in a way similar to the pair-correlation function. To this end the quantity

$$\hat{g}_t(r) = \frac{1}{\hat{\lambda}^2} \frac{\hat{\rho}_t(r)}{c_t} \tag{3.90}$$

is considered, where the subscript *t* refers to the test function *t* used. Note that normalization by the mean of the test function c_t is necessary to yield a value of 1, when a spatial pattern in the points and the marks is absent.

The test function $t_1(m_i, m_j) = m_i\, m_j$ is of particular interest, as it yields the *multiplicatively weighted, pair-correlation function* $g_{mm}(r)$:

$$\hat{g}_{mm}(r) = \frac{1}{\hat{\lambda}^2} \frac{1}{c_t} \frac{1}{2\pi r A} \sum_{i=1}^{n} \sum_{j=1}^{n,\neq} k(\|x_i - x_j\| - r) \times (m_i m_j) \times w_{i,j} \tag{3.91}$$

This summary statistic characterizes the spatial distribution of the "mark mass" rather than the point distribution (Illian et al. 2008, p. 348). The quantity $g_{mm}(r)$ basically acts the same as the pair-correlation function $g(r)$, but weights each point pair additionally by its mark (i.e., the factor $m_i m_j$ in Equation 3.91). If the marks show no spatial correlation, the multiplicatively weighted, pair-correlation function $g_{mm}(r)$ collapses to the pair-correlation function, and if the univariate pattern is a random pattern $g_{mm}(r)$ collapses to the mark-correlation function. The multiplicatively weighted, pair-correlation function $g_{mm}(r)$, therefore, contains signals from potential correlations in the marks and from potential correlations in the spatial distribution of the points. Note that $g_{mm}(r)$ is related to the pair-correlation function derived for objects with finite size and real shape (see Section 3.18).

An interesting example for the application of the multiplicatively weighted pair-correlation function $g_{mm}(r)$ was given in Law et al. (2009). They analyzed the distribution of biomass over space in a 1 ha plot of temperate forest at Rothwald, Austria, to assess the extent to which biomass is decoupled from the spatial pattern of tree locations. The pair-correlation functions revealed aggregation of the tree locations in the forest (Law et al. 2009; Figure 3). They investigated the relationship between the spatial pattern of biomass and trees using the multiplicatively weighted pair-correlation function. For this analysis, they used allometric relationships to convert tree height into biomass, which was then used as a quantitative mark. Comparison of the results of the pair-correlation function $g(r)$ of all trees, the multiplicatively weighted pair-correlation function $g_{mm}(r)$, and the mark-correlation function $k_{mm}(r)$ provided interesting insights into the spatial structure of these forests. While $g(r)$ indicated aggregation of trees up to 20 m, $g_{mm}(r)$ yielded values close to 1, indicating a near random spatial distribution of biomass. In contrast, $k_{mm}(r)$ showed that trees situated close to one another were characterized by below average biomasses (Law et al. 2009; Figure 4). Thus, the spatial pattern of biomass was substantially different from the aggregated pattern of tree locations. A more uniform distribution of biomass (compared to the clustered tree distribution) was reached by closely spaced trees having below average biomass. This compensatory effect was visible up to about 20 m.

3.1.7.2 Univariate Marked K-Functions (Data Type 6)

Cumulative mark-correlation functions can be derived for any test function t as a natural generalization of Ripley's K-functions (Illian et al. 2008, pp. 350–352). Recall that the K-function can be estimated as

$$\hat{K}(r) = \frac{1}{\hat{\lambda}^2}\frac{1}{A}\sum_{i=1}^{n}\sum_{j=1}^{n,\neq}\mathbf{1}(\|x_i - x_j\|, r) \times w_{i,j} \qquad (3.92)$$

where the count function $\mathbf{1}(d, r)$ has a value of 1, if point j is located within distance r of point i, and zero otherwise. As usual, w_{ij} is the edge correction term. *The cumulative mark product function* $K_t(r)$ (Shimatani and Kubota 2004), based on a test function t, can be estimated as

$$\hat{K}_t(r) = \frac{1}{\hat{\lambda}^2}\frac{1}{A}\frac{1}{\hat{c}_t}\sum_{i=1}^{n}\sum_{j=1}^{n,\neq}t(m_i, m_j) \times \mathbf{1}(\|x_i - x_j\|, r) \times w_{i,j}. \qquad (3.93)$$

This approach basically consists of counting all point pairs (i, j) separated by a distance less than r, weighted by the value of the test function $t(m_i, m_j)$ of the respective marks m_i and m_j. Thus, mark-weighted K-functions are the normalized means of the sum of values of the test function t formed by the mark of the typical point and all points in the disk of radius r centered at the typical point. Because of the accumulative nature of the K-functions we find that

$$K_t'(r) = 2\pi r k_t(r)g(r) \qquad (3.94)$$

where $K'_t(r)$ is the derivative of $K_t(r)$ with respect to r (Equation 5.3.40 in Illian et al. 2008). Thus, the mark weighted K-functions depend on both the spatial structure of the unmarked pattern [represented by the pair-correlation function $g(r)$ in Equation 3.94] and the correlation structure of the marks [represented by the mark-correlation function $k_t(r)$].

Finally, it might be desirable to remove the signal of the univariate pattern from the cumulative mark-product function $K_t(r)$ to obtain a *cumulative mark-correlation function*. This can be done by estimating the quantity

$$\hat{k}_t^{cum}(r) = \frac{1}{\hat{c}_t}\frac{\hat{\kappa}_t^{(2)}(r)}{\kappa^{(2)}(r)} = \frac{1}{\hat{c}_t}\frac{\sum_{i=1}^{n}\sum_{j=1}^{n,\neq} t(m_i, m_j) \times \mathbf{1}(\|x_i - x_j\|, r) \times w_{i,j}}{\sum_{i=1}^{n}\sum_{j=1}^{n,\neq} \mathbf{1}(\|x_i - x_j\|, r) \times w_{i,j}}. \qquad (3.95)$$

which is analogous to the approach taken in Equations 3.84 and 3.87. The quantity $\hat{c}_t\hat{k}_t^{cum}(r)$ has a simple interpretation: it is the mean value of the test function for points pairs i and j that are located within distance r of each other.

Figure 3.30 compares the cumulative mark-correlation functions with their noncumulative counterparts for the pattern shown in Figure 3.29a. Clearly, the cumulative nature of the $k_t^{cum}(r)$ glosses over the details of the spatial patterns shown in their noncumulative counterpart $k_t(r)$, and the scales of significant departures from random are often different. The cumulative functions $k_t^{cum}(r)$ are also more likely to show departures from the null model, while the noncumulative $k_t(r)$ shows, in general, scale-dependent effects much more clearly. This is because the cumulative functions are based on a higher number of point pairs, which reduces the effects of stochastic noise. The question as to whether or not to use the noncumulative or the cumulative functions ultimately depends on the biological question: is the contribution of all individuals within the neighborhood r important or is the goal to reveal specific scale effects among individuals separated by a certain distance r?

3.1.7.3 Two Quantitative Marks Attached to a Univariate Pattern (Date Type 7)

The first data structure that yields bivariate mark-correlation functions is a univariate pattern with two quantitative marks m_1 and m_2 (data type 7 in Figure 2.3). An example for this data type is a pattern of trees that hosts two species (or groups of species) of orchids, with the marks representing the number of orchids of each species (or group) on a given tree (Raventós et al. 2011). Stoyan (1987) used another example in which the points were the locations of pine saplings and the marks were their heights and age. The basic interest in analyzing these types of data is to find out whether the two marks show some spatial correlation that depends on the distance r between points. This can be tested by comparing the empirical summary statistics to simulation envelopes arising from a null model that represents the absence of any spatial structure in the two marks.

Depending on the ecological question, different null models need to be used. For example, if the marks are the number of orchids of two species, we may ask whether they are independently distributed over the host trees. In this case, we can condition on the number of orchids of the first species and

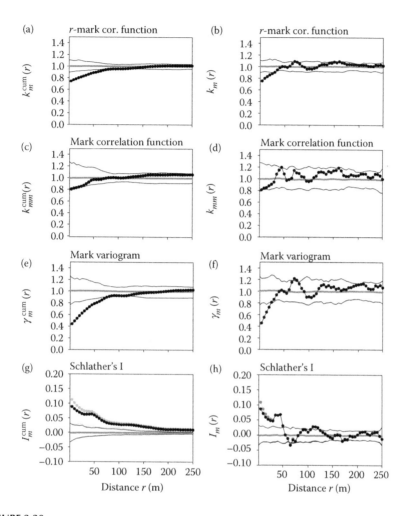

FIGURE 3.30
Cumulative mark-correlation functions vs. noncumulative mark-correlation functions based on the example pattern of the species *Shorea congestiflora* a shown in Figure 3.29a. Left column: cumulative mark-correlation functions. Right column: corresponding noncumulative mark-correlation functions.

shuffle the second mark (i.e., number of individuals of the second orchid species) randomly over the trees of the univariate pattern. However, if we analyze the spatial correlation in marks representing height and dbh of trees, we cannot separate height and dbh, since these are most likely correlated with one another. In this case, we shuffle the vector of marks of the individual trees i, given by (m_{i1}, m_{i2}), randomly over the trees of the univariate pattern. An application of this null model that studied spatial structures in epiphytic orchids caused by hurricane damage is given in Wiegand et al. (2013b). The number of affected orchids of a given species that were located at a host tree was mark m_1, and the corresponding number of non-affected orchids was mark m_2. They used the bivariate mark variogram to test if the number of affected orchids at the focal tree and the number of non-affected orchids at nearby trees tended to be relatively similar or dissimilar compared to those of pairs of trees taken at random.

Summary statistics for the univariate case can be easily modified to perform an analysis of data structures in which two quantitative marks are attached to a univariate pattern. In this case, the mark m_{i1} will refer to the mark of a focal point i and the mark m_{j2} to a second mark for point j, which is located at a distance r from the focal point i. The estimator of the nonnormalized mark-correlation functions therefore yields

$$\hat{c}_t(r) = \frac{\sum_{i=1}^{n} \sum_{j=1}^{n,\neq} t(m_{i1}, m_{j2}) \times k(\|x_i - x_j\| - r) \times w_{i,j}}{\sum_{i=1}^{n} \sum_{j=1}^{n,\neq} k(\|x_i - x_j\| - r) \times w_{i,j}} \tag{3.96}$$

The test function t_1 yields a bivariate mark-correlation function $k_{m1m2}(r)$, which returns the mean product of the marks m_{i1} and m_{j2}. The normalization factor for this test statistic is given by $c_1 = \mu_1\mu_2$, where μ_1 and μ_2 are the mean of the first and second mark, respectively. The test functions t_2 and t_3 are not of interest here because they yield the univariate, r-mark-correlation functions for the first and second mark m_1 and m_2, respectively. The test function t_4 yields a bivariate mark variogram $\gamma_{m1m2}(r)$ that returns the mean of the squared differences of the two marks separated by distance r. However, this test function only makes sense if the values of the two marks are normalized to the same mean, otherwise it will basically depict differences in the numerical values of the two marks. If the marks are normalized, the bivariate mark variogram $\gamma_{m1m2}(r)$ tests whether the two marks of nearby points tend to be relatively similar or dissimilar. The test function t_6 returns a bivariate, distance-dependent correlation coefficient of the two marks m_{i1} and m_{j2} separated by a distance r. Note that in this case Equation 3.88 must be modified to $t_5(m_{i1}, m_{j2}) = [m_{i1} - \mu_1(r)][m_{j2} - \mu_2(r)]$, where $\mu_1(r)$ and $\mu_2(r)$ are the mean of the first and second mark, respectively, given that points i and j are separated by distance r. This Moran's I type summary statistic $I_{m1m2}(r)$ determines, for example, if the dbh and the height of nearby trees are correlated. The normalization constant is given by $c_{t6} = \sigma_{12}$, where σ_{12} is the covariance of the two marks.

Figure 3.31 provides an example of an analysis of a univariate pattern that is augmented with two quantitative marks. It is the pattern of the species $S.$ *congestiflora*, as shown in Figure 3.29a, together with the two marks of size (m_1) and a "constructed mark" (m_2) representing the number of neighbors within 20 m of an individual. The first r-mark-correlation function $k_{m1\cdot}(r)$ is identical with the univariate mark-correlation function of the pattern for the mark size (cf. Figures 3.29b and 3.31b), and indicates that trees separated by distances up to 40 m are smaller than expected. The second r-mark-correlation

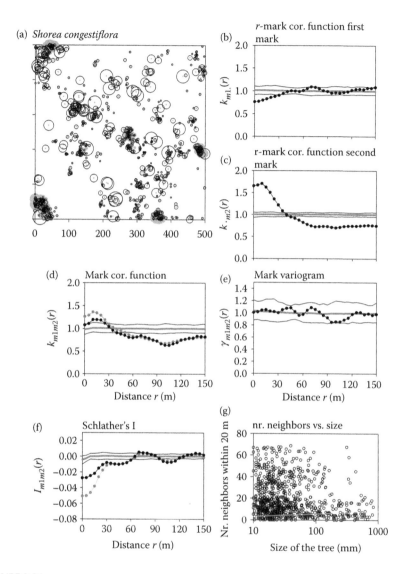

FIGURE 3.31

Bivariate mark-correlation function. (a) Pattern of the species *Shorea congestiflora* a shown in Figure 3.29 together with marks for dbh (m_1; open disks) and number of neighbors within 20 m (m_2; gray disks). Note that the second mark has small values in most cases and is barely visible. (b)–(f) Different bivariate mark-correlation functions analyzing spatial structures in the relationship between the size of a tree and the number of neighbors. (g) Relationship between number of neighbors within 20 m and the size of a tree. In panel (d), we include additional information concerning the product of the two *r*-mark-correlation functions (gray disks) and, in panel (f), we compare Schlather's version of Moran's I mark-correlation function (black) with the common version (gray). The null model in this analysis shuffled the first mark m_1 (i.e., size) randomly over all trees, whereas the constructed mark m_2 (i.e., the number of neighbors within 20 m) remained unchanged.

function $k_{.m2}(r)$ exhibits a strong spatial structure in the number of neighbors within 10 m: for distances up to 40 m we find that two trees separated by distance r have more neighbors than expected (this is due to the strong clustering of the pattern), whereas for distances larger than 40 m two trees separated by distance r have fewer neighbors than expected (Figure 3.31c). The latter happens because trees separated by more than 40 m are often "isolated" trees away from clusters. The bivariate mark-correlation function does not reveal much new information. It basically provides the product of the two r-mark-correlation functions (cf. closed black and closed gray disks in Figure 3.31d). To produce a meaningful variogram, we normalized the marks with their mean value, thus replacing the mark m_1 by m_1/μ_1 and the mark m_2 by m_2/μ_2. When doing this we do not find any significant relationship in the mark differences associated with the distance between trees (Figure 3.31e). However, the Moran's I type summary statistic $I_{m1m2}(r)$ shows a negative correlation between tree sizes and number of neighbors, which indicates that small trees tend to have many neighbors within 20 m.

3.1.7.4 One Qualitative and One Quantitative Mark (Data Type 8)

The second data structure that yields bivariate mark-correlation functions is a univariate pattern with one quantitative mark and one qualitative mark (data type 8 in Figure 2.3). An example for this data structure is a pattern of surviving and dead trees of a given species (qualitative mark), where the size of the trees is also known (the quantitative mark). The basic interest in analyzing these types of data is to find out whether the quantitative mark (e.g., size) shows a spatial correlation with the qualitatively marked points (i.e., surviving and dead). For example, we may expect that dead trees located near surviving trees would be smaller than expected given the over all sizes of surviving and dead trees. This question requires a null model where the qualitative marks (e.g., surviving and dead) are fixed, but the size of the quantitative mark is randomly shuffled over all points (e.g., surviving and dead trees). Alternatively, we may randomly reshuffle the qualitative marks (e.g., surviving vs. dead) over all locations and retain the quantitative mark (e.g., size) as fixed.

Essentially, the same test statistics are used for this data structure as with the univariate mark-correlation functions. The qualitative mark distinguishes between type l and type m points, where the focal type l point i carries the mark m_{il} (e.g., size of surviving tree i) and a type m point j carries mark m_{jm} (e.g., size of dead trees j). The estimator of the nonnormalized mark-correlation functions for this data structure therefore yields

$$\hat{c}_{lm,t}(r) = \frac{\sum_{i=1}^{n} \sum_{j=1}^{n,\neq} t(m_{il}, m_{jm}) \times C_{lm}(x_i, x_j) \times k(\|x_i - x_j\| - r) \times w_{i,j}}{\sum_{i=1}^{n} \sum_{j=1}^{n,\neq} C_{lm}(x_i, x_j) \times k(\|x_i - x_j\| - r) \times w_{i,j}} \tag{3.97}$$

The indicator function $C_{lm}(x_i, x_j)$ evaluates to a value of 1, if point i is a type l point and point j is a type m point. It is zero otherwise. This summary statistic is thus a mixture between a univariate, mark-correlation function (Equation 3.84) and a mark-connection function (Equation 3.79). However, note that the denominator includes the indicator function $C_{lm}(x_i, x_j)$, which selects only point pairs i, j of type l and m, respectively. Thus, the test functions are calculated for pairs of points that fulfill two conditions: the focal point must be of type l and the second point of type m and the points must be separated by distance r.

The test function $t(m_{il}, m_{jm})$ in the numerator of Equation 3.97 can be taken from the set of functions listed in Equation 3.88 and is applied as follows: The test function t_1 yields a bivariate mark-correlation function $k_{m1m2}(r)$, which returns the mean product of the marks of all pairs of type l and m points that are a distance r apart. Because the randomization of this data structure shuffles the marks over the entire unmarked pattern (or randomly shuffles the mark surviving and dead) the normalization constant yields $c_{t1} = \mu_{lm}^2$ where μ_{lm} is the mean of the marks taken over all points of the underlying univariate pattern (i.e., type l and type m points). The test function t_2 yields the mean mark of type l points (e.g., surviving), which are distance r away from a type m point (e.g., dead), whereas t_3 yields the mean mark of a type m point (e.g., dead) located distance r from a type l point (surviving). In both of the latter two cases, the normalization constant is equal to μ_{lm}. Note that there is a subtle difference in the univariate and bivariate r-mark correlation function $k_{m1}(r)$ when it is based on test function t_2. In the univariate case, the condition is that there is a surviving tree at distance r from a surviving tree, whereas in the bivariate case the condition is that there is a dead tree at distance r from a surviving tree.

The test function t_4 yields the mean squared difference of the marks of all pairs of type l and type m points, which are a distance r apart. In the tree example given above, incorporating the quantitative mark for size and the qualitative marks of dead or alive, the bivariate-mark variogram measures whether the sizes of neighboring dead and surviving trees are more similar (low values) or dissimilar (high values) than the sizes of a pair of trees taken at random. The test functions t_5 and t_6 yield a distance-dependent, cross-correlation coefficient of the mark of a pair of points of type l and type m. They must be modified such that

$$t_5(m_{il}, m_{jm}) = [m_{il} - \mu_l][m_{jm} - \mu_m] \qquad (3.98a)$$

$$t_6(m_{il}, m_{jm}) = [m_{il} - \mu_l(r)][m_{jm} - \mu_m(r)] \qquad (3.98b)$$

where μ_l and μ_m are the mean marks of points of type l and type m, respectively, $\mu_l(r)$ is the mean mark of a type l point (e.g., surviving) that is a distance r away from a type m point (e.g., dead), and $\mu_m(r)$ is the mean mark of a type m point (e.g., dead) that is distance r away from a type l point (e.g., surviving). Thus, $\mu_l(r)$ and $\mu_m(r)$ are the two nonnormalized r-mark-correlation functions associated with the test function t_2 and t_3, respectively that arise for this data type. In test function t_5 these conditional means are replaced by the over all means μ_l and μ_m. The normalization constant yields the variance σ_{lm}^2 of the joint pattern of type l and type m points.

Figure 3.32 provides an example of an analysis of a spatial pattern with one qualitative mark (surviving vs. dead) and one quantitative mark (size) using the species *S. congestiflora*. This example illustrates the exploration of distance-dependent spatial patterns between pairs of surviving and dead trees. The r-mark-correlation function $k_{m1\cdot}(r)$, which estimates the mean size of surviving trees (that have a dead tree at distance r), indicates that the size of surviving trees located at distance r from dead trees is significantly smaller than expected (Figure 3.32b; simulation envelopes are produced by reshuffling the mark size over surviving and dead trees). This, however, is mostly a consequence of the univariate structure of surviving trees, as shown

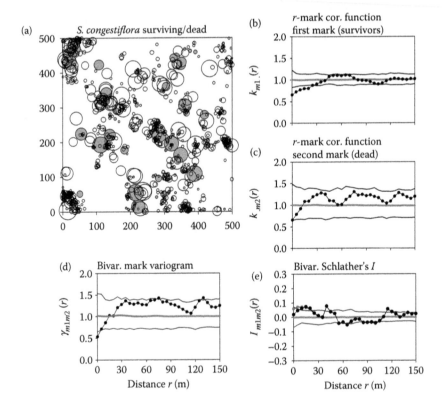

FIGURE 3.32
Mark-correlation function for one qualitative and one quantitative mark. (a) Pattern of surviving (open circles) and dead (closed circles) individuals of the species *Shorea congestiflora*, with a quantitative mark for size (m_1), defined as by dbh, being shown by the size of the circle. (b)–(e) Different bivariate mark-correlation functions analyzing spatial structures with respect to size for pairs of surviving and dead trees located at distance r.

in previous analyses (e.g., Figure 3.29b). However, note that the univariate *r*-mark correlation function of surviving trees (Figure 3.29b) is different from the bivariate *r*-mark correlation function (Figure 3.32b), because in the first case the condition is that the surviving focal tree has a surviving tree at distance *r*, whereas in the second case the condition is that the surviving focal tree has a dead tree at distance *r*. The *r*-mark-correlation function $k_{.m2}(r)$ does not show a significant impact of the proximity of surviving trees on the size of dead trees (Figure 3.32c). The mark variogram shows that nearby (<10 m) surviving and dead trees tend to have similar sizes and more distant pairs (>30 m) tend to have more dissimilar sizes (Figure 3.32d). The same result is depicted by the correlation coefficients that show a tendency to more similar sizes at shorter distances and a tendency to negative correlation in size at greater distances (Figure 3.32e).

3.1.7.5 Bivariate Pattern with One Quantitative Mark (Data Type 9)

Although the case of a bivariate pattern with one quantitative mark seems to be very similar to the case of one qualitative and one quantitative mark, a fundamental difference arises in the randomization of the data structure for the fundamental division. In the case of one qualitative and one quantitative mark, the null model randomizes the marks over the unmarked pattern (either the qualitative or the quantitative mark), whereas for a bivariate pattern with one quantitative mark, the marks must only be randomized within each pattern (i.e., a point of pattern 1 should not receive a mark of a point of pattern 2 and vice versa). Because the two marks in data type 8 are *a posteriori* marks (i.e., they characterize an existing univariate pattern), a point of the univariate pattern could theoretically receive each qualitative mark and one value from any of the quantitative marks. However, if we have a bivariate pattern that carries a quantitative mark, the two univariate component patterns are *a priori* different and therefore we can study potential correlation in their marks only by randomizing the quantitative mark within each component pattern.

We can characterize the bivariate pattern by defining two types of points: type *v* and type *w*. A point *i* of type *v* carries the mark m_{iv} (e.g., size of a tree of species *v*) and a point *j* of type *w* carries the mark m_{jw} (e.g., size of tree of species *w*). The estimator of the nonnormalized, mark-correlation functions of this data structure is, therefore, the same as for data type 8 (i.e., Equation 3.97), except that *l* now stands for type *v* and *m* stands for type *w*. However, the normalization constants of the different test functions differ from those of data structure 8, because the randomization of the data structure shuffles the mark only within data types.

For test function t_1, the normalization constant is given by $c_{t1} = \mu_v \mu_w$, where μ_v and μ_w are the mean marks for points of type *v* and *w*, respectively. The test function t_2 yields the mean mark of type *v* points, taken over all pairs that are a distance *r* apart, where the first point is a type *v* point and the second a type *w* point. In this case, the normalization constant is $c_{t2} = \mu_v$. Note that the analysis with t_2 in this situation is *similar to trivariate random labeling* (see Section 3.1.6.2), which allowed us to determine how the proximity of a point of the second pattern (i.e., type w) influenced the mark of the focal point (i.e., type v). Similarly, test function t_3 allows us to determine how the proximity of a focal point of type v influences the marks of nearby points of type w. Here, the normalization constant is $c_{t3} = \mu_w$. The test function t_4 only makes sense if the values of the two marks are normalized to the same mean, otherwise it basically assesses differences in the numerical values of the two marks. The test functions t_5 and t_6 yield a distance-dependent, cross-correlation coefficient of the marks of a pair of points of type v and type w and is the same as for data type 8 (Equation 3.98). However, in contrast to data type 8, the normalization constant is given by the covariance σ^2_{vw} of the marks of the two patterns.

To illustrate the analysis of this data structure we analyzed data from the large trees (dbh > 10 cm) of the BCI plot (Box 2.6). The bivariate pattern consists of the focal species *Trichilia pallida* (a canopy species) and the individuals of all the other species, which comprise the second pattern. In all cases, we consider individuals with a dbh greater than 10 cm. The quantitative mark in this case is the size of the trees. The bivariate mark-correlation functions, therefore, relate the size of the focal species to the size of individuals of all other species that are located at distance r away from the focal individual. In this example, we are interested in finding out whether there is a correlation between the sizes of the individuals of the focal species *T. pallida* and those of nearby heterospecific trees. We, therefore, randomize the mark size within the focal species, but keep the marks of all heterospecific trees unchanged. This null model assumes that the sizes of *T. pallida* trees are independent of the sizes of nearby heterospecific trees. Note that the univariate summary statistics use the condition that the focal *T. pallida* individual has another *T. pallida* individual at distance r, whereas the bivariate summary statistics use the condition that the focal *T. pallida* individual has a heterospecific tree at distance r.

First, we analyze the spatial correlation structure of the size of the large individuals of the focal species *T. pallida* (Figure 3.33a). The univariate r-mark-correlation function reveals that individuals of this species have a tendency to be smaller than expected, if another *T. pallida* is in the neighborhood (Figure 3.33b). The sizes of individuals located between 5 and 35 m apart are more similar (smaller) than expected by the null model (Figure 3.33c). They are also positively correlated (Figure 3.33d). Thus, individuals located in clumps tend to be smaller than more isolated individuals.

The bivariate r-mark-correlation function, corresponding to the test function t_2, explores whether the size of *T. pallida* individuals depends on the proximity of nearby heterospecific trees. Indeed, we find a significant small-scale effect. *T. pallida* individuals that have a heterospecific neighbor within 10 m are smaller than expected (Figure 3.33e). Interestingly, the r-mark-correlation function corresponding to the test function t_3 (which returns the mean size of heterospecific trees at distance r of individuals of *T. pallida*) reveals that nearby heterospecific trees tend to be larger than expected (Figure 3.33f). At

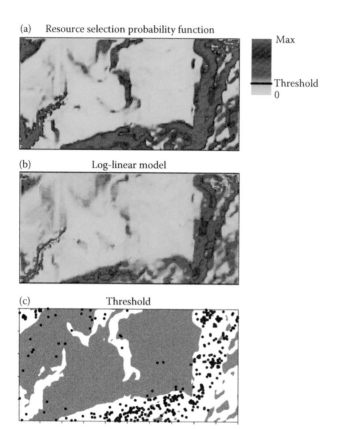

FIGURE 2.27

Habitat model for *Ocotea whitei* within the BCI plot, Panama, for the 2000 census. (a) Probability of occurrence determined by the logistic, resource-selection probability function (Lele 2009). (b) Thinning surface $p(x) = \lambda(x)/\lambda^*$ for the log-linear model; λ^* is the maximum value of $\lambda(x)$. (c) Categorical map showing areas of high suitability ($p(x) > 0.2$; white areas) versus low suitability ($p(x) \leq 0.2$; gray areas), with points of the pattern as black dots.

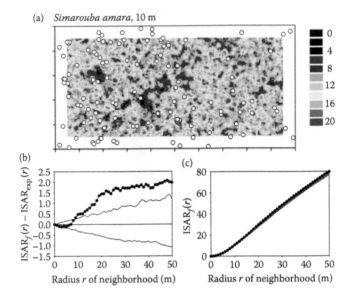

FIGURE 3.23
ISAR for large trees (dbh > 10 cm) at the BCI plot, with *Simarouba amara* used as an example.

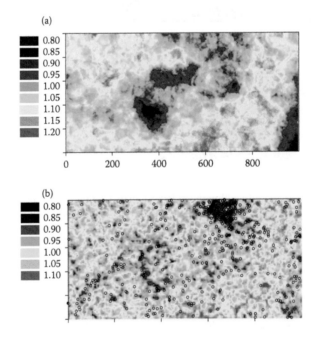

FIGURE 3.24
Landscape maps of phylogenetic neighborhood dissimilarity for large individuals (i.e., dbd > 10 cm) in the 1000 × 500 m² BCI forest dynamics plot, using the species *Gustavia superba* as an example.

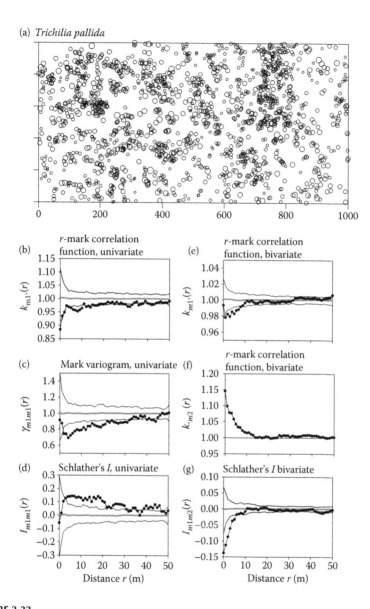

FIGURE 3.33
Mark-correlation function for a quantitatively marked, bivariate pattern derived from the pattern of all trees larger than 10 cm dbh in the BCI plot. The bivariate pattern is given by the individuals of the species *Trichilia pallida* (pattern 1) and the individuals of all other species (pattern 2). The quantitative mark is the dbh of the trees. (a) Map of *T. pallida*, with the area of the disks proportional to basal area. (b–d) Univariate mark-correlation analysis using the independent marking null model. (e–g) Bivariate analysis exploring spatial correlation between dbh of *T. pallida* relative to the dbh of all heterospecific individuals located distance r away. In the null model, dbh of the focal *T. pallida* trees was randomized among locations following independent marking, whereas the dbh of all other (heterospecific) individuals was held constant.

distances of up to 2 m they are approximately 10% larger. Thus, it looks as if individuals of *T. pallida* have a tendency to be located close to larger heterospecific trees. This hypothesis is confirmed by the results of Schlather's correlation function $I_{m1m2}(r)$ (i.e., test function t_6), which show that the sizes of the focal *T. pallida* trees and those of heterospecific individuals are strongly and negatively correlated up to distances of 7 m (Figure 3.33g).

3.1.7.6 Phylogenetic Mark-Correlation Functions

Traditional measures of alpha or beta diversity treat all species as equivalent. For example, the spatially explicit Simpson index (Section 3.1.5.1, Equation 3.70) can be expressed in simplified form as

$$\hat{\beta}(r) = \frac{\sum_{i=1}^{n} \sum_{j=1}^{n,\neq} \mathbf{I}(m_i \neq m_j) k(\|x_i - x_j\| - r)}{\sum_{i=1}^{n} \sum_{j=1}^{n,\neq} k(\|x_i - x_j\| - r)} = \sum_{l,m}^{\neq} p_{lm}(r) \qquad (3.99)$$

where the "test function" $\mathbf{I}(m_i \neq m_j)$ yields 1, if individuals i and j are heterospecific, and 0 if they are the same species. Thus, in this formulation the spatially explicit Simpson index can be viewed as a mark-correlation function that yields the proportion of heterospecific pairs of individuals i–j at distance r. However, it is also the sum of all heterospecific mark connection functions $p_{lm}(r)$.

In Section 3.1.5.3 we generalized the ISAR to a multivariate summary statistic that considers a continuous measure of phylogenetic or functional dissimilarity between species. We can do the same with the Simpson index in a straightforward way, by replacing the function $\mathbf{I}(m_i \neq m_j)$, in Equation 3.99, by a distance matrix $d(a, b)$ that represents the phylogenetic or functional distance between species a and b. The phylogenetic Simpson index $\beta_{phy}(r)$ is then defined as the conditional mean of the test function $d(sp_i, sp_j)$ taken over all pairs of points i and j that are distance r apart:

$$\beta_{phy}(r) = \frac{\sum_{i,j} d(sp_i, sp_j) k(\|x_i - x_j\| - r)}{\sum_{i,j} k(\|x_i - x_j\| - r)} = \sum_{l,m} d(l,m) p_{lm}(r) \qquad (3.100)$$

The kernel function $k(\|x_i - x_j\| - r)$ yields 1, if the distance between points i and j is within the range $(r - dr/2, r + dr/2)$ for a bandwidth of $dr/2$, and 0 otherwise. The phylogenetic distance between species i and j is given by $d(sp_i, sp_j)$, where sp_i and sp_j are the species identifiers of species i and j, respectively.

The nonspatial expectation of the phylogenetic Simpson index β_{phy}^* yields the mean pairwise phylogenetic distance over all pairs of individuals present in the observation window. Comparison with Equation 3.99 shows that the only difference, when compared to the spatially explicit Simpson index, is that the indicator function $\mathbf{I}(m_i \neq m_j)$ is replaced by the distance matrix $d(sp_i, sp_j)$.

The phylogenetic Simpson index $\beta_{phy}(r)$, given in Equation 3.100, contains a signal of both the multivariate spatial structure of the community (captured by the spatially explicit Simpson index $\beta(r)$ shown in Equation 3.99) and a signal of the phylogenetic spatial structure. In fact, if there is no phylogenetic spatial structure, or if all heterospecific species have the same phylogenetic distance (i.e., a star type phylogeny), the phylogenetic Simpson index $\beta_{phy}(r)$ will yield the nonnormalized Simpson index. Thus, the difference between $\beta(r)/\beta^*$ and $\beta_{phy}(r)/\beta^*_{phy}(r)$ is due to the phylogenetic spatial structure embedded in the pattern. To illustrate this, Figure 3.34a compares the normalized spatially explicit Simpson index $\beta(r)/\beta^*$ with the normalized phylogenetic Simpson index $\beta_{phy}(r)/\beta^*_{phy}$ for the spatial pattern of all recruits from the BCI plot (see Figure 3.21a; Box 2.6). We see that the phylogenetic spatial structure is largely driven by the underlying multivariate pattern captured by the Simpson index. However, at distances up to about 20 m, the normalized phylogenetic Simpson index (black dots in Figure 3.34a) is less than the normalized spatially explicit Simpson index (gray line in Figure 3.34a). This indicates that there is phylogenetic spatial structure among recruits, with neighboring recruits tending to be phylogenetically more similar. This is shown more clearly in Figure 3.34b where we divided the normalized

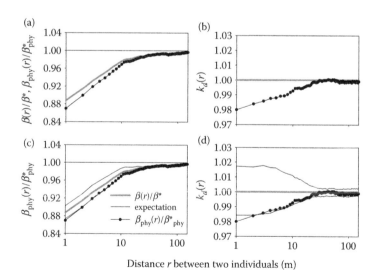

FIGURE 3.34
Phylogenetic mark-correlation function. (a) Comparison of the normalized spatially explicit Simpson index $\beta(r)$ and the phylogenetic Simpson index $\beta_{phy}(r)$ for the spatial pattern of all recruits from the BCI plot (see Figure 3.21). (b) Phylogenetic mark correlation function $k_d(r)$ (c) Same as panel (a), but also including the simulation envelopes resulting from the species shuffling null model. Note that the spatially explicit Simpson index is now the expectation under the null model (gray line) because it corresponds to the case of no spatial phylogenetic structure. (d) Same as panel (b), but now including simulation envelopes. Phylogenies of the tree community in the BCI plot were constructed as in Figure 3.24.

phylogenetic Simpson index $\beta_{phy}(r)/\beta^*_{phy}$ by its expectation under no spatial phylogenetic structure, that is, the normalized spatially explicit Simpson index $\beta(r)/\beta^*$.

To remove the confounding effect of the multivariate spatial structure of the community [captured by the spatially explicit Simpson index $\beta(r)$], we follow the approach already taken in the generalization of the phylogenetic species area relationship PISAR (Section 3.1.5.3; Equation 3.76). This is done by normalizing the phylogenetic Simpson index $\beta_{phy}(r)$ with the spatially explicit Simpson index $\beta(r)$ to yield the phylogenetic mark-correlation function $k_d(r)$:

$$k_d(r) = \left(\frac{\beta_{phy}(r)}{\beta^*_{phy}}\right) \bigg/ \left(\frac{\beta_S(r)}{\beta^*_S}\right) = \frac{1}{c_d} \frac{\beta_{phy}(r)}{\beta_S(r)} \tag{3.101}$$

with $c_d = \beta^*_{phy}/\beta^*_S$ Shen et al. (in press). It can be easily shown that the phylogenetic mark-correlation function $k_d(r)$ can be estimated as

$$\hat{k}_d(r) = \frac{1}{\hat{c}_d} \frac{\displaystyle\sum_{i,j}^{\neq} d(sp_i, sp_j)\mathbf{1}(sp_i \neq sp_j)k(\|x_i - x_j\| - r)}{\displaystyle\sum_{i,j} \mathbf{1}(sp_i \neq sp_j)k(\|x_i - x_j\| - r)} \tag{3.102}$$

Thus, the phylogenetic mark-correlation function $k_d(r)$ uses basically the same estimator as the phylogenetic Simpson index $\beta_{phy}(r)$ (Equation 3.100), the difference being that conspecific pairs are excluded. This is accomplished by the indicator function $\mathbf{1}(sp_i \neq sp_j)$, which selects only heterospecific pairs and results in a value of 1, if the individuals i and j belong to different species, and zero otherwise.

The advantage of the phylogenetic mark-correlation function over the phylogenetic Simpson index $\beta_{phy}(r)$ is that it has the straightforward expectation of $k_d(r) = 1$, when there is no spatial correlation in the phylogenetic structure. However, if the neighbors of individuals tend to be more phylogenetically similar to the focal individual than expected, we have $k_d(r) < 1$ (i.e., phylogenetic clustering). In contrast, we expect $k_d(r) > 1$ (i.e., phylogenetic overdispersion), if the neighbors tend to be more phylogenetically distant species than would be the case, if there were no spatial correlation in the phylogenetic structure.

Figure 3.34c shows the results of an analysis to determine whether the recruits of tree species in the 2005 census of the BCI plot exhibited spatial structure in their phylogenetic relationships. To this end, we contrasted the observed phylogenetic Simpson index to the relationships arising from 199 simulations, where the vector of species names was randomly reshuffled. This null model constrains the entire multivariate pattern and only randomizes the phylogenetic distance matrix $d(sp_i, sp_j)$, representing the case of no phylogenetic spatial structure Shen et al. (in press). The expectation under the null model differs from a value of 1 because the recruits show a strong tendency to conspecific clustering (Figure 3.34c). The analysis shows that there is a tendency for phylogenetic clustering at small distances (<30 m), but the GoF test shows that this tendency was not significant over the range of distances less than 30 m ($P = 0.06$). Figure 3.34c also shows that the spatially explicit Simpson index (bold gray line) is the same as the expectation of the phylogenetic Simpson index under the null model of no phylogenetic spatial structure. Figure 3.34d shows the results of the same analysis, but now we used the phylogenetic mark-correlation function Equation 3.101, which excludes conspecifics. The expectation of the phylogenetic mark-correlation function under no spatial phylogenetic structure is one and, therefore, departures from the expectation are easier to visualize.

To put the phylogenetic mark-correlation function into perspective, we outline the hierarchical structure of plant communities. For example, several plots of tropical forest, arranged for example along environmental gradients over larger distances, may differ in their over all phylogenetic community structure. The latter can be determined by collecting data on the presence/absence (or abundances) of species in each of the plots and using the phylogeny of the regional species pool. Analyses such as these, presented in Kraft et al. (2007) using the program PHYLOCOM (Webb et al. 2008), then allow for quantification of the over all phylogenetic community structure, taking into account the phylogeny of the regional pool. For example, if a plot contains more phylogenetically related species than expected, given the regional pool, the local community is characterized by phylogenetic clustering.

In principle, the individuals present in a given plot could be arranged in quite different ways in space. For example, individuals of phylogenetically similar species may be located close to each other or individuals of phylogenetically dissimilar species could be located close to each other, while the relative abundances of the species in the plot (and therefore the over all phylogenetic community structure) could be exactly the same. Thus, there is an *additional level of complexity hidden in the small-scale placement of individuals with respect to their phylogenetic similarity* that may be manifested for pairs of individuals as correlation between their phylogenetic distance and spatial separation distance *r*. Interestingly, this correlation structure in the small-scale placement of individuals at the plant neighborhood scale is completely independent of the over all phylogenetic community structure of the local community at the plot scale. If the plot is fully mapped, we can use the phylogenetic mark-correlation function to measure how individuals of a local (fully mapped) community are arranged in space with respect to their phylogenetic similarity. This provides a means for hierarchical analyses of phylogenetic structure by decoupling analysis of the over all phylogenetic community structure of a plot (e.g., Kraft et al. 2007) and the smaller-scale spatial phylogenetic structures based on the phylogenetic mark-correlation function.

3.1.7.7 Summary for Mark-Correlation Functions

This chapter has presented the test functions and estimators for mark-correlation functions that correspond to different data types. Mark-correlation functions are the summary statistics for patterns that carry a quantitative mark, such as the size of a tree, or the phylogenetic (or trait) distance between objects, such as two species. Basically, a mark-correlation function looks at pairs of points that are separated by distance r and calculates the mean value of a test function involving the marks. For example, the r-mark-correlation function can estimate the conditional mean size of objects that are separated from another object by distance r. Depending on the structure of the data, the pairs of points are subject to different conditions and only certain combinations of points are selected from the entire marked pattern. While the literature deals mostly with the simplest case of univariate, quantitatively marked patterns (e.g., the pattern of trees of one species and their size), the framework of mark-correlation functions is very flexible. Mark-correlation functions can be used with a variety of test functions, marks, and data structures, which opens up interesting applications in ecology. The framework is especially promising because it also allows for the straightforward spatial analysis of phylogenetic structures and species traits.

The *univariate mark-correlation functions $k_t(r)$* allow us to determine whether there is spatial correlation among marks that depends on the distance between points. It is evident that this permits a deep investigation of the spatial-correlation structure of marks, such as the size of an individual or the phylogenetic distance between two species separated by distance r. In this regard, Law et al. (2009) have demonstrated that the joint use of the pair-correlation function $g(r)$ for univariate patterns, the mark-correlation function $k_{mm}(r)$, and the related, *multiplicatively weighted pair-correlation function $g_{mm}(r)$* can be used to characterize the relationship between the spatial pattern of marks (tree biomass in their case) and the spatial pattern of the individual objects containing the marks. In fact, the multiplicatively weighted pair-correlation function $g_{mm}(r)$ incorporates the influence of both the spatial pattern of the points and the product of the marks of the two points separated by distance r. If the univariate pattern is random, it yields the mark-correlation function, and if the marks are spatially uncorrelated, it yields the pair-correlation function. Analyzing all three summary statistics together allowed Law et al. (2009) to explore whether the spatial distribution of biomass (the mark) was different from the pattern of locations. Indeed, while the locations were clustered, the marks showed inhibition, indicating, in their case, compensatory effects of nearby individuals having lower than average biomass.

The first data structure we presented for bivariate mark-correlation functions involved *a univariate pattern with each point carrying two quantitative marks.* This data structure allows an exploration of the spatial correlation between two properties of the points, such as size and height. It is an especially powerful approach, if a *constructed quantitative mark* is used in the analysis. A constructed mark is a property of the focal point that is constructed based on neighboring points or marks. For example, in Figure 3.31 we used a mark that was constructed from the number of neighbors within 20 m of a focal individual to explore the spatial relationships between the size of the focal tree and the number of neighbors. Indeed, size and number of neighbors were negatively correlated, indicating a density-dependent effect. Other constructed marks could be the total basal area of heterospecific trees located within a given distance (e.g., distance 20 m) from a focal individual or distance to the nearest neighbor. In this case, the *cross-correlation function* $I_{m1m2}(r)$ is a powerful tool to reveal correlations between a quantitative mark and a constructed quantitative mark. However, the Schlather test function t_6 must be used if we want to yield a summary statistic that can be interpreted as a cross-correlation coefficient.

The second data structure presented with two marks carried *one qualitative and one quantitative mark*. Prominent examples of this type of data structure are surviving versus dead trees (the qualitative mark) and a quantitative mark that reflects some aspect of fitness, such as size (Figure 3.32). However, this data structure also has the potential for producing interesting applications when the quantitative mark is a constructed mark. For example, we could explore if dead individuals of a focal species are likely to be associated with competitive interactions by constructing marks from the total basal area (or the total number) of conspecific trees within a certain distance r around the focal individual.

The final data structure we examined involved a *bivariate pattern and one quantitative mark*. This data structure can be used to explore if, for example, the size of a focal species exhibits a correlation with the sizes of neighboring heterospecific individuals. Indeed, in the example we used, we found that the individuals of the species *Trichilia pallida* had sizes that were negatively correlated with the sizes of nearby trees and that *T. pallida* was mostly located close to larger, heterospecific trees.

In summary, mark-correlation functions provide a very flexible tool for analyzing the spatial-correlation structure of quantitative marks and we expect that they will open up new avenues for interesting analyses in ecology. This is especially true when combined with constructed marks or phylogenetic data. Over all, the methods presented in this chapter enable tests of very specific hypotheses regarding the potential drivers of spatial structure in the distribution of marks, when used creatively.

3.1.8 Objects of Finite Size (Data Type 10)

One of the general assumptions of point-pattern analysis is that the objects studied (e.g., plants) are dimensionless points. The point approximation is valid when the size of the object is small in comparison with the spatial scales being investigated. However, when the size of the object is roughly the same order of magnitude as the scale of interest, the point approximation may obscure the real spatial relationships (e.g., Simberloff 1979; Prentice and Werger 1985). For ecologists this can be a real issue as the relationships they are often interested in, for example, the interactions among plants, occur at distances not much greater than the areal extent of the individual objects being investigated (Purves and Law 2002). In these circumstances, the conventional form of point-pattern analysis may suffer from a bias introduced by reducing the scale of the objects to being dimensionless points. For example, when we are interested in exploring the relationships among shrubs to determine whether facilitation or competition is a more important process (e.g., Wiegand et al. 2006), we may be interested in determining if their canopies touch more or less than expected to randomly placed, nonoverlapping shrubs. In such cases, we may need to analyze the spatial pattern of *objects of finite size and real shape*. Although this case falls a bit outside the scope of traditional point-pattern analysis, it is covered by the more general field of stochastic geometry (e.g., Stoyan, Kendall, and Mecke 1995). In fact, some of the tools of point-pattern analysis can be appropriately used to interpret relationships within this additional data category.

Wiegand et al. (2006) introduced an approach for dealing with objects of finite size and irregular shape by extending a grid-based estimator of second-order summary statistics that was introduced in Wiegand and Moloney (2004). The basic idea was to represent objects in a study area by means of a categorical raster map with a cell size smaller than the size of the object (Figure 3.35). One or several adjacent grid cells represent an object, depending on size and shape. These are then included in a map representing several categories such as bare ground, cover of species 1, cover of species 2, and so on (Figure 3.35a). The categorical map is then used to formally construct a point pattern in which the center of the cell is used as a point location and the categories define the types of points (Figure 3.35b). This allows conventional methods of point-pattern analysis to be used. However, consideration of the finite size and irregular shape of objects requires specification of more detail than standard point-pattern analysis. Specifically, rules and options are needed to construct objects from a categorical map, to decide on the rules of overlap among objects, and to randomize the position of the objects.

The aim of an analysis of objects with finite size and irregular shape is to *test simple baseline null hypotheses on the system*. Although this approach corresponds to conventional point-pattern analysis, the explicit consideration of real-world structures (i.e., finite size and irregular shape) prevents an analytical treatment. Analysis of objects with finite size and irregular shape is, therefore, a simulation-based approach for testing specific hypothesis about the spatial dependencies of objects of a system under consideration.

3.1.8.1 Grid-Based Estimators of Second-Order Statistics

The estimation of the univariate K-function in standard point-pattern analysis is based on counting the average number of points located *within distance*

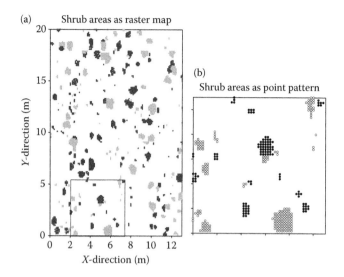

FIGURE 3.35

Typical examples of a bivariate "point" pattern with objects of real size and irregular shape. (a) A 13 × 20 m² plot of semiarid shrubland at Patagonia, Argentina, analyzed in Wiegand et al. (2006), showing the area occupied by three shrub species. The light gray objects are the areas occupied by the dominant shrub species *Mulinum spinosum* and the black objects are the two other dominant species *Adesmia volckmanni* and *Senecio filaginoides*. (b) Approximation of the original raster map as a point pattern for the area within the smaller rectangle of panel (a).

r of the points of the pattern. Similarly, the estimation of the O-ring statistics is based on counting the number of points *at distance r from* the points of the pattern. Thus, the focal point is not counted. However, if we extend the analysis to objects of finite size, which may occupy several cells within a grid, a decision has to be made as to how we should generalize this rule. If we are primarily interested in the spatial structure of the map, the O-ring estimator $O_{lm}(r)$ has the interpretation of being the probability of finding a cell of category m at distance r away from a cell of category l (Wiegand et al. 1999). In the univariate case, $O_{ll}(r)$ has the interpretation of being the probability of finding a cell of the same category at distance r away from a cell of category l. In this case, the estimators of the conventional point-pattern analysis are applied to the point pattern that is formally constructed from the categorical map. The approach presented in Wiegand et al. (2006) does not count the focal cell, but allows the other cells of the focal object to be counted in the analysis. A potential disadvantage of this approach is that it conserves a strong signal from the shape of the objects, since point pairs belonging to the same focal object are considered (Nuske et al. 2009).

If we are primarily interested in the univariate spatial structure of the objects, that is, the probability that a cell of a given category is located at distance r from another cell of the same category, but of a different object, additional modifications are required. Two approaches can be employed.

One approach, which uses only the nearest distance between two objects, was proposed by Nuske et al. (2009). In the second approach, distances between all cells within a focal object are excluded from the analysis. The latter approach is analogous to the standard univariate point-pattern analysis, where the focal point is not counted when estimating the K-function or the pair-correlation function.

When the second approach to analyzing objects of finite size described above is applied, the grid-based estimator for second-order statistics is analogous to the WM estimator presented in Equations 3.13, 3.17, and 3.36. The WM O-ring estimator divides the mean number $\bar{n}^R(r)$ of points located within rings of radius r and width dr, centered on the points of the pattern, by the mean area $\bar{v}(r)$ of the rings; that is, $\hat{O}_{WM}(r) = \bar{n}^R(r)/\bar{v}(r)$. For the grid-based O-ring estimator, rings $R_i^{dr}(r)$ of radius r and width dr are approximated using the underlying grid structure in the calculations. The numerator $\sum_{i=1}^{n_l} \text{Points}_m[R_i^{dr}(r)]$ of the estimator is calculated from the number of type m points, lying within the study region, that occur inside rings centered on the focal points i of the pattern (gray cells in Figure 3.6, with a ring width of one cell). The denominator of the estimator is obtained by calculating the area of the rings $\sum_{i=1}^{n_l} \text{Area}[R_i^{dr}(r)]$ that fall within the observation window, yielding

$$\hat{O}_{lm}^{dr}(r) = \frac{\sum_{i=1}^{n_l} \text{Points}_m[R_i^{dr}(r)]}{\sum_{i=1}^{n_l} \text{Area}[R_i^{dr}(r)]} \tag{3.103}$$

Estimator 3.103 basically counts the number of cells of type m in rings with radius r around cells of type l and divides by the corresponding area of the rings. The estimator of the K-function can be developed in an analogous fashion, yielding

$$\widehat{\lambda_l K_{lm}}(r) = \pi r^2 \frac{\sum_{i=1}^{n_1} \text{Points}_m[C_i(r)]}{\sum_{i=1}^{n_1} \text{Area}[C_i(r)]} \tag{3.104}$$

where the rings $R_i^{dr}(r)$ are replaced by circles $C_i(r)$ with radius r centered on the cells i of the focal pattern l. Here, λ_l represents the intensity of the cells of the focal pattern l (= number of cells of type l divided by number of cells within the observation window). As mentioned above, the estimators given in Equations 3.103 and 3.104 can be evaluated in the univariate case (i.e., $l = m$) in two different ways. First, we may also count pairs of cells that belong to the same object, and second, we may exclude cell pairs belonging to the same object.

For interpretation of the univariate summary statistics, it is worthwhile to note that, if the distance r among objects is relatively large compared to the size of the objects, the resulting pair-correlation and K-functions can be approximated by multiplicatively weighted pair-correlation and K-functions (see Section 3.1.7.1, Equation 3.91). This results from the fact that, if the focal object contains s_i cells and the second object contains s_j cells, the estimator of the pair-correlation function counts the distance between two objects $s_i \times s_j$ times and all $s_i \times s_j$ distances between cells of the two objects are approximately the same. The resulting calculation is essentially the same as the value produced by the multiplicatively weighted pair correlation where the marks are the size of the objects.

3.1.8.2 Additional Features Required for Analysis of Objects of Finite Size

An important difference between the analysis of points versus objects is that objects may or may not overlap whereas points, as one-dimensional objects, cannot overlap if they are approximated by their coordinates (unless they occupy exactly the same location). In particular, randomization of the position of objects of finite size and real shape requires rules that determine whether or not the objects are allowed to overlap. For example, in the grid-based framework as described above, a category (and not a number of points) is assigned to each cell, and overlap of two objects of the same pattern is not allowed. This has several implications. First, the O-ring statistic, the most important summary statistic for studying objects of finite size and real shape, has a slightly different interpretation than in conventional point-pattern analysis. Additional rules are also required for estimation of the O-ring statistic and the K-function. Second, the cell size (i.e., the grain of the map) needs to be adapted to the biological question at hand. For example, if the question requires that small objects be explicitly considered, the cell size must be made small enough to allow individual objects to be represented with at least one cell. Finally, depending on the null hypothesis, it may or may not be allowed for objects of two different types to overlap. This rule will affect data collection and mapping, as well as the application of null models.

3.1.8.3 Algorithm to Define Objects for Randomization

Null models for objects of finite size and irregular shape require randomization of locations of the objects (Wiegand et al. 2006). Therefore, an algorithm must be employed that determines which neighboring cells of a categorical map belong to the same object. One approach uses a "fire spread" (or "flooding") algorithm for this purpose. This algorithm repeats three operations. It starts with selection of random cells. If it encounters a cell i that is of a given category c, the first operation is to place this cell into a queue. The next operation is to select the first element of the queue (in the first step of the algorithm this is cell i), assign it to the object and delete it from the queue. However, the third operation visits the eight (or four) adjacent cells in turn to determine whether they also fall into category c. The neighbors of cell i, which are in category c, are marked and put into the queue. Now the second step of the algorithm starts. Again, the first element of the queue is selected, assigned to the object, deleted from the queue, and its neighbors are examined. If the neighbors are in category c and do not yet belong to the object, they are marked and put into the queue. The third step of the algorithm again selects the first element in the queue and so on. This algorithm is repeated until no cells are left in the queue. Thus, the algorithm starts with a random cell i, which is of category c, and determines all adjacent cells of category c that are connected with cell i via direct neighbors of category c. The algorithm stops if no cell of the object has a direct neighbor of category c, which is not yet assigned to the object. As a result, all adjacent cells of category c are marked and form one object. This algorithm is repeated until all cells of category c are assigned to an object.

One potential problem with this approach is that two objects that touch will be assigned to a single object. This can be avoided by using additional information on the identity of objects that must be collected during the field survey. If it is known which adjacent cells belong to a single object we can represent category c (e.g., one species of shrubs) by several subcategories, say c_1, c_2, and c_3, and assign these "dummy" categories in such a way that two objects, which touch, will still belong to different subcategories. In this way, the software can separate objects of the same category when they touch. This requires slightly more effort in developing the grid-based maps for analysis.

3.1.8.4 Randomization of the Position of Objects

Randomization of points involves assigning random coordinates, whereas randomization of objects of finite size needs to preserve the shape of the object, which may occupy several cells. This can be achieved by rotating and mirroring the object by 0°, 90°, 180°, or 270° (each of the eight variants being equally probable) and then randomly shifting its location in space, as a whole, but not allowing overlap with objects of the same type (Figure 3.36a,b).

Repeated trials are performed for each individual object: if, after being randomly mirrored, rotated, and shifted, an object overlaps with an already distributed object of the same type (or another type if appropriate), or falls partly outside of an irregularly shaped study area, or does not conform with the selected method of edge correction (see below), the trial is rejected. The procedure is repeated until a location is found for all objects.

3.1.8.5 Masking (Space Limitation)

In the analysis of ecological objects, competition for space is often an important ingredient to consider in developing null models for individuals of finite size. We may encounter situations where individuals of a focal species cannot inhabit some areas, for example, those already occupied by other species. In this case, we may ask if the species of interest is randomly distributed,

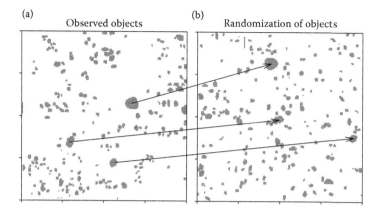

FIGURE 3.36

Randomization of the position of objects using a null model. (a) Pattern of objects with real shape and size. (b) Randomization of objects shown in panel (a), highlighting the change in position for three examples with arrows.

conditioned on the locations of plants of a second species. The categorical map approach facilitates inclusion of such space restrictions in an elegant way. All nonaccessible cells are assigned to a category called "mask" and all cells belonging to the mask are excluded from the study area. In practice, this means that, during the repeated randomization trials of the given species, a trial is rejected if the tentatively placed individual overlaps a cell included in the mask category.

3.1.8.6 Edge Correction

The finite size of objects requires edge correction, since randomly displaced objects may fall partly outside the observation window. In the null model, this would reduce the proportion, λ, of occupied cells and produce a slight (positive) bias toward aggregation. There are three methods that can be used to mitigate this effect (Wiegand et al. 2006). First, randomized objects are not allowed to fall partly outside the observational window. This produces a negative bias toward hyperdispersion since fewer objects of the null model will be distributed close to the border. The second method avoids this problem by treating the observational window encompassing the study region as a torus, that is, the part of an object lying outside the window reappears at the opposite border. However, breaking relatively large objects into two smaller objects produces a slight positive bias. A third method uses the torus correction, but calculates the summary statistics only inside an inner rectangle excluding cells close to the border (guard area). For a guard area wider than the diameter of the largest plants, the biases of the first and the second method disappear, but this may reduce the size of the study rectangle and thus the number of objects considered in the analysis. Therefore, the guard area selected needs to be wider than the diameter of most objects, but still small enough to yield a large enough sample.

3.1.8.7 Univariate Analysis of Objects of Finite Size

In univariate analyses, the fundamental division is provided by the *completely random distribution of objects*, as opposed to aggregation or hyperdispersion. Objects of finite size and irregular shape are, by analogy to CSR, distributed randomly as described above. Depending on the biological question, the available area may be the entire observation window, an irregularly shaped subarea of the observation window, or the area remaining after masking regions of the grid that are hypothesized to constrain the locations of the focal objects, for example, through competitive effects.

Figure 3.37 shows an example of a univariate analysis for a pattern based on the distribution of the shrub species *Mulinum spinosum* (Figure 3.35a). Simulation envelopes were constructed by producing 199 simulations of the null model, which randomized the locations of the plants. We show results of the analysis of two cases, one in which pairs of cells of the focal object were counted (as in Wiegand et al. 2006) and one in which we considered only pairs of cells of different objects. Analysis by the univariate O-ring statistic $O(r)$ gives the probability that a cell located at a distance r away from a cell occupied by a shrub is also occupied by a shrub. Clearly, in the first case, where pairs of cells of the focal object are included in the analysis, $O(r)$ essentially depicts the spatial structure (i.e., size distribution) of the individual shrubs at small scales, which is not of particular interest. This can be seen in Figure 3.37a, which shows that the probability of having a neighbor of the same category (here a *M. spinosum* shrub) declines rapidly with increasing distance from the focal cell, but reaches the expectation given by the over all shrub cover at distances greater than 1 m. The solid gray line in Figure 3.37a gives the expectation of $O(r)$ under the null model and basically captures the geometry of the objects. However, potential departures from the null model are difficult to see because of the steep decline in the value of $O(r)$ at scales up to about 1 m.

Including only distances between different objects in the analysis (open disks in Figure 3.37a) completely removes the signal of the shape of individual objects and places the focus on the spatial relationships among objects. In our example, this reveals a negative departure from the null model at distances less than 30 cm (Figure 3.37b), indicating an additional, subtle effect of regularity at small scales. This information was obscured in Figure 3.37a by including pairs of cells of the focal object in the analysis. However, the effect between shrubs indicated by the analysis may actually be an artifact produced by the method used in mapping the pattern, since shrubs were defined in our example as consisting of adjacent cells of the same type, in effect prohibiting two shrubs from having cells that touched.

The spatial pattern of the focal shrub species may be constrained by the placement of the other shrubs in the system. To test for this potential effect

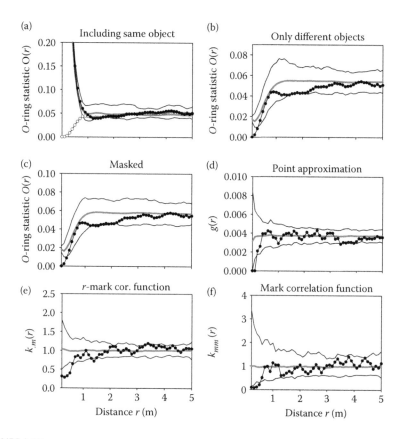

FIGURE 3.37
Univariate "point" pattern analyses for objects of real size and irregular shape. (a) Univariate analysis with $O(r)$ for the shrub species *M. spinosum*, with (closed disks) and without (open disks) including distances among points of the same object in the analysis. Simulation envelopes are produced through randomization of the objects and analysis including points of the same object. (b) Same as panel (a), but excluding points of the same object in the analysis and construction of envelopes. (c) Same as panel (b), but masking out the area occupied by the two other dominant shrub species *A. volckmanni* and *S. filaginoides* as unavailable area for randomizaiton. (d) Standard point-pattern analysis for the shrub species *M. spinosum* where the objects are treated as dimensionless points, with location approximated by the center of mass. (e) Standard marked point-pattern analysis for shrub species *M. spinosum* using the *r*-mark-correlation function treating size as a quantitative mark and using the approach as in panel (d) for location. (f) Same as panel (e), but using the mark-correlation function.

we considered the pattern of the other two species *Adesmia volckmanni* and *Senecio filaginoides* (Figure 3.35a) as a mask, which prohibited relocated shrubs of the focal species from overlapping the areas occupied by the two other shrub species. We obtain basically the same results, with and without masking (cf. Figure 3.37b and c), which indicates that competition for space exerted by the other two dominant shrub species is not an overriding process shaping the spatial pattern of the focal shrub species.

It is instructive to compare the results of the analyses of objects with finite size and irregular shape to those of the point approximation and mark-correlation analysis. Figure 3.37d shows the results of a simple point-pattern analysis, where the objects were approximated by their center of gravity. There is repulsion at very small scales but, in contrast to the results shown in Figure 3.37b, the regularity is caused by the physical size of the objects (i.e., a soft core effect). Next, we marked the point pattern with the size of the corresponding basal areas. The r-mark-correlation function, which returns the (normalized) average size of basal areas at distance r from the focal object, shows that neighboring individuals are smaller, if they are closer than 50 cm to the focal object (Figure 3.37e). The mark-correlation function, which calculated the mean normalized product of the sizes of two basal areas separated by distance r, depicts basically the same significant effect (Figure 3.37f). Thus, either explicitly considering the shape of the shrubs in the analysis or using the size information in a mark-correlation function yields the same result, that is, there is small-scale repulsion among the shrubs.

3.1.8.8 Bivariate Analysis of Objects of Finite Size

In bivariate analyses the fundamental division for objects of finite size is given by *independently distributed patterns*. This provides the separation line between *attraction* and *repulsion* (segregation). The null model testing for independence of two-point patterns assumes that independent stochastic processes created the component patterns. This leads to the idea of using a toroidal shift to produce a null model testing for the independence of the point patterns (see Section 3.1.1). The same approach applies for bivariate patterns of plants of finite size: pattern 1 remains fixed, whereas pattern 2 is randomly shifted as a whole across the study area, using a torus (toroidal shift) to reposition the parts of the pattern that cross an edge.

One aspect should be noted when using the toroidal shift as a null model: it requires that objects of different patterns be allowed to overlap. This may introduce a bias toward repulsion at small scales if the objects are known not to overlap. As an alternative, a univariate null model could be introduced to describe the second-order structure of pattern 2 and, if appropriate, an overlap of plants could be prohibited. In the simplest case, this would be the null model where the plants of pattern 1 remain unchanged, but plants of pattern 2 are distributed at random, with the condition that they are not allowed to overlap plants of pattern 1 (i.e., a completely random distribution of objects of pattern 2 constrained by the objects of pattern 1). Note that one may also use a null model that conditions on the larger scale pattern of the plants of pattern 2 by means of a heterogeneous Poisson process (Sections 2.6.1.2, 2.6.3.3, and Box 2.3). Another option would be to extend the techniques of pattern reconstruction (Section 3.4.3) to objects and create, for the independence null model, stochastic replicate patterns that also conserve intermediate scale correlations in the placement of objects.

Figure 3.38a shows a bivariate pattern composed of the shrub *A. volckmanni* (pattern 1, black) and the pattern of two other dominant shrub species *M. spinosum* and *S. filaginoides* (pattern 2, gray). The bivariate *O*-ring statistic is zero at very small distances, indicating no overlap between the two patterns, and increases up to a distance of 50 cm, before reaching saturation at the expectation

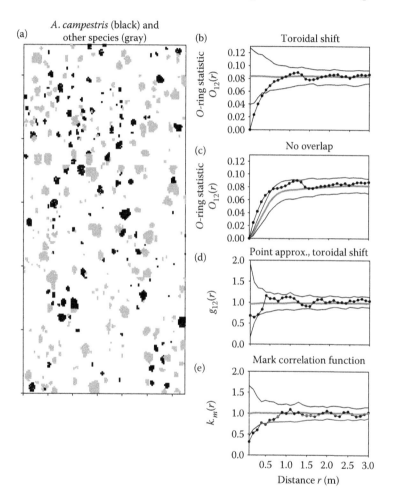

FIGURE 3.38

Analysis of a bivariate "point" pattern for objects of real size and irregular shape. (a) Map of a pattern with *A. campestris*, the focal species, in black and two other dominant shrub species *M. spinosum* and *S. filaginoides* in gray. (b) Bivariate analysis using the toroidal shift null model, keeping the locations of the focal species fixed, but moving the other two. (c) Bivariate analysis using a null model where the locations of the focal species were unchanged, but the individuals of the other two shrub species were randomly relocated, with no overlap allowed. (d) Same as panel (b), but shrubs were approximated by points and the standard, bivariate pair-correlation function was used for analysis. (e) Bivariate analysis with the *r*-mark-correlation function, where shrubs were approximated by points and the size of the shrubs of pattern 2 was used as a quantitative mark.

of independence under the null model (i.e., the probability that a cell is occupied by the second pattern; i.e., *M. spinosum* or *S. filaginoides* shrubs) (Figure 3.38b). As expected, application of the toroidal shift null model exhibits repulsion at small scales (caused by the nonoverlap of the two patterns). At greater distances no significant departures from the null model were observed.

An approach that is more rewarding than using the toroidal shift as a null model is the application of a null model where the *A. campestris* pattern remains fixed and the basal areas of all other shrubs are randomly distributed without overlap. We find that the two groups of shrubs show repulsion at small scales up to 30 cm (Figure 3.38b). This means that the shrubs of the second pattern are located somewhat farther away from *A. campestris* shrubs than would be expected through random placement without overlap. This may be an effect of competition between the two groups of plants.

Using the point approximation and testing for independence with the toroidal shift null model indicates independence between the two patterns. The signal of repulsion between the two groups of shrubs seen in the corresponding analysis with the full shape can only be seen in the point approximation as a nonsignificant tendency (cf. Figure 3.38b and d). This indicates that retaining the size and shape information may be critical in this analysis. The importance of the size information is also outlined by application of the *r*-mark-correlation function, which yields the (normalized) mean size of the shrubs of the second pattern at distance *r* away from shrubs of the focal pattern (Figure 3.38e). We used a null model where the sizes of the shrubs of the second pattern were randomly shuffled among shrubs of this pattern. This analysis revealed that the size of individuals of the second shrub pattern was smaller, if a shrub of the focal pattern was nearby (i.e., nearer than 30 cm). This is an effect of geometry: larger shrubs will not be close together.

In this section, we have presented an extension of conventional point-pattern analysis, where objects are approximated as points in categorical raster maps that can represent objects of finite size and irregular shape. This approach facilitates incorporation of real-world structures into point-pattern analysis and provides a powerful tool for the statistical analysis of objects for which the common point approximation would not be capable of revealing important information. Clearly, mapping plants and characterizing them by their real shapes, instead of treating them as dimensionless points, takes considerably more effort, but is necessary if we are to consider real size and shape effects instead of employing crude point approximations.

⋆⋆ 3.2 Replicate Patterns

Mapping a plant community for the purpose of conducting a point-pattern analysis often involves a tradeoff between sampling one large plot or several smaller plots (Illian et al. 2008, p. 269). Mapping one large plot has the advantage that edge effects will be less severe (i.e., there are fewer plants at the border of the plot impacted by the influence of unknown plants outside the plot), but one large plot may not sufficiently represent the typical conditions occurring within an entire study site. For statistical analysis, it is a common practice to sample several replicate plots under identical conditions within a study site; this can also be done for point patterns. Point-pattern analysis provides methods that allow for the aggregation of results obtained from individual analyses of several replicate plots. Basically, the resulting test statistics of the individual replicate plots are combined to produce an average (Diggle 2003, p. 123; Illian et al. 2008, p. 260f). This is of particular interest when the number of points in each of the replicate plots is relatively low since the envelopes resulting from the analysis of individual plots would be quite wide. By combining the results of several replicate plots into one average test statistic, the sample size is increased and the range of values covered by the simulation envelopes is narrowed. There are several examples in the ecological literature where replicate plots have been combined in a point-pattern analysis (e.g., Riginos et al. 2005; Blanco et al. 2008; De Luis et al. 2008; Raventós et al. 2010, 2011; Jacquemyn et al. 2010).

A straightforward method for combining data from different replicate point patterns would be to place them (artificially) into one large observation window (i.e., combine them into one large data set). This can be done; however, care must be taken to avoid mixing points from one replicate plot with those of another plot in the resulting estimator. We can insure this geometrically by shifting the coordinate systems of the separate replicate plots so that they are "farther apart" than r_{max}, the maximal distance at which the summary statistics are evaluated (Figure 3.39). Additionally, the boundaries of the "observation window" have to be set so that only subplots are considered to be "inside" the observation window. The latter condition is necessary to prevent bias due to edge correction (Figure 3.39). The pair-correlation function can then be estimated using the inhomogeneous pair-correlation function based on the estimator given by Equation 3.31, together with an intensity function $\lambda(x)$ that yields the observed intensity λ_m within each of the replicate patterns m and a value of zero outside (Figure 3.39). In this case, the null models would distribute the points only inside the replicate windows m with their associated intensity λ_m. Although this method may work well for the pair-correlation function, and help us in motivating suitable aggregation formulas, there are actually more direct methods for aggregating the results of individual replicate analyses. The specific formula used in combining results from several replicate plots depends on the estimators of the test statistics used. Illian et al. (2008, p. 261f) provide aggregation formulas for the most important summary statistics. In the following Section 3.2.1, we provide aggregation formulas for univariate patterns, and in Section 3.2.2 we present some

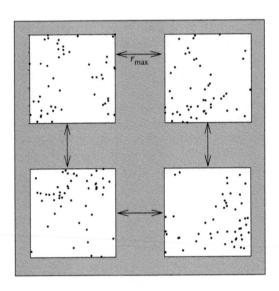

FIGURE 3.39
Naive estimator for deriving a single summary statistic from replicate plots. The four replicate plots are placed within one observational window, but are kept at least distance r_{max} apart, r_{max} being the maximum distance for which we evaluate the summary statistics. The gray shaded area between the plots is excluded from the analysis.

example applications using this approach. Note that it is also relatively straightforward to extend the aggregation formulas for univariate patterns to bivariate and marked patterns.

3.2.1 Aggregation Formula

In developing an aggregation formula for M replicate plots, we can estimate a summary statistic S_m for each plot m. Aggregation of S_m into one summary statistic S can be developed from two basic principles. The first principle is to estimate a weighted mean of the summary statistic S_m of the M individual plots:

$$\hat{S} = \frac{\sum_{m=1}^{M} c_m \hat{S}_m}{\sum_{i=1}^{M} c_m} \tag{3.105}$$

where the c_m are appropriate weights, which depend on the nature of the summary statistic. The second principle for deriving aggregation formulas applies if the estimators of the summary statistic S have a particular functional form, which arise from "adapted" estimators that can be expressed as a ratio of sums over focal points i (we omit the index m for clarity):

$$\hat{S}(r) = \frac{(1/n)\sum_{i=1}^{n} f(x_i, r)}{(1/n)\sum_{i=1}^{n} g(x_i, r)} = \frac{\sum_{i=1}^{n} f(x_i, r)}{\sum_{i=1}^{n} g(x_i, r)} \qquad (3.106)$$

The numerator and the denominator in Equation 3.106 are sums of the functions $f(x_i, r)$ and $g(x_i, r)$, which depend on the focal point i and other points within or at distance r from the focal point. For example, the WM estimator of the O-ring statistic (Equation 3.17) follows this functional form. Here, the numerator is given by $f(x_i, r) = \sum_{j=1, j\neq i}^{n} k_h^B(\|x_i - x_j\| - r)$, which is a factor that returns the number of points falling at distance $r \pm dr/2$ from a focal point i. The denominator $f(x_i, r) = v_{d-1}(W \cap \partial b(x_i, r))$ is the area of the corresponding ring with radius r and width dr centered on point i. The sums are taken over all points i of the focal pattern. For second-order summary statistics the function $f(x_i, r)$ counts the number of points in the neighborhoods of point i, whereas the function $g(x_i, r)$ determines the area of the neighborhood (e.g., Equations 3.17 and 3.36). For mark connection or mark-correlation functions, the function $f(x_i, r)$ determines the sum of a test function involving the marks of focal point i and all other points at distance r from this point, and the function $g(x_i, r)$ counts the number of points in the neighborhood of point i (e.g., Equation 3.84).

Basically, these estimators visit each point i of the pattern and estimate the mean value of the functions $f(x_i, r)$ and $g(x_i, r)$, which summarize neighborhood properties of point i as a function of the radius r. Because the factor $1/n$ appears in both the numerator and denominator and cancels out, the estimator given in Equation 3.106 simply sums up the contributions of all points i of the pattern. We can extend this principle to points of different subplots, which suggests a natural way to aggregate

$$\hat{S}(r) = \frac{\sum_{i=1}^{n_1} f_1(x_i, r) + \cdots + \sum_{i=1}^{n_M} f_M(x_i, r)}{\sum_{i=1}^{n_1} g_1(x_i, r) + \cdots + \sum_{i=1}^{n_M} g_M(x_i, r)} = \frac{\sum_{m=1}^{M} e_m}{\sum_{m=1}^{M} d_m} \qquad (3.107)$$

Each point i contributes in the same way to this estimator. Thus, we can apply a simple aggregation formula by saving, not only the estimate of the summary statistics $\hat{S}_m(r)$, but also the numerator $e_m = \sum_{i=1}^{n_m} f_m(x_i, r)$ and the denominator $d_m = \sum_{i=1}^{n_m} g_m(x_i, r)$ for each plot m. Examples for summary statistics that have the functional form of Equation 3.106 are the O-ring statistic, based on the WM edge correction (Equation 3.17), the K-function based on the WM edge correction (Equation 3.36), the distribution function of the distances to the kth neighbor (Equation 3.47), mark connection functions (Equation 3.80), and non-normalized mark-correlation functions (Equation 3.84). Note that we need to use estimators of $\lambda g(r)$, $\lambda K(r)$, and non-normalized mark-correlation functions, otherwise we would add a weighting factor of

$1/\lambda_m$ or $1/c_l^m$ to the corresponding contribution e_m of subplot m in Equation 3.107. Thus, in this case the results of subplots with low intensity (or low value of c_l^m) would receive too much weight in the combined estimate.

Note that the aggregation principle of Equation 3.107 can be derived from the "naive" method of aggregating replicates shown in Figure 3.39. The data of the M plots are integrated into one data base by arranging them in the way shown in the figure, which guarantees that the distances between points of the different replicate plots are greater than the maximal distance r_{max} to be analyzed. It can be easily verified that the estimator of type 3.108 will result in the estimator given in Equation 3.106, as long as the functions g_m and f_m of replicate plot m do not use points from any other replicate plot. For the O-ring statistic, this is guaranteed if the distance among replicate plots is greater than r_{max}.

3.2.1.1 Intensity

Following Illian et al. (2008, p. 261), the aggregated estimator for the intensity λ of M windows W_m yields

$$\hat{\lambda} = \sum_{m=1}^{M} \hat{\lambda}_m \frac{A_m}{A} = \frac{\sum_{m=1}^{M} \hat{\lambda}_m A_m}{\sum_{m=1}^{M} A_m} = \frac{\sum_{m=1}^{M} n_m}{\sum_{m=1}^{M} A_m} = \frac{n}{A} \qquad (3.108)$$

where $\hat{\lambda}_m$ is the estimate of the intensity of points within the mth window, A_m is the area of the mth window, and A is the sum of the areas of all the windows A_m. If the intensity of all of the m windows were estimated by the natural estimator $\hat{\lambda}_m = n_m/A_m$, we find that the aggregated estimator is given by $\hat{\lambda} = n/A$, where n is the total number of points in all m windows and A is the total area (Equation 3.108). Thus, the aggregation formula follows the recipe given in Equation 3.108. However, in the aggregation formula 3.110, the intensity $\hat{\lambda}_m$ is multiplied by the relative area A_m of plot m. Thus, the aggregation recipe also follows the approach of Equation 3.105, where the area A_m of plot m is the weighting factor c_m. This is reasonable because the number of points will be proportional to the area A_m.

3.2.1.2 Nearest-Neighbor Distribution Function

The aggregation formula for the nearest neighbor (or the kth neighbor) distribution function depends on the estimators used (see Section 3.1.3). If no edge correction is used, the estimators for the m individual plots are weighted with the relative number of points in each window W_m:

$$\hat{D}(r) = \frac{\sum_{m=1}^{M} n_m \hat{D}_m(r)}{\sum_{m=1}^{M} n_m} \qquad (3.109)$$

where n_m is the number of points in the mth window. In this case, the number of points in the mth window provides the weights c_m for Equation 3.105. However, if the border estimator is used (see Section 3.1.3; Equation 3.44), only focal points that are located further away than distance r from the nearest border are used for the evaluation. Therefore, the aggregation formula becomes

$$\hat{D}(r) = \frac{\sum_{m=1}^{M} n_{r,m} \hat{D}_m(r)}{\sum_{m=1}^{M} n_{r,m}} \tag{3.110}$$

where the $n_{r,m}$ is the number of points inside a reduced window W_m, with only points farther than distance r to the nearest border included as focal points (Illian et al. 2008, p. 261).

We can also apply the second aggregation principle to a calculation of nearest-neighbor distribution functions, since the border or Hanisch estimator (Equation 3.47) has the functional form of Equation 3.106. For example, with the Hanisch estimator we can define $g(x_i, r) = \mathbf{1}(d_{ik} < \beta_i) \times w(d_{ik})$ and $f(x_i, r) = \mathbf{1}(0 < d_{ik} \leq r) \times g(x_i, r)$ and rewrite Equation 3.45 as

$$\hat{D}^{k,H}(r) = \frac{\sum_{i=1}^{n} \mathbf{1}(0 < d_{ik} \leq r) g(x_i, r)}{\sum_{i=1}^{n} g(x_i, r)} = \frac{e}{d} \tag{3.111}$$

Here, d_{ik} is the distance to the kth nearest neighbor and β_i is the distance from the point i at location x_i to the nearest border. We can then use the aggregation formula

$$\hat{D}^{k,H}(r) = \frac{\sum_m e_m}{\sum_m d_m} \tag{3.112}$$

if we save the numerators e_m and denominators d_m of estimators for the m replicate plots. Note that both aggregation receipts (i.e., Equations 3.105 and 3.107) yield basically the same results because the denominator of Equation 3.112 is an estimator of the intensity of points in W_m.

3.2.1.3 Spherical Contact Distribution

The spherical contact distribution $H_s(r)$ yields the probability that the first neighbor of a univariate pattern is distance r away from an arbitrary test location t. This can be conceptualized as a bivariate nearest-neighbor distribution function $D_{tp}(r)$ that gives the proportion of test points t which has the nearest point of pattern p within distance r. In this case, we can use the preceding aggregation formulas for the distribution function to the nearest neighbor. Illian et al. (2008, pp. 204, 261) propose a slightly different estimator and aggregation formula based on the weighted mean (Equation 3.105). Their weighting factor is given by the relative area of the mth replicate plot A_m/A.

3.2.1.4 Pair-Correlation Function

Illian et al. (2008) and Law et al. (2009) proposed an aggregation formula for the pair-correlation function that considers a weight (c_m) based on the isotropizised set covariance $\bar{\gamma}_W(r)$ (Section 3.1.2.1; Equation 3.7):

$$\hat{g}(r) = \frac{\sum_{m=1}^{M} \bar{\gamma}_{W_m}(r)\hat{g}_m(r)}{\sum_{m=1}^{M} \bar{\gamma}_{W_m}(r)} \tag{3.113}$$

This is a reasonable approach for the case that all windows W_m have the same shape, as the weight reduces to the relative area of the mth window due to the fact that $\bar{\gamma}_W(r)$ is proportional to the area A of the window (Equation 3.7). However, this weight also provides an additional correction if the shape of the windows W_m differ. Regarding this, Law et al. (2009) provide an interpretation of Equation 3.113: "The isotropized set covariance gives a larger weight to estimates of the pair-correlation function in square windows than in rectangular windows at larger distances. This reflects the fact that a larger number of points with larger distances can be expected in a square window than in a rectangular window yielding a better-quality estimation at these distances, p. 625." Note that the formula given in Equation 3.113 is adapted to the case of Ohser edge correction (Equation 3.21) because it eliminates the individual edge correction terms for replicates m given by the isotropizised set covariance $\bar{\gamma}_{W_m}(r)$ of window W_m and replaces them by their average.

If we use the WM estimator of the O-ring statistic (Equation 3.17) we can apply the aggregation formula (3.107)

$$\hat{O}_{WM}(r)$$
$$= \frac{\sum_{i=1}^{m_1}\left(\sum_{j=1}^{m_1,j\neq i} k_h^B(\|x_i^1 - x_j^1\| - r)\right) + \cdots + \sum_{i=1}^{n_M}\left(\sum_{j=1}^{n_M,j\neq i} k_h^B(\|x_i^M - x_j^M\| - r)\right)}{\sum_{i=1}^{m_1} v_{d-1}(W_1 \cap \partial b(x_i^1,r)) + \cdots + \sum_{i=1}^{n_M} v_{d-1}(W_M \cap \partial b(x_i^M,r))}$$

$$\tag{3.114}$$

where the superscripts refer to the plots. The estimator of the O-ring statistic can then be combined with the estimator of the intensity (Equation 3.108) for the final estimate of the pair-correlation function.

3.2.1.5 K-Function

Diggle (2003, p. 123) and Illian et al. (2008, p. 262) propose the following aggregation formula for the K-function:

$$\hat{K}(r) = \frac{\sum_{m=1}^{M} n_m \hat{K}_m(r)}{\sum_{m=1}^{M} n_m} \tag{3.115}$$

This is the weighted average of the K-functions for the individual windows W_m, with weights provided by the relative proportion of points in W_m. However, this weighting recipe is somewhat inconsistent with that of the pair-correlation function (Equation 3.113). Following the same philosophy as for the pair-correlation function, the cumulative isotropized set covariance

$$c_m = \frac{2\pi \int_0^r t \bar{\gamma}_{W_m}(t) dt}{\pi r^2} \tag{3.116}$$

can be used as the weighting factor c_m.

If we use the WM estimator of $\lambda K(r)$ given in Equation 3.36, we can apply the aggregation formula (3.107) with a similar approach to the one used in calculating the pair-correlation function. We then obtain

$$\lambda K^{WM}(r) = \pi r^2 \frac{\sum_{i=1}^{n_1} \sum_{j=1}^{n_1, \neq} \mathbf{1}(\|x_i^1 - x_j^1\|, r)) + \cdots + \sum_{i=1}^{n_M} \left(\sum_{j=1}^{n_M, \neq} \mathbf{1}(\|x_i^M - x_j^M\|, r) \right)}{\sum_{i=1}^{n_1} v(W_1 \cap b(x_i^1, r)) + \cdots + \sum_{i=1}^{n_M} v(W_M \cap b(x_i^M, r))}$$

$$\tag{3.117}$$

Again, the estimator of $\lambda K(r)$ can be combined with the estimator of the intensity (Equation 3.108) to estimate the K function.

3.2.1.6 Mark-Connection Functions and Mark-Correlation Functions

The estimators of mark-connection functions (Equations 3.80 and 3.83) and those of non-normalized mark-correlation functions (Equations 3.84, 3.96, 3.97, 3.99, and 3.102) all have the functional form of Equation 3.100. It is, therefore, straightforward to apply the aggregation formula of Equation 3.107, but, in doing this, the numerator and denominator of the estimators must be saved separately for each plot.

3.2.2 Examples of the Application of Aggregation Formulas

In this section, we provide several practical examples of the application of the aggregation formulas. The basic application is, of course, to combine the results of several replicate plots into a single summary statistic and produce the associated simulation envelopes. However, we also point to the fact that the aggregation formulas can be used to implement very specific null models that otherwise would require very specialized software.

3.2.2.1 Increasing the Power of GoF Test

One of the principle applications of the aggregation formulas is to combine the results of replicate plots with a low number of points to produce more sensitive analyses, since the low number of points within individual plots generally results in wider simulation envelopes. The combined data act to improve the results of a point-pattern analysis by yielding a single summary statistic with narrower simulation envelopes, due to increased sample size. To illustrate this, we consider an example of nine 250×250 m^2 plots, with each pattern containing $n = 50$ points (Figure 3.40). The patterns were created with a Thomas process (Box 2.5) consisting of 13 clusters with a radius of 44 m. The expectation of the pair-correlation function under this point process is shown as a gray line in Figure 3.40a,b. Figure 3.40a presents the analysis of an individual plot (plot 1), which has very wide simulation envelopes arising from the 199 simulations of the CSR null model. Additionally, the pair-correlation function of plot 1 fluctuates somewhat erratically. Although the observed pair-correlation function lies, at some spatial scales, marginally outside of the simulation envelopes, the GoF test (see Section 2.5.1.2) does not detect a significant departure from the CSR null model over the distance interval of 1–50 m. The P-value of the test yields $P = 0.12$. These results are a consequence of the high degree of stochasticity, which does not allow a separation of the real signal of clustering from the stochastic fluctuations. As a consequence, five of nine plots did not yield significant departures from the CSR null model, although they were all created with a point process that incorporated an explicit cluster mechanism. However, when aggregating the results of the nine plots with Equation 3.114, the estimate of the pair-correlation function approximates the expected pair-correlation function of the underlying Thomas process reasonably well, and the simulation envelopes are considerably narrower (Figure 3.40b). Because the sample size was much larger for the combined analysis ($n = 450$), there is a clear departure from CSR (Figure 3.40b).

3.2.2.2 Intra- versus Interspecific Interactions

The aggregation formulas are not only an effective way of combining the results of several replicate plots; they can also be used for specific applications that would otherwise require very specific software. One example for such an application is the study of mortality in multi-species communities,

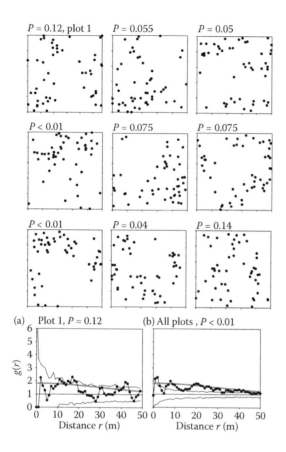

FIGURE 3.40
Combining replicates in a point-pattern analysis using nine 250×250 m^2 plots with 50 points each, created by a Thomas process with a cluster radius of 44 m and on average 13 parents. (a) Pair-correlation function results for plot 1 (black dots), with the thin black horizontal line indicating the expectation under CSR and the thick gray line the expectation for the Thomas process. (b) Same as panel (a), but with the combined results of all nine plots. The P-value associated with each plot is from the GoF test applied to the data in the plot for 199 simulations of the CSR null model.

where we may wish to test the "segregation hypothesis" (Pacala 1997; see also Chapter 1). This hypothesis states that most plants in communities that show a particular segregation pattern (i.e., intraspecific aggregation and interspecific segregation) will compete at short distances with con-specifics, which prevents (or retards) competitive exclusion and contributes to coexistence (Pacala 1997; Murrell et al. 2001).

One possibility for testing this hypothesis is to use data from repeated censuses of a plant community. In this case, each individual is characterized by its spatial position and species identifier, as well as a mark indicating whether it is alive or dead. We use the subscript 1 to refer to dead individuals and subscript 2 to refer to those surviving. The task is to determine whether dead plants, as well as surviving plants, are in fact mostly surrounded by conspecifics, as suggested by the segregation hypothesis. To this end, we decompose the over all neighborhood density of dead plants (subscript 1) around dead plants (subscript 2) [i.e., the O-ring statistic $O_{11}(r)$] and of surviving plants around surviving plants [i.e., the O-ring statistic $O_{22}(r)$] into two components, one that considers only conspecific pairs [$O_{ii}^{\text{intra}}(r)$] and one that considers only heterospecific pairs [$O_{ii}^{\text{inter}}(r)$].

To derive the estimators of these quantities, we first need to calculate for each species the four possible neighborhood densities $O_{11}(r)$, $O_{12}(r)$, $O_{21}(r)$, and $O_{22}(r)$, considering both dead and surviving plants. The subscript 1 refers to dead individuals and subscript 2 to surviving individuals. In all cases, the first subscript indicates the focal pattern and the second subscript represents the pattern that is assessed around the focal pattern. For a given species s, $O_{11}(r)$ is the density of dead plants at distance r away from other dead plants; $O_{22}(r)$ is the density of surviving plants at distance r away from other surviving plants; $O_{12}(r)$ is the density of surviving plants at distance r away from dead plants; and $O_{21}(r)$ the density of dead plants at distance r away from surviving plants. For a given species s, the estimator of the bivariate $O_{ij}(r)$ $(i, j = 1, 2)$ can be written in simplified notation as

$$\hat{O}_{ij}^{s}(r) = \frac{\sum_{k=1}^{n_i^s} \text{Points}_j[R_k^s(r)]}{\sum_{k=1}^{n_i^s} \text{Area}[R_k^s(r)]} \tag{3.118}$$

where the summation is done over the n_i^s type i points of species s, $R_k^s(r)$ is a ring segment with radius r and width h centered at a point k (which is of species s and focal type i) that overlaps the observation window W. The operator Points$_j[X]$ counts the number of points of type j inside an area X, and the operator Area$[X]$ determines the area of X.

Given the previous definitions, the first task is to derive an estimator for the over all neighborhood density function of conspecifics that integrates over all species S. This can be accomplished by considering the neighborhood density $\hat{O}_{ij}^{s}(r)$ for different species s as replicates. Thus, the replication here is given by different species that are located within the same plot rather than by one species found in several plots. By pooling the neighborhood densities for the S species s, we obtain, by using aggregation formula 3.107, the over all (community average) intraspecific neighborhood density of type j points at distance r from type i points:

$$\hat{O}_{ij}^{\text{intra}}(r) = \frac{\sum_{k=1}^{n_i^1} \text{Points}_j[R_k^1(r)] + \cdots + \sum_{k=1}^{n_i^S} \text{Points}_j[R_k^S(r)]}{\sum_{k=1}^{n_i^1} \text{Area}[R_k^1(r)] + \cdots + \sum_{k=1}^{n_i^S} \text{Area}[R_k^S(r)]} \tag{3.119}$$

Note that the estimator given in Equation 3.119 looks only at conspecific pairs of points within the community.

The next task is to derive the corresponding estimator $O_{ij}^{\text{inter}}(r)$ that considers only heterospecific pairs of points. This can be done by estimating the neighborhood density function $O_{ij}^{\text{all}}(r)$ for all individuals in the plot, regardless of species, and subtract the estimator for $O_{ij}^{\text{intra}}(r)$. By maintaining the same notation as used in the estimator of $O_{ij}^{\text{intra}}(r)$ we find:

$$\hat{O}_{ij}^{\text{all}}(r) = \frac{\left(\sum_{k=1}^{n_i^1} \text{Points}_j[R_k^1(r)] + \text{mix}_{\neq 1}\right) + \cdots + \left(\sum_{k=1}^{n_i^S} \text{Points}_j[R_k^S(r)] + \text{mix}_{\neq S}\right)}{\sum_{k=1}^{n_i^1} \text{Area}[R_k^1(r)] + \cdots + \sum_{k=1}^{n_i^S} \text{Area}[R_k^S(r)]}$$

$$(3.120)$$

where the symbol $\text{mix}_{\neq 1}$ refers to the count of all points at distance $r \pm dr/2$ of points of species 1 that are heterospecific (these pairs of points were disregarded in $O_{ij}^{\text{intra}}(r)$). The symbol $\text{mix}_{\neq S}$ refers to the count of all points at distance $r \pm dr/2$ of points of species S that are heterospecific. In a more simplified notation we thus obtain:

$$\hat{O}_{ij}^{\text{all}}(r) = \frac{\text{Points} + \text{mix}}{\text{Area}} \qquad (3.121)$$

$$\hat{O}_{ij}^{\text{intra}}(r) = \frac{\text{Points}}{\text{Area}} \qquad (3.122)$$

$$\hat{O}_{ij}^{\text{inter}}(r) = \frac{\text{mix}}{\text{Area}} = \hat{O}_{ij}^{\text{all}}(r) - \hat{O}_{ij}^{\text{intra}}(r) \qquad (3.123)$$

$\hat{O}_{ij}^{\text{all}}(r)$ can be easily estimated by pooling the individuals of all species, distinguishing only between surviving and dead individuals, and applying Equation 3.118. $\hat{O}_{ij}^{\text{intra}}(r)$ can be estimated by joining the results of the analysis obtained for individual species with Equation 3.119.

We exemplify this method with a reanalysis of data presented in Raventós et al. (2010). In this study, a long-term data set was analyzed that comprised fully mapped seedling emergence and subsequent survival in a Mediterranean gorse shrubland after experimental fires. The experimental fires killed all plants and restarted the system though the germination processes. The analyses involved surviving and dead individuals of the four most common species *Cistus albidus*, *Helianthemum marifolium*, *Ulex parviflorus*, and *Ononis fruticosa* obtained one, two, three, and nine years after the fire (Raventós et al. 2010). Figure 3.41 shows results from the first year after fire for the "fire only" treatment (corresponding to Figures 2A and 2I in Raventós et al. 2010). We used

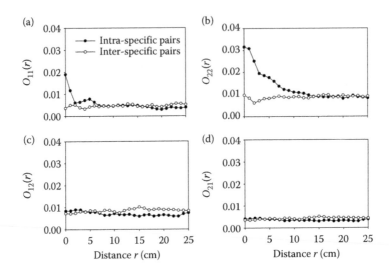

FIGURE 3.41

Use of the aggregation formula applied to the bivariate O-ring statistic to determine the density of inter- and intraspecific pairs of dead and surviving plants in a plant community. The example is a reanalysis of data provided in Raventós et al. (2010), corresponding to their figures 2A and 2I (first year, fire treatment). (a) Dead around dead. (b) Surviving around surviving. (c) Surviving around dead. (d) Dead around surviving.

Equation 3.122 to estimate the neighborhood densities of intraspecific pairs of plants and Equation 3.123 to calculate the neighborhood densities for interspecific pairs of plants. The intraspecific neighborhood density $O_{22}(r)$ of surviving plants around surviving plants (Figure 3.41b, closed disks) showed strong aggregation. This is indicated by a neighborhood density at short distances r that is approximately three times higher than at greater distances. However, the interspecific neighborhood density $O_{22}(r)$ of surviving plants around surviving plants (Figure 3.41b, open circles) exhibits almost no dependence on distance. Thus, the neighborhood density of surviving plants around conspecific surviving plants is much higher than the neighborhood density of surviving plants around heterospecific surviving plants. The same pattern is shown by dead individuals (Figure 3.41a), but is somewhat weaker. Interestingly, the mixed pairs of surviving and dead plants showed no clustering for interspecific and intraspecific pairs (Figure 3.41c,d). Thus, surviving plants (and to a lesser extent dead plants) occur in intraspecific clusters, but the clusters of different species rarely overlap. These results point to positive intraspecific association among surviving plants (Raventós et al. 2010).

3.2.2.3 Nonparametric Estimation of a Pollen Dispersal Kernel

This example is given by a study by Niggemann et al. (2012) who estimated pollen dispersal kernels and contrasted the observed kernels to those expected under a random mating null model. The study analyzed a complex quantitatively marked pattern and the aggregation formulas were used to implement specific null models that would otherwise require specialized software. Seeds were harvested from female black poplar trees (*Populus nigra* L.) and genetic paternity analysis was used to determine the pollen donor (i.e., male parent) for each harvested seed from among all male poplar trees in the study area. (The poplar population comprised a dense patch and a few scattered individuals further away.) The basic interest in analyzing this data type was to find out if pollen dispersal occurred in a spatially correlated way (i.e., pollen of nearby males had a higher probability of reaching a female than pollen from males located further away) and to quantify the distance dependence. To answer these questions, parametric dispersal kernels are usually fitted to such data sets (e.g., Robledo-Arnuncio and García 2007); however, for assessment of uncertainty with respect to the random mating null model (no distance dependence), or any other model, it is desirable to calculate nonparametric dispersal kernels. This can be done within the framework of quantitatively marked point patterns and mark-correlation functions (Section 3.1.7).

The data structure involves mother trees m, potential father trees f, and marks m_{mf}, which are the number of seeds of mother tree m fathered by male f (Figure 3.42). The null model assumes random mating, where all potential pollen donors f have the same probability of fathering a seed of mother m. To determine how the observed values of the marks m_{mf} depend on the distance r to the mother m, Niggemann et al. (2012) developed a bivariate mark-correlation function:

$$\hat{k}_{mf}(r) = \frac{1}{\hat{c}_f} \frac{\sum_{m=1}^{n_m} \sum_{f=1}^{n_f} t(m_{mf}) k(\|x_f - x_m\| - r)}{\sum_{m=1}^{n_m} \sum_{f=1}^{n_f} k(\|x_f - x_m\| - r)} \tag{3.124}$$

where n_m and n_f are the total number of mother trees and potential pollen donors, respectively; $t(m_{mf}) = m_{mf}$ is the test function with associated normalization constant c_f (the mean of the mark m_{mf} taken over all $f–m$ pairs of female and male trees); x_f and x_m are the location of the pollen donor f and the mother tree m, respectively; and $k(\|x_f - x_m\| - r)$ is the box kernel (Equation 3.15; Figure 3.7a) with bandwidth h. The kernel function $k()$ basically defines a ring with radius r and width $2h$ centered on the mth mother tree, and the sum $\sum_{f=1}^{n_f} k(\|x_f - x_m\| - r)$ counts the number of potential pollen donors f within such a ring (Figure 3.42a).

The interpretation of the mark-correlation function (3.124) is straightforward. All mother trees m are visited in sequence, the marks m_{mf} of all potential donors f are counted at distance r from the mother m [i.e., the inner sum $\sum_{f=1}^{n_f} t(m_{mf}) k(\|x_f - x_m\| - r)$] and the number of potential donors f at distance r from the mother tree m are counted [i.e., $\sum_{f=1}^{n_f} k(\|x_f - x_m\| - r)$]. Thus, $k_{mf}(r)$

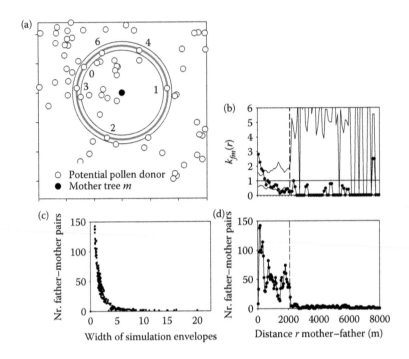

FIGURE 3.42
Nonparametric estimation of pollen dispersal kernels. (a) Data structure for estimation of pollen dispersal kernel, which consists of one focal female tree m (closed disk), a ring of distance r
and width $2h$ around the focal female, and potential pollen donors f (open disks) lying in the
ring, marked by the number of seeds m_{mf} they sired with the female m (only for trees inside
the ring). (b) Estimate of the observed pollen dispersal kernel and simulation envelopes of the
random mating null model. (c) The relationship between the width of the simulation envelope
and the number of father–mother pairs. (d) The number of father–mother pairs for a ring of
distance r. (Adapted from Niggemann, M. et al. *Journal of Ecology* 100: 264–276.)

is proportional to the average number of seeds sired by male trees located at
distance r from a representative mother tree and is therefore proportional to
the pollen probability density $f(r)$ at distance r from the location of the dis-
persing father tree (i.e., the dispersal kernel; Clark et al. 1999). Note that no
edge correction is required for mark-correlation functions (Illian et al. 2008).
    Because the mark m_{mf} depends on the mother tree m, an implementation
of the random mating null model must randomize the mark individually for
each female. In practice, the null model randomly shuffles, for a given female
parent m, the marks m_{mf} (i.e., number of seeds of female parent m fathered by
male f) over all males f. This procedure must be independently repeated for
each female parent m. Implementation of this null model would, therefore,
require specialized software. However, when using the aggregation formula for mark-correlation functions (based on Equation 3.107), the analysis
can be done using the standard procedure for mark-correlation functions.

To run the analysis, the data for each mother are formally treated as replicates and the data set comprises the coordinates of the mother (this is the focal pattern) and the coordinates of all male trees with the attached mark m_{mf} (i.e., number of seeds of female parent m fathered by male f). Conducting a separate analysis for each female guarantees that the simulation of the random mating null model keeps the identity of the mother tree for the marks m_{mf}. The analysis for mother tree m yields the (nonnormalized) estimator

$$\frac{\sum_{f=1}^{n_f} t(m_{mf}) k(\|x_f - x_m\| - r)}{\sum_{f=1}^{n_f} k(\|x_f - x_m\| - r)} \tag{3.125}$$

Subsequent aggregation of the replicates (i.e., females), using the principle in Equation 3.107, yields Equation 3.124 where the normalization constant $\hat{c}_{mf}$ becomes $\hat{c}_{mf} = \sum_{m=1}^{n_m} \sum_{f=1}^{n_f} t(m_{mf})/n_m n_f$. The resulting dispersal kernel estimated using distance intervals of 100 m and a ring width of $h = 200$ m is shown in Figure 3.42b. There is a clear signal of distance dependence at distances less than 400 m, but at larger distances the simulation envelopes become quite wide. This is especially the case for distances greater than 2000 m, where the number of data points within each ring (i.e., the number of father–mother pairs) becomes very low (i.e., below 10; Figure 3.42d). Clearly, the data at these distances are not sufficient, prohibiting the discrimination of any reasonable kernel function and the result is completely overpowered by stochasticity. While this simple fact becomes clear when looking at the simulation envelopes, it is not that clear in "traditional" kernel fitting. While lack of data in fitting long-distance dispersal kernels is a well-known problem, it seems to be not fully recognized how serious this problem is. The approach of point-pattern analysis in deriving nonparametric dispersal kernels and the associated simulation envelopes can also be used to estimate the data demand for a certain level of uncertainty (i.e., width of simulation envelopes). In our case, we would need roughly more than 80 father–mother pairs within a given ring of distance r to make reasonably good inferences. With less data, the envelopes become quite wide (Figure 3.42c).

3.2.2.4 Blowouts and Waterholes

Blanco et al. (2008) used the aggregation formula for the O-ring statistic (Equation 3.119) to implement a specific null model. Their study was concerned with an assessment of the impact of grazing on soil erosion processes prevailing in arid rangelands. Blowouts are depressions or hollows formed by wind erosion on a preexisting sand deposit in vegetation-stabilized dune fields. In arid zones, they are among the most common aeolian erosional

landforms in dune landscapes. Vegetated parabolic dunes are leeward extensions of blowouts. Plants colonize, and may eventually stabilize the trailing
arms of the dunes, which are formed by the advancing dune apex. Grazing,
however, can contribute to destabilization of dunes. The study by Blanco
et al. (2008) investigated the neighborhood density of blowouts around waterholes in a vegetated dune field of Península Valdés, in the Patagonia region
of Argentina. Their goal was to analyze changes induced by grazing on soil
erosion processes. The spatial pattern of blowouts was used as an indicator
of erosion intensity. They used the neighborhood density $O_{12}(r)$ of blowouts
(i.e., type 2 points) at distance r away from waterholes (i.e., focal points of
type 1) as a summary statistic to identify scales at which the spatial pattern
of blowouts was significantly aggregated around water points (Figure 3.43a).
The hypothesis was that large numbers of livestock concentrating around
waterholes might destroy vegetation cover, resulting in dune reactivation
and consequently generating blowouts. Accordingly, the impact of livestock
on soil erosion processes should be greatest in sites near waterholes relative
to those farther away from waterholes (i.e., a utilization gradient). This gradient should result in aggregation of blowouts near waterholes.

The study area comprised eight paddocks, four of which were lightly
grazed, four heavily grazed, and each paddock contained one waterhole. The

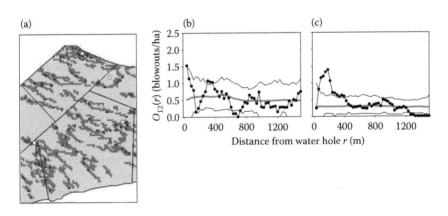

FIGURE 3.43
Use of the aggregation formula for the O-ring statistic in implementing specific null models.
(a) Partial map of the study area, showing two complete paddocks. Each paddock has one
water hole (open circle), the area of the dune crests are indicated by darker gray, and the blowouts are shown as black squares. (b) O-ring analysis for lightly grazed paddocks. In a first step,
an analysis was done for each paddock separately, using techniques for irregularly shaped
study areas. The neighborhood density $O_{12}(r)$ of blowouts at distance r away from the water
holes was determined and the null model randomized the blowouts (black squares) over the
dune crests (dark gray areas). The results of the four analyses for the individual paddocks were
then combined using the aggregation formula for the O-ring statistic. Analysis of separate paddocks was necessary because the sheep cannot move from one paddock to another. (c) Same as
panel (b), but for the heavily grazed paddocks. (Adapted from Blanco, P.D. et al. 2008. *Rangeland
Ecology and Management* 61: 194–203.)

paddocks had an irregular shape but were adjacent (Figure 3.43a). Because sheep could not move among paddocks the grazing treatment within one paddock had no influence on the dune crests and the formation of blowouts of neighboring paddocks. As a consequence, Blanco et al. (2008) divided the study area into four replicate plots for each treatment and conducted separate analyses for each paddock. The waterhole of a paddock (open disks) was the focal point and the blowouts within ring segments centered on the respective paddock were counted (black squares). Consequently, the null model distributed the blowouts randomly on dune crests of a given paddock (which are the areas susceptible to erosion; dark gray areas). However, because sheep could not move from one paddock to another, the null model needed to randomize the blowouts only inside paddocks.

Similar to the example in the preceding section, the analysis was first done separately for each paddock, and then the results of the four replicate paddocks that were lightly grazed and the four paddocks that were heavily grazed were combined using the principle of Equation 3.107. Aggregation yielded an estimator similar to that shown in Equation 3.119. As expected, the lightly grazed paddocks showed only a weak distance dependence of blowouts around waterholes (Figure 3.43b), whereas the heavily grazed paddocks showed a strong attraction of blowouts up to 400 m away from waterholes, with a clear peak in the neighborhood density at 200 m, which for sheep is the walking distance away from a water hole (Figure 3.43c).

3.3 Superposition of Point Processes

Point patterns in ecology may be the result of the independent superposition of two or more different mechanisms. A good example illustrating this can be provided by examining the effects of animal dispersal on the seed distributions of tropical trees. As Wiegand et al. (2009) note, this may have consequences for the resulting spatial pattern of recruits: "Howe (1989) pointed to two contrasting seed deposition patterns of animal dispersed tree species: scatter-dispersal and clump-dispersal. The behavior of frugivores such as small birds or bats that regurgitate, spit, or defecate seeds singly may lead to isolated recruits. Conversely, large frugivores that defecate seeds in masses may cause aggregated recruit patterns (Howe 1989). Additionally, secondary dispersal by central-place foragers such as ants may increase clumping of seed deposition by depositing seeds from a wide area within nest or its refuse piles (Passos and Oliveira 2002). Scattered vs. clustered seed deposition may occur for plant species with a diverse assemblage of dispersers, but can even be caused by two different behavioral patterns in a single frugivore species (e.g., Russo and Augspurger 2004), p. E107." Thus, point patterns may be a superposition of two (or more) component patterns. In the example of recruits of tropical tree species given by Wiegand et al. (2009), we may expect a random component pattern generated by scatter-dispersal and a clustered component pattern caused by clump-dispersal.

Calculating the summary statistics of superposition patterns is a powerful way to better understand the nature of complex spatial patterns. Illian et al. (2008, p. 371) provide formulas for the superposition of summary statistics. Under certain conditions, the summary statistics that result from the superposition of the two-point processes can be calculated, if the summary statistics of the two independent point processes are known. In the following sections, we present the superposition formulas for important summary statistics and provide examples.

3.3.1 Intensity Function

The superposition of two independent (but not necessarily homogeneous) point processes with intensities $\lambda_1(x)$ and $\lambda_2(x)$ yields the sum of the intensity functions of two individual intensity functions:

$$\lambda(x) = \lambda_1(x) + \lambda_2(x) \tag{3.126}$$

3.3.2 Second-Order Statistics

If two-point processes are independent, stationary, and isotropic, the superposition of the two pair-correlation functions $g_1(r)$ and $g_2(r)$ yields

$$g(r) = \frac{\lambda_1^2 g_1(r) + 2\lambda_1\lambda_2 + \lambda_2^2 g_2(r)}{(\lambda_1 + \lambda_2)^2} \tag{3.127}$$

The above formula can be understood based on product densities. Consider two disjoint disks of infinitesimal magnitude dx_1 and dx_2 centered on location x_1 and x_2, respectively. The product density $\rho^{(2)}(r)\, dx_2 dx_1 = (\lambda_1 + \lambda_2)$ $g(r)\, dx_1 dx_2$ gives the probability that one point of a given point process occurs within the first disk and a second point occurs within the second disk. In the superposition pattern we have four possibilities: (i) the first point (at x_1) is of type 1 and the second point (at x_2) is of type 1; (ii) the first point is of type 1 and the second point is of type 2; (iii) the first point is of type 2 and the second point is of type 1; and (iv) the first point is of type 2 and the second point is of type 2. If the patterns are stationary, the (partial) product density of type 1 points around type 1 points yields $\lambda_1\lambda_1 g_1(r)$, that of type 1 points around type 2 points yields $\lambda_1\lambda_2 g_{12}(r)$, that of type 2 points around type 1 points yields $\lambda_2\lambda_1 g_{21}(r)$, and that of type 2 points around type 2 points yields $\lambda_2\lambda_2 g_1(r)$. Because the two patterns are independent we have $g_{12}(r) = g_{21}(r) = 1$, and summing up all four cases and dividing by λ^2 yields Equation 3.127.

Integrating Equation 3.127 yields the superposition of the two *K*-functions $K_1(r)$ and $K_2(r)$:

$$K(r) = \frac{\lambda_1^2 K_1(r) + 2\lambda_1\lambda_2\pi r^2 + \lambda_2^2 K_2(r)}{(\lambda_1 + \lambda_2)^2}$$ (3.128)

If the two component processes are homogeneous Poisson processes with $g_1(r) = g_2(r) = 1$ and $K_1(r) = K_2(r) = \pi r^2$ it can be easily verified that the superposition process yields $g(r) = 1$ and $K(r) = \pi r^2$.

3.3.3 Spherical Contact Distribution

If two-point processes are independent, stationary, and isotropic, the superposition of the two spherical contact distribution functions $H_{s,1}(r)$ and $H_{s,2}(r)$ yields

$$H_s(r) = 1 - [1 - H_{s,1}(r)][1 - H_{s,2}(r)]$$ (3.129)

where $H_{s,1}(r)$ and $H_{s,2}(r)$ are the spherical contact distributions of the two-component processes. The formula can be understood with the following arguments. The quantity $1 - H_s(r)$ is the probability that a test point has no neighbor of pattern 1 within distance *r* and no neighbor of pattern 2 within distance *r*. If the two patterns are independent, this probability can be separated into the product of the two separate probabilities, that is, the test point has no neighbor of pattern 1 within distance *r*, which yields $1 - H_{s,1}(r)$, and the probability that the test point has no neighbor of pattern 2 within distance *r*, which yields $1 - H_{s,2}(r)$. This leads directly to Equation 3.129.

3.3.4 Nearest-Neighbor Distribution Function

If two-point processes are independent, stationary, and isotropic, the superposition of the distances to the nearest neighbors of the two distributions $D_1(r)$ and $D_2(r)$ yields

$$D(r) = 1 - \frac{1}{\lambda_1 + \lambda_2}(\lambda_1(1 - D_1(r))(1 - H_{s,2}(r)) + \lambda_2(1 - H_{s,1}(r))(1 - D_2(r)))$$

(3.130)

The formula can be understood analogously to that of the spherical contact distribution when considering the probability $1 - D(r)$ that there is no point of the superposition pattern within distance *r*. In this case, a type 1 point has no type 1 neighbor and no type 2 neighbor within distance *r*. The probability that a type 1 point has no type 1 neighbor within distance *r* is given by $[1 - D_1(r)]$.

If we think of a type 1 point as a test point for the spherical contact distribution of point type 2, the quantity $(1 - H_2(r))$ gives the probability that the type 1 (test) point has no type 2 neighbor within distance r. Thus, if the patterns are independent, the joint probability yields $[1 - D_1(r)][1 - H_2(r)]$. The argument for type 2 points is analogous. If we now sum up the respective probabilities for all points, Equation 3.130 arises. To do this we need to weight the contribution of each pattern with its respective intensity (e.g., the contribution to the distribution function due to type 1 points needs to be weighted by λ_1).

3.3.5 Examples of the Superposition of a Thomas Process with a Random Pattern

Application of formula 3.127 for the independent superposition of a Thomas process, with $g_1(r)$ comprising a proportion c of all points of the joined pattern, with a random pattern yields

$$g_{1+2}(r) = 1 + \frac{c^2}{\rho} \frac{\exp(r^2/4\sigma^2)}{4\pi\sigma^2} \tag{3.131}$$

where c is the proportion of the points associated with the Thomas process. We can understand Equation 3.131 by recalling that the pair-correlation function of the Thomas process (Box 2.5) is given by

$$g_1(r) = 1 + \frac{1}{\rho} \frac{\exp(r^2/4\sigma^2)}{4\pi\sigma^2} \tag{3.132}$$

and the pair-correlation function for a random pattern is $g_2(r) = 1$. Thus, the resulting pair-correlation function 3.131 is exactly the same as that of the component Thomas process (3.132), except that the parameter ρ which characterizes the density of cluster centers results in the superposition pattern $\rho_{1+2} = \rho/c^2$. This means that the pair-correlation function alone cannot distinguish between the pure Thomas process and the superposition of a Thomas process with a random pattern. However, the fit with the pair-correlation function will overestimate the real number of clusters by a factor of $1/c^2$, where c is the proportion of the points belonging to the cluster component pattern.

To illustrate the values of the summary statistics under superposition we use the pattern shown in Figure 3.44, which was generated by a Thomas process within a 500×500 m^2 grid (with cluster radius of 10 m and 100 randomly distributed cluster centers). We then superposed different numbers of random points with this pattern. As expected, the values of the pair-correlation function decrease with the addition of random points (Figure 3.44b). The spherical contact distribution also shows a clear and strong response to the addition of random points. Because the random points fill gaps, the probability of finding a point of the superposition pattern within distance r from a test location increases substantially with the addition of random points

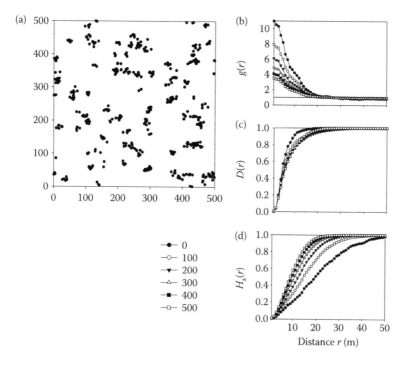

FIGURE 3.44

Impact of the superposition of increasing numbers of random points with a Thomas process consisting of 500 points. (a) Map of the Thomas process used in the analysis. (b) Pair-correlation function $g(r)$. (c) Distribution function $D(r)$ of nearest-neighbor distances. (d) The spherical contact distribution $H_s(r)$. The numbers in the legend refer to the number of random points added to the map in panel (a) for the corresponding analyses.

(Figure 3.44d). The distribution function $D(r)$ of the distances to the nearest neighbor shows a more complex behavior. First, when adding 100 random points, $D(r)$ decreases at larger scales, because many random points will be located within the gaps of the cluster component pattern and will have their nearest neighbor at larger distances. However, addition of more than 100 random points does not change $D(r)$ much, once all the gaps are filled. In this case, additional random points will only rarely be the nearest neighbor of an existing point of the superposition pattern. This indicates that using the spherical contact distribution as an additional summary statistic, to complement the pair-correlation function, can provide important additional information on the nature of the pattern.

Figure 3.45 shows a similar situation, but now the number of points is fixed ($n = 500$) and a certain number of points of the cluster pattern are removed and replaced by random points. As expected, the pair-correlation function decreases steadily. When random points replace 80% of the original clustered points, the signal of clustering almost disappears (Figure 3.45b). The change

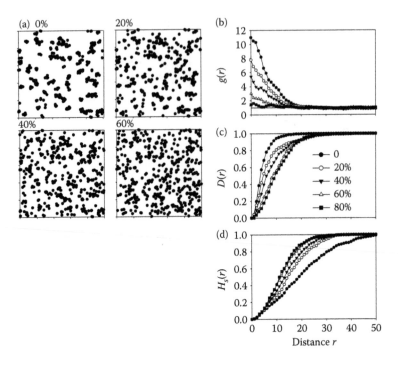

FIGURE 3.45
Impact of randomly displacing points of a Thomas process. (a) Maps of the superposition of a Thomas process (top left), with a random pattern where 20%, 40%, and 80% of the points of the original cluster pattern are replaced by random points. Panels (b)–(d) show the resulting summary statistics.

in shape of the distribution function $D(r)$ of the distances r to the nearest neighbor is also interesting (Figure 3.45c). A low proportion of random points (i.e., 20%) affects mostly the values of $D(r)$ at intermediate distances r (say between 7 and 12 m). This is due to the replacement of points from clusters (which usually have short nearest-neighbor distances) by random points. The random points are mostly placed in gaps of the cluster pattern and therefore yield larger distances to the nearest neighbor. However, higher proportions of random points do not change $D(r)$ at distances greater than 18 m, but occur mostly at smaller distances (say 5 m). This is because many small neighbor distances disappear due to removal of points from the cluster pattern.

The spherical contact distribution exhibits a more continuous response to replacement of clustered points by random points. In all cases, the maximal change occurs at distances between 15 and 35 m (Figure 3.45d). This is understandable because the gaps of the cluster pattern are continuously filled by random points. As a result, the spherical contact distribution is a better summary statistic to use in supplementing the pair-correlation function than is the nearest-neighbor distribution function.

3.3.6 Examples of Superposition of Two Thomas Process

Another important case of the superposition of patterns is given by the independent superposition of two different Thomas processes. The pair-correlation function can be constructed from the parameters for each Thomas process, that is, σ_1, σ_2, ρ_1, and ρ_2, and information regarding the proportion of points associated with each process (Stoyan and Stoyan 1996):

$$g_{1+2}(r) = 1 + \frac{c^2}{\rho_1} \frac{\exp(r^2/4\sigma_1^2)}{4\pi\sigma_1^2} + \frac{(1-c)^2}{\rho_2} \frac{\exp(r^2/4\sigma_2^2)}{4\pi\sigma_2^2} \quad (3.133)$$

In this case, c is the proportion of points associated with Thomas process 1. If the two Thomas processes are the same, the random superposition yields

$$g_{1+2}(r) = 1 + \left(\frac{c^2 + (1-c)^2}{\rho} \right) \frac{\exp(r^2/4\sigma^2)}{4\pi\sigma^2} \quad (3.134)$$

For the case where $c = 0.5$ (i.e., both Thomas processes contain the same number of points), we find that $c^2 + (1-c)^2 = 0.5$, thus the resulting pair-correlation function yields a Thomas process with double the number of cluster centers (which is indeed the case since the clusters of a Thomas process are independently placed). Wiegand et al. (2007c, 2009) present applications of the superposition of Thomas processes to recruits of tropical trees. We will treat these and similar examples in detail in the following chapter on univariate point patterns (Section 4.1.4).

⋆⋆ 3.4 Toolbox

This chapter presents different techniques and methods that are especially useful in a variety of applications, but not specific to any given data type. They include general methods for detecting clusters, the analysis of irregularly shaped observation windows, and pattern reconstruction.

⋆⋆ 3.4.1 Cluster Detection Algorithm

Identifying and characterizing clusters is a task that is of interest in several applications of point-pattern analysis (e.g., Coomes et al. 1999; Plotkin et al. 2002). Here, we present a simple algorithm that can be used to detect clusters and subsequently describe their properties, such as the "center of mass" or the number of points belonging to each cluster.

The purpose of a cluster algorithm is to join together ecological objects idealized as points into successively larger clusters, using a measure of distance between objects. In this section, we present a point-pattern-based approach of cluster analysis (Sneath and Sokal 1973) that identifies cluster centers and points belonging to these clusters. Initially, we have a univariate point pattern given as a map, with the positions of points within an observation window W. Each initial point is called a "cluster," even if it is initially composed of a single point. In later steps, the points are replaced by clusters, which contain several of the original points of the pattern. Each cluster is characterized by the number of points it contains and by the cluster center, which is the center of gravity of all points of the cluster.

In the first step, the algorithm determines the pair of cluster centers that have the smallest distance among all pairs of cluster centers. The points of the two clusters are then amalgamated into one new cluster, and the two original clusters are deleted. Different rules can be used to amalgamate clusters. Here, we use a rule that maintains the identity of all points, but uses the "center of gravity" of all points of the cluster as coordinates of the new cluster. The algorithm proceeds by searching for the pair of cluster centers of the updated pattern that has the shortest separation distance among all cluster centers. It then amalgamates these two clusters and uses all points belonging to the new "cluster" for determining the new mean position (which is the center of gravity). In this way, the number of clusters is reduced by one in each step. The same procedure is repeated as long as cluster centers can be found that have a separation distance less than a predefined maximal distance r_c, which is the basic parameter of the algorithm.

Depending on the value of r_c, the size of the clusters will vary because the distance r_c is the minimal distance between two cluster centers. Larger values of r_c yield fewer, but larger clusters. Note that this cluster algorithm uses an amalgamation rule called "unweighted pair-group centroid" because the coordinate (or centroid) of a cluster is the center of gravity of all points contained in the cluster (Sneath and Sokal 1973). Cluster analysis is not restricted to point patterns, it can be used for any objects for which we can define a distance or similarity index. Many other cluster algorithms exist that differ in their amalgamation rules and distance definition. For example, the distance between two clusters may be defined as the distance between the two nearest or furthest points of the two different clusters, or as the average distance between all pairs of objects in the two clusters. Additionally, the number of objects contained in the clusters may be used to weight the distance measure. However, the unweighted pair-group centroid generally works well for the purpose of identifying clusters in point patterns.

The output of the cluster algorithm is a list of cluster centers and points that belong to each corresponding cluster. This information can be used to characterize the pattern with respect to the distribution of cluster sizes and the frequency of points within a given distance to the cluster center. This information can also be used to construct a classification tree. Finally, for specific applications it may be important to replace the artificial cluster centers by a point of the pattern. For example, we may use the point of the cluster, which is closest to the cluster center. In this way, each cluster can be replaced by a representative point of the cluster for analysis.

One application of the cluster detection algorithm is *de-clustering* of a pattern for nonparametric intensity estimation (Box 2.4). The intensity function reflects the impact of first-order effects on the position of points. For example, if the points are trees of a given species in a tropical forest, the intensity function may provide a good description of habitat suitability: some areas of the plot will be better suited for the tree species and have a higher probability of hosting a tree than other areas. This leads to locally elevated point densities. However, other mechanisms, such as clumped or limited seed dispersal, which are independent of habitat suitability, may partly override the signal of habitat suitability and create additional areas of elevated point densities. Such second-order clustering is not caused by better environmental conditions, but by a mechanism that deposits many seeds at a given location. Due to a "mass effect" a location may host more recruits than expected by its habitat suitability. Thus, second-order clustering may "contaminate" nonparametric estimates of the intensity function. This phenomenon has been known for a long time and is an important issue in studies using parametric estimation techniques to develop habitat and species distribution modeling (e.g., Dorman et al. 2007). The main issue, in the context of parameter estimation, is that second-order clustering causes a form of pseudo-replication, because nearby points contain similar information.

The cluster detection algorithm can be used to remove strong small-scale clustering. This is an important step before applying a nonparametric kernel estimate of the intensity function, if it is suspected that the strong clustering is due to factors other than localized patterns of favorable habitat conditions. In the first step, one should analyze a plot of the pair-correlation function to identify an initial value of the cluster detection parameter r_c (Figure 3.46a). Several values of r_c should be examined to find the appropriate parameter value. After application of the cluster algorithm, clusters are replaced by the point of the cluster that lies closest to the cluster center. In this way, each cluster is replaced by one representative point. We call this de-clustering.

Figure 3.46 shows an example of this approach. The full pattern is shown in Figure 3.46b, and the resulting pair-correlation functions for the parameters $r_c = 0, 1, 2, 3, 4$, and 5 m are shown in Figure 3.46a. The pair-correlation function of the original pattern (open squares in Figure 3.46a) shows strong clustering at small distances, with a neighborhood density at 1 m being 25 times as high as expected under a random pattern. Most of the small-scale clustering in the pair-correlation functions is removed only when minimal distances of $r_c = 4$ (open triangles) or 5 m (closed squares) are used in declustering. The sequence of Figure 3.46c through f shows how the declustering works as the maximal distance r_c is increased. Indeed, the small clusters are still visible for parameter $r_c = 2$ and 3 m (Figure 3.46c,d). Ideally, the selection

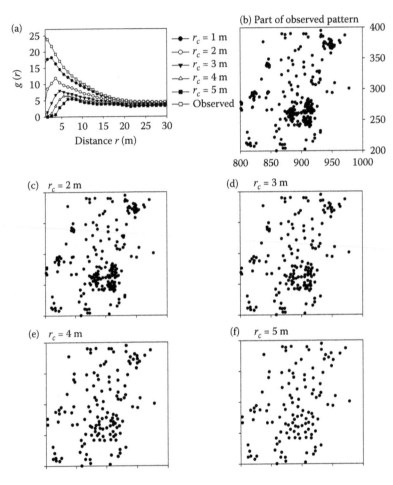

FIGURE 3.46
Application of a de-clustering algorithm. (a) The pair-correlation functions of the observed and de-clustered patterns, varying the parameter r_c, the maximal distance between clusters. (b) Observed pattern. (c–f) De-clustered patterns with different parameter values for r_c. See Section 3.4.1 for a detailed explanation.

of the parameter r_c should be guided by biological information on the scales at which the environment typically varies and on the scale of potential clustering mechanisms. Otherwise, one may remove too much clustering from the pattern and underestimate the variability in habitat suitability.

3.4.2 Analyzing Irregular Observation Windows

In most applications of point-pattern analysis the observation window W has a rectangular shape, or sometimes a circular shape. The selection of regular shapes is not entirely driven by esthetic considerations, but is more a function of the practical issue of edge correction, which is especially important for calculating the pair-correlation and K-functions. For example, the popular Ripley edge correction (Section 3.1.2.1) is based on a weighting factor that is given by the circumference of a disk with radius r divided by the circumference of the disk that overlaps W. Relatively straightforward analytical formulas exist for calculating the Ripley weights for rectangular and circular observation windows. However, their calculation is nontrivial if the observation window has an irregular shape. An analytically tractable method was not introduced until 1999, when Goreaud and Pélissier (1999) presented a method that approximated the real shape of an irregularly shaped observation window W either by a polygonal observation window or by removing a polygonal area inside a rectangle.

Because of the difficulties produced by having irregularly shaped observation window, most observation windows for point-pattern data are placed so that they can retain a rectangular shape. However, as outlined in Section 2.6.3.2, it might be necessary to exclude some areas from an initially rectangular observation window to retain a homogeneous pattern (e.g., Figures 2.25, 2.28, 2.30). In other cases, the area has an irregular shape that cannot be avoided, such as the paddocks in the example of blowouts around water holes (Figure 3.43). In the following, we present two methods that can deal with edge correction for pair-correlation and K-functions and irregular observation windows. We also recommend that the interested reader examine the study by Goreaud and Pélissier (1999), which presents an extension of the Ripley edge correction method (Section 3.1.2.1; Figure 3.2) approximating observation windows by polygons.

3.4.2.1 Grid Approximation

Perhaps the simplest method in considering an irregular observation window is to use an underlying grid and represent the irregular observation window by small cells (Figure 3.47; Wiegand and Moloney 2004). Each cell can be either in or out of the observation window. The area of a ring or circle located inside the observation window W is then approximated by the underlying grid (Figure 3.47). We can then estimate the various point-pattern statistics accounting for edge effects in a straightforward fashion. For example, the estimators can account for the number of points lying inside the rings or circles around a focal point and also determine the number of cells within the circle or ring lying inside W. The latter information can be used in calculating an edge correction empirically. In the case of the O-ring statistic, we can rewrite Equations 3.13 and 3.118 to yield

$$\hat{O}_{ij}(r) = \frac{\sum_{k=1}^{n} \text{Points}_j[R_k(r)]}{\sum_{k=1}^{n} \text{Area}[R_k(r)]} \tag{3.135}$$

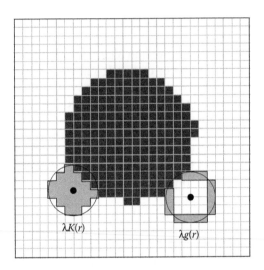

FIGURE 3.47
Approximation of an irregular observation window by an underlying grid. The figure shows part of an observation window where a circular patch is excluded. The cells that belong to the observation window are shown as white and those excluded as black. Locations that will be used to estimate two types of point-pattern statistic are approximated by the underlying grid (darker gray) as shown in the figure.

where $R_k(r)$ is a ring with radius r around point k, which is approximated by the underlying grid (Figure 3.47). The operator Points$_j[X]$ counts the number of points of type j in X, and the operator Area$[X]$ counts the number of cells in X. The corresponding estimator of the K-function yields

$$\lambda K_{ij}(r) = \pi r^2 \frac{\sum_{k=1}^{n} \text{Points}_j[C_k(r)]}{\sum_{k=1}^{n} \text{Area}[C_k(r)]} \tag{3.136}$$

where $C_k(r)$ is a circle of radius r around point k, which is approximated by the underlying grid (Figure 3.47). This method allows for efficient approximation of the WM estimators for $\lambda g(r)$ and $\lambda K(r)$ (3.17, 3.36), if the cell size is sufficiently small. However, the calculation of the estimator will be quite slow, if the grid is too fine. Figure 3.48a shows a random pattern inside an irregular observation window where a circular area is missing in the center. We approximated this irregular shape with an underlying grid with a mesh width of 0.5 m. Figure 3.48b shows the results of an analysis with 199 simulations of the CSR null model and the pair-correlation function with a ring width of 2 m.

3.4.2.2 Using Inhomogeneous Functions

A second approach for analyzing irregularly shaped observation windows is provided by the estimators of the inhomogeneous pair-correlation and

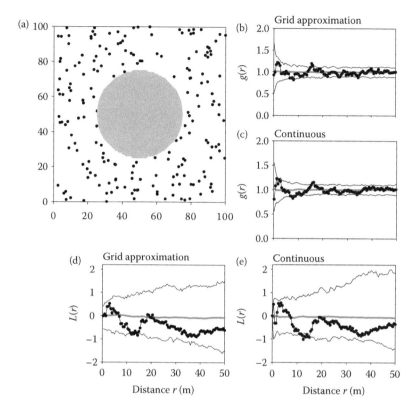

FIGURE 3.48

Analyses accounting for irregularly shaped observation windows. (a) A random pattern where a central circle is excluded. (b) Grid approximation of the pair-correlation function (dots), with a mesh size of 0.5 m and ring width of 2 m. A CSR null model distributing points only within the irregular observation window was used. (c) Same as panel (b), but applying the methods for inhomogeneous pair-correlation functions using the WM estimator with a ring width of 2.5 m. (d) The L-function, using a grid approximation as in panel (b). (e) The L-function, using the approach as in panel (c).

K-functions, based on the Ohser or WM edge correction methods (Equations 3.29, 3.31, 3.41, 3.42). This approach represents the irregularly shaped observation window by using the intensity function $\lambda(x)$: if the location is inside the observation window, we have $\lambda(x) = \lambda$ (where λ is the intensity of the pattern inside the irregularly shaped observation window), otherwise we have $\lambda(x) = 0$. Clearly, with this approach we will still approximate the intensity function with an underlying grid, but this grid can be made as fine grained as the resolution of the data and does not slow down the estimation as with the grid-based estimator above. Figure 3.48c shows the results of this approach, which uses an underlying grid with a mesh width of 0.5 m for the intensity function and a ring width of 2.5 m in calculating the WM estimator

(Equation 3.31). There is basically no difference to the grid-based approach. We also calculated the *L*-function with both approaches (Figure 3.48d,e) and found essentially the same results.

★★ 3.4.3 Pattern Reconstruction

For a number of applications in the study of point patterns, it is useful to generate patterns with predefined properties. For example, null models produce patterns with known properties by holding certain aspects of a pattern fixed, while randomizing other aspects of the data (see Section 2.4). This is especially difficult in the case of bivariate or multivariate patterns. The null model of independence in bivariate patterns should conserve the observed univariate structure of the two component patterns, but only break their (potential) association. Although this has been recognized for quite a while (see, e.g., Lotwick and Silverman 1982 or Berman 1986), no satisfying solution has yet been found. The early solution proposed to solve this problem was the toroidal shift (Lotwick and Silverman 1982; Figure 2.4), where one of the two component patterns is shifted entirely by adding a fixed random vector to each coordinate and then wrapping it on a torus. The toroidal shift null model has two potential problems: first, it maintains most of the observed interpoint distances exactly, with no stochastic variability. Second, it may produce additional artifacts at the edges, were the patterns are wrapped. In contrast, the independence null model should produce stochastic replicates of one or both component patterns that have the same statistical properties as the observed pattern. Another nonperfect solution suggested to achieve the desired properties is to use parametric, point-process models to fit the second component pattern and use realizations of the fitted, point process as a null model. This method works well if the point process represents a good fit to the pattern, which is, in general, not guaranteed. *Pattern reconstruction provides a solution to developing a bivariate null model of independence* (Wiegand et al. 2013a), because it can produce stochastic replicate patterns, which share predefined properties of the observed pattern. The predefined properties are summary statistics of the observed pattern. The nonparametric algorithm of pattern reconstruction can reproduce patterns that match the selected summary statistics as closely as desired. Thus, pattern reconstruction can be used for generating null model patterns.

Another important application of pattern reconstruction is assessment of the degree of information captured by a given summary statistic or a combination of summary statistics (Wiegand et al. 2013a). In Section 3.1, we presented many different summary statistics that capture certain aspects of the spatial structure of point patterns. Although the interpretation of most of the summary statistics is relatively straightforward, it is not clear how well a given summary statistic (or several summary statistics together) describe the potentially complex properties of point patterns. In other words, if we use only the pair-correlation function to summarize the properties of a given point pattern, do we miss other properties of the pattern that may be important for our ecological question? We will certainly do so in many cases. For example, it is well known that several different point processes (such as cluster processes) have the same pair-correlation function. In this case, additional summary statistics, such as the distribution function to the nearest-neighbor distances or the spherical contact distribution, are required to differentiate from among the various candidate processes (e.g., Section 3.3.5). In particular, the adequate description of complex point patterns may require the use of several summary statistics simultaneously (e.g., trees in tropical forests can exhibit clustering at multiple scales; Condit et al. 2000; Wiegand et al. 2007c, 2009). While it is common practice for experienced spatial statisticians to employ multiple summary statistics (e.g., Illian et al. 2008, p. 214), ecological studies using point-pattern analysis evaluate their data mostly with only one or two. However, it should be noted that important properties of the spatial patterns may remain undetected, if nonappropriate summary statistics are used, and the inferences made would be much weaker. Thus, the question arises as to how many, and which, summary statistics are needed to fully describe the properties of a point pattern. One elegant way of answering this is to use pattern reconstruction.

The *purpose of pattern reconstruction* is to create stochastic replicates of observed patterns that closely approximate predefined summary statistics of the observed pattern. For pattern reconstruction, we use an inverse approach originally proposed in materials science (Rintoul and Torquato 1997; Torquato 1998). This approach uses optimization techniques to reconstruct a point pattern only with the limited information provided by a given combination of summary statistics. To this end, a variant of the simulated annealing algorithm (Kirkpatrick et al. 1983) is used to generate patterns that minimize the deviations between the summary statistics of the observed and reconstructed patterns. This idea has been translated by Tscheschel and Stoyan (2006) into point-pattern analysis and is illustrated in detail in Illian et al. (2008, p. 407ff). The following sections follow largely Wiegand et al. (2013a).

3.4.3.1 Annealing Algorithm

Here, we describe the pattern reconstruction algorithm outlined in Tscheschel and Stoyan (2006) and Illian et al. (2008), but expand it for non-stationary patterns by conditioning on the intensity function $\lambda(x)$ (Wiegand et al. 2013a). The reconstruction method is a variation of simulated annealing (Kirkpatrick et al. 1983) that generates a series of patterns by trial and error, approaching the values of the summary statistics of the observed patterns more closely in each simulation step (Yeong and Torquato 1998).

Structural information on the observed pattern φ is measured by a set of I functional summary statistics $f_i^{\varphi}(x)$ (with index $i = 1...I$), where the variable

x represents distance r or neighborhood rank k. During each simulation step t, we also estimate the corresponding $f_i^{\psi_t}(x)$ of the simulated pattern ψ_t. This allows us to determine the deviation between $f_i^{\varphi}(x)$ and $f_i^{\psi_t}(x)$ by means of the "partial energy"

$$E_i^{\varphi}(\psi_t) = \sqrt{\frac{1}{n_i}\sum_{b=1}^{n_i}[f_i^{\varphi}(x_b) - f_i^{\psi_t}(x_b)]^2} \qquad (3.137)$$

where the variable x is evaluated at n_i discrete values x_b. We divide by the number of measurements n_i and use the square root to obtain a direct measure of mean error. To combine the I partial energies E_i^{φ} into a total energy $E_{\text{total}}^{\varphi}$, we need to normalize the values because their absolute values may vary greatly or depend on the intensity λ. One approach to normalization is to give all summary statistics approximately the same importance in the reconstruction. In practice, one has to select the weights w_i so that the different summary statistics yield approximately the same value of E_i^{φ}, if the observed and simulated patterns are in a good agreement. By using the weights w_i, the total energy at simulation step t yields

$$E_{\text{total}}^{\varphi}(\psi_t) = \frac{\sum_{i=1}^{I} w_i E_i^{\varphi}(\psi_t)}{\sum_{i=1}^{I} w_i} \qquad (3.138)$$

The reconstruction of a pattern φ starts with a random pattern ψ_0 that has the same number of points as φ. In each simulation step t, a randomly selected point is tentatively removed and a new point with random coordinates is proposed instead. This new point is accepted if the total energy $E_{\text{total}}^{\varphi}(\psi_t)$ decreases, that is, $E_{\text{total}}^{\varphi}(\psi_t) < E_{\text{total}}^{\varphi}(\psi_{t-1})$, otherwise another new point is considered (Tscheschel and Stoyan 2006). Thus, the new pattern ψ_t is slightly more similar to the observed pattern than the previous pattern ψ_{t-1}. The algorithm proceeds until a maximum number of steps is reached (e.g., 40,000 or 80,000) or until the total energy is less than a predefined threshold (e.g., 0.005) (Tscheschel and Stoyan 2006). An additional stopping rule may be added to avoid conducting a large number of steps in a configuration that shows no improvement. An index that controls the improvement in the fit relative to the number of iteration steps already conducted is the proportion of accepted points. If this proportion drops below a given threshold, the simulation stops. This "stochastic improvements-only" algorithm is able to find local minima with very small total energies and each simulation will end up in a different local minimum (the absolute minimum would be the observed pattern φ). We, therefore, obtain stochastic replicate patterns with predefined properties provided by the I summary statistics $f_i^{\varphi}(x)$ of the observed pattern φ (Tscheschel and Stoyan 2006).

Because the pattern reconstruction algorithm requires many simulation steps, an efficient calculation of the summary statistics is necessary. As proposed by Tscheschel and Stoyan (2006) and Illian et al. (2008, p. 414), only the part of an estimator that was affected by the exchange of one point is updated. Note that there is no need to use edge correction for the estimators of the pair-correlation function and the K-function. Although this is counterintuitive at first, it becomes understandable if we consider Ohser edge correction (Equation 3.21), which accomplishes edge correction by multiplying the naive estimator $\hat{g}''(r)$ of the pair-correlation function by a factor $c(r)$, characterizing the expected bias due to edge effects. In this case, the estimator of the pair-correlation function can be written as $\hat{g}(r) = c(r)\hat{g}''(r)$, where the correction factor depends only on distance r and the geometry of the observation window, but not the pattern. However, because both, the estimator of the pair-correlation function of the observed pattern φ and that of the reconstructed pattern ψ_t share the same edge correction factor $c(r)$, it can be factored out in Equation 3.137. This means that the edge correction factor $c(r)$ plays only the role of a weight when combining the partial energy of all summary statistics in Equation 3.138. Because the factor $c(r)$ increases with distance r, it has the effect of putting slightly more weight on the larger values of $\hat{g}''(r)$ in the estimation of $E^{\varphi}_{\text{total}}(\psi_t)$. However, this will have little effect on the reconstruction. Thus, reconstructions based on the naive estimator of $g(r)$ and $K(r)$ will produce the same results as reconstructions using the Ohser estimator. However, using the Ripley estimator will produce slightly different patterns than the Ohser or the WM estimator because the numerical values of $K(r)$ may differ (e.g., Figure 3.14). Note that we used the pair correlation function $g(r)$ not directly for pattern reconstruction, but the related quantity $2\pi r dr\, \lambda g(r)$ which gives the mean number of points at rings with radius r and width dr around the points of the pattern. We used this quantity instead of $g(r)$ because it varies over a much reduced range, especially if the pattern shows strong small-scale aggregation (e.g., Figure 3.49).

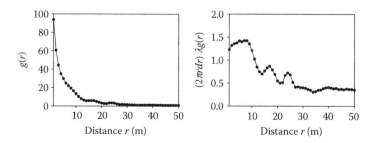

FIGURE 3.49
Pair-correlation function $g(r)$ and the transformation $2\pi r dr\lambda g(r)$ used for pattern reconstruction for the highly clustered species *C. insignis* (shown as open disks in Figure 3.57a).

3.4.3.2 Reconstruction of Homogeneous Patterns

A homogeneous pattern is composed of stochastic repetitions of a typical point configuration within an entire observation window and the variability around the typical point configuration is only due to small stochastic fluctuations. An important aspect of the typical point configuration, the average number of points within or at distance r from the typical point, is captured by second-order summary statistics. Therefore, the pair-correlation function (and/or the K-function) will capture important aspects of homogeneous patterns. However, even homogeneous patterns may require additional summary statistics to describe the variability in the typical point configuration around the mean. For example, the same pair correlation function may occur for a pattern with all the points having approximately the same number of neighbors within distance r as there would be for a pattern where, for example, most points have only a small number of neighbors, but a few points have many neighbors. The variability in the number of neighbors can be captured in the reconstruction by incorporating the nearest to kth neighbor distribution functions in the algorithm.

The impact of these considerations is illustrated in Figure 3.50, where we reconstructed a pattern that resulted from an independent superposition of a homogeneous cluster pattern (196 points) and a homogeneous hyperdispersed pattern (197 points) (Figure 3.50a). If we reconstruct the pattern using only the pair-correlation function, the resulting pattern is a sort of "average" pattern that yields a perfect reconstruction of the pair-correlation function (Figure 3.50b), but lacks the pronounced clusters and hyperdispersion of the observed pattern (Figure 3.50a). Consequently, the distribution function $D(r)$ of the distance to the nearest neighbor of the reconstructed pattern shows strong departures from the observed function because it does not represent the mixture of a clustered pattern with a hyperdispersed pattern (Figure 3.50g). However, if we reconstruct the pattern based only on the distribution function $D(r)$ of the distance to the nearest neighbor, we obtain a perfect fit of $D(r)$ (Figure 3.50h), but the reconstruction lacks the strong clusters of the original pattern, although it yields a satisfying representation of the hyperdispersed part of the pattern (Figure 3.50c). However, using the pair-correlation function and the distribution function $D(r)$ together, we obtain a reconstruction that shows both the pronounced clusters and the hyperdispersion (Figure 3.50d). It also now matches both summary statistics used in reconstruction quite well (Figure 3.50i). It is notable that a similarly good reconstruction can be obtained by using the first 50 distribution functions $D_k(r)$ of the distance to the kth neighbor (Figure 3.50e). This is reasonable to expect, since the $D_k(r)$ taken together contain more information than the pair-correlation function and the K-function and we have the relationship

$$\lambda K(r) = \sum_{k=1}^{\infty} D_k(r) \qquad (3.139)$$

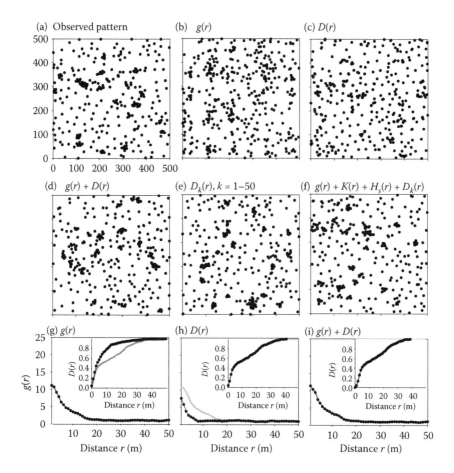

FIGURE 3.50
Pattern reconstruction. (a) Observed pattern, being the random superposition of a cluster pattern created with a Thomas process (with parameter $\sigma = 5$ m and 25 clusters) and a regular pattern, where points separated by distances below 15 m occurred only rarely. (b–f) Patterns reconstructed with different combinations of summary statistics, as indicated. (g) Observed (gray bold line) and reconstructed (closed disks) pair-correlation function for the case where the reconstruction used only the pair-correlation functions as a summary statistic as shown in panel (b). (h) Same as panel (g), but reconstruction was only done with the distribution function $D(r)$ of the distance to the nearest neighbor. (i) Same as panel (g), but reconstruction was done with $g(r)$ and $D(r)$. For estimators of the summary statistics refer to Wiegand et al. (2013a).

In the observed pattern we find that the 40th nearest neighbor was at least 50 m away (which means that $D_k(r) = 0$ for $k > 40$ and $r < 50$ m). Therefore, the K-function is completely determined by the first 40 $D_k(r)$ up to distances of 50 m. Finally, using several summary statistics together [i.e., $g(r)$, $K(r)$, $H_s(r)$, and $D_k(r)$ with $k = 1,...50$], the reconstruction does not improve much (Figure 3.50f). This suggests that each pattern may have, depending on its degree of complexity, an "optimal" set of summary statistics. We will investigate this issue in a later section (Section 3.4.3.6).

Wiegand et al. (2013a) found that homogeneous patterns are usually well reconstructed by using the pair-correlation function $g(r)$, the spherical contact distribution $H_s(r)$, and some of the distribution functions $D_k(r)$ of the distance to the kth neighbor. Note that they weighted the $D_k(r)$ in Equation 3.138 such that all the $D_k(r)$ taken together were weighted to the same degree as each of the other summary statistics. Thus, the $D_k(r)$ were functionally treated in the reconstruction as one summary statistic (see also Tscheschel and Stoyan 2006). The reason for the good performance of the combination of $g(r)$, $D_k(r)$, and $H_s(r)$ in reconstructing homogeneous patterns is that the pair-correlation function controls the average neighborhood density, the distribution functions of the distance to the kth neighbor control the variability of the neighborhood density (i.e., the small-scale clustering that determines the number of neighbors individual points have within distance r), and the spherical contact distribution controls the "gaps" in the pattern.

3.4.3.3 Impact of Spatial Scale on Reconstruction

Successful pattern reconstruction depends not only on the selection of appropriate summary statistics, but also on the scales over which the summary statistics are evaluated. If a pattern contains nonrandom structures at distances greater than say 50 m, but the summary statistics (as in Figure 3.51) capture only information up to 50 m, the reconstruction will not be able to correctly reassemble the pattern at distances greater than 50 m. To demonstrate this point we generated 100 reconstructions of the pattern shown in Figure 3.50a, based on the pair-correlation function and the distribution functions of the kth neighbor $D_k(r)$ for $k = 1 \ldots 50$, using information from the summary statistics at distances up to $r = 50$ m. Figure 3.51a shows the results for the quantity $(2\pi r dr)\, \lambda g(r)$, which was used instead of the pair-correlation function as the functional summary statistic $f_i(r)$ in Equation 3.137. For distances r smaller than 50 m, the values were virtually identical to the observed pattern, but at greater distances there was increasing variability. This is because the reconstruction algorithm did not use information at the larger scales (i.e., $r > 50$ m) and, as a consequence, stochastic effects influenced the reconstruction more with increasing scale. However, because the pattern was homogeneous and showed no larger-scale structures (except stochastic ones) the departures are relatively small and the observed function is the same as the average of the 100 reconstructions (Figure 3.51a). We find a similar result for the L-function (Figure 3.51d), but because the L-function emphasizes larger-scale effects, the stochastic effects at distances greater than 50 m result in increasingly wider envelopes, although the mean value of all reconstructions is close to the value observed for the actual pattern. The same was true for the mean

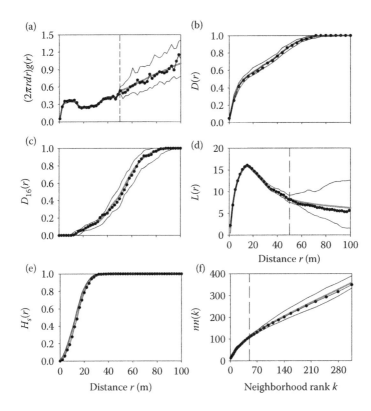

FIGURE 3.51
Pattern reconstruction of a homogeneous pattern. We conducted 100 reconstructions of the pattern shown in Figure 3.50a based on the pair-correlation function and the distribution functions to the kth neighbor $D_k(r)$ for $k = 1\ldots50$, using the information from the summary statistics up to distances of $r = 50$ m. The panels show the observed summary statistics of the pattern (closed disks) with simulation envelopes being the minimal and maximal values of the 100 reconstructions (solid lines) and their mean value (gray bold line).

distance to the kth neighbor, which fits very well up to distances of 50 m, but is subject to stochasticity for the larger neighbor ranks (Figure 3.51f). As expected, the distribution function $D(r)$ to the nearest neighbor is well reconstructed (Figure 3.51b) and even the distribution functions to the 16th neighbor $D_{16}(r)$ are well matched in all reconstructions (Figure 3.51c). The spherical contact distribution $H_s(r)$, which was not used for reconstruction, is only slightly overestimated (Figure 3.51e).

The results shown in Figure 3.51 indicate that a homogeneous pattern can be reconstructed well, even with limited distance information. This is a consequence of being homogeneous and having no larger-scale dependencies. At larger distances we find only differences due to random stochasticity, rather than to larger-scale dependencies.

3.4.3.4 Reconstruction of Heterogeneous Patterns

Although the summary statistics used in pattern reconstruction are developed for homogeneous patterns, they can also be used to represent certain aspects of heterogeneity. As a consequence, the pattern reconstruction algorithm is also able to reconstruct heterogeneous patterns to some extent. First, second-order summary statistics such as $K(r)$ or $g(r)$ that capture the average number of points in a neighborhood can be used to describe large-scale clustering. When used in a reconstruction, they will produce some sort of "average" large-scale clustering, but will have problems in correctly reproducing areas void of points or with low-point density. Second, the nearest-neighbor distribution functions $D_k(r)$ can quantify certain aspects of heterogeneous patterns, which are missed by the pair-correlation function. For example, if the pattern comprises areas of low point density (or isolated points), some points will have their kth nearest neighbor at large distances r. This will be accounted for by the $D_k(r)$, which will saturate only at large distances r [i.e., $D_k(r) < 1$ also for large values of r]. This will force the pattern reconstruction algorithm to produce low-density areas, with isolated points, and create heterogeneities, if the neighborhood rank k and the maximal distance r_{max} over which $D_k(r)$ is evaluated are large enough. Finally, the spherical contact distribution $H_s(r)$ is useful for characterizing an additional aspect of heterogeneous patterns not captured by the pair-correlation function or the $D_k(r)$. Because $H_s(r)$ measures the distribution of the sizes of gaps in a pattern, it is able to force the pattern reconstruction algorithm to produce areas void of points.

Several summary statistics of a different nature [such as $g(r)$, $D_k(r)$, and $H_s(r)$] can be used together to recreate a heterogeneous spatial structure, but only up to the maximal spatial scale r_{max} considered in the calculation of the summary statistics. This is demonstrated in Figure 3.52, where we generated a simple heterogeneous pattern by enlarging the homogeneous pattern shown in Figure 3.50a to include a void area of the same size (Figure 3.52a). We use this to explore whether, when used together, the pair-correlation function, the spherical contact distribution, and the distribution functions to the kth neighbor are able to produce a reasonable reconstruction of this strongly heterogeneous pattern. Indeed, the pattern reconstruction algorithm deals very well with this pattern. Figure 3.52b shows one reconstruction. Because the intensity was not fixed, the void areas of the observed and the reconstructed pattern do not overlap, but the two patterns show the same typical structure internally, with small clusters and an otherwise hyperdispersed pattern. This demonstrates the power of the "homogeneous" summary statistics to depict heterogeneities in the intensity function.

Some reconstructions, especially patterns with large-scale gradients in the intensity, may be incomplete, as illustrated in Figure 3.53. The example pattern used for reconstruction is characterized by a strong heterogeneity

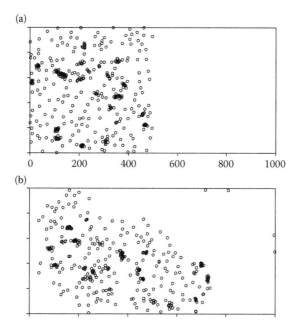

FIGURE 3.52
Pattern reconstruction of a heterogeneous pattern. (a) The observed pattern is the same as in Figure 3.50a, but enlarged by a void area from x-coordinates 500 to 1000. (b) Pattern reconstructed using the pair-correlation function, the spherical contact distribution, and the distribution functions to the *k*th neighbor $D_k(r)$ for $k = 1...50$ for distances up to 200 m.

in the intensity, with a curved band of high-point density in a pattern otherwise characterized by low-point density or an absence of points (Figure 3.53a). The points inside the high, point-density area exhibit a strong cluster structure, with larger clusters appearing to be composed of smaller clusters. The reconstruction of this pattern with the summary statistics $K(r)$, $g(r)$, $H_s(r)$, and $D_k(r)$ (with $k = 1,...50$) for distances up to 200 m are in good agreement with the small to intermediate scale spatial structures in the original pattern (Figure 3.53b). Although the clusters in the reconstruction resemble those of the original pattern closely, these elements were not assembled in the same way as the original pattern (Figure 3.53b). Instead, they were distributed differently within the plot, leaving approximately the same area void of points, but did not form the characteristic band shown by the original pattern. The reconstruction in Figure 3.53b has a distinctly heterogeneous appearance, which again confirms that summary statistics developed for homogeneous patterns can generate heterogeneous patterns. However, the different small- to medium-scale elements are structured in a somewhat different fashion than those of the original pattern. Clearly, the pattern reconstruction algorithm is not able to recreate larger-scale

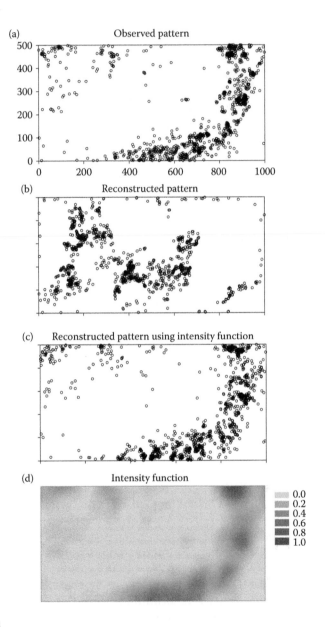

FIGURE 3.53

Pattern reconstruction of a heterogeneous pattern. (a) Original pattern that contains a high-intensity band extending from the bottom center of the plot to the top right. (b) Reconstruction using the summary statistics $g(r)$, $K(r)$, $H_s(r)$, and $D_k(r)$ ($k = 1$–50) taken over the distance interval of 0–200 m. (c) Reconstruction as panel (b), but the intensity function shown in panel (d) was also used. (d) Estimate of the intensity function after de-clustering with bandwidth 50 m.

gradients such as the high-density band in the observed pattern (Figure 3.53a). To capture this component of the pattern, we must incorporate the intensity function in the reconstruction.

3.4.3.5 Reconstructions Using the Intensity Function

Pattern reconstruction can be conditioned on the location-based intensity function $\lambda(x)$ of the observed pattern, if it is critical to conserve the large-scale structure (e.g., smooth gradients in intensity such as those in Figure 3.53a). However, the estimate of the intensity function should not include spatial structure at small scales, so that the reconstruction can capture smaller-scale features that are independent of the intensity. Intensity functions with this property can be estimated, for example, by using a nonparametric kernel estimate that averages over the small-scale structures up to a bandwidth of h (Box 2.4). However, for patterns with strong small-scale clustering, this approach is somewhat problematic because the strong clustering is a small-scale, autocorrelated phenomenon that can bias the intensity estimate. For example, small clusters may be due to some inherent clustering mechanisms and not caused by a locally elevated intensity function (e.g., better environmental conditions). Therefore, strong small-scale clustering should be removed before the intensity function can be estimated nonparametrically from the observed pattern (see Section 3.3.1).

Once the intensity function is estimated (Figure 3.53d), the reconstruction algorithm starts with an initial pattern based on a heterogeneous Poisson process in which a random point is accepted with a probability proportional to $\lambda(x)$. Similarly, any further tentative point is selected in accordance with the intensity function $\lambda(x)$. Conditioning on $\lambda(x)$ forces the reconstructed pattern to have the same large-scale properties as the original pattern, but the spatial structure at smaller scales is entirely driven by the partial energies of the particular summary statistics selected for pattern reconstruction. This is illustrated by the pattern shown in Figure 3.53c, which now reproduces both the small-scale characteristics of the pattern [represented by the summary statistics $g(r)$, $K(r)$, $H_s(r)$, and $D_k(r)$] as well as the large-scale gradient structure of the pattern [represented by the intensity function $\lambda(x)$]. The large-scale properties of the reconstructed pattern approximate those of the original pattern well, but the typical small-scale structures appear at stochastically displaced locations (cf. Figure 3.53a and c).

3.4.3.6 Ranking of Summary Statistics

Statisticians often recommend the use of multiple summary statistics in analyzing point patterns (e.g., Illian et al. 2008, p. 214), since each statistic characterizes a somewhat different aspect of the pattern. However, as noted above, it is a widespread practice in ecological applications to use just one, or occasionally two, summary statistics. Very little is known about the loss of information that ensues from this practice. As a consequence, Wiegand et al. (2013a) conducted a systematic study to determine which summary statistics(s) are required to accurately characterize the spatial structure of a point pattern. This involved the investigation of three fundamental issues. First, is the utilization of different summary statistics to some extent redundant? While the natural recommendation would be to use combinations of nonredundant summary statistics, it is not clear *a priori* if the same ranking (and specific combination) of summary statistics would apply for all patterns. A second issue was to determine to what extent such a ranking would be influenced by pattern idiosyncrasies. Finally, real-world patterns often show some aspects of nonstationarity and it would be useful to determine if one, or several, stationary summary statistics taken together can be used to describe key properties of a nonstationary pattern. For example, the density of points may be influenced by environmental covariates or the properties of local point configurations may depend on location.

Wiegand et al. (2013a) used an inverse approach to pattern reconstruction, as originally proposed in materials science, to explore the three issues described above (Rintoul and Torquato 1997; Torquato 1998). Their approach involved the reconstruction of a point pattern using only the limited information provided by a given combination of summary statistics. The degree to which the reconstructed pattern matched the original pattern was then assessed by examining how well the reconstructed pattern reproduced the summary statistics of the observed pattern for statistics not used in the reconstruction (Torquato 1998). They applied this approach to eight simulated point patterns that were homogeneous and to eight, complex, point patterns given by the locations of tree species in the BCI plot (Box 2.6). The patterns obtained from the tree species could exhibit some heterogeneity. These patterns were selected to span a wide range of possible spatial structures.

The summary statistics used in this test were $g(r)$, $K(r)$, $D(r)$, $D_k(r)$, $H_s(r)$, and $E(r)$. In addition, the intensity function $\lambda(x)$ was used to constrain the reconstruction of the (possibly heterogeneous) tree point patterns. The summary statistic $E(r)$ is related to the nearest-neighbor distribution function $D(r)$ and yields the probability that a ring of radius r and width dr around a point of the pattern contains no point. Wiegand et al. (2013a) used 32 combinations of the six summary statistics $g(r)$, $K(r)$, $D(r)$, $D_k(r)$, $H_s(r)$, and $E(r)$ in exploring the ability to reconstruct the 16 spatial patterns. For each combination of summary statistics c, they calculated an index $M_c(i)$ that measured the information contained in this combination with respect to summary statistic i

$$M_c(i) = \frac{1}{8} \sum_{\varphi=1}^{8} \frac{1}{10} \sum_{rep=1}^{10} I[E_i^\varphi(\psi^{rep}) < E_i^{match}] \tag{3.140}$$

The indicator function $I(\ )$ yields a value of 1 if the condition is true and zero otherwise. The "partial energy" $E_i^\varphi(\psi^{rep})$ quantifies the deviation of the observed pattern φ from one reconstruction ψ^{rep} produced by the summary statistic i (Equation 3.137). If the partial energy was smaller than a predefined threshold E_i^{match}, the observed φ and simulated pattern ψ^{rep} agreed in the summary statistic i. The threshold E_i^{match} was the largest partial energy $E_i^{\varphi}(\psi)$ observed in reconstructions ψ where the summary statistic i was used. If summary statistic i was used for reconstruction, it always matched the threshold. When a summary statistic was not used for reconstruction, it matched if the deviation yielded $E_i^{\varphi}(\psi) < E_i^{\text{max}}$. The index $M_c(i)$ represents the proportion of 80 reconstructions (of eight different patterns φ and their 10 replicate reconstructions), based on a combination of c summary statistics, that yielded a match for summary statistic i. Finally, the six "partial" matches $M_c(i)$ of the different summary statistics i were averaged to obtain the final index $M_c = \sum M_c(i)/6$ that measures the information contained in a given combination c of summary statistics.

Wiegand et al. (2013a) found a surprising amount of consistency between the 32 indices M_c estimated for the eight simulated patterns and the eight tree patterns (Figure 3.54). The combination of summary statistics that did well for the homogeneous patterns also did well for the partly heterogeneous (real-world) patterns. The different summary statistics were also redundant, to a certain extent (Figure 3.54). The index M_c should have a value of $sc/6$, with sc being the number of summary statistics used for reconstruction

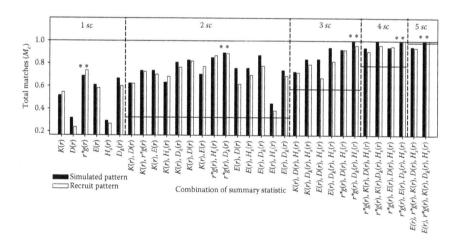

FIGURE 3.54
Redundancy and ranking of the different combinations of summary statistics. The figure shows the index M_c that measures the information contained in a given combination of summary statistics for cases of 1, 2, 3, 4, and 5 summary statistics. Pattern redundancy occurs if the value of M_c is smaller than $sc/6$ (indicated as horizontal lines), with sc being the number of summary statistics of the combination. Filled bars: simulated (homogeneous) patterns, open bars: observed patterns of trees at BCI (possibly heterogeneous). The best combination for a given number of summary statistics is indicated by a star. (Modified after Wiegand, T., F. He, and S.P. Hubbell. 2013a. *Ecography* 36: 92–103.)

(horizontal lines in Figure 3.54), if each statistic used in the reconstruction had captured completely independent information.

The pair-correlation function carried the most information on spatial structure, if it was used alone for pattern reconstruction; in this case, approximately 70% of the six summary statistics [i.e., $g(r)$, $D(r)$, $D_k(r)$, $H_s(r)$, and $E(r)$] were matched on average (Figure 3.54). As expected, the two nearest-neighbor statistics $D(r)$ and $H_s(r)$, when considered alone, contained very little information, and $D_k(r)$ carried notably more information than $D(r)$. Interestingly, for the homogeneous patterns the 50 nearest-neighbor distribution functions $D_k(r)$, when considered together, yielded a match almost as good as that of the pair-correlation function $g(r)$ ($\Delta M_c = 0.021$). However, for the point patterns of trees at BCI, the performance of $g(r)$ was clearly better ($\Delta M_c = 0.14$). The summary statistic that, together with the pair-correlation function, contained the most information was $D_k(r)$, with $M_c = 0.89$ for both pattern types. $g(r)$ together with $H_s(r)$ yielded only slightly poorer matches ($M_c = 0.87$). The combination of three summary statistics that captured the most information was $g(r)$, $D_k(r)$, and $H_s(r)$, yielding $M_c = 1$ for the homogeneous patterns and $M_c = 0.96$ for the recruits pattern. Finally, the combination of four summary statistics $g(r)$, $D_k(r)$, $H_s(r)$, and $E(r)$ yielded a perfect match with $M_c = 1$ for both pattern types. Thus, we found a consistent ranking for the summary statistics, with respect to average pattern match, of first $g(r)$, second $D_k(r)$, third $H_s(r)$, and fourth $E(r)$.

Wiegand et al. (2013a) repeated the analysis of Figure 3.54 to explore if the above ranking was also robust with respect to pattern idiosyncrasies. However, instead of looking at the average match among the eight patterns, they conducted the analysis separately for each pattern φ. A given combination of summary statistics c was defined as appropriate, if the "partial" index $M_c(\varphi)$ shown in Equation 3.140 was greater than 0.97. This result means that a combination c contained more than 97% of the information contained in all six summary statistics together. They found that the minimal number of summary statistics required for a satisfying match varied between one and four. In some cases, only one summary statistic [$g(r)$ or $K(r)$] was needed to reconstruct a pattern satisfactorily. These were patterns that contained little spatial structure. However, more complex patterns required four summary statistics to yield satisfying reconstructions. Thus, pattern idiosyncrasy plays an important role in the selection of summary statistics. Some patterns contain little spatial structure, which can be quantified with one or two summary statistics, whereas other patterns contain more structure that can only be captured by employing a larger number of summary statistics. Even so, the sequence of useful summary statistics was, in most cases, defined by the ranking $g(r)$ first, $D_k(r)$ second, $H_s(r)$ third, and $E(r)$ fourth.

Including the intensity function in pattern reconstruction improved the over all match M_c for all combinations c of summary statistics (Figure 3.55). The improvement was greatest for combinations of summary statistics that captured relatively little of the spatial structure [such as $D(r)$ or $H_s(r)$], but low for the best combinations of summary statistics [such as $g(r)$, $g(r) + D_k(r)$,

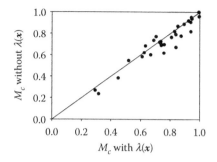

FIGURE 3.55
A comparison of the information captured by pattern reconstruction between the cases where the intensity function is not included and where it is included. We estimated the M_c indices that measure the information contained in a given combination c of summary statistics for the tree recruits from BCI using 32 combinations of summary statistics. The figure shows the resulting M_c indices using (x-axis) and not using (y-axis) the intensity function $\lambda(x)$. (Modified from Wiegand, T., F. He, and S.P. Hubbell. 2013a. *Ecography* 36: 92–103.)

$g(r) + D_k(r) + H_s(r)]$ that already captured a large degree of spatial structure (Figure 3.55). This result indicates that the summary statistics $g(r)$, $K(r)$, $D(r)$, $D_k(r)$, $H_s(r)$, and $E(r)$ capture properties of the pattern, which are largely independent of the intensity function. This is understandable because these summary statistics describe mostly small- to medium-scale structures, whereas the intensity function captures the large-scale structures.

One summary statistic that is able to capture larger-scale properties of a pattern is the mean distance $nn(k)$ to the kth neighbor. Wiegand et al. (2013a), therefore, analyzed the match in $nn(k)$ for three different cases: (i) reconstruction of eight real patterns from the BCI plot (represented by the recruits of eight different species) without use of the intensity function; (ii) reconstruction of the same eight patterns using the intensity function; and (iii) reconstruction of eight simulated homogeneous pattern, without the intensity function. The results were interesting. The eight real patterns, when reconstructed without the intensity function, generally yielded a poor fit for the mean distance $nn(k)$ to the kth neighbor (Figure 3.56a). For most combinations of summary statistics, only two of the eight patterns showed a good match. Those were patterns with little spatial structure. For most of the other patterns, the combinations of the six summary statistics $g(r)$, $K(r)$, $D(r)$, $D_k(r)$, $H_s(r)$, and $E(r)$ were not able to yield a satisfying fit in $nn(k)$. Thus, these summary statistics were not able to capture the larger-scale properties of the pattern. However, when the intensity function was included in the reconstruction, there were combinations of summary statistics that yielded satisfying reconstruction of $nn(k)$ for all eight real patterns (Figure 3.56b). Thus, the large-scale and small- to intermediate-scale properties of the patterns were largely separated. Interestingly, in most cases reconstructions of the eight simulated homogeneous patterns yielded

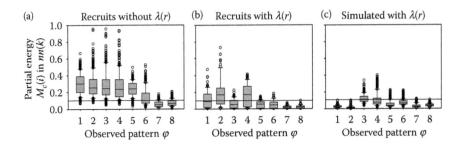

FIGURE 3.56

Box plots of the partial energies for the mean distance $nn(k)$ to the kth neighbor estimated for reconstructions of eight recruit patterns and eight simulated homogeneous patterns shown in the supplementary material Appendix 1 of Wiegand et al. (2013a). (a) Partial energies for real patterns of recruits from BCI reconstructed without the intensity function. (b) Partial energies for real patterns of recruits from BCI reconstructed with the intensity function. (c) Partial energies for simulated homogeneous patterns reconstructed without the intensity function. The horizontal line indicates the acceptance threshold E_i^{match} for the summary statistic $nn(k)$. Reconstructions with partial energies below the threshold are satisfying. (Modified after Wiegand, T., F. He, and S.P. Hubbell. 2013a. *Ecography* 36: 92–103.)

a good fit to $nn(k)$, even if no intensity function was used for reconstruction (Figure 3.56c). This is understandable because these patterns do not contain large-scale structure and therefore a suitable combination of the summary statistics $g(r)$, $K(r)$, $D(r)$, $D_k(r)$, $H_s(r)$, and $E(r)$ will yield reconstructions that also match $nn(k)$.

In summary, the study by Wiegand et al. (2013a) found clear answers to the three issues explored. First, a certain degree of *redundancy* exists among summary statistics and the redundancy depends on the complexity of the pattern analyzed. Patterns with near random structures can be described well by only one summary statistic, but four or five summary statistics are needed to capture the properties of patterns with more complex spatial structures. Second, Wiegand et al. (2013a) found a clear and *robust ranking of summary statistics* in the order of utility of first $g(r)$, second $D_k(r)$, third $H_s(r)$, and fourth $E(r)$. These four summary statistics were able to capture the spatial structure of simulated stationary patterns and were also able to recover small- to medium-scale heterogeneity in real-world patterns. However, larger-scale properties of nonstationary patterns could only be recovered when using the intensity function $\lambda(x)$ for pattern reconstruction. The ranking of summary statistics was remarkably robust with respect to pattern idiosyncrasies, in the sense that patterns with little spatial structure could be quantified with the first one or two summary statistics in the ranking, whereas patterns with increasingly complex spatial structures required addition of further summary statistics in the sequence defined by the ranking. Finally, Wiegand et al. (2013a) found that *stationary summary statistics can indeed describe key aspects of heterogeneous patterns*. The spherical contact distribution and nearest-neighbor distribution functions play important roles in describing gaps and in capturing heterogeneous elements, such as presence of isolated points, or areas of low point density. However, *the intensity function is needed to characterize larger-scale structures of heterogeneous patterns*. In general, spatial analysis will require the use of several summary statistics at the same time to avoid flawed inferences that leave important properties of a pattern undetected.

3.4.3.7 Pattern Reconstruction as a Null Model

We now provide two examples for the analysis of bivariate patterns, where we test for independence using pattern reconstruction. In testing independence, the interest is in the relationship between the two patterns and not in the spatial structure of the composite pattern. Therefore, a null model used in analyzing the pattern needs to condition on the univariate spatial structures of the two component patterns, but must also break apart the possible spatial dependence between the two. Pattern reconstruction can be used to generate this null model. For this purpose, we keep the first component pattern unchanged but randomize the second pattern based on pattern reconstruction. The reconstructions are stochastic realizations of the second component pattern that show all typical structures of the original pattern, but at somewhat displaced locations. The advantage of this method is that it conserves the detailed spatial structure of the original pattern without producing potential artifacts of the torus shift null model. Additionally, pattern reconstruction is more flexible in capturing potentially complex spatial structures than parametric reconstruction methods that use models, such as the Thomas process.

The first example investigates independence of two patterns with complex univariate structures. The patterns represent recruits of two tropical tree species *Cecropia insignis* and *Miconia argentea* as found at the BCI plot (Figure 3.57) (Box 2.6). Recruits are all individuals that exceeded the size threshold of 1 cm diameter at breast height during the preceding five years and therefore entered the census for the first time. They were usually saplings up to 3 m in height. The two species used in this example are typical light-demanding gap species, which exhibit a complex superposition of a random pattern (of isolated recruits) with a cluster pattern, where very strong small-scale clustering was nested within larger clusters. These univariate spatial patterns were analyzed in detail in Wiegand et al. (2009).

The analysis with the pair-correlation function and the null model based on pattern reconstruction of *M. argentea* shows that the recruits of the two species are strongly attracted at scales up to 15 m (Figure 3.57b). Thus, the neighborhood density of *M. argentea* recruits around *C. insignis* recruits is significantly greater than expected under independence. For neighborhoods of 2 m, the neighborhood density is more than 20 times higher than expected under independence (Figure 3.57b). The bivariate distribution function of nearest-neighbor distances $D_{12}(r)$ shows a significant attraction affect up to 40 m. The nearest *M. argentea* recruits were up to 40 m closer to *C. insignis*

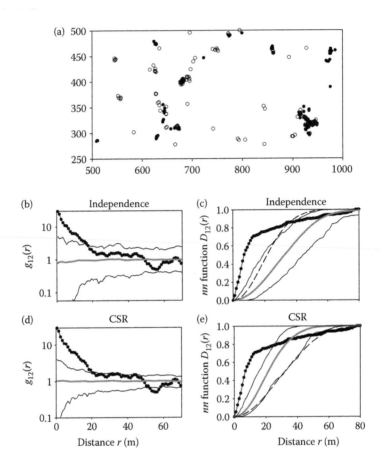

FIGURE 3.57
Pattern reconstruction and the independence null model. (a) Map of two patterns representing recruits of the species *Cecropia insignis* (focal species, black) and *Miconia argentea* (open disks) from the 2000 census at BCI. The patterns have been analyzed in detail in Wiegand et al. (2009). The panel shows the bivariate pattern for the upper right part of the study area. (b) Bivariate pair-correlation analysis using the using pattern reconstruction for *M. argentea* as the independence null model. Note the logarithmic scale of the y-axis. (c) Same as panel (b), but for the nearest-neighbor distribution function $D_{12}(r)$. The dashed line is the expectation under CSR. (d) Same as panel (b), but using CSR as the null model for *M. argentea*. (e) Same as panel (d), but for the nearest-neighbor distribution function $D_{12}(r)$. The dashed line is the expectation under independence.

recruits than expected by independence. The independence null model only agreed with the observed pattern at distances greater than 60 m (Figure 3.57c). The strong attraction was probably caused by canopy gaps shared by the recruits of the two light-demanding species.

Comparison of the results of independence with those of the CSR null model (cf. Figure 3.57d and b) shows that the CSR null model underestimates the variability in point configurations and yields simulation envelopes in the pair-correlation function that are too narrow. This is understandable because CSR cannot produce the strong clustering observed, since the CSR null model destroys the clusters. In contrast, pattern reconstruction with the independence null model may produce many configurations where clusters of *M. argentea* accidently overlap those of *C. insignis*. Similarly, configurations where the clusters of the two species accidently segregate are unlikely under CSR, but are likely under independence. As a consequence, the independence null model yields wider simulation envelopes. While CSR and independence yield the same expectations under the pair-correlation function, the expectations of the nearest-neighbor distribution function $D_{12}(r)$ differ for the two null models (cf. Figure 3.57c and e). Under CSR the nearest *M. argentea* neighbor of *C. insignis* trees is usually much closer than under the independence null model. The reason for this is that the CSR null model does not maintain the observed clustering (which concentrates points in space) and therefore fills the space more thoroughly, thereby placing the *M. argentea* individuals closer to those of *C. insignis*.

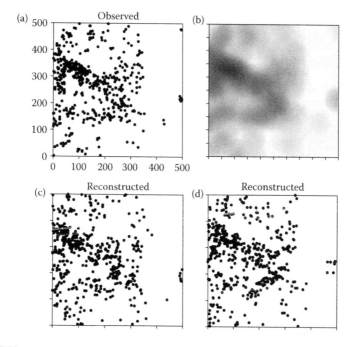

FIGURE 3.58
Pattern reconstruction using the intensity function. (a) Map of large individuals (dbh > 10 cm) of the species *Shorea trapezifolia* from the Sinharaja forest in Sri Lanka. (b) Intensity function of *S. trapezifolia* estimated with a nonparametric kernel estimate using a bandwidth of $R = 50$ m, after slight de-clustering using parameter $r_c = 3$ m for the minimal distance between cluster centers. (c) and (d) Reconstructions using the intensity function in panel (b).

The second example illustrates how pattern reconstruction can help to assess independence even in the case of heterogeneous patterns. The data are large individuals (i.e., dbh > 10 cm) of the two species *M. tetrandra* (focal species) and *Shorea trapezifolia* in the Sinharaja forest of Sri Lanka (Figures 3.58a, 3.59a). The plot contains strong topographic structuring and both species occur predominately in the western half of the plot. Thus, the

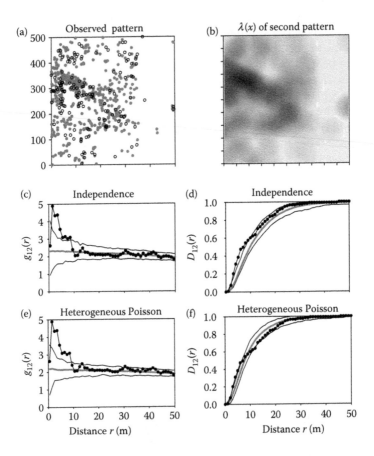

FIGURE 3.59

Pattern reconstruction using the independence null model under heterogeneity. (a) Map of patterns of large individuals (dbh > 10 cm) of the species *Mastixia tetrandra* (focal species, open disks) and *Shorea trapezifolia* (reconstructed pattern, gray closed disks) from the Sinharaja forest. (b) Intensity function of *S. trapezifolia* estimated with a nonparametric kernel estimate using a bandwidth of $R = 50$ m after slight de-clustering using parameter $r_c = 3$ m for the minimal distance between cluster centers. (c) Bivariate pair-correlation function and the independence null model for pattern reconstruction of *S. trapezifolia* using the intensity function shown in panel (b). (d) Same as panel (c), but for the nearest-neighbor distribution function $D_{12}(r)$. (e) Same as panel (c), but incorporating the corresponding heterogeneous Poisson null model based on the intensity function shown in panel (b). (f) Same as panel (e), but for the nearest-neighbor distribution function $D_{12}(r)$.

pair-correlation function is clearly larger than the expectation under independence (i.e., a value of 1; Figure 3.59a) and testing for independence, as done in the previous example, would most likely depict the effects of habitat association of the two species (i.e., they both co-occur largely within the same area of the plot), but provides no answer to the question of whether there is potential attraction or repulsion of the individuals of the two species within a local neighborhood. To do this we need to condition on the observed large-scale structure of the second pattern (i.e., that of *S. trapezifolia*). This can be done by using a heterogeneous Poisson process based on a nonparametric intensity estimate of the second pattern (Figure 3.58b), which then randomizes the individuals of this pattern only locally. Here we selected a bandwidth of $R = 50$ m, which is somewhat larger than the interaction range among trees. However, pattern reconstruction using the intensity function allows us to additionally condition on the small-scale structure of the second pattern. In this case, the null model patterns have the same intensity function and show the same small-scale structures as the observed pattern, but they are slightly displaced relative to the observed pattern (Figure 3.58c,d).

Figure 3.59c shows the results of the analysis for the pair-correlation function. At distances of 2–9 m we observe a clear attraction of the two species. At larger distances, the observed pair-correlation function agrees with that expected under the independence null model implemented with pattern reconstruction. Thus, both species co-occur in the same restricted area of the plot (i.e., the heterogeneity) and show additional attraction. This is also evident from visualizing the patterns (Figure 3.59a); there are small, shared clusters, where both species co-occur. The analysis using only the heterogeneous Poisson process (which does not maintain the observed small-scale structure of the pattern of *S. trapezifolia*) shows basically the same result, but produces a smaller level of attraction in the null model at small distances (Figure 3.59e). However, the nearest-neighbor distribution functions show fewer differences between the two null models. However, the pattern reconstruction null model suggests stronger attraction at distances of 2–9 m. Thus, the nearest neighbors under the heterogeneous Poisson null model (which does not maintain the observed clustering) are in general somewhat closer than under the corresponding pattern reconstruction null model that conserves the observed clustering.

4

Examples

In this chapter, we present a systematic collection of examples and additional methods, which we find useful, structured around different data types. For each data type, we present the most important null models and point-process models that are relevant in an ecological context. In contrast to Chapters 2 and 3, in which we only briefly treated methods for some data types, such as multivariate and marked point patterns, we present here a more detailed and comprehensive description of methods that can be used to answer ecological questions related to the different data types.

4.1 Analysis of Univariate Patterns

Univariate point patterns are the simplest data type in spatial point-pattern analysis: they involve only one type of ecological object without additional marks. Early examples of point-pattern analysis were mostly concerned with determining whether a pattern was random as opposed to clustered or hyperdispersed. The points of clustered patterns (Figure 4.1a) have more neighbors in close proximity than expected for a random pattern (i.e., the null model of complete spatial randomness (CSR); also called the homogeneous Poisson process), and points of a regular or hyperdispersed pattern (Figure 4.1d) have fewer neighbors in close proximity than expected. Most of the methods discussed in this book explicitly consider local neighborhood structure around the point of a pattern. This allows for a subtle, scale-dependent assessment of the small-scale spatial autocorrelation structure of univariate patterns. For example, many ecological questions require assessment of the range of clustering or hyperdispersion. The range of aggregation in the pattern shown in Figure 4.1a is approximately 30 m. As a consequence, the pair-correlation function $g(r)$ produces values greater than the expectation $g(r) = 1$ under CSR for distances of 1–30 m (Figure 4.1b). In this case, interpoint distances below 30 m are more frequent than expected by CSR. In contrast, in the case of hyperdispersion short interpoint distances occur less frequently than expected under CSR. The pattern shown in Figure 4.1d contains no interpoint distances less than 20 m. As a consequence, the pair-correlation function yields a value of $g(r) = 0$ for distances r less than 20 m (Figure 4.1e).

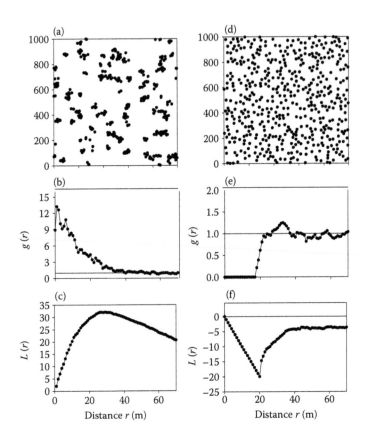

FIGURE 4.1
Examples of clustered and hyperdispersed patterns. (a) Map of a clustered pattern created with a Thomas process composed of 100 randomly distributed clusters $\sigma = 10$ m, with each of the 100 clusters consisting of, on average, 5 points. (b) Pair-correlation function of the clustered pattern (closed disks) and expectation under CSR (horizontal line). (c) Same as panel (b), but for L-function. (d) Hyperdispersed pattern created with a hard-core process, with random points distributed in sequence within the observation window. Each new point in the sequence was accepted if it was located at least 20 m away from all other points already placed. (e) Pair-correlation function of the hyperdispersed pattern (closed disks) and expectation under CSR (horizontal line). (f) Same as panel (e), but for L-function.

In addition to providing basic information on the characteristics of univariate patterns, the shape of the summary statistics can contain even more information that can be used to clearly describe the characteristic features of an observed point pattern. Using this information to account for the stochastic nature of the underlying point processes, we can test whether the observed point pattern agrees or departs from simple null models or more detailed point-process models. In the following, we describe and summarize the analysis of several null models and point-process models that have proven useful in ecological applications. We illustrate these examples with analyses that follow the basic steps in conducting a point-pattern analysis outlined in Sections 2.3 through Section 2.6 (i.e., we present summary statistics, consider appropriate null models, compare null models with data and take into account heterogeneous patterns if necessary).

4.1.1 Homogeneous Poisson Process (CSR)

The homogeneous Poisson process (Box 2.2) is characterized by two fundamental properties: (i) the intensity λ of the process is a constant (i.e., there is no effect of the environment on the placement of the points), and (ii) the points of the process are independently distributed (i.e., there are no point interactions). Because this point process accomplishes a complete randomization of the pattern, it is often referred to as "complete spatial randomness" (CSR). The CSR null model can be used for a variety of purposes; the most important purposes are to determine whether the pattern contains spatial structure that can be distinguished from pure stochastic effects (i.e., it is used as fundamental division; Section 2.4.1), and to determine whether the pattern exhibits indications of heterogeneity.

4.1.1.1 Is the Pattern a Random Pattern?

The first question to ask in a point-pattern analysis is whether the pattern contains spatial structure that can be distinguished from the pure stochastic effects expected for the given data type. To answer this question, we use the idea of the "fundamental division" as a null model (see Section 2.4.1). The basic approach is to condition on all the features of the data that are not of current interest for the data type (e.g., the spatial-correlation structure of the univariate component patterns of a bivariate pattern), but randomize what is of interest (e.g., the association between two component patterns). In the case of univariate patterns, the fundamental division is provided by the homogeneous Poisson process (Section 2.6.1), which basically relocates the points of the pattern to random locations within the observation window, independent of the positions of already relocated points. The only constraint on the data type is the number of points within the pattern, which needs to be conserved in the null model simulations.

   Figure 4.2a shows a CSR pattern composed of 500 points. No systematic spatial structures are visible; some points are very close to each other and some areas of the observation window show larger gaps. Clearly, these characteristics are strongly determined by the stochastic nature of the process producing the pattern, which can cause departures of the empirical summary statistics from the theoretical expectations that can be estimated analytically for the homogeneous Poisson process (Table 2.2). For instance, in the example pattern the values of the pair-correlation function show small departures from the expected value of 1 (Figure 4.2b), that is, some interpoint distances are accidently more frequent than others. A similar observation

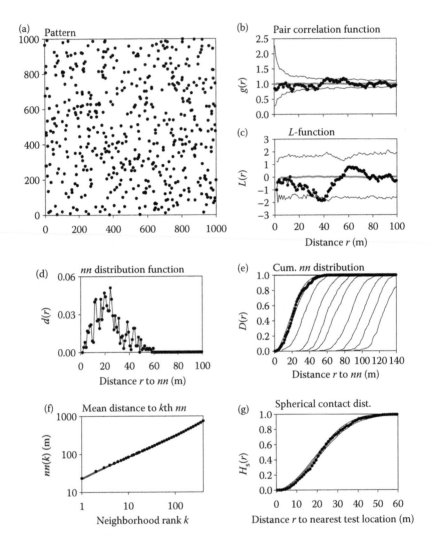

FIGURE 4.2
Examples of summary statistics for a random point pattern. (a) Map of a pattern created with a homogeneous Poisson process (Box 2.2) comprising 500 points within a 1000 × 1000 m sample domain. (b)–(g) Various summary statistics applied to the pattern in panel (a), with simulation envelopes produced by a CSR null model. (e) Sequence of 10 distribution functions $D^k(r)$ of the distances to the kth neighbor with $k = 1, 2, 4, 6, 8, 12, 16, 20, 25$, and 30, with $D^1(r)$ being indicated by solid dots.

can be made for the *L*-function, which also exhibits departures from the expected value of 0 (Figure 4.2c).

The important issue here is to determine whether these *departures from the expected value are already an indication of an underlying spatial structure or whether they can arise by pure chance* (i.e., are only due to stochastic effects within the homogeneous Poisson process). To address this issue, we have to evaluate the match between observed data and data generated by the homogeneous Poisson process. We initially do this using the simulation envelope approach outlined in Section 2.5.1.1 and shown in Figure 2.17. In the example, we generate 199 realizations of a homogeneous Poisson process and calculate the summary statistics for the observed data and for each of the 199 simulated data sets. Figure 4.2b shows the resulting simulation envelopes for the pair-correlation function. The simulations envelopes in this case are composed of the 5th lowest and 5th highest value of the 199 simulations, that is, 95% of the simulated values lie between the envelopes. The values of the observed pair-correlation function lie fully within the simulation envelopes, thus indicating that the fluctuations of the pair-correlation function are within the range of values that would occur by pure chance alone. This diagnosis can be confirmed by the GoF test (Section 2.5.1.2), which yields a p-value of 0.7 for a distance interval of 1–50 m. Similar results were obtained for the L-function (Figure 4.2c; $p = 0.16$), the distribution function $D(r)$ of the distances to the nearest neighbor (Figure 4.2e; $p = 0.48$), the mean distances to the kth neighbor $nn(k)$ (Figure 4.2f; $p = 0.81$), and the spherical contact distribution $H_s(r)$ (Figure 4.2g; $p = 0.21$). Thus, as expected, the pattern shown in Figure 4.2 shows no indication that it contains a nonrandom spatial structure; all departures of the empirical summary statistics from the expected values are fully compatible with stochastic effects.

Note that the simulation envelopes of the pair-correlation function at short distances r are substantially wider than at larger distances (Figure 4.2b). This is a general phenomenon and a consequence of different numbers of point pairs at different distances. Since the region assessed by the pair-correlation function around the "typical point" scales at a rate of r, the effect of stochasticity is stronger at shorter distances r because the sample size (i.e., point pairs located at this distance interval) is smaller. Similarly, the effect of stochasticity will be stronger if the pattern has fewer points. As a consequence, wider simulation envelopes will be produced by patterns with fewer points. Thus, our ability to detect nonrandom spatial structures in the data will be low if the pattern has only a few points.

4.1.1.2 Is the Pattern Homogeneous?

The homogeneous Poisson process can also be used to detect certain aspects of heterogeneity in the pattern. Especially useful for this are the pair-correlation function and the L-function, which exhibit distinct shapes if the pattern shows larger-scale heterogeneity due to gradients in point density or large areas void of points (Section 2.6.2.2, Figure 2.28). If the pattern is homogeneous (i.e., the properties of the pattern and especially the intensity λ are the same at all locations of the observation window) and if the observation window is sufficiently large, the value of the pair-correlation function must approach a value of 1 asymptotically for distances r larger than the range of point interactions. This can be seen clearly in Figure 4.1b, in which the range of clustering is approximately 30 m. Similarly, the *L*-function must approach a value of 0 at large distances *r*. Because of its cumulative nature, the *L*-function may decline only slowly to zero, as shown in Figure 4.1c. The decline should begin approximately at the distance where the pair-correlation function approaches a value of 1. If the pattern shows larger-scale heterogeneity, the pair-correlation function and the *L*-function will be above the simulation envelopes for distances clearly larger than the range of interpoint interactions (Figure 2.22). Getzin et al. (2008) used this property successfully to assess potential heterogeneity in the spatial distribution of stems of all mature adult trees (dbh > 15 cm) in two Douglas-fir forests on Vancouver Island, Canada (Figure 2.23).

Large-scale heterogeneity can also influence the values of the pair-correlation and *L*-functions at shorter distances. For patch or gradient-type heterogeneities, we demonstrated in Section 2.6.2.2 that the pair-correlation function typically produces constant values that are greater than one at shorter distances and the *L*-function typically shows a linear increase at increasing scales. For a patch-type heterogeneity, consisting of a CSR pattern confined to an area covering a proportion *c* of the observation window, we found that at shorter distances the pair-correlation function yielded a value of $g(r) = (1/c)$ and the *L*-function was given by $L(r) = r[(1/c)^{0.5} - 1]$. Additionally, we found that the mean distance to the *k*th neighbor always yielded a value that was $-\log(c)/2$ smaller than the expectation of the corresponding CSR null model, if $\log(nn(k))$ is plotted against $\log(k)$ on a double-logarithmic scale. These results clearly demonstrate that large-scale heterogeneities can influence the value of the second-order (and nearest neighbor) summary statistics at short distances *r* and that these statistics are, in general, shaped jointly by small-scale interactions and by large-scale heterogeneity.

The examples in Figure 2.22a,b show patterns containing a simple gradient and patch heterogeneity, respectively, with no interpoint interactions. In both cases, we observe that the pair-correlation function (Figure 2.22c,d) and the *L*-function (Figure 2.22e,f) produce evidence of aggregation at short distances, much below the scales of the underlying heterogeneities; that is, the pair-correlation function produces a constant value $g(r) = (1/0.75) = 1.33$ (Figure 2.22c) and the *L*-function produces a linear increase at increasing values of *r*, that is, $L(r) = r[(1.33)^{0.5} - 1] = 0.155 * r$ (Figure 2.22e). As a result, we observe *"clustering at all scales"* in Figure 2.22e,f, which is a spurious result induced by the underlying large-scale heterogeneity of the pattern. Figure 2.22h shows that the mean distance $nn(k)$ to the *k*th neighbor also exhibits the typical behavior expected under patch heterogeneity: on a double-logarithmic scale the observed values show a constant difference of $-\log(c)$ from the value expected under the corresponding CSR null model. The observed value of the difference (0.058) is close to the expected, theoretical difference of $\log(c) = 0.062$ for $c = 0.75$ (Figure 2.22h). However, this is not shown by the gradient heterogeneity (Figure 2.22g).

The effects of large-scale heterogeneity on the values of the second-order and nearest-neighbor summary statistics at small scales are somewhat counterintuitive, because one might naively expect that large-scale heterogeneity in the intensity function would influence the values of summary statistics only at broad scales, but leave the values at shorter scales untouched. As demonstrated above, this is not the case. As a consequence, the effect of *"virtual aggregation"* has been frequently overlooked in the literature and has often resulted in the spurious interpretation of clustering at all scales in patterns affected by large-scale heterogeneity. For a more detailed treatment of this effect of virtual aggregation see Section 2.6.3.1 (cf. Wiegand and Moloney 2004).

4.1.2 Heterogeneous Poisson Process

The heterogeneous Poisson process (HPP) modifies the first fundamental property of the homogeneous Poisson process; the intensity function is no longer treated as a constant λ, but now becomes a function of location x, that is, $\lambda(x)$. A non-constant intensity function can arise due to the impact of the environment on the placement of points: some areas of the observation window may be less suitable, and therefore have a lower value of $\lambda(x)$, while other areas may be more suitable, and have a higher value of $\lambda(x)$. However, other effects of population dynamics may also cause situations where points are not homogeneously distributed. For example, dispersal and recruitment limitation may cause the patchy distribution of a species even in a homogeneous habitat, if propagules are rare and do not disperse sufficiently far to reach all suitable sites. Similar situations may arise if alien species invade suitable habits outside their native range or if suitable habitat remains unoccupied because of population contraction. Thus, *spatial effects of internal population dynamics* that are unrelated to environmental factors may also shape the intensity function. If appropriate, such factors should be considered when estimating the intensity function.

The second property of the homogeneous Poisson process, that is, that the points of the process are independently and identically distributed, is also valid for an HPP. Thus, the heterogeneous Poisson point process is completely determined by the intensity function $\lambda(x)$ and the number of points n. To simulate a realization of an HPP, random coordinates within the plot are tentatively proposed, but only retained with probability $\lambda(x)/\lambda^*$ where λ^* is the maximal value of $\lambda(x)$.

An HPP has several important applications in the analysis of univariate patterns; the most important is that it can account for the effect of a non-constant intensity function caused by environmental effects and possibly the effects of internal population dynamics on the placement of points. In theory, incorporating an HPP in the analysis of a point pattern allows the study of the "pure" effects of point interactions in systems structured by an underlying large-scale trend, given that we are able to estimate the intensity function satisfactorily. The principle difficulty in the application of this null model is, in fact, the estimation of the intensity function $\lambda(x)$. Depending on the ecological question at hand, a variety of approaches are available for this task. In broad terms, the intensity function can be either estimated non-parametrically (Section 2.6.3.3) or parametrically (Section 2.6.2.9). In the non-parametric case, the estimation of the intensity function is only based on the data through the application of smoothing techniques, whereas in the parametric case additional information on environmental covariates is used to reconstruct the intensity function $\lambda(x)$. In the following, we propose using a combination of both approaches.

4.1.2.1 HPP: Nonparametric Intensity Estimate to Avoid Virtual Aggregation

One solution for avoiding the problem of virtual aggregation, caused by large-scale heterogeneity, is to apply HPP, with a nonparametric estimate of the intensity function (HPPnpe), to an analysis (see Figure 4.3). The HPPnpe null model is, in a way, a generalization of the homogeneous Poisson process. Recall that the homogeneous Poisson process basically relocates the points of the pattern randomly within the entire observation window using the same intensity throughout. The HPPnpe of the intensity function uses basically the same approach as the standard, univariate CSR null model. However, instead of relocating the points of the pattern to random locations within the entire observation window W, the HPPnpe null model displaces the points of the pattern only within a neighborhood of radius R around their original locations. While the homogeneous Poisson process effectively removes all spatial structures in a pattern at all scales, the HPPnpe process removes only structures at scales smaller than the displacement radius R, because it only randomizes within this radius. This effectively conditions the randomized pattern on the observed spatial structure at scales $r > R$.

How does an HPP, employing a nonparametric intensity estimate, factor out the effect of virtual aggregation, if the pattern shows larger-scale heterogeneity? This can be understood as follows: The assumption of larger-scale heterogeneity means mathematically that the intensity function $\lambda(x)$ at location x has very similar values within distance $r < R$, thus $\lambda(x) \approx \lambda(x + e^*r)$, where e is a vector of length 1. Thus, relocated points will land at locations with approximately the same intensity $\lambda(x)$ as their original locations (see circle in Figure 4.3a). As a result, the null model will track the intensity function represented in the underlying structure of the data. This null model is thus a classic null model that holds certain aspects of the data fixed (here the number of points n and the spatial structure at scales greater than R), but randomizes other aspects of the data (spatial structure at scales less than R). This means also that the HPPnpe null model will detect only departures from randomness caused by mechanisms that operate at scales below R.

It is clear that the parameter R (also called the bandwidth) plays an important role in the HPPnpe null model. We need to select a radius R that is slightly greater than the expected range of point interactions (Wang et al.

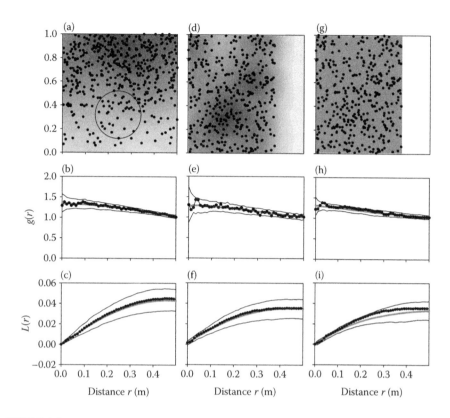

FIGURE 4.3
Application of HPP with nonparametric intensity estimation to factor out the effects of virtual aggregation. The heterogeneous patterns are shown in panels (a), (d), and (g), together with their intensity functions. In panels (a) and (d) we used a nonparametric intensity estimate based on an Epanechnikov kernel with a bandwidth of $R = 0.2$, and in panel (g) we used an intensity function with a value of λ in the gray-shaded area and a value of zero otherwise. The circle in panel (a) has a radius of $R = 0.2$ to illustrate the maximum relocation radius of the null model. The panels in the middle and bottom row show the observed summary statistics and simulation envelopes of HPP lined up under the patterns they analyze.

2010; Wiegand et al. 2012), since the null model should reveal the effects of interpoint interactions, but condition on all other effects (mostly environmental heterogeneity). The analysis may also be repeated with several values of R, if the expected interaction range is uncertain. If we choose a radius for R that is too large, the approximation of $\lambda(x) \approx \lambda(x + e^*R)$ will not hold and the displaced points will not be placed at locations with equivalent values of the intensity function. As a consequence, the corresponding null model will not fully remove the effects of larger-scale heterogeneity. It is also clear that this method cannot separate the effects of point interactions and heterogeneity, if the scales at which both operate are too similar. For example, if small-scale edaphic factors cause variation in the intensity function at scales less than or equal to the scales at which interpoint interactions occur, the effects of environmental heterogeneity will not be removed effectively by the null model. Thus, the HPP with a nonparametric estimate of the intensity function is explicitly based on the *assumption of separation of scales*, which means that point interactions operate over short distances and that smooth variation in the intensity function occurs at broader scales.

To formalize this null model, we need to calculate the intensity function $\lambda(x)$ that results from random displacement of the points of the pattern within a small neighborhood R around their observed locations. Clearly, all points of the pattern that are located closer than distance R from location x have, in principle, the chance to be displaced by the null model to lie within the infinitesimally small disk dx centered at x. Additionally, all of these points will have the same probability of being located within the small disk dx centered at x, because they are displaced to a random location within their R neighborhood. Thus, we can estimate $\lambda(x)$ by counting the number of points within distance R of x and then divide by the corresponding area ($=\pi R^2$) around x that can contribute points. If a neighborhood with radius R only partially overlaps the observation window W, we divide the number of points by the area that falls inside of W.

In technical terms, the method of estimating the intensity function $\lambda(x)$ just described corresponds to a nonparametric kernel estimate, using a "box kernel" (see Box 2.4; Section 2.6.2.1) with bandwidth R (Stoyan and Stoyan 1994). The box kernel has the disadvantage of producing somewhat rugged surfaces (Figure 2.20b). For this reason, other kernel functions are usually used. For example, the Epanechnikov kernel (Stoyan and Stoyan 1994), which we introduced earlier (Box 2.4), counts points at shorter distances somewhat more than points at larger distances (but still within distance R), providing smoother intensity surfaces than the box kernel (e.g., Figure 2.20c).

Figure 4.3 shows how the HPP with nonparametric estimate of the intensity function factors out the effects of large-scale heterogeneity. The observation window is the unit square, with coordinates ranging from zero to one. The pattern shown in Figure 4.3a is subject to a smooth gradient heterogeneity that increases from zero at the bottom to a maximal value at the top. Relocating the points randomly within neighborhoods of radius $R = 0.2$ only moves the points to locations with a similar value of the intensity function (indicated by the circle in Figure 4.3a). As a consequence, the simulation envelopes of the pair-correlation function shift upwards and completely enclose the observed pair-correlation function (Figure 4.3b). The analysis, therefore, indicates that the pattern does not exhibit small-scale interactions at distances below 0.2. The same result can be found for the L-function (Figure 4.3c). The patch type heterogeneity (Figure 4.3d) can also be successfully factored out with a displacement distance of $R = 0.2$ as shown in Figure 4.3e,f. However, in this case it would be possible to delineate a subarea of the observation window with constant intensity, defining an intensity function $\lambda(x)$, which has a value of $\lambda^* = \lambda/c$ in the occupied area (c is the proportion of the observation window covered by the patch), and a value of 0 outside, in the area void of points (Figure 4.3g). Using this intensity function for an HPP yields a similar result (Figure 4.3h,i).

4.1.2.2 HPP and Separation of Scales

As mentioned in the previous section, the assumption of separation of scales is an important issue in using the HPP based on a nonparametric intensity estimate (HPPnpe). To illustrate this issue in more detail we consider two examples. The first is a pattern with patch heterogeneity. The second is a pattern from the 2000 census at the BCI plot (Box 2.6), consisting of all living individuals of the species *Ocotea whitei* for which we have already estimated the intensity function based on topographic variables (Figure 2.27). The first example represents an extreme case of patch heterogeneity, formed by a CSR pattern in the left half of the observation window and an area void of points in the right half (Figure 4.4a). The typical scale of the heterogeneity of this pattern is given by the smallest width of the patch, which is 50 m (Figure 4.4a). Thus, a circle with radius $R > 25$ m does just fit fully within the patch, indicating that the typical scale of the patch heterogeneity is about 25 m. An appropriate displacement radius R (i.e., bandwidth) for the HPPnpe used in analyzing the pattern should, therefore, be somewhat smaller than this. Indeed, if we use a displacement distance of $R = 40$ or $R = 30$ m, the effect of the large-scale heterogeneity is not completely removed and, as a consequence, the pair-correlation function exhibits a departure from the null model at scales of $r < R$ (Figure 4.4b,c). When using a displacement radius of $R = 20$ m, we find that, for distances $r = 1$–20 m, the pair-correlation function is fully within the simulation envelope (Figure 4.4d).

Based on the above argument for determining the maximum, suitable, displacement radius R, we hypothesize that a radius of $R = 70$ m should be appropriate for removing the large-scale patch heterogeneity shown by the species *O. whitei* (Figure 4.4e). Indeed, for radius $R = 70$ we find that the pair-correlation function is well within the simulation envelopes for distances of $r = 30$–70 m (Figure 4.4g). In contrast, we find a strong effect of aggregation at distances below 30 m, where neighborhood densities are up to 18 times higher than expected under the heterogeneous Poisson null model. A displacement radius of $R = 50$ m yields a similar result (Figure 4.4h). However, selecting a value of $R = 30$ m already violates the assumption of separation of scales, because at this scale the displacement radius is too close to the scale of clustering (Figure 4.4i).

In summary, to verify that the separation of scales holds, and to determine the optimal displacement radius R, we recommended that a range of displacement radiuses R, ranging from the maximal scale of point interactions R_{inter} to the scale of the heterogeneity R_{hetero}, be tested. There is a sufficient separation of scales if the pair-correlation function is outside the simulation envelopes for distances $r < R_{inter}$, but inside the envelopes for a wide interval $R_{inter} < r < R$ (such as in Figure 4.4d,g,h).

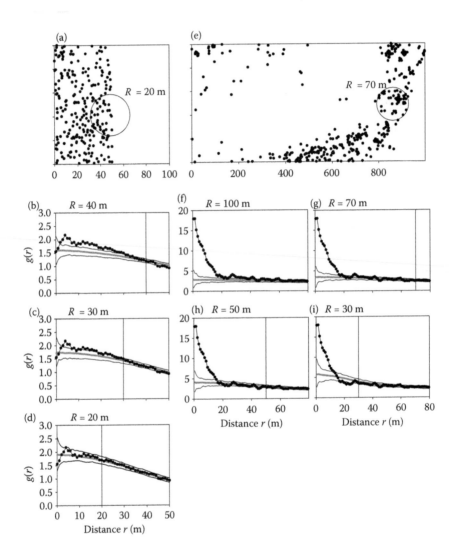

FIGURE 4.4
Separation of scales and HPP with nonparametric intensity estimation. (a) Pattern with patch heterogeneity, which is CSR inside the left half of the observation window. (b)–(d) Results of analyses using HPP with different bandwidths R, indicated by vertical lines in the figure, for pattern shown in panel (a). (e) Pattern of all living individuals of *Ocotea whitei* in the BCI plot from the 2000 census. (f)–(i) Results of analyses using HPP with different bandwidths R for the pattern shown in panel (e).

4.1.2.3 HPP Based on Parametric Intensity Estimates

One of the principal objectives of using an HPP is to explicitly separate the effects of environmental dependency from those of point interactions. The most biologically satisfying approach for removing the impact of environmental dependency is to use an HPP with an intensity function based on environmental covariates (see Section 2.6.2.9). However, this approach faces several problems that one needs to bear in mind. First, if the available covariates do not fully describe the environmental dependency of the pattern (i.e., one or more important environmental variables are missing), we may only obtain an approximation of the intensity function and will not be able to fully factor out the effects of heterogeneity in the pattern. This problem will be indicated, for example, by departures of the pair-correlation function from the null model at large distances r. Second, even if our habitat model includes all the important environmental variables, it may not reflect the true intensity function, if a species exhibits the effects of dispersal limitation (or other mechanisms of internal population dynamics). The latter issues may cause some areas of suitable habitat to be unoccupied in the observation window. Under these circumstances, we will most likely see the effects of dispersal limitation (or other mechanisms of internal population dynamics) as a signal of heterogeneity in the pattern at broader scales. Because of the two effects potentially influencing the pattern, the approach of using a habitat model as an intensity function does not always allow us to reveal the small-scale effects of point interactions, since the heterogeneity may not be fully removed. However, this approach does quantify the contribution of the currently known environmental variables in determining the spatial structure of the pattern, providing a starting point for further analysis.

To demonstrate the use of parametric intensity estimates, we will consider a pattern from the 2000 census at the BCI plot (Box 2.6), consisting of all living individuals of the species *O. whitei*, for which we have already estimated the intensity function based on topographic variables (Figure 2.27). Figure 4.5a shows the observed pattern with a strong patch type heterogeneity, and Figure 4.5c shows one realization of the HPP based on the habitat model shown in Figure 2.27a. It is evident that this HPP does not match the observed intensity of the species well. As a consequence, the pair-correlation function lies above the simulation envelopes at all distances up to the maximal distance of 100 m analyzed (Figure 4.5d). The expectation of the null model (i.e., solid gray line in Figure 4.5d) yields a value of 1.3 at a distance of $r = 100$ m, compared to a value of 2 for the observed pattern. This indicates that a substantial part of the heterogeneity of the pattern remains unexplained by the habitat model.

We combined the approaches of parametric and nonparametric intensity estimates to overcome the somewhat poor fit from applying the parametric approach alone, as just described. This was done by adding a nonparametric intensity estimate, with a bandwidth of $R = 50$ m (IntensityR50), as an additional variable to the habitat model. The resulting model includes two significant environmental variables (i.e., topographic wetness index TWI and slope), as well as the nonparametric intensity variable (IntensityR50) (Table 4.1). The nonparametric intensity variable does not dominate the habitat model, but supplements the environmental variables. Figure 4.5b shows the intensity estimate based on the combined habitat model, which now presents

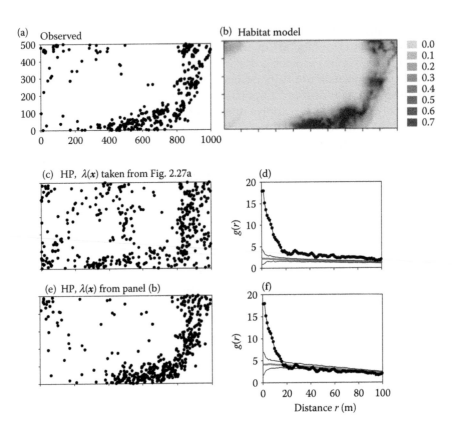

FIGURE 4.5
Application of HPP with parametric intensity estimate. (a) Map of an observed pattern consisting of all living individuals of the species *Ocotea whitei* from the 2000 census of the BCI plot. (b)
Map of the habitat model incorporating the nonparametric estimate of the intensity function
as a variable (Table 4.1). (c) One realization of HPP based on the habitat model shown in Figure
2.27a, which did not include the nonparametric estimate of the intensity function as a variable
(Table 2.4). (d) Pair-correlation analysis using the habitat model shown in Figure 2.26b. (e) Same
as panel (c), but using the habitat model in panel (b). (f) Same as panel (d), but using the habitat
model shown in panel (b).

a suitable compromise between the effects of the large-scale patch, as characterized by intensityR50, and intermediate-scale effects, produced by the
environmental variables. The realizations of the HPP, based on the intensity
function shown in Figure 4.5b, reassemble the larger-scale structures of the
observed pattern quite well (cf. Figure 4.5a,e), but cannot show smaller-scale
clustering. In addition, the expected pair-correlation function for this HPP
approximates the observed pair-correlation function for distances between
20 and 100 m (Figure 4.5f), although it falls slightly below the simulation
envelopes at some distances. At distances below 20 m, the pair-correlation
function reveals substantial clustering in the data not represented by the heterogeneous Poisson null model (Figure 4.5f). These results indicate that the

TABLE 4.1

Results of the Habitat Model for the Species *Ocotea whitei* Analogous to Those Shown in Table 2.4, but with the Addition of a Nonparametric Estimate of the Intensity Function, Based on a Bandwidth of $R = 50$ m (IntensityR50)

	Estimate	Std. Error	z Value	Pr(>|z|)
(Intercept)	−3.17***	0.43	−7.461	≪0.001
TWI	−0.263***	0.029	−9.119	≪0.001
VDist.Chann	−0.168	0.180	−0.931	0.351857
Slope	0.063***	0.018	3.577	0.000348
Aspect	0.0025	0.0013	1.95	0.051234
IntensityR50	1418***	117.7	12.043	≪0.001

Note: The additional variable IntensityR50 is introduced to consider internal effects of population dynamics (e.g., dispersal limitation) not accommodated by the environmental variables. The habitat model is based on the logistic resource selection probability function (Lele 2009). The Hosmer and Lemeshow goodness-of-fit (GOF) test for the model yields a *P*-value 0.5657 and we obtain an AUC of 0.872.

***P < 0.001.

small-scale clustering of *O. whitei*, which we attribute to point interactions, has a range of up to 20 m, whereas the observed pair-correlation function is approximated at broader scales by an HPP.

4.1.3 Null Model of Pattern Reconstruction

In Section 3.4.3, we presented a technique of pattern reconstruction that uses simulated annealing to generate stochastic replicates of observed patterns, which are constrained by one or more of the observed summary statistics. This approach to pattern reconstruction allows the selection of one or more summary statistics and then generates patterns that produce almost identical fits to the same summary statistics as the observed pattern. One of the most important applications of this form of pattern reconstruction is its use as a null model for testing independence in bivariate patterns (Section 3.4.3.7, Figure 3.57). In addition, pattern reconstruction also has interesting applications for the study of univariate patterns.

The application of pattern reconstruction for univariate patterns is based on the fact that the pattern reconstruction algorithm basically searches for the simplest (or "most random") pattern that is compatible with the constraints imposed by the summary statistics to be matched (Wiegand et al. 2013a). This is, in a way, analogous to the concept of maximum entropy (Volkov et al. 2009) that basically states that the system will tend to be in a state of maximal disorder, given a set of constraints. This allows us to determine, for example, whether the second-order summary statistics are sufficient to characterize the spatial structure of an observed pattern, or if additional summary statistics are required. Wiegand et al. (2013a) showed that second-order summary statistics may, in general, not be sufficient to capture the spatial structure of "real world" univariate patterns, and that the number of summary statistics required for this may vary. Some patterns with near random structure are already well described by one summary statistic whereas other patterns, with more complex spatial structures, may require four or five summary statistics (see also Section 3.4.3.6). This has important applications when parametric point-process models, such as cluster processes, are fit to the data (see the next chapter). In this context, the essential question is whether the summary statistics used for fitting the process are informative and can describe the spatial structure of the observed pattern sufficiently well. This can be explored beforehand by using pattern reconstruction (of course, this will also be revealed by testing the realizations of the fitted point-process model with other summary statistics not used for fitting). To this end, we reconstruct the pattern, using only the summary statistics applied in fitting the point-process model (e.g., the K-function and the pair-correlation function), and employ the resulting patterns as a null model, allowing us to identify departures of the pattern from other summary statistics not used in fitting the model (e.g., the spherical contact distribution or nearest-neighbor statistics). Note that pattern reconstruction is a nonparametric method and will, therefore, usually provide a better fit than parametric point-process models, which are constrained by their parametric functional form.

As a first example of this approach, we consider the complex spatial pattern of all recruits of the species *Cecropia insignis* obtained during five censuses from 1985 to 2005 at the BCI plot (Figure 4.6) (Box 2.6). This pattern has been analyzed in detail in Wiegand et al. (2009) using cluster point processes fit to the second-order summary statistics $K(r)$ and $g(r)$. In this context, recruits are defined as individuals growing past a size threshold of 1 cm dbh during a census period. The recruit pattern is complex, containing a strong signal of small-scale aggregation (Figure 4.6a). The reconstructions based on the second-order summary statistics $K(r)$ and $g(r)$ are able to reproduce the strong clustering, but visual inspection of the patterns suggests that the observed pattern contains more structure than a pure second-order structure. For example, the observed pattern shows somewhat larger gaps than the reconstructions (cf. Figure 4.6a,c). This visual diagnosis is formalized by looking at the fit of the different summary statistics. As expected, the pair-correlation function is almost perfectly reproduced (Figure 4.6b). In contrast, the spherical contact distribution $H_s(r)$ of the data departs substantially from that of the pattern reconstructed using the null model (Figure 4.6d). The observed $H_s(r)$ is clearly below the simulation envelopes, indicating that the gaps in the reconstructed pattern are too small. The reason for this is shown by $D(r)$, the distribution function to the nearest-neighbor (Figure 4.6e), for which the observed values lie above the simulation envelopes. This result indicates that the reconstructed patterns produce too many "isolated" points, which fill the gaps left by the clusters. Thus, we cannot expect that a fit with second-order summary statistics alone will yield a good fit of the observed recruit pattern.

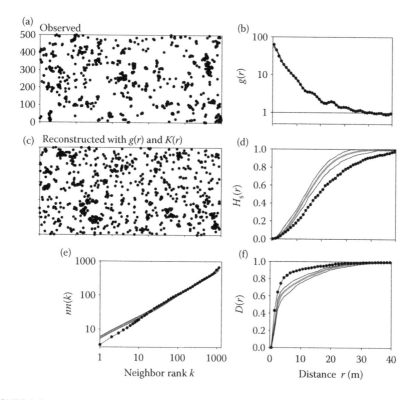

FIGURE 4.6

Pattern reconstruction as a null model. In this example, we examine how well second-order statistics [i.e., $K(r)$ and $g(r)$] capture the spatial pattern of all recruits of the species *Cecropia insignis* during the 1985–2005 census period for the BCI plot (recruits are defined as individuals that have a dbh ≥ 1 cm for the first time for a given census). (a) Observed pattern. (b) Pair-correlation function (closed disks) for the observed pattern, including simulation envelopes being the lowest and highest values of the pair-correlation function for 99 patterns based on reconstructions using the second-order summary statistics $K(r)$ and $g(r)$ (black line). The simulation envelopes are narrow because the pair-correlation function was used for the reconstruction. (c) Example of one reconstructed pattern. (d)–(f) Same as panel (b), but for summary statistics not used for pattern reconstruction. The simulation envelopes are wider because the summary statistics were not used for the reconstruction.

For a second example, we examine the pattern of all recruits of the shade-tolerant understorey species *Faramea occidentalis* from the 2005 BCI census (Box 2.6), as analyzed in Wiegand et al. (2013a). This species also exhibits a complex spatial structure (Figure 4.7a; clustering at two critical scales, see Section 4.1.4.5). However, this case is already very well defined by second-order summary statistics. Figure 4.7c shows one example for a reconstructed pattern. In this case, visual comparison of the reconstructed and the observed patterns suggests a good match. This is confirmed by the different summary statistics. First, as expected, the pair-correlation function is, in

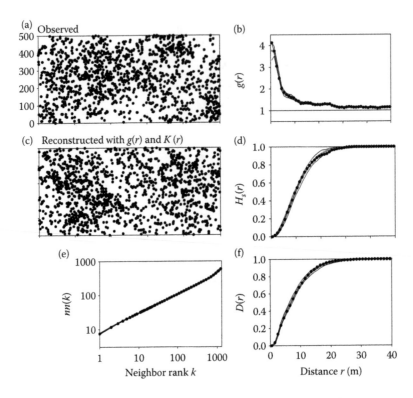

FIGURE 4.7
Pattern reconstruction as a null model. In this example, we examine out how well second-order statistics [i.e., $K(r)$ and $g(r)$] capture the spatial pattern of all recruits of the species *Faramea occidentalis* using the 2005 census from the BCI plot. (a) Observed pattern. (b) The observed pair-correlation function (closed disks) for the observed pattern, including the lowest and highest value of the pair-correlation function for 99 patterns based on reconstructions using the second-order summary statistics $K(r)$ and $g(r)$ (black line). (c) Example of one reconstructed pattern. (d)–(f) Same as panel (b), but for summary statistics not used for pattern reconstruction.

general, reproduced very well (Figure 4.7b). The spherical contact distribution (Figure 4.7d), the distribution function to the nearest neighbor (Figure 4.7f), and the mean distances to the kth neighbor (Figure 4.7e) are also extremely well matched by the reconstructed patterns, although they were not used for the reconstruction. The recruits of the species *F. occidentalis* are therefore a good example of the "most random" pattern that is possible under the constraints of the observed (complex) second-order structure. Reconstructions of this pattern based only on $K(r)$ and $g(r)$ show no departures from the range of summary statistics tested here. It would be very interesting to investigate why the pattern of recruits of some species, such as the shade-tolerant understorey species *F. occidentalis*, show only a "second-order spatial structure," whereas other species, such as the large-sized gap species *C. insignis*

(Figure 4.6), exhibit additional components of spatial structure not accommodated by the second-order characteristics alone.

4.1.4 Poisson Cluster Point Processes

Point processes producing elements of clustering are the most important parametric point processes for practical applications in ecology. For this reason, we will give them a fairly detailed treatment. Cluster producing processes are commonly used as null models in a large number of contexts. For example, they are useful for characterizing observed patterns of spatial clustering (e.g., Plotkin et al. 2000; Potts et al. 2004; John et al. 2007; Shen et al. 2009; Wang et al. 2011), exploring the spatial structure of observed patterns (e.g., Batista and Maguire 1999; Wiegand et al. 2007c, 2009), relating cluster properties of models fit to species properties (e.g., Seidler and Plotkin 2006), or deriving theoretical expectations for species turnover and distance-decay under species clustering (Morlon et al. 2008). A clustered pattern typically contains areas of elevated point density (i.e., clusters) that contrast with areas of low point density or even areas void of points (e.g., Figure 4.6a). As a first approximation in characterizing a homogeneous cluster pattern, we may develop the following point-process model (Figure 4.8): we assume that the pattern consists of a certain number $n_\rho = A\rho$ of randomly and independently distributed "clusters," where A is the area of the observation window and ρ the intensity of the pattern of the cluster centers (Figure 4.8a). A pattern is then constructed following two stochastic rules: (i) the number of points S belonging to a given cluster follows a certain probability distribution p_s: $s = 0, 1, 2, \ldots$, and (ii) the position of points in the cluster, relative to the cluster

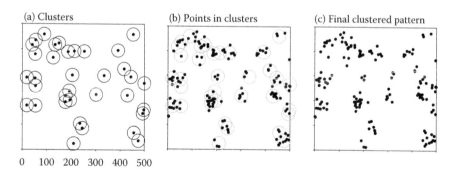

(a) Clusters

(b) Points in clusters

(c) Final clustered pattern

0   100   200   300   400   500

FIGURE 4.8

Schematic representation of the construction principle for a Poisson cluster process. (a) A number of n_ρ randomly and independently distributed "clusters" (with centers following CSR) are distributed within the observation window. The typical size of the clusters is indicated by circles. (b) Each cluster contains S points governed by a probability distribution p_s, and the position of points in the cluster relative to the cluster center are independently and identically distributed according to a bivariate probability density function $h(.)$. (c) The resulting Poisson cluster pattern.

center, are independently and identically distributed according to a bivariate probability density function $h(.)$ (Diggle 2003, p. 64). In addition, the function $h_2(r)$ is given by

$$h_2(r) = \int h(x)h(x - r)\,dx \qquad (4.1)$$

which yields the probability that two points of the same cluster are distance r apart. The function $h_2(r)$ is used in the following to derive analytical expressions for the second-order summary statistics of the cluster processes. Note that different clusters will potentially overlap, which hinders a direct identification of the points that belong to a given cluster (Figure 4.8b,c).

The point-process model described above is called the *Poisson cluster* or *Neyman-Scott process* and was introduced by Neyman and Scott (1958). The resulting realizations of this point process are homogeneous and isotropic, if $h(.)$ is radially symmetric. The intensity of the process is given by $\lambda = \rho\mu$, where μ is the mean number of points per cluster and ρ is the intensity of the pattern of the cluster centers. Following Diggle (2003, p. 65), we can calculate the pair-correlation function of the Poisson cluster process as

$$g(r) = 1 + \frac{\rho}{\lambda^2} E[S(S - 1)]h_2(r) \qquad (4.2)$$

where S is the number of points a given cluster contains [i.e., $E(S) = \mu$] and $E[S(S - 1)]$ is the expectation of $S(S - 1)$, which is determined by the probability distribution p_s associated with the process. For points that are randomly assigned to a cluster according to a *Poisson distribution*, we find that $E[S(S - 1)] = \mu^2$ and $g(r) = 1 + h_2(r)/\rho$, where $\lambda = \rho\mu$.

The most important examples of the Neyman-Scott process are the so-called *Thomas processes*, introduced by Thomas (1949). For these processes, the locations of the point in a given cluster, relative to the cluster center, follow a *bivariate Gaussian distribution* $h(r, \sigma)$ with variance σ^2. The Thomas process has the convenient property that the function $h_2(r)$ in Equation 4.2 has a simple analytical form, because the convolution of a normal distribution is again a normal distribution. As we will see in the next sections, this property allows the Thomas process to be used as a basic construction unit for more complex cluster processes that include several scales of clustering. Another Neyman-Scott process, with a more complex analytical solution for the pair-correlation function, is the so-called *Matern process*. In this process, the points of a given cluster are randomly placed within a disk of radius R around the cluster center (Stoyan and Stoyan 1994, p. 311). The selection of the "kernel function" $h(r, \sigma)$ of the Neyman-Scott process depends on the objective of the analysis. In some cases, the Matern process may match the underlying biology better than the Thomas process, whereas the Thomas process may be preferred in other cases.

The expectation $E[S(S-1)]$ may differ from μ^2, if we select a distribution function p_s other than the Poisson distribution. For example, if we select a *negative Binominal distribution* with "clumping parameter" k and mean μ we find that $E[S(S-1)] = \mu^2 f_k$ and $g(r) = 1 + f_k\, h_2(r)/\rho$ where $f_k = (1+k)/k$. "Clumping" here refers to the way the points are distributed within the clusters. If the parameter k is small, there are very few points in a large number of clusters and a few clusters with a large number of points (see insets of Figure 4.11b,f,j). Thus, the negative Binominal distribution is appropriate for cases in which the variance in the number of points per cluster is greater than would be expected for the Poisson distribution. As the magnitude of k increases, the variance among clusters decreases. In fact, the negative Binominal distribution approaches the Poisson distribution as $k \to \infty$ and $f_k = (1+k)/k \to 1$.

The Poisson distribution and the negative Binomial distribution yield pair-correlation functions of exactly the same structure, since their pair-correlation functions are

$$g(r) = 1 + h_2(r)/\rho \qquad \text{Poisson}$$
$$g(r) = 1 + [(1+k)/k]\, h_2(r)/\rho \quad \text{Negative binominal} \tag{4.3}$$

The implication of this is that we cannot distinguish between these two-point processes, based on second-order summary statistics alone. If the number of points per cluster S is "clumped" (i.e., follows a negative Binominal distribution with parameter k), we can obtain the same pair-correlation function as under a Poisson distribution, if we increase the number of clusters ρA (by increasing ρ) by the factor $f_k = (1+k)/k$. (Note that the pair-correlation function contains the factor f_k/ρ.) Thus, the Thomas process based on a negative Binominal distribution needs more clusters to yield the same pair-correlation function as compared to the Thomas process using a Poisson distribution. The underlying reason for this is that increased clumping in the distribution of points over the clusters (i.e., decreasing value of k) yields an elevated number of "empty" clusters that need to be included to compensate for the clumping. The probability that a cluster is empty is given by $p_{\text{Bin}}(S=0,k) = (1+\mu/k)^{-k}$ which, for large k, approximates the probability $p_{\text{Pois}}(S=0) = e^{-\mu}$ for the Poisson distribution.

Illian et al. (2008, p. 375) and Diggle (2003, p. 66) provide general formulas for the distribution function $D(r)$ of the nearest neighbor and the spherical contact distribution $H_s(r)$ for the Poisson cluster process:

$$D(r) = 1 - [1 - H_s(r)]J(r) \tag{4.4}$$

$$H_s(r) = 1 - \exp(-\rho C(r)) = 1 - \exp(-\lambda\pi r^2 C^*(r)) \tag{4.5}$$

where $J(r)$ and $[1 - H_s(r)]$ in Equation 4.4 are the probabilities that no point within the typical cluster and no point from a neighboring cluster, respectively, are located within distance r of an arbitrary point of the typical cluster. $H_s(r)$ is the spherical contact distribution and the function $J(r)$ is the J-function introduced by Van Lieshout and Baddeley (1996), which is defined to be a combination of $D(r)$ and $H_s(r)$ as $J(r) = [1 − D(r)]/[1 − H_s(r)]$ (Equation 4.4).

Given the preceding definitions, Equation 4.4 can be understood by knowing that $[1 − D(r)]$ represents the probability that the typical point of the Poisson cluster pattern has no neighbor within distance r. In this case, the typical point must match two conditions; first, it has no neighbor from its own cluster within distance r [with associated probability $J(r)$]; and second, it has no neighbor from a different cluster within distance r [with associated probability $1 − H_s(r)$], which gives us $1 − D(r) = [1 − H_s(r)] J(r)$, yielding Equation 4.4.

The spherical contact distribution $H_s(r)$, as defined in Equation 4.5, is determined in the following fashion: $C(r)$ is the mean value of a quantity $C_i(r)$ taken over all clusters i, where $C_i(r)$ is the area of overlap of all disks of radius r centered on the points of cluster i (Figure 4.9). For a random pattern we find that $H_s(r) = 1 − \exp(−\lambda \pi r^2)$ (Table 2.2). In this case, $C_i(r)$ is simply the area πr^2 of a disk of radius r for all "clusters" i, since there is only one point in each cluster. In a

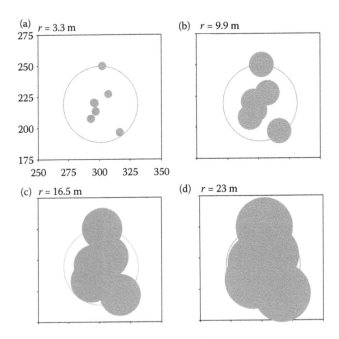

FIGURE 4.9
Schematic representation of the calculation of the spherical contact distribution $H_s(r)$ for a Poisson cluster process based on the quantity $C(r)$. We show a cluster i composed of six points and the corresponding quantity $C_i(r)$ for different distances r. $C_i(r)$ is the area of the overlap of all discs of radius r centered at the points of the cluster i (gray area). The quantity $C(r)$ is the average taken over the $C_i(r)$ of all clusters i.

Poisson cluster process, this area has to be replaced by the overlap of areas centered on the points belonging to cluster i (Figure 4.9). In other words, the disk i of the CSR pattern has to be replaced by the corresponding area that arises from the overlap of all disks around the points that belong to cluster i. Estimating $C(r)$ is an elementary, but difficult, problem (e.g., Stoyan and Stoyan 1994, p. 313). Analytical solutions are available only for specific cases (e.g., Morlon et al. 2008).

We can rescale $C(r)$ to yield $C^*(r) = C(r)/(\pi r^2 \mu)$, where μ is the mean number of points in the typical cluster and πr^2 the area of a circular sampling area with radius r. Therefore, $C^*(r)$ is a correction factor in Equation 4.5 that ranges between zero and one and accounts for the degree of clustering at the sampling scale r (Morlon et al. 2008). It yields one for a pattern without clustering (i.e., CSR) and its value decreases with increasing clustering.

Many different cluster processes can be obtained by using different distributions for p_s and $h(.)$. A particularly interesting feature of the Poisson cluster process is that the distribution p_s of the number of points S per cluster enters only as a constant (i.e., the expectation $E[S(S - 1)]$; Equation 4.2). Because of this, we can always find Thomas processes that have different distributions p_s, which govern the distribution of the number of points among the clusters, but have the same pair-correlation function and the same function $h(.)$. This means that second-order statistics will not allow us to infer the distribution p_s of the number of points S per cluster. However, it is evident that the use of other summary statistics, such as $D(r)$ or $H_s(r)$, can be helpful in differentiating among these cases. Thus, in general, we may need the information from additional summary statistics to determine the distribution p_s of the number of points S per cluster.

4.1.4.1 Interpretation of Poisson Cluster Processes

Sometimes the Poisson cluster processes are interpreted as "parent–offspring" processes, where the offspring are randomly scattered around the parents according to a "dispersal kernel," which is determined by the probability density function $h(.)$. This interpretation can be made if the pattern of interest really represents offspring (e.g., recruits). In Section 4.1.4.4, we present a Thomas process for the pattern of offspring that can be applied if the pattern of the parents is known. However, we cannot expect, in general, a direct relationship between the "real" dispersal kernel and the pattern of the adult population. This is because the resulting pattern of present-day adults is the accumulated outcome of many generations of offspring and, if the species exhibits dispersal limitation, we may have simultaneous expansion and contraction of clusters that can create spatial signatures quite different from those expected from the dispersal kernel. Nevertheless, very short, mean dispersal distances should result in strongly clustered adult patterns and long-ranging dispersal may cause larger-scale clustering. For example, a study by Seidler and Plotkin (2006) found that the parameter σ, which defines the size of clusters (Equation 4.7), was correlated with the dispersal capacity of tropical tree species. In general, however, we have to consider the Poisson cluster process to be a phenomenological description of the cluster characteristics of the pattern, not a mechanistic explanation that would provide a direct link to the underlying processes. For a recent approach used in fitting ecological process models to spatial patterns of tree species at the BCI plot (Box 2.6) see Detto and Muller-Landau (2013). The Poisson cluster process and other more complex cluster processes, based on the Thomas process, are best used as a "benchmark" for processes with known structure. They are best at characterizing the basic properties of the aggregation structure of the patterns, but do not necessarily give direct insight into the real nature of the processes producing a pattern (Wiegand et al. 2007c, 2009). The parameters of these processes are thus summary indices of the cluster properties of the pattern. They allow a detailed description of the properties of the observed pattern, which is a prerequisite for deriving hypotheses on the underlying processes actually driving the pattern.

4.1.4.2 Thomas Process

The Thomas process (Thomas 1949) is an algorithm that produces relatively realistic, clustered point patterns based on the simple stochastic construction principles outlined above (Figure 4.8). Generally, it consists of a bivariate probability density function $h(\cdot)$ that governs the location of points in a given cluster, relative to the cluster center. Usually, this is a *symmetric bivariate Gaussian distribution* $h(r, \sigma)$ with variance σ^2, which is commonly applied together with a *Poisson distribution* that governs the distribution of the number of points per cluster, but there are exceptions (e.g., Tscheschel and Stoyan 2006). The most important exception is the generalization of the Poisson distribution (which determines the random allocation of offspring within the clusters) to a *negative binominal distribution* (which produces a "clustered" distribution of points among the clusters) (Section 4.1.4). The negative Binominal distribution, through the parameter k, characterizes cases in which the variance in the number of points per cluster is greater than would be expected for a Poisson distribution of points among the clusters. As already noted in Section 4.1.4, plugging in the negative binominal distribution for p_s does not change the functional form of the pair-correlation function (Equation 4.2) relative to the Poisson distribution. This means that second-order properties of the pattern alone do not allow us to make statements on the way the points are distributed over the clusters. However, it is evident that changing the distribution of points over the clusters will change the nearest-neighbor summary statistics. In the case of a small value of the parameter k, we obtain a few clusters with many points and many clusters with a few points. In this case, a pattern generated with a parameter $k < 1$ will have substantially more isolated points than a pattern with the same pair-correlation function, but generated by a Poisson distribution for p_s (i.e., $k \to \infty$). This will be depicted by the distribution function $D(r)$ of the distances to the nearest neighbor. Such situations are of particular relevance in ecology because they point to an additional mechanism that governs the distribution p_s. Generalizing p_s by using the negative binominal distribution, instead of the Poisson distribution, allows us to characterize a much broader range of aggregated spatial patterns.

The *first step in the construction of a Thomas process* is the generation of a random pattern of cluster centers. Once the coordinates of the cluster centers are determined, the individual clusters are constructed. To this end, the points of the pattern may be, in the simplest case, randomly assigned to the clusters in which the number of points that belong to a given cluster follows a Poisson distribution with mean $\mu = \lambda/\rho$. Note that this also means that a given cluster may be empty (i.e., there is no point assigned to the cluster); if the points are randomly distributed over the clusters, the probability that a given cluster will be empty is given by $p_s(S = 0, \mu) = \exp(-\mu)$ and the probability of having just one point in a cluster is $p_s(S = 1, \mu) = \mu\exp(-\mu)$. It is important to keep this construction principle in mind, when interpreting the parameters that were fit by a Thomas process for a given data set.

In addition to the previous considerations, we may also encounter patterns where the fit with the Thomas process indicates that the number of clusters is higher than the number of points in the cluster pattern, implying that many clusters contain no points. Although cases in which there are more clusters than points in the process (i.e., $\rho \gg \lambda$) are at first somewhat counterintuitive, it makes perfectly sense since this case arises when most points are isolated (i.e., assigned to clusters with only one point), although there are occasionally two or three points that are located close together. The latter is an important class of cluster processes that can be characterized quite well by the Thomas process. For example, if the mean number of points per cluster is $\mu = 0.25$, the Poisson distribution predicts that 78% of the clusters will be empty, 19% will contain one point, and 3% will contain more than one point. Thus, if the pattern comprises 500 points we expect 1558 empty clusters [$= 500 * \exp(-0.25)$], 389 clusters with only one point [$= 500 * 0.25 * \exp(-0.25)$], and 111 clusters with more than one point. Thus, 78% ($= 389/500$) of the points are isolated. By using the Poisson distribution for p_s, we can further estimate that 19% of the points ($= 95$) will have one close by neighbor (i.e., belong to a cluster with two points) and 3% ($= 16$) will be in clusters with three or more points. Thus, the Thomas process is able to capture a wide range of possible cluster configurations.

The *second step in the construction of the Thomas process* is the determination of the locations of the points that belong to a given cluster. For the Thomas process, the location of the points in a given cluster, relative to the cluster center, follows a bivariate Gaussian distribution $h(r, \sigma)$ with variance σ^2, as indicated above. We can interpret $h(r, \sigma)$ as being the *"dispersal kernel"* with respect to the (unknown) locations of the cluster centers. In cases in which we analyze the pattern of the offspring, the fitted parameters of the Thomas process allow us to infer the underlying dispersal kernel.

Thomas processes have a big advantage in that their pair-correlation functions can be calculated analytically. This allows the Thomas process to be fit to a given point pattern by fairly standard statistical tools. As a consequence, the Thomas process is commonly used in ecological applications. Based on Equations 4.2 and 4.3 we can easily calculate the pair-correlation function of the Thomas process. Using the symmetric bivariate normal distribution with variance σ^2 we find that

$$h_2(r,\sigma) = h(r,\sqrt{2}\sigma), \tag{4.6}$$

If the distribution p_s is a Poisson distribution, we know that $E[S(S-1)] = \mu^2$, and if the distribution p_s is a negative Binominal distribution, with clumping parameter k and mean μ, we know that $E[S(S-1)] = \mu^2(k+1)/k$. Thus, with $\mu = \lambda/\rho$ the pair-correlation function of the Thomas process becomes

$$g(r,\rho,\sigma) = 1 + \frac{f_k}{\rho}\frac{\exp(-r^2/4\sigma^2)}{4\pi\sigma^2} \tag{4.7}$$

where $f_k = (k+1)/k = 1$ for the Poisson distribution. To better understand the estimation of more complex "multi-generation" Thomas processes, we provide in the next sections a detailed estimate of the pair-correlation function for the Thomas process. We will use the terms *"parent"* to refer to the cluster centers and *"offspring"* (or type 1 point) to refer to the points belonging to a given cluster.

The estimation of the pair-correlation function is based on the *product density* $\rho^{(2)}(r) = \lambda^2 g(r)$, where $\rho^{(2)}(r)\, dx_1 dx_2$ is the probability of finding two offspring lying within small disks dx_1 and dx_2, separated by distance $r = |x_1 - x_2|$, and centered on locations x_1 and x_2, respectively (see Section 3.1.2.3). Because the two offspring may be produced either by the same parent or by different parents, we need to account for the contribution of these two cases (Figure 4.10a). First, the locations of the two offspring are clearly independent (remember that the parents are CSR) if they were produced by different parents. In this case, the contribution to the product density is λ^2, since the probability of finding one offspring within disk dx_1 at location x_1 is given by λdx_1. However, if the two offspring were produced by the same parent, we have to multiply the intensity of the parents ρ by the expected number of ordered pairs of offspring ($E[S(S-1)] = \mu^2 f_k$ in the case of a negative Binominal distribution), and by the probability $h_2(r, \sigma)$ that two offspring of the same parent are at a distance r apart:

$$\rho^{(2)}(r) = \lambda^2 + \rho E[S(S-1)]h_2(r,\sigma) \tag{4.8}$$

(a) Simple Thomas process

(b) Simple bivariate parent-offspring
Thomas process

(i) Different parents   (ii) Same parent

(i) Different parents (ii) Same parent

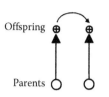

(c) Bivariate parent–offspring Thomas process with clustered parents

(i) Type 1 and 2 points   (ii) Type 1 parent   (iii) Type 1 and 2 points from
from different   of type 2 point   different parents but
grandparents   same grandparent

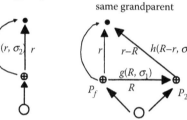

(d) Nested double-cluster
Thomas process

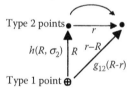

FIGURE 4.10
Estimation of the pair-correlation functions of the different Thomas processes. Curved arrows represent parent–offspring relationships and the curved arrows the relationship of interest. (a) Simple Thomas process where the offspring is clustered around unknown parent points. (b) Simple bivariate parent–offspring Thomas process where the offspring is clustered around known parent points. (c) Bivariate parent–offspring Thomas process with known and clustered parent points. (d) Nested double-cluster Thomas process where the offspring is clustered around unknown parent points that themselves follow a simple Thomas process.

With $\mu = \lambda/\rho$ and $E[S(S-1)] = \mu^2 f_k$ we find that

$$\rho^{(2)}(r) = \lambda^2 \left[1 + \frac{f_k}{\rho} h_2(r,\sigma) \right] = \lambda^2 \left[1 + \frac{f_k}{\rho} h(r,\sqrt{2}\sigma) \right] \qquad (4.9)$$

Equation 4.9, in combination with Equation 4.6, yields the formula given in Equation 4.7, since $\rho^{(2)}(r) = \lambda^2 g(r)$.

As shown in Equation 4.7, the Thomas process has four parameters, the intensity λ of the process, which determines the number of points of the pattern ($n = \lambda A$), the intensity ρ of the cluster centers, the parameter σ, which determines the spatial extent of a cluster, and the parameter k of the negative binominal distribution [or alternatively $f_k = (k + 1)/k$)] that governs the clumping of the points over the clusters. Using Equation 4.7, we can calculate the neighborhood density of points at very short distances r around the typical point of the pattern, as $O(r) = \lambda\, g(r) \approx \lambda(1 + (f_k/\rho)/(4\pi\sigma^2))$, since $\exp(-r^2/4\sigma^2) \approx 1$. Thus, if the number of clusters ($= \rho A$) decreases (for a fixed number of points), the neighborhood density $O(r)$ of the pattern increases at small distances r, because the points are now concentrated in fewer clusters. The neighborhood density $O(r)$ of the pattern will also increase if the variance σ decreases, since the variance determines the location of the points relative to the cluster center. With a lower variance driving the pattern, the points will be closer together. Additionally, we will also see increases in the neighborhood density $O(r)$ at small distances r when there is a higher degree of clumping of points over the clusters, as indicated by smaller values of k and larger value of f_k. In this case, a few clusters will have many points, but many clusters will have very few points. Because the intensity ρ of clusters and the quantity f_k appear in Equation 4.7 as a quotient, they can compensate for each other in yielding identical pair-correlation functions. To maintain the same structure of the pair-correlation function for different values of k, the intensity of clusters ρ must also be altered, since Equation 4.7 contains the factor $f_k = (k + 1)/k$.

In the standard Thomas process, the distribution of points relative to cluster centers is governed by the symmetric bivariate normal distribution with variance σ^2. Thus, it makes sense to define the radius of the "typical" cluster of this point process as the distance for which 86% of all the points are located within the cluster center. For the symmetric bivariate normal distribution, this distance is given by $r_C = 2\sigma$. Thus, the approximate area covered by the "typical" cluster is given by $A_C = \pi r_C^2 = 4\pi\sigma^2$. Note that this definition of cluster size is not directly related to the specific properties of individual clusters (e.g., the number of points belonging to a given cluster; Plotkin et al. 2002), but is instead based on the (stochastic) construction principle of the clusters. In fact, most of the clusters can only be appropriately called "clusters," in a literal sense, if the average number of points per cluster is high (say $\mu > 3$). Otherwise, a majority of the "clusters" would only contain a single point. Thus, care has to be taken to not interpret the fitted parameters too literally.

The value of the pair-correlation function at $r = 0$ is related to the probability that a given point of the pattern has a neighbor very close by. As already seen above, this is a useful property of the Thomas process that characterizes the overall degree of clustering. It yields

$$g(r = 0, \sigma, \rho) = 1 + \frac{f_k}{\rho}\frac{1}{4\pi\sigma^2} = 1 + \frac{f_k}{\rho}\frac{1}{A_C} \tag{4.10}$$

Thus, the area of the cluster, the number of clusters, and the clumping of points over the clusters determine the overall clustering of a pattern in the same way. This is consistent with the intuitive idea that suggests that clustering may increase if there are fewer clusters, fewer clusters with more points, or if the area covered by individual clusters is smaller. However, the rate of decline in the value of the pair-correlation function with separation distance r depends only on the cluster size A_C:

$$g(r,\sigma,\rho) = 1 + \frac{f_k}{\rho A_C}\exp(-\pi r^2/A_C) \tag{4.11}$$

as this is proportional to the area of the disk with radius r measured in units of cluster size A_C.

We can illustrate the issues being discussed here by examining the effects of changing k, while holding all other parameters constant. For example, Figure 4.11a shows one realization of a Thomas process with 626 points, a cluster size 2σ of 12 m, and 50 clusters ($\rho = 0.0002$). Thus, the mean number of points per cluster is $\mu = 12.5$. The small inset of Figure 4.11b shows the distribution of the number of points per cluster, which ranges from 6 to 22 points, but is in most cases close to the mean (i.e., 10–14). As a consequence, the clusters all have more or less the same number of points and the pattern looks very "orderly." As a demonstration of the impact of increased clumping of points among clusters, we generated two additional realizations of the Thomas process containing the same general parameter values as before, but different values of the parameter k. Thus, the original parameter ρ, which fixed the number of clusters based on the Poisson distribution for p_s to 50, needed to be increased by a factor of f_k to yield the same pair-correlation function (Equation 4.9).

In the first example, shown in Figure 4.11e, we have a realization of the Thomas process based on a negative binominal distribution with parameter $k = 1$. To maintain the same pair-correlation function, the corresponding number of clusters must be $f_k = (k + 1)/k = 2$ times the number of clusters of the Thomas process, based on a random distribution of points over the clusters (i.e., 100 clusters). Because the variance in the number of points per cluster increases, the pattern looks more diverse, with many isolated points (i.e., no cluster associated), many clusters with a few points, and a few clusters with many points (small inset of Figure 4.11f, indicating that there are 17 empty clusters and 13 with one point). For the second example, a realization of a Thomas process with parameter $k = 0.1$ is shown in Figure 4.11i. For this realization of the Thomas process, 11 times the number of clusters [$f_k = (k + 1)/k = 11$] (i.e., 550 clusters) is required to yield the same pair-correlation function as the Thomas process, based on the Poisson distribution for p_s (i.e., Figure 4.11a). In this case, 422 of the 550 clusters (77%) are empty and 42 have only one point (i.e., isolated points). In contrast, there are 2 large clusters containing 29 and 31 points (inset Figure 4.11j).

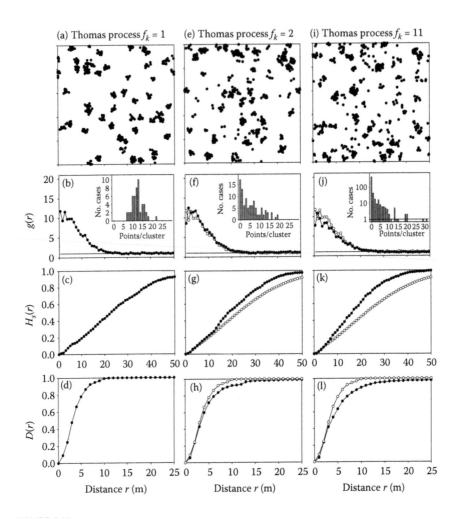

FIGURE 4.11
Simulated Thomas processes with 626 points based on the same pair-correlation function, but different distribution functions p_s for the number of points per cluster. The cluster size was in all cases $2\sigma = 12$ m and $\rho A / f_k = 50$ (a) Realization for a Poisson distribution for p_s with $\rho A = 50$ clusters containing 626 points. (e) Realization for a negative binominal distribution for p_s using a parameter of $k = 1$ ($f_k = 2$), with $50 * f_k = 100$ clusters. (i) Realization for a negative binominal distribution for p_s using a parameter of $k = 0.1$ ($f_k = 11$), with $50 * f_k = 550$ clusters. The other panels show, as solid disks, the pair-correlation function, the spherical contact distribution $H_s(r)$ and the distribution function $D(r)$ of the distances to the nearest neighbor for the patterns occurring within the same column. The open disks show, for comparison, the summary statistics for the Thomas process based on the Poisson distribution. Insets in panels (b), (f), and (j) show the observed distribution of the number of points per cluster.

While the pair-correlation function is maintained across these three examples, there are changes in the spherical contact distribution $H_s(r)$ (Figure 4.11g,k) and the distribution function $D(r)$ of the distances to the nearest neighbor (Figure 4.11h,l). If the number of points per cluster is clumped (i.e., $f_k > 1$), the gaps in the pattern become smaller, which is indicated by $H_s(r)$ for the binomial distribution lying above those for the Poisson distribution (Figure 4.11g,k; open disks). This is because there are more isolated points and clusters with fewer points than for the Poisson distribution. Because of the increased number of isolated points, the values of $D(r)$ for the Thomas process with a negative binominal distribution are below those of the corresponding case with a Poisson distribution (i.e., $f_k = 1$; Figure 4.11h,l; open disks). In sum, although we cannot infer the distribution p_s of the points over the clusters based on the pair-correlation function (or other second-order statistics), we can extract this information using $H_s(r)$ and $D(r)$ as additional summary statistics.

4.1.4.3 Fitting a Thomas Process to the Data

For some point-process models we can derive analytical formulas for important summary statistics. For example, with the Thomas process we specified the pair-correlation function (Equation 4.7). In this case, we can fit the point process directly to the summary statistics of an observed pattern without the need to simulate the point process. The standard approach is to use the minimum contrast method (Section 2.5.2.1; e.g., Stoyan and Stoyan 1994; Diggle 2003; Illian et al. 2008, p. 450f). In short, one summary statistic (typically the pair-correlation function or the K function) is estimated from the observed point pattern, and the model is then fit by finding the optimal parameter values for the model that yield the closest match between the theoretical and empirical curves. However, as suggested in Wiegand et al. (2007c), we recommend fitting both the g-function and the L-function simultaneously to the data (for details see Section 2.5.2.1). Theoretically, the g and the L functions contain the same information and a fit using either g or L should yield the same parameter estimates. In practice, however, we found improved results by fitting both simultaneously. The explanation for this is that the g-function is especially sensitive at shorter distances r, but approaches the asymptote of $g = 1$ more quickly at increasing distances r, as compared to the slower approach by L to its asymptote of 0. In contrast, the cumulative L-function is not very sensitive at small distances, but is sensitive at greater distances. Optimizing both the g- and the L-function (Equation 2.12), in general, produces a more balanced fit than using either function alone.

A potential problem when using the L-function for parameter estimation is that the L-function is cumulative and has a "memory" (Wiegand et al. 2007c, 2009). To understand this, we first discuss the behavior of the pair-correlation function. For example, if a pattern was generated by a point process that contains two critical scales of clustering (this point process will be introduced in the following Section 4.1.4.5 in detail), the contribution of the small-scale clustering in the pair-correlation function fades away at larger distances. For example, the gray line in Figure 4.12d shows only the contribution of large-scale clustering in the pair-correlation function, which at distances $r > 10$ m is identical with the pair-correlation function of the point process that incorporates both small- and large-scale clustering (closed disks in Figure 4.12d). However, at distances smaller than 10 m we observe the effect of the small-scale clustering, as indicated by strongly elevated values of $g(r)$. That means that we can safely fit a simple Thomas process (Equation 4.7) to the pair-correlation function of a pattern with two scales of clustering, if we do this selectively for distances $r > r_0 = 10$ m (i.e., only over the range of large-scale clustering). In this case the simple Thomas process will reveal the correct characteristics of the observed large-scale clustering (Figure 4.12d) but, of course, cannot make statements about the small-scale clustering. However, fitting the simple Thomas process for the entire distance interval of 1 to 50 m will produce a biased estimate (Figure 4.12b), because this point process does not incorporate a mechanism for small-scale clustering and will, therefore,

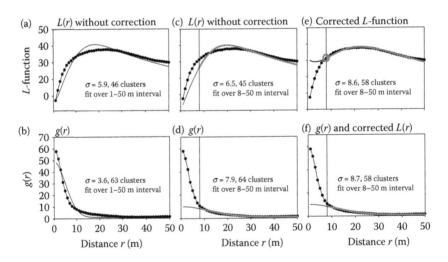

FIGURE 4.12
Memory of the cumulative L-function as a source of problems when fitting a Thomas process to point patterns with additional small-scale effects. (a) Fit of a pattern generated by a double-cluster Thomas process using the L-function for scales of 0–50 m. The fit is not satisfying. (b) Same as panel (a), but using the pair-correlation function. (c) Same as panel (a), but the L-function was fit only over the distance interval 8–50 m. Again, the fit is not satisfying. (d) Same as panel (c), but using pair-correlation function. The fit is satisfying. (e) Fit with the transformed L-function that removes the memory for the distance interval 8–50 m. The gray open disk shows how the transformation attaches the fitted L-function to the value observed at distance $r = 8$ m. The fit is now satisfying. (f) Fit that minimized the contrast of the transformed L-function and g-function simultaneously over the distance interval 8–50 m. The fit is satisfying. In all the panels, the closed disks are the results of analyzing the observed pattern and the gray line is the best fit. The vertical lines in (c)–(f) indicate the distance interval at which the fit was conducted.

yield some sort of average between small- and large-scale clustering. In summary, the noncumulative pair-correlation function $g(r)$ has the convenient property that a pattern at small distances (i.e., small-scale clustering at distances $r < r_0$) does not influence the shape of $g(r)$ at larger distances. This allows for application of the simple Thomas process (Equation 4.7) for distances $r > r_0$ to selectively reveal the large-scale clustering.

In a more general sense, the above considerations mean that we can use the pair-correlation function to fit a point process selectively for larger distances r, if the pattern shows additional small-scale structure not accommodated by the point process. However, this is not directly possible with the K- or L-function. For example, Figure 4.12c shows the best fit with the L-function over distances $r > 10$ m (gray line) to the observed L-function (closed disks), which yields biased estimates. The same happens if we fit over the entire range from 1 to 50 m (Figure 4.12a). Thus, fitting the Thomas process with the L-function to a pattern that exhibits additional small-scale clustering produces biased estimates of the parameters of the cluster process (Stoyan and Stoyan 1996) and leads to the observation that the parameter estimates are sensitive to the upper limit r_{max} to which the L-function is fit (e.g., Batista and Maguire 1998; Plotkin et al. 2000). To overcome these problems, Wiegand et al. (2007c) developed a transformation of the K-function that removes the memory. The underlying idea is simple and is based on the fact that the K-function at larger distances r incorporates the values at smaller distances r:

$$K(r) = \int_{t=0}^{r} g(r)2\pi t\, dt = \int_{t=0}^{r_0} g(r)2\pi t\, dt + \int_{t=r_0}^{r} g(r)2\pi t\, dt = \underbrace{K(r_0)}_{\text{memory}} + [K(r) - K(r_0)] \quad (4.12)$$

To remove the memory, we fit the transformation $K_t(r) = \hat{K}(r_0) + K(r) - K(r_0)$, in place of the theoretical K-function of the point process, for distances $r > r_0$, where $\hat{K}(r_0)$ is the value of the observed K-function at distance r_0, $K(r)$ is the theoretical K-function of the point process, and $K(r_0)$ is the value of the theoretical K-function at distance r_0. In other words, we replaced the theoretical value $K(r_0)$ by the observed value $\hat{K}(r_0)$ in Equation 4.12. Thus, we find $K_t(r_0) = \hat{K}(r_0)$ which means that the values of the observed and (transformed) theoretical K-function coincide at distance r_0 (Figure 4.12e; open disk). When we base the L-function of the Thomas process on the transformed K-function $K_t(r)$ during the fitting procedure we obtain a satisfying fit (Figure 4.12e). The reason for this is that the transformation removes the memory and, therefore, allows us to selectively fit the larger-scale features of the pattern at distances $r > r_0$.

4.1.4.4 Bivariate Parent–Offspring Thomas Processes

When data for the spatial pattern of offspring and parents are available, we may be interested in characterizing the parent–offspring relationship in detail. If the offspring are clustered around the parents, we can model the parent–offspring relationship as a *simple bivariate parent–offspring Thomas process*. As an example of this approach, we can consider a study by Jacquemyn et al. (2007), who analyzed data on the spatial distribution of the adults and seedlings of the orchid *Orchis purpurea*. The particular interest in their study was to determine how dispersal distances, as estimated by parental analysis, related to the observed scales of seedling-adult association.

The simplest point process, based on a Thomas process, that can describe parent–offspring relationships is a generalization of the Neyman-Scott process for univariate patterns, as described in Section 4.1.4 and Figure 4.8. However, because the point process should describe a bivariate pattern, it is natural to define the spatial pattern of the cluster centers (i.e., Figure 4.8a) as pattern 1 (i.e., the parent pattern), and the resulting spatial pattern of the univariate Neyman-Scott process as pattern 2 (i.e., the offspring). The task is now to derive the probability of finding a type 2 point (= an offspring) at distance r away from a type 1 point (= a parent). The parameters for this type of cluster process are the intensity λ_1 of pattern 1, the intensity λ_2 of pattern 2, and the variance σ_2^2 of the distribution function $h(r, \sigma)$ that governs the location of the type 2 points in a given cluster, relative to the associated cluster center (i.e., a type 1 point).

The calculation of the pair-correlation function of the simple bivariate parent–offspring Thomas process is analogous to the derivation of the Thomas process previously described (Section 4.1.4.2). Here, we must consider two cases of type 1–type 2 point pairs that are separated by distance r (Figure 4.10b): (i) the case in which the type 2 point belongs to the cluster associated with the type 1 point and (ii) the case in which the type 2 point does not belong to the cluster associated with the type 1 point. In the second case, the type 1 and type 2 points are distributed independently of one another (because cluster centers are independently placed) and we multiply the intensity of type 1 points (= λ_1) by the intensity of type 2 points (= λ_2). Thus, the contribution of case (ii) to the product density yields $\lambda_1\lambda_2$. In case (i), in which the type 2 point belongs to the cluster associated with the type 1 point, we multiply the intensity of type 1 points (= λ_1) by the expected number of offspring per cluster ($\mu_2 = \lambda_2/\lambda_1$) and the probability that an offspring is at distance r from the parent [= $h(r, \sigma_2)$], resulting in the product $\lambda_2 h(r, \sigma_2)$. We now add the two cases together and divide the resulting product density by $\lambda_1\lambda_2$ to yield the bivariate pair-correlation function of the process:

$$g_{12}(r, \sigma_2, \lambda_1) = 1 + \frac{1}{\lambda_1} h(r, \sigma_2) \qquad (4.13)$$

Note that the bivariate pair-correlation function for this process is not dependent on the distribution function p_s of the number of offspring per parent (i.e., the parameter k for the situation where p_s is from a negative Binominal distribution).

Figure 4.13 shows a bivariate pattern composed of 35 parent points (pattern 1) and 157 offspring (pattern 2) that were constructed based on the simple bivariate parent–offspring Thomas process described above (Figure 4.8b). We fit the parameters ρ and σ to the (univariate) offspring pattern using the minimum contrast method (see previous section) and simulated the realizations of this point process using the location of the known parent points as cluster centers. Thus, in contrast to the simulations of the standard Thomas process, where for each realization new cluster centers are used, the cluster centers now remain the same (= pattern 1) for all realizations of the null model. The fit to the model output yields 34 parents, which is very close to the actual 35 parents used in constructing the pattern, and a parameter value for σ of 12.75, which yields an approximate cluster radius of 25 m (the parameter for simulating the process was $\sigma = 13.3$). Thus, the fit with the Thomas process recovered the parameters used for simulating the point process quite well and the simulations of the point process, using the known parent locations, are in agreement with the summary statistics of the observed offspring

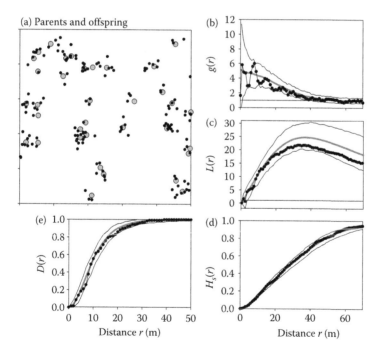

FIGURE 4.13
Analysis of a simulated pattern based on a simple bivariate parent–offspring Thomas process. (a) Map of offspring (closed black disks) and parents (=cluster centers; gray disks). (b)–(e) Univariate summary statistics for the observed offspring pattern, with simulation envelopes produced by the simple bivariate parent–offspring Thomas process generating the pattern shown in panel (a). Cluster centers were the same observed parent locations for all simulations of the null model.

pattern (Figure 4.13b–e). Figure 4.13 also illustrates the degree of stochastic-ity that is inherent in this type of point process, even though the realizations of the point process used exactly the same parent point positions.

Clearly, the use of Equation 4.13 is only valid if pattern 1 (i.e., the parents) is CSR, which is a quite restrictive assumption. It would be desirable to generalize this point-process model to encompass a wider class of parent patterns. This can be done by assuming that the parents follow a Thomas process (Wiegand et al. 2007c; Jacquemyn et al. 2007). In this case, the pattern of the parents is itself clustered and the cluster centers of the parent pattern are given by the pattern of the "grandparents." The *bivariate parent–offspring Thomas process with clustered parents* (i.e., a grandparent–parent–offspring process) assumes that the parent pattern (pattern 1) follows a Thomas pro-cess and that points of pattern 2 are offspring of the points of pattern 1. The parameters of this cluster process are given by ρ_1 and σ_1 of the Thomas pro-cess for the parents (= pattern 1), the intensity λ_1 of the parents, which is also the intensity of the clusters of pattern 2 (= ρ_2), the intensity λ_2 of the offspring (= pattern 2), and the variance σ_2^2 of the distance between parents and offspring. To be slightly more general, we also consider that the two, dis-tribution functions p_{s1} and p_{s2} can be negative binominal distributions, with parameters k_1 and k_2 and means μ_1 and μ_2, respectively.

We can calculate the product density $\rho_{12}^{(2)}(r)$ for the more complex process with clustered parents in a manner analogous to the calculation of the pair-correlation function for the simple Thomas process (Section 4.1.4.2). Here, we need to include three cases for the relationship between parent and off-spring in the formula: (i) the parent and offspring points trace back to differ-ent grandparents (Figure 4.10c); (ii) the offspring is a direct descendant of the parent (Figure 4.10c); and (iii) the parent and offspring are descendants of the same grandparent point, but the offspring is not a direct descendent of the parent (Figure 4.10c, case iii). In case (ii) the calculation is simple: as before, we multiply the intensity of type 1 parent points (= λ_1) by the expected num-ber of offspring ($\mu_2 = \lambda_2/\lambda_1$) and the probability that an offspring is distance r away from the parent [= $h(r, \sigma_2)$]. The contributions of the first (i) and third (iii) cases (i.e., where the offspring is not the direct descendent of the focal type 1 parent point) can be handled together. To this end, we have to con-sider all possible locations l_2 of the actual, unknown parent point (a second type 1 point; Figure 4.10c) relative to the focal type 1 point. In Figure 4.10c (case iii) we therefore have the focal parent (named P_f) and the second parent (named P_2), which is an unknown (vector) distance $\mathbf{R}$ away from the focal parent P_f, and the offspring of the focal parent P_f which is distance r away from P_f. Thus, we have three points involved in the estimation, and because the second parent point P_2 can be located anywhere in the plane we have to consider all possible locations of P_2 in our estimation. This means that we have to integrate over all possible configurations of the two parent points and the offspring as shown in Figure 4.10c (case iii). Thus, we need to multi-ply the probability of finding a second type 1 point (i.e., point P_2) at a location

l_2 (which is located an unknown vector distance $\mathbf{R}$ away from the first type 1 point P_f) [= $\lambda_1^2 g(R, \sigma_1)$], by the expected number of offspring ($\mu_2 = \lambda_2/\lambda_1$) and the probability that the type 2 offspring is a distance $|R - r|$ away from the second type 1 point [= $h(R - r, \sigma_2)$]. This term is then integrated over all possible vector distances $\mathbf{R}$:

$$\rho_{12}^{(2)}(r)$$

$$= \lambda_1\mu_2 h(r,\sigma_2) + \int \lambda_1\lambda_1 g(R,\sigma_1,\rho_1)\mu_2 h(R - r,\sigma_2)dR$$

$$= \lambda_1 \left[\frac{\lambda_2}{\lambda_1}(h(r,\sigma_2) + \lambda_2 \int \left[1 + \frac{f_{k1}}{\rho_1} h(R,\sqrt{2}\sigma_1) \right] h(R - r,\sigma_2)dR \right]$$

$$= \lambda_1\lambda_2 \left[\frac{1}{\lambda_1} h(r,\sigma_2) + 1 + \frac{f_{k1}}{\rho_1} h\left(r,\sqrt{2\sigma_1^2 + \sigma_2^2}\right) \right] \tag{4.14}$$

Note that $f_{k1} = (k_1 + 1)/k_1$, $h_2(r,\sigma) = h(r,\sqrt{2}\sigma)$, $\mu_2 = \lambda_2/\lambda_1$, and $\lambda_1 = \rho_2$. Thus, the bivariate pair-correlation function $g_{12}(r)$ for the bivariate Thomas process, with clustered parents, yields

$$g_{12}(r,\rho_1,\sigma_1,\rho_2,\sigma_2) = 1 + \underbrace{\frac{1}{\rho_2} h(r,\sigma_2)}_{\text{Offspring–own parent}} + \underbrace{\frac{f_{k1}}{\rho_1} h(r,\sqrt{2\sigma_1^2 + \sigma_2^2})}_{\substack{\text{offspring–other parent} \\ \text{but same grandparent}}} \tag{4.15}$$

The first term in Equation 4.15 (i.e., the 1) corresponds to the relationship between a type 2 and a type 1 point that traces back to different grandparents points and are therefore independent. The second term corresponds to the case in which the type 2 point is a direct offspring of the type 1 point (governed by the distribution function $h(r)$). The third term corresponds to the case in which the type 1 and type 2 points go back to the same grandparent, but the type 2 point is not a direct offspring of the type 1 point. The third term contains an interaction between the two different scales of clustering, as indicated by the variance term $2\sigma_1^2 + \sigma_2^2$. If the parents exhibit a random pattern (i.e., $\sigma_1^2 \to \infty$), Equation 4.15 collapses to a simple bivariate parent–offspring Thomas process (Equation 4.13), since the third term in Equation 4.13 disappears (i.e., there are no parents clusters). If type 2 points are independent of their parents (i.e., $\sigma_2^2 \to \infty$) Equation 4.13 collapses to $g_{12}(r) = 1$ as expected for independence.

Figure 4.14 shows an example of a parent–offspring Thomas process with clustered parents. The parents are clustered using the same pattern shown in Figure 4.13. The 626 offspring points are randomly located around the parents with a bivariate normal distribution, with parameter $\sigma = 3.2$. Simulating this process based on the observed parent locations provides a good fit to the summary statistics for the offspring pattern (Figure 4.14b,c). Figure 4.14b

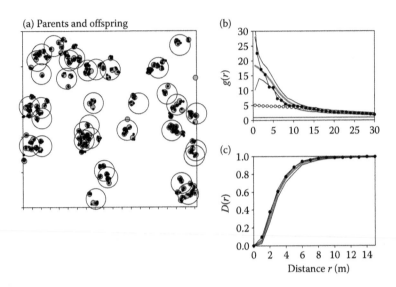

FIGURE 4.14
Analysis of a simulated pattern based on a bivariate parent–offspring Thomas process, where the parents are clustered following a Thomas process. (a) Map of offspring (closed black disks), clustered parents (= cluster centers; gray disks with size indicating the cluster radius) and CSR parents of the parents (open disks with size indicating the cluster radius). (b)–(c) Univariate summary statistics for the observed offspring pattern, with simulation envelopes produced by the bivariate parent–offspring Thomas process generating the pattern shown in panel (a), with the parents of the parents being in the same location for all simulations. The open disks in panel (b) represent the expected pair-correlation function for the cluster pattern of the parents.

also shows the expected pair-correlation function for the clustered pattern of the parents (gray disks), which drives the larger-scale pattern (i.e., distances $r > 13$ m). Comparison of the parent pair-correlation function with the observed pair-correlation function of the offspring clearly shows the effect of the additional small-scale clustering around the parents.

4.1.4.5 Generalized Thomas Process with Two Nested Scales of Clustering

In the last section we introduced a point process where the parents follow a Thomas process and the offspring are distributed around the clustered parents following a bivariate normal distribution. Considering only the final pattern of the offspring, we have a pattern with *two nested scales of clustering*. The nested construction principle is principally the same as for a Thomas process, where the cluster centers are built upon an initial CSR process, except that in this case the cluster centers are built upon a Thomas process. These types of point processes are often referred to as "multi-generation processes" (Diggle 2003), where small clusters are located within larger clusters (Figure 4.14a). In the example shown in Figure 4.14, there are 626 points distributed over 157 small clusters with radii of $r_C = 6.4$ m. Thus, there are on average $\mu = \lambda/\rho = 4$ points within a small cluster.

We will now calculate the pair-correlation function of the *nested double-cluster Thomas process*. The parameters of this cluster process are given by the parameters ρ_1 and σ_1 of the Thomas process of the parents (= pattern 1), the intensity λ_1 of the parents, which is also the intensity of the clusters of pattern 2 (i.e., $\lambda_1 = \rho_2$), the intensity λ_2 of the offspring (= pattern 2), the variance σ_2^2 of the distance between parents and offspring, and the parameters of the negative binominal distributions k_1 and k_2, for the large and small clusters, respectively. Note that the (clustered) parent pattern is in general unknown and that we need to infer the parameters for the parent pattern and the offspring pattern only from the univariate pattern of offspring.

The estimation of the product density of the nested double-cluster Thomas process is based on the pair-correlation function of the bivariate parent–offspring Thomas process with clustered parents, as derived above (Equation 4.15). It also requires a consideration of three types of points: grandparent points, type 1 points (parents), and type 2 points (offspring). The task here is to derive the probability that a type 2 offspring point (second point) is located at a distance r away from an initial, arbitrarily chosen type 2 point (focal point) (Figure 4.10d). This problem can be reduced because we already know the probability of finding a type 2 point at distance r away from a type 1 point. The second type 2 point can, therefore, be considered to be the offspring of the type 1 point (Figure 4.10d). Therefore, the relationship of the second type 2 point to the type 1 point is governed by the distribution function $h(r, \sigma_2)$. We also know that the relationship of a type 1 point to the focal type 2 point is governed by the pair-correlation function $g_{12}(r)$ of the bivariate parent–offspring Thomas process with clustered parents (Equation 4.15). To estimate the product density we therefore need to integrate over all possible distances $\mathbf{R}$ between type 1 points and the first type 2 point (Figure 4.10d):

$$\lambda_{22}(r) = \int \lambda_2 g_{12}(R - r) h(R, \sigma_2) dR \qquad (4.16)$$

Integrating Equation 4.16, using Equation 4.15 for $g_{12}(R - r)$, and considering $E[S_2(S_2 - 1)]$, yields

$$g_{22}(r, \rho_1, \sigma_1, \rho_2, \sigma_2) = 1 + \underbrace{\frac{f_{k2}}{\rho_2} h(r, \sqrt{2}\sigma_2)}_{\text{Same parents}} + \underbrace{\frac{f_{k1}}{\rho_1} h(r, \sqrt{2\sigma_1^2 + 2\sigma_2^2}}_{\substack{\text{Different parents} \\ \text{but same grandparent}}} \qquad (4.17)$$

The first term (i.e., 1) corresponds to the case in which the two type 2 points have different grandparents; since grandparents in this case are independently placed, the parents are also independent, and the contribution of this case to the pair-correlation function yields one. The second term corresponds to the case in which the type 2 points share the same type 1 (parent) point. In this case, the grandparent is not of interest and the contribution is the same as for a simple Thomas process where the offspring are from the same parent (i.e., the second term in Equation 4.8) and there is no effect of nested clustering. The third term corresponds to the case in which the two type 2 points have the same grandparent but different parent points. Here, the nested clustering has an effect, which can be seen in the sum of the variances of the two normal distributions that correspond to the two levels of clustering in this point process. Note that a negative binomial distribution for the distribution of type 1 and type 2 points over their respective parents and type 1 clusters only impacts the intensity of the corresponding cluster centers.

Because of the previous considerations, the formula for the pair-correlation function of the univariate double-cluster (offspring) pattern (Equation 4.17) is basically the same formula as for the bivariate offspring–parent pattern (Equation 4.15). The only differences are that the parameter σ_2, which describes the size of the small clusters, is now multiplied by a factor of $\sqrt{2}$ (which is due to the fact that we do not account for parent to offspring relationships, but only offspring to offspring) and the factor f_{k2} is added (due to the distribution of type 2 points over their own small clusters).

Equation 4.17 collapses to the pair-correlation function for the Thomas process (Equation 4.7), if there is no large-scale clustering (i.e., pattern 1 follows a random pattern with $\sigma_1^2 \to \infty$), and it collapses further to the g-function of CSR, that is, $g(r) = 1$, if the offspring pattern 2 is not clustered around the parents (i.e., $<\sigma_2^2 \to \infty$). Also, if both scales of clustering are the same (i.e., $\sigma_1 = \sigma_2$), the pair-correlation function yields

$$g_{22}(r, \rho_1, \sigma, \rho_2) = 1 + \left(\frac{f_{k2}}{\rho_2} + \frac{f_{k1}}{\rho_1} \right) h(r, \sqrt{2}\sigma) \tag{4.18}$$

which is the same as that of a simple Thomas process (Equation 4.7), but the intensity ρ/f_k of cluster centers is $(\rho_1 \rho_2)/(f_{k2} \rho_1 + f_{k1} \rho_2)$. Thus, without knowledge of the pattern of the parents (which is generally unknown) we will not be able to distinguish this nested process from a simple Thomas process.

Figure 4.15 shows the application of a nested double-cluster Thomas process to the spatial pattern of 626 small saplings (i.e., 1–5 cm dbh) of the species *Shorea congestiflera* from the Sinharaja tropical forest in Sri Lanka (Figure 4.15a). This pattern has been analyzed in detail by Wiegand et al. (2007c). Note that we cannot use second-order statistics to estimate the parameters k_1 and k_2, which characterize the degree of aggregation of points among clusters. Instead, we use an indirect approach by applying a double-cluster process, fit for different values of k_1 and/or k_2. This is used to determine which parameter values yield the best estimates, relative to the observed pattern, for the spherical contact distribution and the distribution function of the distances to the nearest neighbor.

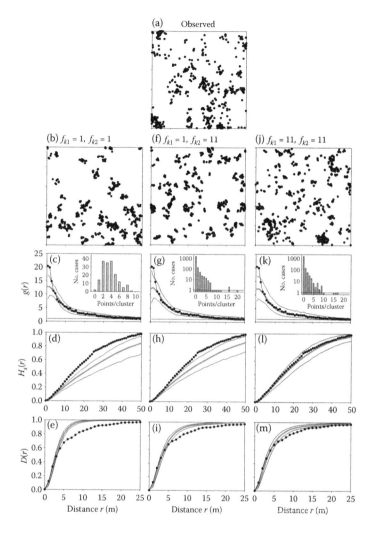

FIGURE 4.15
Fit of a double-cluster Thomas process to an observed pattern of small saplings of the species *S. congestiflera* from the Sinharaja plot as analyzed in Wiegand et al. (2007a). (a) Observed pattern, which was fit with the pair-correlation and L-functions, yielding 33 large clusters with a size of $2\sigma_1 = 26$ m, and 168 small clusters with a size of 7.3 m. (b) One realization of the double-cluster process, fit with a Poisson distribution of the points over the clusters. (c) Pair-correlation function of the observed pattern and simulation envelopes of the fitted double cluster process, with $f_{k1} = 1$ and $f_{k2} = 1$ (i.e., Poisson distributions for p_{s1} and p_{s2}). (d) Same as panel (c), but for the spherical contact distribution. (e) Same as panel (d), but for the distribution function of the distances to the nearest neighbor. (f) One realization of the double cluster process fit with a negative binominal distribution of points over the small clusters with parameter $k_2 = 0.1$ ($f_{k1} = 1$, $f_{k2} = 11$). (g)–(i) Summary statistics and simulation envelopes for the cluster process in panel (f), with $f_{k1} = 1$ and $f_{k2} = 11$. (j)–(m) Same as panels (f) through (i), but for a cluster process with $f_{k1} = 11$, $f_{k2} = 11$. The small insets in panels (c), (g), and (k) show the distribution of the number of points over the small clusters.

Separation of the scales of clustering (i.e., $\sigma_1^2 \gg \sigma_2^2$ in Equation 4.17) suggests a convenient approach to fit the four parameters of the double-cluster process to the pair-correlation function and L-function. In the first step of the analysis, we fit the value of $\sigma_{\text{sum}} = \sqrt{\sigma_1^2 + \sigma_2^2}$ and the intensity ρ_1 of the larger-scale clustering process using a simple Thomas process (Equation 4.7). In this case, only scales r that are greater than the range r_{02} of small-scale clustering are considered. The value of r_{02} can be determined by fitting the simple Thomas process for several plausible values of r_{02} (e.g., Figure 4.12d,e). If the value of r_{02} is too small, the Thomas process does not yield a good fit (Figure 4.12b). Using this approach, we find a suitable value of $r_{02} = 15$ m for the pattern of small *S. congestiflera* saplings. In the second step of the analysis, we used the full double-cluster model, incorporating the parameter estimates of σ_{sum}^2 and ρ_1 obtained in the first step within the model, and obtain a fit for σ_2^2 and ρ_2, the two remaining, unknown parameters of the small-scale cluster process. After fitting the parameters, using the methods described in Section 4.1.4.3, we obtained the following estimates for the general structure of the process: σ_{sum} of 13.5, an estimate of 33 large clusters with radii of 26 m and an estimate of 168 small clusters with radii of $2\sigma_2 = 7.3$ m. Estimates of the distribution of mean number of points were $\mu_1 = 19.2$ saplings located in the large clusters, 5.1 small clusters located in large clusters, and $\mu_2 = 3.7$ saplings located in the small clusters.

In general, the double-cluster process fit the pair-correlation function very well, although there was evidence for additional clustering at distances below 3 m, as indicated by the observed pair-correlation function lying above the simulation envelopes at scales below 3 m (Figure 4.15c). When randomly assigning the centers of the smaller clusters and the saplings within these centers to the larger clusters (i.e., $f_{k1} = 1$ and $f_{k2} = 1$; Figure 4.15b), we find that the corresponding point process yields gaps that are too large, as indicated by a positive departure from the spherical contact distribution $H_s(r)$ (Figure 4.15d). Also, there are substantially more isolated points, as indicated by a negative departure of $D(r)$ from the simulation envelopes at scales greater than 3 m (Figure 4.15e). This diagnosis is confirmed by comparison of one realization of the point process (Figure 4.15b) with the observed pattern (Figure 4.15a). The simulated pattern exhibits not enough interspersed, isolated saplings (or clusters of one or two saplings), as is characteristic of the observed pattern.

We now explore how a negative binominal distribution, which governs the distribution of saplings in the small clusters, changes and possibly improves the fit of $D(r)$ and $H_s(r)$. We use a value of $k = 0.1$ for the small clusters, which corresponds to a value of $f_{k2} = 11$. Thus, instead of 168 small clusters we now have 1845 small clusters. The 626 saplings are, in turn, distributed within the 1845 small clusters following a negative binominal distribution (inset Figure 4.15g). As expected, a clumped distribution of saplings among the small clusters changed the distribution function $D(r)$ to be closer to the observed $D(r)$ (Figure 4.15i; as compared to the case of $f_{k1} = 1$ and $f_{k2} = 1$ in Figure 4.15e).

However, the point process still exhibits departures from the observed pattern (Figure 4.15i). The spherical contact distribution $H_s(r)$ is closer to the observed $H_s(r)$ (cf. Figure 4.15d,h), but still clearly lies outside the simulation envelopes. Selecting larger values for f_{k2} did not improve the fit. Using the case of $f_{k1} = 11$ and $f_{k2} = 1$, where the distribution of small clusters occurring within large clusters followed a negative Binominal distribution, produced similar results, although $D(r)$ was closer to the observed values, but $H_s(r)$ yielded a poorer fit (results not shown).

Finally, we use a negative binominal distribution for both the large and small clusters, with parameter $k = 0.1$ (i.e., $f_{k1} = 11$ and $f_{k2} = 11$). In this case, we have 359 large clusters and 1845 small clusters (inset Figure 4.15k). The combined effect of the non-Poisson distribution of large and small clusters now yields a spherical contact distribution which is in good agreement with the data (Figure 4.15l) and a distribution function $D(r)$ which is now, at most distances, within (or very close to) the simulation envelopes (Figure 4.15m). Increasing the values of f_{k1} and f_{k2} did not substantially improve the fit, as a small departure in the distribution function $D(r)$ remains. Thus, the pattern of small saplings of the species *S. congestiflera* is largely compatible with a double-cluster Thomas process with negative binominal distributions for the number of small clusters and points within large and small clusters, respectively. The parameter k of the negative binominal distribution is approximately $k = 0.1$, which produces a strongly skewed distribution (i.e., insets in Figure 4.15k). The process that creates the large and small clusters must, therefore, have a non-Poisson component that produces many clusters with one or two points, but also some clusters with a very high numbers of points.

4.1.4.6 Independent Superposition of CSR within a Double Cluster Process

In the previous section we demonstrated how modeling the distribution of points in clusters (i.e., the distributions p_{s1} and p_{s2}) can enhance the fit of the double-cluster process. The simplest version of the Thomas process assumes a random allocation of points over the clusters (i.e., a Poisson distribution). However, this assumption may not apply to real data, as we found for the pattern of small saplings of the species *S. congestiflera*. Generalizing the Poisson distribution by using a negative binominal distribution allows us to model a "clumped" distribution of points over the clusters, such that some clusters have substantially more points than expected by a random distribution and other clusters have substantially fewer points than expected. This yields an increased number of "isolated" points.

Wiegand et al. (2007c, 2009) suggested an alternative point process that can produce isolated points. They observed that, in analyzing the spatial patterns of saplings and recruits of several tropical tree species, the pair-correlation function of a double-cluster Thomas process yields a good fit to the data. However, it was also observed that the distribution function $D(r)$ of distances to the nearest neighbor yielded a poor fit, with many more isolated points occurring than predicted by the "pure" double-cluster Thomas process. They also found that the patterns could be approximated reasonably well by applying an independent superposition of a random pattern on a "pure" double-cluster Thomas process (i.e., one with $f_{k1} = 1$ and $f_{k2} = 1$). The superposition is an alternative way of modeling distribution functions p_{s1} and p_{s2} that differ from a pure Poisson distribution. The double-cluster Thomas process with $f_{k2} > 1$ and/or $f_{k1} > 1$ will only rarely generate situations where isolated points approximate a CSR pattern, independent of the clustering process, since this would require many large clusters having exactly one small cluster consisting of exactly one point. The double-cluster Thomas process with $f_{k2} > 1$ and/or $f_{k1} > 1$ can instead produce situations in which many small clusters have exactly one point. For example, the insets of Figure 4.15 g,k show that this process comprises about 130 small clusters with only one point. The double cluster Thomas process with $f_{k2} > 1$ and/or $f_{k1} > 1$ could, therefore, be approximated by an independent superposition of a "pure" double-cluster Thomas process (i.e., one with $f_{k1} = 1$ and $f_{k2} = 1$) and a simple Thomas process with variance $2\sigma_1^2 + 2\sigma_2^2$.

Wiegand et al. (2007c, 2009) considered a cluster process with two critical scales of clustering that arises from an independent superposition of a nested double-cluster Thomas process (Equation 4.17) and a CSR pattern. Denoting p_C as the proportion of the points belonging to the nested double-cluster component Thomas process and using the formulas for independent superposition (Equation 3.127 in Section 3.32) yields the pair-correlation function:

$$g(r, \sigma_1, \rho_1, \sigma_2, \rho_2) = 1 + p_C^2 \frac{1}{\rho_2} h(r, \sqrt{2}\sigma_2) + p_C^2 \frac{1}{\rho_1} h(r, \sqrt{2\sigma_1^2 + 2\sigma_2^2}) \quad (4.19)$$

The pair-correlation function of the superposition Equation 4.19 yields exactly the same functional form as that of a nested double-cluster process Equation 4.17. Instead of the factors f_{k1} and f_{k2} that appear in Equation 4.17, we have now the factor p_C^2. If a cluster pattern has isolated points (i.e., $p_C < 1$) the cluster component of the pattern comprises fewer clusters than expected by the fit with the pair-correlation function. This is because, in the case of $p_C < 1$, we need to decrease the values of ρ_1 and ρ_2 to maintain the pair-correlation function. Thus, the intensities of cluster centers of the double-cluster component process are given by $\rho_1 = \rho_1^* p_C^2$ and $\rho_2 = \rho_2^* p_C^2$, where ρ_2^* and ρ_1^* are the corresponding estimates for the superposition pattern when using the pair-correlation function for fitting the pattern (Wiegand et al. 2007c).

As an example of using an independent superposition of a nested double-cluster Thomas process (Equation 4.17) with an additional CSR pattern, we again consider the pattern of the small saplings of the species *S. congestiflera* (Figure 4.15a). The fitted parameters of the double-cluster process, without superposition, yield $A\rho_1^* = 33$, $A\rho_2^* = 168$, $2\sigma_1 = 26$ m, and $2\sigma_2 = 7.3$ m (see the previous section). As expected, adding 40 random points (i.e., $p_C = 586/626 = 0.88$) produces an improvement in the fit of the spherical contact distribution $H_s(r)$, but there are still gaps that are too large (results not shown). The distribution function $D(r)$ of the distances to the nearest neighbor also improved slightly (results not shown). If we add 80 random points, the pattern produces a good approximation to the values of $H_s(r)$ (Figure 4.16g), but the observed values of $D(r)$ for distances of 6 to 12 m (Figure 4.16h) are significantly

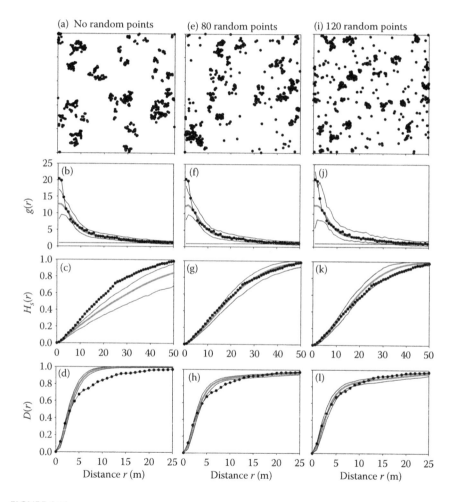

FIGURE 4.16
Fit of a superposition of a random pattern and a double-cluster Thomas process, based on the observed pattern of small saplings of the species *S. congestiflera* analyzed in Figure 4.15. (a) One realization of the double-cluster pattern fit without random points. (b) Pair-correlation function (closed disks) of the observed pattern, with simulation envelopes of the model fit without random points, which produced panel (a). (c) Same as panel (b), but for the spherical contact distribution. (d) Same as panel (b), but for the distribution function of the distances to the nearest neighbor. (e) One realization of the superposition pattern with 80 random points. (e)–(h) Summary statistics and simulation envelopes for the superposition pattern with 80 random points. (i)–(l) Same as panels (e) through (h), but for 120 random points.

below the values produced by the point-process model. Thus, the superposition process predicts clustering to be too strong at this distance range, since the nearest-neighbor distances produced are generally too short relative to the observed pattern. However, increasing the number of random points further (to 120) yields a poorer fit in $H_s(r)$, as the number of gaps with larger diameters tend to be too few (Figure 4.16k). The introduction of the random points also results in a poor prediction of the observed pattern by the point-process model at neighbor distances >15 m in a positive direction, as it produces too many isolated points (Figure 4.16l). As expected, we observe a similar fit to the pair-correlation function in all cases (Figure 4.16b,f,j). Taking the results of this and the previous example (Figure 4.15) together, we find that small saplings of species *S. congestiflera* exhibit a complex cluster pattern, with an uneven (nonrandom) distribution of points in the clusters. This could be due to a superposition of a random pattern and a "pure" double-cluster Thomas process, with a Poisson distribution of points over the clusters, or it could be due to a clumped distribution of points over the clusters, as would be consistent with a negative binominal distribution. Although the point process based on the negative binominal distribution approximated the nearest-neighbor distances somewhat better than the superposition process, we cannot provide conclusive evidence for one or the other point process being more realistic in producing the observed pattern.

4.1.4.7 Independent Superposition of Two Thomas Processes

The point process presented in the previous section is not the only possibility for generating univariate patterns with two scales of clustering. Stoyan and Stoyan (1996) proposed a somewhat simpler point process, with two scales of clustering, resulting from the independent superposition of two simple Thomas processes. The parameters of this process are given by ρ_1 and σ_1 of the first Thomas process and the ρ_2 and σ_2 of the second Thomas process. Denoting the relative intensities of these processes by p_1 and p_2, respectively, and using the formulas for independent superposition (Equation 3.127 in Section 3.3.2) yields

$$g_{1+2}(r,\sigma_1,\rho_1,\sigma_2,p_2) = 1 + \underbrace{(1-p_1)^2 \frac{1}{\rho_2} h(r,\sqrt{2}\sigma_2)}_{\text{Contribution of Thomas process 2}} + \underbrace{p_1^2 \frac{1}{\rho_1} h(r,\sqrt{2}\sigma_1)}_{\text{Contribution of Thomas process 1}} \quad (4.20)$$

Comparison of Equation 4.20 with the pair-correlation function of the nested double-cluster process, Equation 4.17 shows that both have the same functional form. Thus, the two point processes cannot be distinguished based on their second-order properties. However, the superposition process does not show the interaction between the two nested scales of clustering as seen in Equation 4.17 [i.e., $h(r,\sqrt{2\sigma_1^2 + 2\sigma_2^2})$], but only the contribution of the first Thomas process [i.e., $h(r, \sqrt{2}\sigma_1)$]. This is because the process is not nested; the two cluster processes are independently superposed. The distribution function $D(r)$ of the distances to the nearest neighbor should, therefore, help to distinguish between these two alternative point processes. For the same parameters, the distances to the nearest neighbor will be generally smaller under the nested double-cluster point process than under the independent superposition.

4.1.5 Cox Processes

Cox processes are a broad class of point-process models that can characterize clustered point patterns and are frequently used in ecological applications. The Neyman-Scott and Thomas processes treated in the previous sections can also be embedded within the framework of Cox processes. The basic idea of a Cox process is simple: it is an extension of the heterogeneous Poisson process (HPP). Recall that HPPs are entirely driven by an intensity function $\lambda(x)$ and that realizations of HPPs are all based on the same intensity function $\lambda(x)$. The Cox process generalizes this by treating the intensity function $\lambda(x)$, not as a constant, but as a realization of a stochastic process $\Lambda(x)$. This point process is therefore *"doubly stochastic"* (Diggle 2003, p. 68), since it is formed by an HPP with a stochastic intensity function.

Thomas processes can also be Cox processes. Following the construction principle of the Thomas process, the points of a realization of a Thomas process (offspring) are randomly distributed according to a symmetric bivariate kernel function $h(r, \sigma)$ around a pattern of cluster centers that follow a CSR distribution. For one realization of the Thomas process, we can estimate the expected intensity of offspring that results from this point process. Because each parent has a random number of offspring, which is equivalent to the offspring being randomly assigned to clusters, and a given offspring is placed following $h(r, \sigma)$ relative to the location of the cluster center, the resulting intensity function is proportional to the sum of the different bivariate kernel functions $h(r, \sigma)$ that are centered on the cluster centers (Figure 4.17). If the pattern of cluster centers is the same for all realizations, we obtain an HPP. However, if the pattern of cluster centers is itself a stochastic process (i.e., CSR), we obtain a more general Cox process. From these considerations it becomes clear that Cox processes can be used in a unified framework to characterize both the impact of environmental covariates on the intensity function and clustering. Section 6.4 of the book of Illian et al. (2008, pp. 379–386) provides a great deal of details on Cox processes, which may be of use to the interested reader.

4.1.5.1 Construction Principle of Cox Processes

In the first step of the hierarchical construction of a realization of the Cox process, an intensity function $\lambda(x)$ is generated as a realization of a stochastic

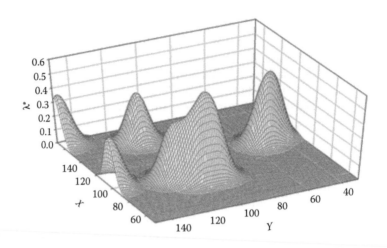

FIGURE 4.17
Part of the (normalized) intensity function $\lambda^*(x)$ that results from the superposition of two-dimensional, symmetric normal distributions with parameter $\sigma = 13$ m centered at the locations of a CSR pattern.

process $\Lambda(x)$. In a second step, the intensity function $\lambda(x)$ is used to generate one realization of the corresponding HPP. It is clear that many different point processes can be generated with this construction principle. The first- and second-order properties of the Cox process can be estimated based on the properties of $\Lambda(x)$. If $\Lambda(x)$ is stationary, the intensity is given by

$$\lambda = E[\Lambda(x)] \tag{4.21}$$

and, in the isotropic and stationary case, the second-order product density is given by

$$\rho^{(2)}(r) = E[\Lambda(x)\Lambda(y)] = \lambda^2 + \gamma(r) = \lambda^2[1 + \gamma(r)/\lambda^2] \tag{4.22}$$

where $r = |x - y|$, and $\gamma(r) = \text{Cov}[\Lambda(x)\ \Lambda(y)]$ (Diggle 2003, p. 69). The analogy between the Cox process and the Neyman-Scott process becomes evident when looking at Equation 4.23, which presents the general formula for the product density of the Neyman-Scott process (see Equation 4.2):

$$\rho^2(r) = \lambda^2\left[1 + \frac{\rho}{\lambda^2}E[S(S-1)]h_2(r)\right] \tag{4.23}$$

The covariance function $\gamma(r)$ of the Cox process is analogous to the expression $\rho E[S(S-1)]h_2(r)$ in the Neyman-Scott process (Equation 4.23). The equivalence between these two point processes can be shown by using a bivariate probability density function $h(.)$ and constructing a stochastic process $\Lambda(x)$ as

$$\Lambda(x) = \mu \sum_{i=1}^{n_p} h(x - X_i) \qquad (4.24)$$

where the X_i are the coordinates of n_p points of a homogeneous Poisson process that provide the cluster centers and μ is the mean number of points per cluster. One realization of the process $\Lambda(x)$ looks like Figure 4.17 and is the superposition of bivariate probability density functions $h(\cdot)$ centered on locations X_i. For different realizations of $\Lambda(x)$, the X_i are stochastic realizations of a homogeneous Poisson process and are therefore, in general, different. If intensity functions generated by Equation 4.24 are used to generate the patterns, they basically allocate the offspring randomly to the parents. This corresponds to a Poisson distribution in the Neyman-Scott process.

Another important Cox process is the so-called *log-linear Cox process* that is used to characterize environmental heterogeneity. In this case, the intensity function is usually not stochastic (i.e., it is not a doubly stochastic process) and a natural choice for modeling the intensity function $\lambda(x)$, which is dependent on k environmental covariates $v_1(x), v_2(x), \ldots, v_k(x)$, is a log-linear model of the form:

$$\log(\lambda(x)) = c_0 + c_1 v(x) + \cdots + c_k v_k(x) \qquad (4.25)$$

with coefficients c_i and environmental variables $v_i(x)$. This point process is basically a "habitat model" that can be used to estimate the intensity function of an HPP from environmental covariates (see Sections 2.5.2.2 and 2.6.2.9).

Cox processes provide a flexible mechanism that allows the integration of several components into one unifying framework, for example, environmental heterogeneity and clustering. An especially relevant example is given in the theoretical study of Waagepetersen and Guan (2009), which was then applied by Shen et al. (2009) and Wang et al. (2011, 2013) within an ecological context. The underlying problem is that both environmental heterogeneity and internal clustering mechanisms can create locally elevated point densities. It is inherently difficult to differentiate between these two alternative explanations of the observed spatial patterns. One promising way for resolving this problem is to consider environmental heterogeneity and clustering together in a point-process model. Cox processes provide a suitable framework for accomplishing this. The basic idea is to interpret a Thomas process that represents the "pure effect" of clustering as a Cox process (i.e., Equation 4.24). This is done by multiplying the stochastic process $\Lambda_T(x)$, which is used to generate the intensity function $\lambda(x)$ for the homogeneous Thomas process, with a log-linear Cox process that represents the "pure effect" of environmental heterogeneity

$$\Lambda(x) = \exp(c_0 + c_1 v_1(x) + \cdots + c_k v_k(x))\mu \sum_{i=1}^{n_p} h(x - X_i) \qquad (4.26)$$

In Equation 4.26, the c_j are coefficients, the $v_j(x)$ are environmental variables depending on location x and the X_i are the n_p points of a homogeneous Poisson process that serve as cluster centers. The expected intensity function $\lambda(x)$ of this point process (i.e., $E[\Lambda(x)]$) is proportional to $\exp(c_1 v_1(x) + \cdots + c_k v_k(x))$. Waagepetersen and Guan (2009) proposed a two-step procedure to estimate the parameters c_j of the "habitat model" and the parameters μ and σ of the Thomas process, using a bivariate Gaussian distribution that determines the spread of the points of the pattern around the cluster centers X_i. In a first step, the regression coefficients c_j of the log-linear model are fit. In a second step, the duality between the Cox process and Thomas process is used to fit the parameters μ and σ of the Thomas process, using minimum contrast estimation (Section 2.5.2.1). However, this is somewhat more complicated, because the resulting Thomas process is an inhomogeneous Thomas process (see e.g., Section 2.6.3.5 and Figure 2.32). Therefore, the approach of Waagepetersen and Guan (2009) involves estimation of the inhomogeneous K function and/or the inhomogeneous pair-correlation function, where the estimate of the intensity function obtained in the first step is plugged into the analysis.

4.1.5.2 Examples of the Cox Process: Parent–Offspring Relationships

The first example we present for the Cox process takes advantage of the duality between a bivariate parent–offspring Thomas process and an HPP, by reanalyzing the simulated bivariate pattern shown in Figure 4.13a. The pattern was simulated based on the simple bivariate parent–offspring Thomas process (Equation 4.13) and the bivariate data comprises the parent pattern (pattern 1; e.g., adult orchids) and the offspring pattern (pattern 2; e.g., orchid seedlings). The problem of fitting a bivariate parent–offspring Thomas process to such data is that the parent pattern may not follow the assumptions of the bivariate parent–offspring Thomas process (i.e., it can be described by a CSR process or a Thomas process). However, the Cox process using the observed parent locations for construction of the intensity function is not plagued by this problem.

We tested several HPP to fit the (in general) unknown parameter σ of the symmetric bivariate kernel function $h(r, \sigma)$. In our case, the value of the parameter was $\sigma = 13.3$. For example, a value of $\sigma = 10$ m yields parent–offspring clustering that is too tight and does not fit the observed summary statistics well (Figure 4.18a–c). However, a parameter $\sigma = 13$ m produces a good fit for both the observed parent–offspring relationship (Figure 4.18d–f) and the univariate offspring pattern (Figure 4.18g–i). Increasing the value of σ would

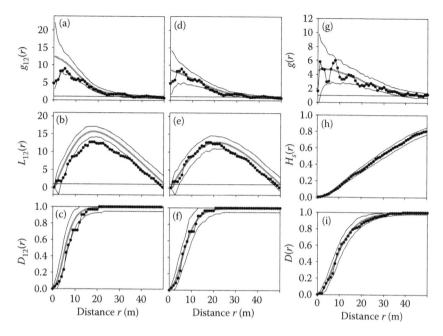

FIGURE 4.18

Analysis of a simulated pattern (shown in Figure 4.13a) using a simple, bivariate, parent–off-spring, Thomas process represented as a Cox process. The parameter used for simulating the pattern used here as an example was $\sigma = 13.3$ m. (a)–(c) Using patterns generated with a bivari-ate parent (pattern 1) and offspring (pattern 2) process for a parameter $\sigma = 10$ m as the null model to be contrasted with the observed pattern. (d)–(f) same, but for parameter $\sigma = 13$ m. (g)–(i) Univariate summary statistics for the offspring pattern and the heterogeneous Poisson null model with parameter $\sigma = 13$ m.

result in a pattern with clustering by the offspring around the parents that is too weak. In general, an appropriate value for σ can be easily determined by varying σ and calculating the GoF for the different summary statistics that are produced.

The duality between the parent–offspring Thomas process and HPP has been used in several ecological studies. A special advantage of this approach is that the associated HPP provides a flexible framework for integrating sev-eral hypotheses concerning, for example, the interaction between environ-mental heterogeneity, disperser activity, and seed dispersal. HPPs also allow the intensity functions expected from different mechanisms (e.g., habitat suitability for offspring or a seed shadow for primary dispersal) to be com-bined into one compound intensity function. This structure can therefore be used to test the relative contribution of several mechanisms to the observed parent–offspring pattern.

Rodríguéz-Perez et al. (2012) used the representation of the Thomas process as an HPP to understand the effects of a seed disperser (the lizard *Podarcis lilfordi*) on the bivariate relationship between the distribution of juveniles and adults of the shrub species *Daphne rodriguezii*, which is endemic to Menorca Island (Balearic Islands, W. Mediterranean Sea). The HPP provided a unifying framework to assess the relative impact of (i) habitat suitability, (ii) disperser activity, and (iii) proximity of adults on the distribution of juvenile shrubs. In a model exploring the habitat hypothesis for juveniles and adults, Rodriguez-Perez et al. (2012) randomly redistributed the observed *D. rodriguezii* plants with a heterogeneous Poisson distribution based on an intensity function derived from a previously developed habitat suitability map. As a result, the local plant density was proportional to the habitat suitability. They also characterized lizard home ranges using habitat suitability maps (Rodríguez-Pérez et al. 2012). This was done to implement the seed disperser hypotheses for juveniles and adults. The corresponding point-process model was analogous to the one used in studying the habitat hypothesis and generated patterns with local plant densities proportional to the disperser's habitat use.

In their examination of the third "proximity of adults" hypothesis, Rodriguez-Perez et al. (2012) assumed that the juvenile distribution depended directly on the adult pattern. Consequently, the locations of juveniles were governed in the corresponding point-process model by a kernel function centered on the observed locations of the adult plants. For this, a symmetric bivariate Gaussian distribution with variance σ^2 was used (Jacquemyn et al. 2007). Superposition of the various kernel functions yielded a map proportional to the expected juvenile density that was then used, as above, to generate patterns, where the density of juveniles depended on the distance from adults. They systematically varied the value of the unknown parameter σ and searched for the value that produced a reasonable fit. Because all three point processes used the same framework of probability maps (i.e., HPPs), Rodríguez-Pérez et al. (2012) also contrasted the plant distributions to point-process models that were based on different combinations of the probabilities of habitat suitability for plants, dispersers, and the seed shadows. To this end, the different intensity functions were combined by using the geometric mean of the superposition. In the first set of analyses, they investigated the three independent hypotheses individually. They found that hypothesis (iii) (proximity of adults) provided the best fit among the individual hypotheses. In contrast, the point-process model, which combined hypothesis (ii) (lizard activity) with hypothesis (iii) (proximity of adults), improved the fit of the data over hypothesis (iii) alone. Adding the habitat suitability to the other two (i.e., i + ii + iii) did not improve the fit any further. Thus, the proximity of adults was the most important factor determining the pattern of the juveniles of the shrub species *D. rodriguezii*, but the activity of the unique seed disperser *P. lilfordi* also left a detectable signal in the juvenile—adult relationship. The activity of the seed disperser may also contribute to the proximity of adult shrubs hypothesis, since the dispersers defecate frequently when they move on or close to shrubs for feeding.

4.1.5.3 Examples of the Cox Process: Clustering and Habitat Association

Inhomogeneous Thomas processes (Equation 4.26) can model the impact of environmental dependency and clustering simultaneously. This makes them especially attractive as models for ecological applications. As an example, we continue the analysis of the pattern of all living individuals of the species *O. whitei* from the 2000 census at the BCI plot (Box 2.6) (Figure 4.19a), for which we have already estimated the intensity function based on topographic variables and a variable representing the nonparametric intensity estimate (Figure 4.5; Table 4.1). Following the approach of Waagepetersen and Guan (2009), we fit the parameters of the Thomas process to the estimate of the inhomogeneous *K*-function and the pair-correlation function. The inhomogeneous Thomas process, using the intensity function shown in Figure 4.5b, provides a good fit to the inhomogeneous *K*- and pair-correlation functions (Figure 4.19c,d). The estimate of cluster size yields $2 * \sigma = 9.6$ m. A realization of the fitted point process is shown in Figure 4.5e. The observed inhomogeneous *K*-function and pair-correlation function are fully within the simulation envelopes of the fitted Cox process (Figure 4.19c,d) and the distribution

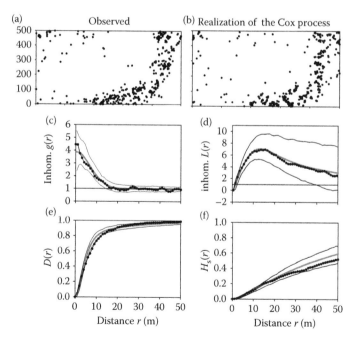

FIGURE 4.19
Use of Cox processes to estimate cluster parameters and environmental dependency for a heterogeneous pattern of all living individuals of the species *Ocotea whitei* from the 2000 census at the BCI plot. (a) Observed pattern. (b) One realization of the fitted Cox process based on the intensity function shown in Figure 4.5b. (c)–(f) Summary statistics and simulation envelopes of the fitted Cox process.

function $D(r)$ of the distance to the nearest neighbor shows only a minor departure from the simulation envelopes (Figure 4.19e). Additionally, the spherical contact distribution $H_s(r)$ is fully within the simulation envelopes (Figure 4.19f). Thus, this point process produces a good approximation to the observed data.

4.1.6 Point Processes Modeling Hyperdispersion

While clustering corresponds to situations where the neighborhood density of points is greater than expected, hyperdispersion corresponds to the opposite case, where the neighborhood density at short distances is less than the expectation for a random (CSR) point process. Thus, points of a hyperdispersed pattern tend to "avoid" each other. There are two different mechanisms that can create hyperdispersed patterns. The hyperdispersion of points may be the consequence of the *finite size of ecological objects*, as discussed in detail in Section 3.1.8, or it may be the consequence of *interactions among ecological objects*. In the extreme case of a *"hard core"* pattern, a minimal distance of at least r_0 separates all the points in the pattern. The term hard core is due to the fact that the pattern behaves as if it were produced by randomly placed, nonoverlapping disks of radius $r_0/2$. The centers of the disks are treated as if they are the points of the pattern. In contrast, *"soft core"* patterns are produced when the nonoverlapping disks have a range of sizes, with the centers of smaller disks potentially lying closer to each other than the centers of larger disks.

In some cases, hyperdispersion arises because of the finite size of ecological objects. In these circumstances, the common assumption that ecological objects can be approximated as points is not valid. This is particularly true if the scales of interest are of the same order of magnitude as the physical size of the objects. An example of this was presented in an earlier Chapter (Section 3.1.8.7), where we discussed an example in which the physical size of the shrubs in a plant community in Patagonia, Argentina, interfered with the analysis of shrub interactions (Figure 3.37). In this case, a null model such as CSR, which ignores the size of the objects, may not be able to reveal the "real," and possibly subtle, interactions producing the pattern (Wiegand et al. 2006). As a consequence, the null model should be modified to account for the observed sizes of the ecological objects. If the objects are explicitly mapped, methods that consider the finite size and real shape of the objects can be used (Section 3.1.8). However, in many cases, a complete mapping of the shape of each object within the observation window is not possible and only the mean size, or a typical size distribution, of the species may be known. In this case, using a model employing hard or soft-core point processes may produce a suitable approximation for removing the bias produced by the finite size of the objects.

If hyperdispersion arises because of ecological interactions, not because of the physical size of the objects, it is of special interest to *extract information regarding the possible nature of the interactions from the spatial pattern.* This may lead to the identification of the range of interactions or the functional form of distance dependence among objects, which may in turn result in the identification of the underlying ecological processes. There are many examples of ecological processes that can give rise to hyperdispersed patterns. Competition for limited resources among animals and plants, a major process limiting local densities, can produce hyperdispersed spatial patterns. For example, Bourguignon et al. (2011) used the L-function to study the spatial pattern of nests of the termite *Anoplotermes banksi* and found that established nests were overdispersed at a short distances, whereas young nests had a random or clumped distribution.

As an example for hyperdispersed patterns, Figure 4.20b–d shows the results of an analysis of the spatial pattern of large trees (i.e., dbh > 20 cm) found in the 1995 census of the BCI forest (Box 2.6) (Figure 4.20a). The pattern is typical of a soft-core pattern, with hyperdispersion up to distances of 9 m. This is clearly shown by the pair-correlation function (Figure 4.20b), which at short distances (e.g., $r = 1$ m) is substantially below the expectation of the CSR null model [i.e., $g(r) = 1$]. The neighborhood density at a distance of 1 m, for example, is only 67% as large as expected under CSR. The neighborhood density increases almost linearly with distance and reaches the expectation of $g(r) = 1$ for CSR at 9 m, the range of the soft-core process. $D(r)$, the distribution function of the distances to the nearest neighbor, also clearly shows a pattern of hyperdispersion, with its values lying below the simulation envelopes for distances of 2 to 6 m (Figure 4.20d). This result indicates that nearest neighbors at these distances are fewer than expected under CSR. Interestingly, almost every tree of this size class has a nearest neighbor within 9 m (Figure 4.20d), again consistent with the range of the soft-core process. Also of interest is that this pattern does not contain any indications of heterogeneity; at distances greater than 20 m, the L function remains fully inside the simulation envelopes of the CSR null model (Figure 4.20c).

The pair-correlation function for the large trees at BCI has a shape typical of a soft-core point process (Figure 4.20b). Values at short distances are clearly greater than zero, indicating that the crowns of the trees overlap to some extent, and two trees may occasionally be located close to one another, but less frequently than expected under CSR. The different sizes of the trees also contribute to the observed pattern of hyperdispersion. For example, the pattern changes if we only analyze trees with dhb > 50 cm (Figure 4.20e). Trees of this size class are somewhat less likely to lie close to one another than trees with dbh > 20 cm (Figure 4.20f). There are also clear indications that trees larger than 50 cm dbh are heterogeneously distributed, as indicated by the L-function (Figure 4.20g).

The simplest approach to quantifying the various aspects of hyperdispersion is to contrast the pattern (if it is homogeneous) to the null model of the homogeneous Poisson process. This can be used to determine the scales

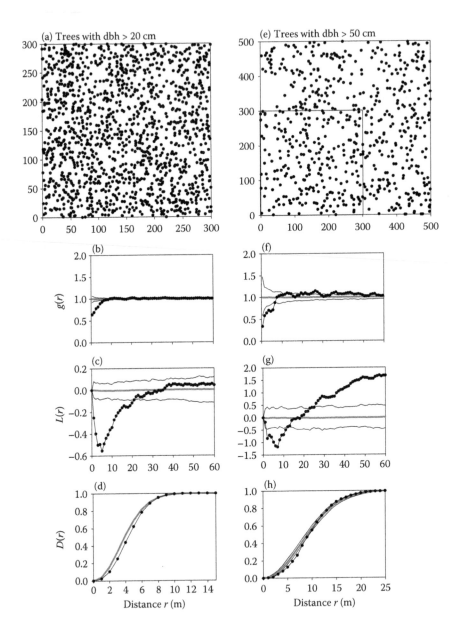

FIGURE 4.20
Hyperdispersed patterns. (a) Map of all large trees with dbh > 20 cm for the 1995 census of
the BCI plot. Note that we only show a 300 × 300 m subsection of the original plot. (b)–(d)
Application of the CSR null model to the pattern shown in panel (a). (e) Same as panel (a), but
only including trees with dbh > 50 cm. The inset square indicates the 300 × 300 m subsection
shown in (a). Note that we show only the western 500 × 500 m subsection of the plot. (f)–(h)
Application of the CSR null model to the pattern shown in panel (e).

of hyperdispersion. We followed this approach in the example of Figure 4.20. However, in some cases it is of interest to more precisely characterize hyperdispersed patterns by using specific point-process models. Different approaches exist for this purpose. Statisticians may prefer to apply the so-called inhibition processes (e.g., Diggle 2003, p. 72f; an ecological example is given in Picard et al. 2009). Or, if a precise description of the functional form of the point interactions is aimed for, a *Gibbs process* might provide the best model. As an alternative, ecological modelers may prefer *individual-based simulation models* that explicitly model how spatiotemporal birth and death processes depend on neighborhood interactions (e.g., Grimm and Railsback 2005: their Section 6.7).

4.1.6.1 Simple Inhibition Processes

Hard core point patterns, where any two points maintain a minimal distance r_0, but do not incorporate further departures from CSR, are also called *simple inhibition processes* (e.g., Diggle 2003, p. 72). An important descriptor of these processes is the packing intensity τ

$$\tau = n\pi(r_0/2)^2/A \tag{4.27}$$

which gives the proportion of the observation window (with area A) covered by the n nonoverlapping disks with diameter r_0 (Diggle 2003). Clearly, if the packing intensity τ is high, no further disks may be added to the pattern.

Patterns following simple inhibition processes can be generated in several ways. One approach is to remove all pairs of points that have a separation distance shorter than r_0 from an existing CSR pattern. Another is to use a sequential algorithm, where a series of CSR points is generated, but the only ones accepted are those located farther away than distance r_0 from any point produced earlier in the sequence (e.g., Figure 2.11a). In the latter case, it is clear that not all inhibition point processes with parameters n and r_0 exist in a given observation window of size A, because only a limited number of nonoverlapping disks can be squeezed into an area A. Thus, the packing intensity τ must be sufficiently small. The area within distance r_0 from a point can be called its *zone of influence* (ZOI) because this is the area that cannot be penetrated by another point.

4.1.6.2 Gibbs or Markov Point Processes

Hard core processes are often too simple to characterize inhibition processes producing observed point patterns in ecology. As shown in Figure 4.20, interactions between individuals may not result in total inhibition of other points within the zone of influence. Instead, interactions among points may make it only less likely that a point is located within the zone of influence of another point, rather than impossible. Within this context, an interesting task is to determine the type of interactions underling point patterns, such as the pattern observed for large trees (i.e., dbh > 20 cm) in the BCI forest (Figure 4.20).

A Markov point process generalizes simple inhibition processes by making more detailed assumptions about the point interactions within the zone of influence of the focal points x_i. Two points interact if their distance is $r < r_0$. The interaction structure of this type of point process is then characterized by means of the so-called *pair-potential function* $\Psi(r)$, which is a function of the distance $r = \|x_i - x_j\|$ between two points x_i and x_j. The pair-potential function was developed in physics and measures the "potential energy" caused by the interaction among pairs of points x_i and x_j, as a function of their separation distance (Illian et al. 2008, p. 140). It can take values within the range of $-\infty < \Psi \leq \infty$.

A *Gibbs process* (with a fixed number of points n) within a given observation window is then defined by a *location density function* $f_n(x_1, \ldots x_n)$. This function is a measure of the likelihood of a particular spatial configuration of points at locations $x_1, \ldots x_n$ under a given Gibbs process, defined by the pair-potential function Ψ. Interactions are assumed to be symmetric, and hence the location density function does not depend on the order of points. For a Gibbs process, with a fixed number of points and pairwise interactions, the location density function can be given by

$$f_n(x_1, \ldots, x_n) = \exp\left(-\frac{1}{2} \sum_{i,j}^{i \neq j} \psi\left(\|x_i - x_j\|\right)\right) / Z_n \qquad (4.28)$$

where Z_n is a normalization constant that ensures that f_n is a probability density function, and $\psi(\|x_i - x_j\|)$ is the pair-potential function. The exponent of Equation 4.28 is basically the sum of the potential energies $\psi(\|x_i - x_j\|)$ produced by the interactions of all pairs of points x_i and x_j and is referred to as the total energy (where each i–j pair is only counted once). The location density function given in Equation 4.28 arises in physics from the principle of maximal entropy.

To yield a more intuitive interpretation, Equation 4.28 can be rewritten by using a quantity called the *pairwise interaction function* $h(r) = \exp(-\Psi(r))$. The pairwise interaction function yields a value of 1 if the two points do not interact, since in this case $\Psi(r) = 0$. However, pairs of points located at distance r exhibit hyperdispersion if $h(r) < 1$ and clustering if $h(r) > 1$ (Illian et al. 2008, p. 143). The location density function 4.28 can be expressed in terms of the pairwise interaction function $h(r)$ as

$$f_n(x_1, \ldots, x_n) = \prod_{i=1}^{n-1} \prod_{j=i+1}^{n} h\left(\|x_i - x_j\|\right) / Z_n \qquad (4.29)$$

To understand how a Gibbs process works we have to consider that it cannot be simulated in a direct manner, as would be the case for an HPP. Simulation of a Gibbs process requires indirect methods, because the Gibbs process is defined only by means of a high-dimensional probability density function $f_n(x_1, \ldots x_n)$, which returns the likelihood of a given point configuration. The likelihood, in turn, is related to the assumed interaction structure of pairs of points governed by the pairwise interaction function $h(r)$. Each point interacts with its closer neighbors, creating a network of interactions that may involve most of the points in a pattern. For this reason, the Gibbs process is simulated by trial and error, starting with a suitable initial pattern. Points are removed at random and are replaced only by points that are likely to occur, given the remaining point configuration and the probability density function $f_n(x_1, \ldots x_n)$.

Point patterns $x_1, \ldots x_n$ that yield high values of the location density function $f_n(x_1, \ldots x_n)$ are more likely to occur than those with lower values, as dictated by the Gibbs process defined by the pairwise interaction function $h(r)$. For example, a *Gibbs hard-core process* with a fixed number of points has the following pairwise interaction function

$$h(r) = \begin{cases} 0 & \text{for} \quad r \leq r_0 \\ 1 & \text{for} \quad r > r_0 \end{cases} \tag{4.30}$$

If only one point pair is located closer than distance r_0, the probability density function (Equation 4.29) yields a value of 0, which means that the configuration is impossible. As a consequence, a hard-core pattern arises. The packing intensity τ (Equation 4.27) must be sufficiently small for the process to exist.

An extension of the Gibbs hard-core process (with fixed number of points) is the so-called *Strauss process* (with fixed number of points; Illian et al. p. 141). In this case, the pairwise interaction function for a point x_j inside the zone of influence of a point x_i yields a value γ greater than zero, but less than one:

$$h(r) = \begin{cases} \gamma = \exp(-\beta) & \text{for} \quad r \leq r_0 \\ 1 & \text{for} \quad r > r_0 \end{cases} \tag{4.31}$$

β is the corresponding value of the pair-potential function for points with $r \leq r_0$. In this case there is a finite probability that point x_j is located within the zone of influence of point x_i. Using Equations 4.28 and 4.31, the location density function of this Strauss process yields

$$f_n(x_1, \ldots, x_n) = \exp\left(-\beta \frac{1}{2} \sum_{i,j}^{i \neq j} \mathbf{I}\left(\|x_i - x_j\| \leq r_0\right)\right) / Z_n$$

$$= \exp(-\beta n_2(r_0)) / Z_n = \gamma^{n_2(r_0)} / Z_n \tag{4.32}$$

where the quantity $n_2(r_0)$ counts the number of pairs of points that are located closer than distance r_0 (Illian et al. p. 141). Clearly, the case of $\gamma = 0$ in Equation 4.31 yields the hard core process described above (i.e., $\beta = \infty$), cases of $0 < \gamma < 1$ (i.e., $0 < \beta < \infty$) represent a type of nonstrict inhibition, and the case $\gamma = 1$ (i.e., $\beta = 0$) yields the homogeneous Poisson process, where points do not interact. The pair-correlation function for the Strauss process with a fixed number of points can be approximated as (Dixon 2002)

$$g(r) = \begin{cases} \gamma & \text{for} \quad r \leq r_0 \\ 1 & \text{for} \quad r > r_0 \end{cases} \tag{4.33}$$

The Strauss process assumes that the interaction strength of two points is always the same when the second point x_j is located inside the zone of influence of a focal point x_i, irrespective of their separation distance r. However, a more realistic characterization of interactions between two points can be attained by using the *overlap of the zone of influence* of the two points to calculate the strength of the interaction. This model of competition is popular in plant ecology (e.g., Weiner et al. 2001) and is based on making the strength of the interaction proportional to the area of overlap of the zones of influence $O(r, r_0)$ for the two competing points. The corresponding pairwise interaction function proposed by Penttinen (1984) is

$$h(r) = \begin{cases} \exp(-\theta O(r, r_0)) & \text{for} \quad r \leq r_0 \\ 1 & \text{for} \quad r > r_0 \end{cases} \tag{4.34}$$

where θ is a model parameter. For a distance of $r = 0$ we find $h(r) = 0$, and for distances $r \leq r_0$ the interaction strength declines with increasing distance r. If $\theta = 1$, the pairwise interaction function approaches a value of 1 smoothly as the distance r approaches r_0. The overlap area for $r < r_0$ can be calculated as

$$O(r, r_0) = (r_0^2 \arccos(r/r_0) - r\sqrt{r_0^2 - r^2})/2$$

Many other models of the pairwise interaction functions are possible, for example, Illian et al. (2008, p. 142) list several.

It is clear from the preceding description that the Gibbs processes cannot be simulated in as straightforward a fashion as the Neyman-Scott cluster process, the HPP, or Cox process, just to name a few. As mentioned above, more complex simulation methods are required (Illian et al. 2008, p. 144). The probability of different possible patterns, composed of n points $x_1, \ldots x_n$ within the observation window, must be evaluated based on the location density function $f_n(x_1, \ldots x_n)$. In theory, we could simulate a large number of patterns and select the realization of the Gibbs process that is the most likely, given the location density function. However, this approach is not feasible in practice because of the high number of possible point configurations. For this reason, more "intelligent" approaches are required that work in a way similar to the algorithm used for pattern reconstruction (Section 3.4.3.1). In one such approach, points of an initial pattern are deleted and replaced by randomly drawn points, which are accepted if the new point configuration becomes more likely, given the location density function $f_n(x_1, \ldots x_n)$.

Algorithms for simulating a Gibbs process generally start with a suitable, initial pattern of n points. In general, the criterion for suitability is that the initial pattern is in fact possible, given the location density function. For example, in the case of a Gibbs hard core process, the initial distribution of points cannot have interpoint distance less than r_0, because this would lead to a value of $f_n(x_1, \ldots x_n) = 0$. A suitable initial condition for the Gibbs hard-core process could be provided by a regular point pattern distributed on the nodes of a lattice. Next, one of the points x_k is deleted (all points have the same probability of being deleted) and replaced by a new tentative point x_k^t, which is randomly placed within the observation window through a CSR process. The tentative point x_k^t is then accepted with a probability that is proportional to the product of the pairwise interaction functions of the tentative point x'_k relative to the $n - 1$ "old" points $x_1,..., x_{k-1},..., x_{k+1}, \ldots x_n$ (Illian et al. 2008, p. 146):

$$\varphi(x_k^t) = \exp\left(-\sum_{i}^{i \neq k} \psi\left(\left\| x_k^t - x_i \right\| \right) \right) \tag{4.35}$$

The algorithm proceeds by replacing points until it reaches the "stationary regime." In this case, the resulting patterns are statistically similar and can be regarded as samples from the underlying Gibbs process. This approach is called the Markov chain Monte Carlo (MCMC) method where the states are the possible point configurations $x_1, \ldots x_n$ of n points in the observation window. The probability that a new configuration is selected depends only on the current configuration (i.e., a Markov chain). MCMC methods are generally used in situations where it is difficult to directly sample from multi-dimensional, probability distributions and can be traced back to Metropolis et al. (1953). The algorithm described above is a "birth-and-death" algorithm, but other algorithms are also possible (for a detailed treatment of such simulation techniques in point-process statistics see the book of Møller and Waagepetersen 2004).

In the case of the Gibbs hard core process (Equation 4.30), the function $\varphi(x_k^t)$ of Equation 4.35 yields zero if a new point penetrates one of the zones of influence of the remaining $n - 1$ points, otherwise it yields one. In the case of the Strauss process (Equation 4.31), the function $\varphi(x_k^t)$ of Equation 4.35 yields one if the tentative point is located outside the zone of influence of the $n - 1$ "old" points, and $\varphi(x_k^t) = \exp(-\beta n_2^t)$ if the tentative point is located within the zone of influence of n_2^t of the original points. Thus, the probability that the tentative point is accepted is lower if it is placed within the zone of influence of old points. This is a reasonable assumption for competition or other forms of repulsive interactions among points. In practice, a suitable upper bound M [i.e., a value larger than the values of $\varphi(x_k^t)$] is selected for $\varphi(x_k^t)$ and a random number u between 0 and M is drawn from a uniform distribution in assessing the potential placement of a point. The new point is selected if $\varphi(x_k^t) \geq u$. For the Strauss process a suitable upper bound is $M = 1$.

It often makes sense to use a point process that constrains the number of points to n, the number of points occurring in an observed pattern, when analyzing and characterizing the pattern. However, in some cases, such as the study of plant population dynamics, the number of objects is usually not constant in time and therefore is best represented as a random variable. Thus, when the objective of the study involves dynamic aspects, one may wish to generalize the Gibbs process with a fixed number of points to a *Gibbs process with a random number of points*. Methods for these point processes are described in Illian et al. (2008, their Section 3.6.3). Finally, one may wish to use *stationary Gibbs process*, which are not defined for an observation window of a given size A, but are based on the process occurring within the entire two-dimensional plane. The latter approach is suitable for describing very large point patterns and is also presented in Illian et al. (2008, their Section 6.6). Because closed-form expressions for summary statistics like the pair-correlation function are usually not available for Gibbs processes, fitting a Gibbs process model to observed data require, in general, complex simulation approaches. Illian et al. (2008, their Sections 3.6 and 6.6) provide a detailed overview of the techniques used with Gibbs processes.

In many respects, Gibbs processes closely resemble aspects of population dynamics. The birth-and-death simulation algorithm described above is basically a "zero-sum" model (i.e., a dying plant is immediately replaced by a new one), where plants die randomly (i.e., a random point of the pattern is removed) and the establishment of a new plant depends on interactions with plants located in the neighborhood of the "candidate" position. If the zero-sum assumption is removed and birth and dead occur independently, a Gibbs process with a random number of points may already represent a realistic characterization of population dynamics. However, note that this class of point process neglects an important aspect of plant population dynamics, the development of the size of a plant. Growth can be considered by using *marked Gibbs processes* incorporating marks, such as the size of a plant. This type of model can then be used to study the birth, dead, and growth of plants in a dynamic way (e.g., Renshaw and Särkkä 2001). These models are structurally very similar to *spatially explicit, individual-based simulation models*, which are used by ecologists to study the spatio-temporal dynamics of plant populations and communities. Grimm and Railsback (2005, their Section 6.7) provide a thorough overview of this type of model. Examples for such models are a treeline model that addresses the spatiotemporal dynamics of *Pinus uncinata*, the dominant tree species in most treeline ecotones in the Pyrenees (Wiegand et al. 2006; Martinez et al. 2011), and a zone-of-influence simulation model by Weiner and Damgaard (2006) used to investigate the relationship between size asymmetric competition and size-asymmetric growth.

4.2 Analysis of Bivariate Patterns

While univariate analysis aims to characterize the spatial distribution pattern of a given type of point, *bivariate analysis is concerned with the characterization of the spatial relationship between two types of points.* The major interest is in exploring the small-scale interaction structure between the two types of points. The fundamental division of bivariate patterns is given by the *independence* of the two patterns, as opposed to *attraction* or *repulsion*. If the two component patterns show attraction, we find more points of pattern 2 within the neighborhood of points of pattern 1 than expected under independence of the two component patterns. In this case, the two patterns show a positive interaction. Conversely, if the two component patterns show repulsion, we find fewer points of pattern 2 within the neighborhood of points of pattern 1 than expected. If repulsion occurs at distances greater than the local neighborhood, this is generally referred to as *segregation*. Most techniques presented in this book explicitly consider the local neighborhood distribution of pattern 2 around the points of pattern 1 and therefore allow for a subtle, scale-dependent assessment of the spatial relationship between the points of the two patterns. However, interactions between the two types of points are not the only source of departures from independence. Similar to the case of univariate patterns, first-order effects (i.e., heterogeneity of the patterns) may also produce segregation or "mixing" of the two patterns, even if there are no interactions between the two types of point.

In general, analysis of bivariate patterns is more complicated than that of univariate patterns, because each univariate component pattern can have a complex spatial structure, whereas the points of the two univariate patterns may or may not show complex interactions across a range of scales. For example, the two component patterns of the bivariate pattern shown in Figure 4.21a are each hard-core patterns that exhibit hyperdispersion up to distances of 4 m (Figure 4.21b,c), but the two patterns are independent of one another (Figure 4.21d). Thus, existence of spatial structure in the univariate component patterns does not mean that there is also an interaction between the two patterns. However, we may also encounter cases where the two univariate patterns do not show any spatial structure (i.e., they follow CSR), but the bivariate pattern does. An example for a case of bivariate repulsion with random univariate component patterns is given in Figure 4.21e–h.

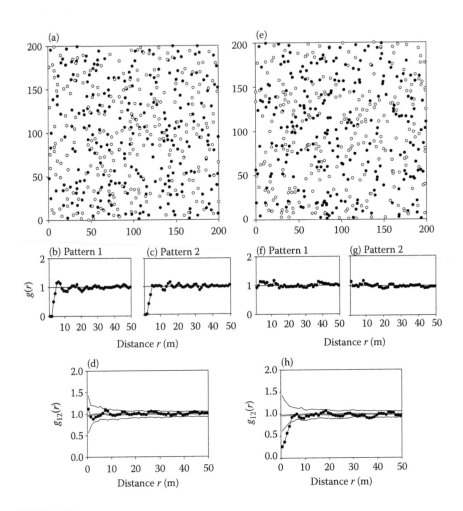

FIGURE 4.21
Bivariate patterns. (a) Map of an independent bivariate pattern composed of two independent, hyperdispersed, component patterns, created by a hard core process with a separation distance of 5 m for any point within the same component pattern. (b)–(c) Pair-correlation analysis for the patterns in panel (a). (d) Test of independence for the two univariate component patterns in panel (a), based on 199 simulations of using a toroidal shift. (e) Map of a bivariate pattern where one pattern (pattern 1) was produced by a CSR process and the second pattern (pattern 2) was also CSR, but with an additional constraint. Tentative points of pattern 2 were randomly placed, but only accepted with a probability that increased linearly from zero at distance $r = 0$ m to one at distance 5 m with respect to the distance to the nearest neighbors of pattern 1. (f)–(g) Pair-correlation analysis for the patterns in panel (e), indicating that the component patterns are univariate CSR. (h) Test of independence for the two patterns in panel (e), based on 199 simulations of using a toroidal shift.

This example clearly indicates that a lack of univariate structure does not automatically imply that there is no bivariate spatial structure.

As stated earlier, independence is the *fundamental division for bivariate patterns* and is examined through the randomization of the characteristic features of the data type (i.e, the relationship between the two patterns), but must also be constrained by all other features of the data. What this means, in practice, is that the pattern must be conditioned on the structure of the two point processes that created the two component patterns and only questions regarding the interaction between the two processes are explored. Therefore, point-process models exploring the fundamental division for bivariate patterns must maintain the univariate spatial structure of the two patterns, but break apart any dependence between them (Dixon 2002). Thus, stochastic replicates of one of the component patterns must be generated without regard to the actual locations of the points of the other pattern, such that the stochastically produced points exhibit the same spatial characteristics (e.g., strong clustering) as the component pattern they are modeling. In the following, we present different approaches for constructing the null model of independence. This includes a consideration of several specific null models for bivariate patterns that are developed to address specific questions, and a discussion of various methods to analyze heterogeneous bivariate patterns.

4.2.1 Testing for Independence

Testing for independence in the spatial patterns of two species is an important issue in community ecology. For example, McGill (2010) reviewed six theories of biodiversity and showed that they were all based on the assertion that individuals are placed without regard to individuals of other species. Empirical evidence for independent species placement is scarce, although models assuming no species interactions are successful in predicting diversity patterns (e.g., Hubbell 2001; Morlon et al. 2008). The assertion that species are placed independently is also inconsistent with the vast amount of literature in ecology devoted to the study of species interactions (Chesson 2000b). The tools of point-pattern analysis, if employed properly, provide a suitable mechanism for exploring these issues using data collected in the field.

As we have stressed elsewhere in this book, one of the key issues in testing for the independence of two patterns is the maintenance of univariate spatial structures in the component patterns. To understand this issue more fully let us consider an example of two clustered, but independent point processes. In this case, it is quite likely that clusters of patterns 1 and 2 accidently overlap. Figure 4.22a shows such a realization of two independent cluster processes. We observe that the bivariate pair-correlation function $g_{12}(r)$ yields values larger than one (Figure 4.22d). In analyzing this pattern for independence, we run into a problem if we employ CSR as the null model for pattern 2, distributing its points randomly and independently of the points of

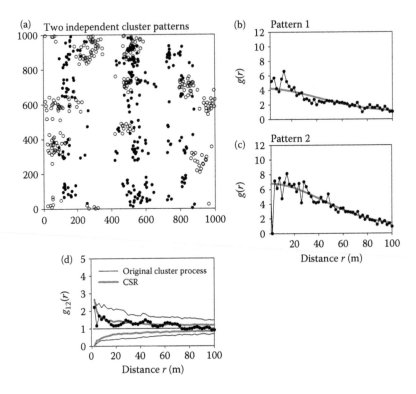

FIGURE 4.22
Bivariate patterns II. (a) Map of an independent bivariate pattern with two component patterns comprising 250 points, generated by Thomas cluster processes with parameters $A * \rho = 15$ clusters and $\sigma = 31$ m. (b)–(c) Univariate pair-correlation functions of the two component patterns (closed disks) and fits with a Thomas process (gray lines). (d) Bivariate pair-correlation function (closed disks) of pattern in panel (a), with simulation envelopes for CSR (solid gray lines) and for the original Thomas process used to generate pattern 2 (solid lines).

pattern 1, which are held fixed. The issue is that the CSR process cannot recreate the overlapping clusters seen in the pattern, because CSR does not include a mechanism of clustering. As a consequence, the CSR null model will yield values for the bivariate pair-correlation function $g_{12}(r)$ that are generally below the observed values, producing a spurious signal of attraction (Figure 4.22d; $p = 0.015$). The underlying reason for this is that the CSR-based null model does not capture the full variability of spatial arrangements that are possible under two independent cluster processes and, therefore, yields simulation envelopes that are too narrow. This is shown in Figure 4.22d where we conducted additional simulations of the original cluster process of pattern 2 as the null model. The simulation envelopes of the original cluster process (black lines) are much wider that those produced by the CSR null model (gray bold lines).

In the above example we used simulated data, where the underlying point process was known. This allowed us to use the correct null model for independence (i.e., the simple Thomas cluster process for pattern 2). However, the "true" point processes are generally unknown and *accounting for the observed autocorrelation structure in the point patterns in testing for independence is a difficult problem.* This was recognized very early on in the analysis of bivariate patterns (e.g., Lotwick and Silverman 1982), but was only partly solved before Jacquemyn et al. (2012a,b) and Wiegand et al. (2013a) proposed pattern reconstruction as a mechanism for generating null model patterns for testing independence.

Implementations of the independence null model keep the first pattern unchanged and randomize the second pattern in a way that preserves the important characteristics of its univariate structure. This can be done by generating the null model of pattern 2 using either nonparametric or parametric techniques. Nonparametric implementations include the toroidal shift null model and the pattern reconstruction null model, whereas parametric implementations fit parametric point processes to the observed pattern and use the realizations of the fitted process as a null model. Both methods are explained in the following.

4.2.1.1 The Toroidal Shift Null Model

An early solution for implementing an independence null model for bivariate patterns was proposed by Lotwick and Silverman (1982) in their seminal 1982 paper; a paper that also introduced the bivariate *K*-function. In this paper, they developed the nonparametric, toroidal shift method, which moved one of the component patterns in its entirety, while holding the second pattern fixed in space. In practice, this null model treats the study plot as a torus where the upper and lower edges of the observational window are connected, as are the right and left edges (Figure 4.23). The pattern re-assembled after a toroidal shift approximates the observed univariate spatial structure, because it conserves most of the interpoint distances within the shifted pattern. However, one problem of the toroidal shift is that it may artificially split apart larger structures (e.g., clusters), when they lie close to the border. This can result in artificially abrupt boundaries or other artifacts and as a result does not produce a pure, stochastic replicate of the observed pattern.

Examples of analyses using the toroidal shift null model for homogeneous patterns are shown in Figure 4.21. The bivariate pattern shown in Figure 4.21a is composed of two independent patterns, which are hyperdispersed at scales below 4 m, and random at larger scales (Figure 4.21b,c). Clearly, it would be difficult to tell by visual inspection that the two patterns are independent, but we can clearly see this in the pair-correlation function using the toroidal shift null model (Figure 4.21d). Figure 4.21e shows an example for a homogeneous bivariate pattern where each component pattern follows CSR, but where the points of the two components show repulsion at short distances. As before, differences between the two bivariate patterns shown in

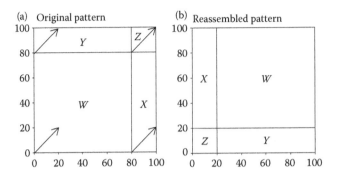

FIGURE 4.23

Toroidal shift null model. (a) Original pattern within a 100 × 100 m observation window. The arrows indicate a random shift by the vector (20 m, 20 m), which is added to the coordinates of the pattern. (b) Rearrangement of the shifted pattern, following toroidal geometry. The areas X, Y, and Z, which are shifted to lie outside the observation window, are reallocated following the rules of toroidal geometry. Points now lying within zones X and Z (to the right of the observation window) are relocated within the observation window by subtracting a value of 100 from the x-coordinate. Points lying within zones Y and Z (above the observation window) are relocated by subtracting 100 from the y-coordinate.

Figure 4.21a and e are difficult to judge just by eye. We recognize somewhat more "order" in the pattern shown in Figure 4.21a, but this is caused by the univariate hyperdispersion of the patterns. Visual judgment of the bivariate relationship is very difficult, especially since the elements of hyperdispersion occur only over short distances compared to the scale of the observation window. As expected by the construction of the patterns, the toroidal-shift null model shows that the patterns are not independent, since the pattern exhibits repulsion at shorter scales (Figure 4.21h).

4.2.1.2 Parametric Point-Process Models

A second possibility for a test of independence is to first conduct univariate analyses of the two component patterns. This can provide a mechanism for finding suitable parametric models that will be able to produce the desired stochastic replicates of the observed pattern. In this case, one of the goals is to maintain the most important features of the univariate structures. If a null model is found that does a good job of characterizing the univariate spatial structure of one of the component patterns, one may use this null model to randomize the pattern, while keeping the pattern of the other component pattern unchanged. This is quite similar to the approach used with the toroidal-shift null model. We used a parametric method for the bivariate pattern shown in Figure 4.24a. Both patterns are clustered (Figure 4.24b,c) and were created independently with a simple Thomas processes.

Fitting the Thomas process to the data of pattern 2 produced the parameters $A\rho = 15$ for the number of clusters and $\sigma = 31$ m for the distribution of

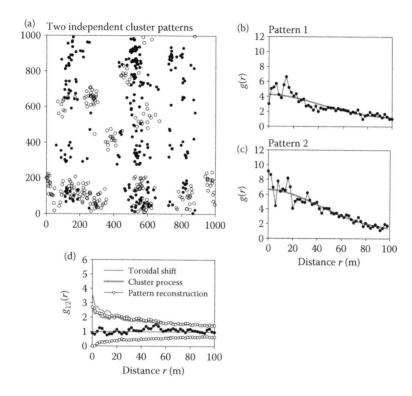

FIGURE 4.24
Bivariate patterns III. (a) Map of an independent bivariate pattern where the two component patterns comprise 250 points and were generated with a Thomas cluster process with parameters $A * \rho = 15$ clusters and $\sigma = 31$ m. (b)–(c) Univariate pair-correlation functions of the two component patterns (closed disks) and the fits for a Thomas process (gray lines). (d) Tests of significance using different implementations of the independence null model.

the pattern around the focal points of the process. $A = 1000$ m $\times$ 1000 m was the area of the observation window. The clusters of pattern 2 had an approximate diameter of 124 m. Application of the parametric null model indicates that the patterns are independent (Figure 4.24d; gray bold lines). Again, this diagnosis would be difficult to judge by eye as some of the clusters accidently overlap and others are segregated. We also estimated the simulation envelopes for the toroidal shift implementation of the independence null model (black lines in Figure 4.24d); they were similar to those of the parametric null model, since the toroidal shift null model produces reasonable approximations of the Thomas process. However, the toroidal shift model does overestimate the variability at smaller scales; the upper simulation envelope is somewhat wider at scales less than 20 or 30 m (Figure 4.24d). This is probably an artifact due to the fact that the toroidal shift freezes the second pattern and leaves most of its inter-point distances unchanged.

4.2.1.3 Pattern Reconstruction Null Model

A third possibility to test for independence is to use nonparametric pattern reconstruction techniques (Section 3.4.3.7) to generate patterns that have the same spatial structure as one of the component patterns. The reconstructions are stochastic realizations of one of the component patterns, which closely approximate several summary statistics of the observed pattern that were selected *a priori*. The reconstructions were used as a null model for the second pattern. As before, the first pattern was kept unchanged. The advantage of this method is that it very closely approximates the detailed spatial structure of the original pattern, without producing potential artifacts like those that occur in the toroidal-shift null model. Pattern reconstruction is also more flexible in capturing the potentially complex, univariate spatial structures than are the parametric reconstruction methods. The latter are severely constrained by their functional form in generating spatial structures. The question, however, is how much of this detail really matters with respect to the simulation envelopes.

We show two examples where pattern reconstruction is used as a null model. The first example is the pattern generated by a Thomas cluster process, which was already analyzed in Figure 4.24a. The reconstruction used the summary statistics $g(r)$, $D(r)$, $K(r)$, and $H_s(r)$. Figure 4.24d demonstrates that the simulation envelopes of this null model (open disks) coincide very well with those of the cluster process (gray lines). This result was expected because the data were generated with a Thomas process and the fitted Thomas process should do a good job in capturing the spatial structure of the data.

The second example investigates the spatial association of two "real world" patterns, which contain complex univariate structures. The two component patterns represent the recruits of two typical light demanding tree species in the tropical forest at BCI (Box 2.6), *Cecropia insignis* and *Miconia argentea* (Figure 4.25a). In this case, recruits are defined to be all individuals that crossed a size threshold of 1 cm dbh at some time during the 5 years prior to the census used for the analysis. For both species, the recruits show complex superposition patterns consisting of a random pattern independently superimposed on a nested, double-cluster pattern (Wiegand et al. 2009).

The analysis using a null model based on the pattern reconstruction of *M. argentea* shows that the recruits of the two species are strongly attracted at distances up to 15 m (Figure 4.25b). Thus, the neighborhood density of *M. argentea* recruits around *C. insignis* recruits is significantly greater than expected through independence. At neighborhoods of 2 m the neighborhood density is more than 20 times higher than expected under independence (Figure 4.25b). The bivariate distribution function $D_{12}(r)$ of the nearest-neighbor distances exhibits significant attraction up to 40 m (Figure 4.25c). The strong attraction is probably caused by the occurrence by the recruits of these two light-demanding species in shared canopy gaps. Comparison with

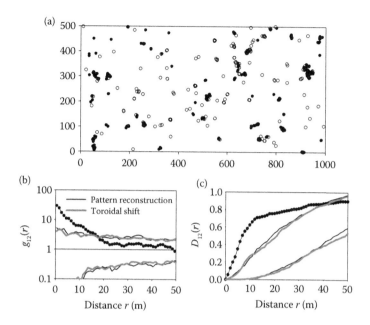

FIGURE 4.25

Test of independence for a bivariate pattern. (a) Map of two component patterns of a bivariate pattern representiing all recruits of the light demanding tree species *Cecropia insignis* (closed disks, pattern 1) and *Miconia argentea* (open disks, pattern 2) for the 2005 census in the tropical forest at BCI. (b) Bivariate analysis using the pattern reconstruction implementation of the independence null model. Simulation envelopes of the pattern reconstruction null model are shown as solid black lines and those for the toridal shift implementation as solid, gray lines. Note the logarithmic scale for the y-axis. (c) Same as panel (b), but for the distribution function $D_{12}(r)$ of the distances to the nearest neighbor.

the toroidal-shift null model (gray lines) shows no differences in the simulation envelopes. This is understandable because the typical spatial structures in these two patterns are relatively small-scale and the patterns show no apparent gradients in intensity. Thus, toroidal shift and pattern reconstruction are equivalent.

4.2.2 Other Null Models for Bivariate Analysis

4.2.2.1 Homogeneous Poisson Processes

A null model randomizing one of two component patterns using a homogeneous Poisson process is not suitable for producing a fundamental division for bivariate patterns if the goal is to test for independence. Because this is a frequent issue in studies of bivariate patterns, we use the example of the recruits of two tropical tree species presented in Figure 4.25 to illustrate this point further. Figure 4.26 compares the results obtained through the CSR null model with that obtained through pattern reconstruction. CSR

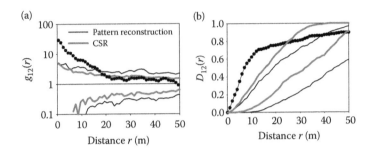

FIGURE 4.26
Independence vs. CSR for the bivariate pattern of recruits of the two species *Cecropia insignis* and *Miconia argentea* (see Figure 4.25a). (a) Bivariate analysis using the pattern reconstruction implementation of the independence null model. Note the logarithmic scale of the y-axis. (b) Same as (a) but for the bivariate distribution function $D_{12}(r)$ of the distances to the nearest neighbor.

underestimates the variability in possible spatial configurations that can arise through independent superposition of these two strongly clustered patterns, resulting in simulation envelopes that are too narrow for the pair-correlation function (Figure 4.26a). The distribution function $D(r)$ of the distances to the nearest neighbor, using CSR as the null model, yields shorter nearest-neighbor distances than the pattern reconstruction null model (cf. Figure 4.26b). This is because the recruits of *C. insignis* are strongly clustered. When the clustering is conserved, the likelihood that recruits will be close to the individuals of the first species is, on average, lower than that compared to a case where the recruits of *C. insignis* are randomly distributed. In other words, randomly relocated recruits of *C. insignis* will be spread over a larger area compared with recruits that are relocated in a clustered manner, and therefore they will be located closer to the individuals of the first species.

Nonetheless, the CSR null model can be used in addressing some specific questions for bivariate patterns. For example, Wiegand et al. (2007a, 2012) and Wang et al. (2010) studied pair-wise distribution patterns of large trees in forest communities. As explained in more detail in the next sections, they used the CSR null model to randomize the second component pattern to study how trees of a second species were distributed within local neighborhoods of a focal species. This allowed them to determine whether trees of the second species occurred, on average, more (or less) frequently within the neighborhoods of the focal species than expected by chance alone. Knowing how two species co-occur in smaller neighborhoods is important because it informs us as to how often plants of two species are in direct contact with each other, and how this relationship changes with neighborhood radius. This provides basic information related to different hypotheses on species coexistence (e.g., the segregation hypothesis; Pacala 1997; Raventós et al. 2010). For instance, departures from the CSR null model may be caused by different sources of patterning, that is, (i) species interactions, (ii) the univariate pattern of one of the species (e.g., strong aggregation preempting space), and (iii) environmental heterogeneity. If a given species pair shows departures from the CSR null model we know that there are processes acting that make their spatial distribution structure different from one that occurs by pure chance alone.

4.2.2.2 Classification Scheme for Co-Occurrence Patterns

In Section 4.2.1, we discussed the use of the CSR null model in exploring how plants of a second species were distributed within the local neighborhoods of a focal species. This provided a picture as to how frequently the two different species were in close contact, providing the opportunity to interact. This is of special interest in plant community ecology, and the application of this type of analysis to all pairs of species in a plant community can produce deep insights into the spatial co-occurrence structure of the community, especially if we can classify co-occurrence patterns in a general framework. We present such a framework in the following.

In the simplest case, where all component patterns are homogeneous, we have basically three types of co-distribution patterns: (i) *attraction*, where we find, on average, more individuals of pattern *j* within the neighborhoods of pattern *i* than expected by chance alone; (ii) *repulsion*, where we find fewer individuals of pattern *j* within the neighborhoods of pattern *i* than expected by chance alone; and (iii) *independence*, where there is no departure from the null model. Because we ask about the distribution of individuals of species *j* within neighborhoods around individuals of species *i*, the bivariate *K* function $K_{ij}(r)$ is the summary statistic of choice to diagnose membership within one of these three co-occurrence patterns. $K_{ij}(r)$ is proportional to the mean number of individuals of pattern *j* occurring within neighborhoods of radius *r* around individuals of pattern *i*, and is, therefore, perfectly suited for this purpose.

For more complex cases, which involve heterogeneous component patterns, we need to apply a slightly different approach than for the simple homogeneous case. For example, in most natural communities, such as forests, we may observe (first-order) heterogeneity in the individual patterns, which means that some species are more (or less) likely to be found within certain areas of the observation window than in others. As a consequence, the two patterns *i* and *j* may exhibit intensity functions $\lambda_i(x)$ and $\lambda_j(x)$ that are positively or negatively correlated. In the case of a positive association in the intensities, both patterns are found within the same areas of the observation window (Figure 4.27c). Wiegand et al. (2007a) called this pattern "*mixing*," because the patterns mix within the same areas. However, if the two intensity functions are negatively correlated, the two patterns will be found within different areas of the observation window (Figure 4.27a). Wiegand et al. (2007a) called this pattern "*segregation*" because the distribution patterns tend to segregate into different areas of the observation window.

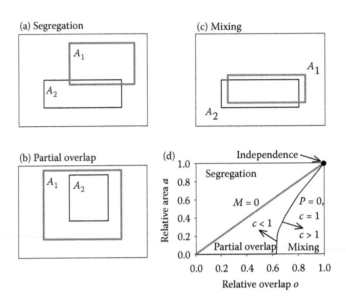

FIGURE 4.27
Classification scheme for a simple patch heterogeneity, where pattern 1 and pattern 2 are CSR patterns inside patches A_1 and A_2, respectively, with no points occurring outside the patches. (a) *Segregation*: there is only a little overlap between patches of patterns 1 and 2. (b) *Partial overlap*: pattern 2 is located totally within pattern 1, whereas pattern 1 has some individuals with many pattern 2 neighbors and some individuals with few pattern 2 neighbors. (c) *Mixing*: patches of patterns 1 and 2 overlap to a large extent, covering only a limited proportion of the observation window. (d) Isoclines for $M = 0$ (bold gray line) and $P = 0$ (solid line) in the o–a parameter space, with the isocline for $P = 0$ being associated with a parameter value of $c = 1$, where $c = \lambda_2 \pi r^2$. The two isoclines delineate areas of segregation, partial overlap, and mixing, dependent on the three parameters o, a, and c. The $P = 0$ isocline moves left if $c < 1$ and right if $c > 1$. Independence is only possible for $o = 1$ and $a = 1$ (closed disk). See Section 4.2.2.3 for a detailed account of the figure.

While in some respects segregation and mixing correspond to the attraction and repulsion of homogeneous patterns, heterogeneous patterns can also exhibit additional modes of patterning. In the case of mixing, most individuals of the focal pattern i will have a large number of individuals of pattern j in their neighborhoods and, in the case of segregation, we will find that most individuals of a focal pattern i will have few, if any, neighbors of pattern j. However, as shown in Wiegand et al. (2007a), there are two other cases that we need to consider. These arise when there is more heterogeneity in the distribution of pattern j individuals with respect to the neighborhoods of pattern i individuals than for the cases of mixing and segregation.

We can have a pattern of "*partial overlap*" when pattern i is partially overlapped by pattern j (Figure 4.27b). Within the overlap area, the individuals of pattern i will have a larger number of pattern j neighbors than for a completely random pattern, but outside of the overlap area they will have fewer neighbors. In contrast, we can have a *"type IV"* pattern, where individuals of pattern *i* are highly clustered and pattern *j* somewhat overlaps clusters of pattern *i*. As a result, the mean number of pattern *j* neighbors for a pattern *i* point is small, but the probability of having the nearest neighbor of pattern *j* within distance *r* of a pattern *i* point is greater than expected. This is because a few pattern *j* individuals are in fact the nearest neighbor of a large number of pattern *i* individuals.

The above considerations suggest that a second summary statistic is required to capture the effects of heterogeneity in the distribution of type *j* points with respect to type *i* points. Wiegand et al. (2007a) proposed using the bivariate distribution function $D_{ij}(r)$ of the distances to the nearest neighbor for this purpose. $D_{ij}(r)$ characterizes the distribution of distances from type *i* points to the nearest type *j* point and is therefore able to distinguish, together with the *K* function, among the four cases described above. Additionally, the summary statistics $K_{ij}(r)$ and $D_{ij}(r)$ are a good choice for classifying different types of bivariate associations, since they express fundamentally different properties of bivariate point patterns.

Wiegand et al. (2007a) developed a classification scheme characterizing patterns with respect to the four categories described above, using a two-dimensional classification space based on the two summary statistics $K_{12}(r)$ and $D_{12}(r)$, where the subscripts 1 and 2 now refer to the focal pattern (= 1) and the secondary pattern (= 2), the latter being examined within the neighborhoods around the individuals of the focal pattern. This scheme can be applied to each neighborhood distance *r* of interest. The null association, indicating no departure from the null model, forms the origin of the scheme. This is done by subtracting the theoretical values under the (e.g., CSR or independence) null model from the two statistics and log-transforming the *K*-function, to weight positive or negative departures from the null model in the same way. The expectations of the two summary statistics under both the CSR and independence null models yield $D_{12}^E = 1 - \exp(-\lambda_2 \pi r^2)$ and $K_{12}^E(r) = \pi r^2$, where $\lambda_2 = n_2/A$ is the overall intensity of pattern 2 within the observation window and the "*E*" superscript means "expected by the null model for no spatial patterning." The two axes of the scheme are, therefore, defined as

$$\hat{P}(r) = \hat{D}_{12}(r) - (1 - e^{-\lambda_2 \pi r^2})$$
$$\hat{M}(r) = \ln(\hat{K}_{12}(r)) - \ln(\pi r^2) \tag{4.36}$$

with the hat symbol indicating a value estimated from the data.

The *two-axis scheme* includes the "no association" category and the four fundamental categories of bivariate association. In the case of "segregation" (type I), both the average number of neighbors within distance *r* and the proportion of nearest neighbors within distance *r* are fewer than expected [i.e. $\hat{M}(r) < 0$ and $\hat{P}(r) < 0$]. In the case of "mixing" (type III), both are greater than expected [i.e. $\hat{M}(r) > 0$ and $\hat{P}(r) > 0$]. In the case of "partial overlap" (type II), the mean number of individuals of pattern 2 within neighborhoods of radius r around individuals of pattern 1 is greater than would be expected according to the null model [i.e. $\hat{M}(r) > 0$], but the probability that an individual of pattern 1 has no nearest neighbor of pattern 2 is less than expected [i.e. $\hat{D}_{12}(r) < 0$]. This configuration is only possible for heterogeneous patterns, if some individuals of pattern 1 are surrounded by a large number of individuals of pattern 2 within neighborhood r, but others are surrounded by few (or no) individuals of pattern 2 (Figure 4.27b). Finally, for type IV, individuals of pattern 1 are highly clustered and individuals of pattern 2 overlap the clusters of pattern 1. As a result, the mean number of pattern 2 neighbors is less than expected $\hat{M}(r) < 0$, but the probability of having the nearest neighbor of pattern 2 within distance r is greater than expected [i.e. $\hat{D}_{12}(r) > 0$].

For each neighborhood r, the bivariate association of a given pair of patterns can be classified into one of the five categories "no association," "segregation," "partial overlap," "mixing," and "type IV" (Figure 4.28a,b). To get an overview of how the relative frequencies of the different co-distribution types among species pairs change with neighborhood radius r, one may count the number of cases for the different distances r (i.e., Figure 4.28c,d). The simplest approach for accomplishing this is to determine whether the values of the two observed summary statistics $D_{12}(r)$ and $K_{12}(r)$ are outside the simulation envelopes of the null model for neighborhood r. If both observed summary statistics are located within the simulation envelopes, the "no association" type is assigned to the pattern pair and, if at least one of the summary statistics is located outside the simulation envelope, one of the four remaining types is assigned as explained above. However, this approach is prone to problems associated with multiple testing, because the assessment is repeated for a set of different neighborhoods r. It may, therefore, happen that a GoF test indicates global fit of the null model over a given distance interval, however the observed summary statistics may still be outside the simulation envelopes for individual distances r. Wiegand et al. (2012), therefore, proposed eliminating, in an initial step, all pattern pairs for which the GoF test conducted over the entire interval r of interest did not detect significant departures from the null model. These pattern pairs were assigned to the "no association" category for all neighborhoods r.

The GoF test is conducted for the two summary statistics, but the L-function is used instead of the K-function to stabilize the variance. Because the two summary statistics $D_{12}(r)$ and $L_{12}(r)$ are used simultaneously to assess departures from the null model, the GoF test should be conducted for each summary statistic with an error rate of 2.5% which yields an approximate error rate of 5% for both summary statistics taken together. Alternatively, if the two summary statistics are tested with a 5% error rate, the combined test would yield an error rate of approximately 10%.

4.2.2.3 Classification Scheme Applied to a Simple Patch Type Heterogeneity

To get a feeling as to how the different types of heterogeneity influence the distribution of a bivariate pattern within the classification scheme spanned by the $P(r)$ and $M(r)$ axes, we utilize the approach presented in Wiegand et al. (2007a) and approximate the values of the $P(r)$ and $M(r)$ axes for a simple patch-type heterogeneity (Figure 4.27). The cases we explore represent a rather extreme case of the distribution of two patterns. Patterns 1 and 2 each occupy a rectangular patch within the observation window and overlap to a different degree in the three patterns shown in Figure 4.27. The patterns follow CSR inside their respective patches and no points lie outside of a patch. In calculating the summary statistics, we assume that the neighborhood r is small relative to the areas A_1 and A_2 of patterns 1 and 2, respectively. We also ignore edge effects.

The bivariate K-function yields the mean number of type 2 neighbors within distance r of the individuals of pattern 1 divided by $\lambda_2 = n_2/A$, the overall density of pattern 2 within the observation window. To approximate the K-function for the patch heterogeneities shown in Figure 4.27a–c we need to know the proportion o of the area covered by pattern 1 that pattern 2 overlaps [i.e., $o = (A_1 \cap A_2)/A_1$], and the relative area a (with respect to the area A of the observation window) that is occupied by pattern 2 [i.e., $a = A_2/A$]. The two indices o and a characterize the heterogeneity of the bivariate pattern. If o is small, there is only a small amount of overlap between the two patches. If o is one, pattern 2 completely overlaps pattern 1. If a is small, the local density of pattern 2 is very high and if $a = 1$ we find that pattern 2 covers the entire observation window and, consequently, completely overlaps pattern 1, which means $o = 1$.

The M classification axis is relatively easy to calculate, since only the individuals in pattern 1 that lie within the overlap area have neighbors of pattern 2. Also, the local density of individuals of pattern 2 within their own patch is given by λ_2/a. Given these considerations, we find that

$$K_{12}(r) \approx (o/a)\pi r^2 \tag{4.37}$$

$$M(r) \approx \ln(o/a) \tag{4.38}$$

This approximation holds if the neighborhood is smaller than the size of the patches. Thus, bivariate patterns of the form depicted in Figure 4.27 are located in the upper half of the scheme if $o > a$, and in the lower half of the scheme if $o < a$, since this leads to $M > 0$ and $M < 0$, respectively. If both patterns occupy the entire observation window, we have no heterogeneity and, as expected, we find that $K_{12}(r) \approx \pi r^2$ and $M(r) \approx 0$. We also find that $o = a = 1$ and $M(r) \approx 0$, if pattern 2 occupies the entire observation window. When both patterns occupy the same patch, we find $o = 1$ and $M(r)$ ranges from a value of 0 [when the patch covers the entire observation window (i.e., $a = 1$) to fairly high values when the relative area a occupied by pattern 2 is small. The two parameters o and a may also scale such that increasing overlap o is compensated for by an increase in the relative area a occupied by pattern 2. The values of the axis $M = \ln(o/a)$ may range between $M = -\infty$ ($o = 0$; complete segregation) and $M = \infty$. The latter occurs if both patterns occupy the same subarea (i.e., $o = 1$) and the subarea is infinitesimally small (i.e., $a = 0$).

With arguments analogous to those employed for the approximation of the K function, we can approximate the probability that an individual of pattern 1 has a nearest neighbor of pattern 2 within distance r as

$$D_{12}(r) \approx o[1 - \exp(-(\lambda_2/a)\,\pi r^2)] \tag{4.39}$$

Equation 4.39 arises since only a proportion o (approximately) of individuals of pattern 1 have nearby neighbors of pattern 2, which occur within their own patch at a local density of λ_2/a. Using the abbreviation $c(r) = \lambda_2 \pi r^2$, where $c(r)$ yields the expected number of points of pattern 2 under CSR (or independence) within a disk of radius r, we can write $D_{12}(r) \approx o - o\,e^{-c(r)/a}$ and the values of the P-axis yield

$$P(r) \approx o(1 - e^{-c(r)/a}) - (1 - e^{c(r)}) \tag{4.40}$$

For small values of $c(r)$ we find that $P(r) \approx c(r)[o/a - 1]$. Thus, a patch-type, bivariate pattern is located on the right side of the scheme (i.e., $M > 0$) if $o > a$, and on the left side (i.e., $M < 0$) if $o < a$. Thus, in this situation we have only segregation ($o < a$) or mixing ($o > a$). The more interesting situations are, therefore, the cases where $c(r)$ is larger. In the case of total segregation (i.e., $o = 0$) and a large value of $c(r)$, we find that $P(r) \approx -1$. In the case of complete overlap of pattern 1 by pattern 2 ($o = 1$) and $a < 1$, we find that $P(r) \approx e^{-c(r)}$.

Equations 4.38 and 4.40 provide analytical approximations for the values of the two axes $P(r)$ and $M(r)$ of the scheme for the case of simple patch heterogeneity shown in Figure 4.27. This approximation is governed by three parameters, the proportion o of the area covered by pattern 1 that pattern 2 overlaps, the relative area a (with respect to the area A of the observation window) that is occupied by pattern 2, and the expected number $c(r)$ of points of pattern 2 under CSR within a disk of radius r. To better understand how these three parameters determine classification of a given patch-type pattern into the mixing, segregation and partial overlap categories, we calculated the $M(r) = 0$ and $P(r) = 0$ isoclines in the o–a parameter space (Figure 4.27d). Remember that, for a given neighborhood radius r, we have mixing if $M(r) > 0$ and $P(r) > 0$, segregation if $M(r) < 0$ and $P(r) < 0$, partial overlap if $M(r) > 0$ and $P(r) < 0$, and no patterning if $M(r) = 0$ and $P(r) = 0$. Using Equation 4.38, we find that $M(r) = 0$ is only possible for $o = a$. Thus, the bold gray line in Figure 4.27d indicates the points in the o–a parameter space where $M(r) = 0$ (i.e., the M-isocline). We find $M(r) > 0$ in the lower right half of the parameter space shown in Figure 4.27d, because here $o > a$, and $M(r) < 0$ in the upper left half of the parameter space where $o < a$. The black line shows the points in the parameter space where Equation 4.40 predicts a value of $P(r) = 0$ (i.e., the

P-isocline). Because the solution $P(r) = 0$ of Equation 4.26d is also dependent on the value of the third parameter $c(r) = \lambda_2 \pi r^2$, we have a family of isoclines where each indicates the solution of $P(r) = 0$ for a different neighborhood radius *r*. In Figure 4.27d, we show the solution for $c(r) = 1$ (black line). For (o, a) points located to the left of the *P*-isocline, we find values of $P(r) < 0$ and, for points located to the right of the *P*-isocline, we have $P(r) > 0$.

With the above information we can determine the areas in the *o–a* parameter space that are associated with independence, mixing, partial overlap, and segregation. Independence [i.e., $P(r) = 0$, $M(r) = 0$] is given if the *P*- and *M*-isoclines meet in *o–a* parameter space, as shown in Figure 4.27d. There is only one solution for independence given by $o = 1$ and $a = 1$. As expected, this corresponds to a homogeneous pattern where both types of points occupy the entire observation window. Segregation [i.e., $P(r) < 0$, $M(r) < 0$] arises if $o < a$. This means that bivariate patterns of the patch-type show segregation if the relative area *a* occupied by pattern 2 is greater than the relative overlap *o* between the two patterns (i.e., $o/a < 1$). Partial overlap [$P(r) < 0$ and $M(r) > 0$] or mixing [$P(r) > 0$ and $M(r) > 0$] arise if $o > a$, but the division line between mixing and partial overlap (i.e., the *P*-isocline in Figure 4.27d) depends on the third parameter $c(r)$. Remember that $c(r) = \lambda_2 \pi r^2$ yields the expected number of points of pattern 2 under CSR within a disk of radius *r*. If $c(r)$ becomes smaller, for example, by reducing the neighborhood radius *r*, the *P*-isocline moves toward the *M*-isocline (shown as a bold gray line) and the *a–o* parameter space of partial overlap becomes smaller, and finally completely disappears for $c(r) = 0$. However, if $c(r)$ becomes greater than one, the *P*-isocline moves to the right and the *a–o* parameter space of partial overlap becomes larger and that of mixing becomes smaller. For large values of $c(r)$ the mixing parameter space disappears entirely. The latter yields the interesting observation that a patch-type bivariate pattern may change its categorization, as the neighborhood radius *r* is increased. For small values of *r* and $o > a$ [i.e., $K(r) > \pi r^2$], the association will be categorized as mixing, but with an increasing neighborhood radius, the category will switch to one of partial overlap. A type IV categorization ($P > 0$ and $M < 0$) does not occur under the assumptions of our approximation; it may only arise if there are strong second-order effects.

4.2.2.4 The Classification Scheme Applied to Recruits at the BCI Forest

For a second illustration of the classification scheme, we use data for the recruits at the BCI forest (Box 2.6) from the 2005 census (e.g., the example in Figure 4.25). We included only the 76 species that were represented by more than 50 individual recruits, analyzing all 5600 possible bivariate species pairs using the CSR null model for the second species. Figure 4.28a shows the location of the [$M(r)$, $P(r)$] point for a neighborhood of $r = 5$ m for each of the 5600 species pairs in the two-dimensional *M–P* scheme. Species pairs that did not differ significantly from the CSR null model are marked with

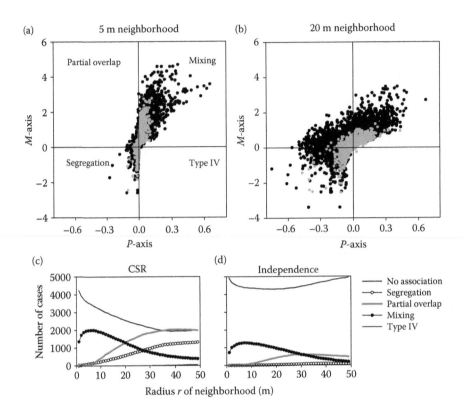

FIGURE 4.28
The classification scheme for recruits of the 2005 census at the BCI forest. We included all species with more than 50 individuals classified as recruits (yielding a total of 76 species) for $76 \times 75 = 5600$ bivariate pairs. (a) Location of all 5600 species pairs within the P–M scheme for a 5 m neighborhood. Pairs with significant departures from the CSR null model are shown as black disks and pairs without significant departures from the CSR null model as gray disks. Most significant pairs are of the mixing type. (b) Same as panel (a), but for a 20 m neighborhood. (c) Dependence of the number of different co-distribution types (with respect to the CSR null model) on the neighborhood radius r. Only pairs with significant GoF test were counted for the significant co-distribution types. (d) Same as panel (c), but for a pattern reconstruction implementation of the independence null model.

a gray color. As expected, these pairs are located close to the origin of the scheme. We find that, for the 5 m neighborhood, the majority of the species pairs are categorized as "mixing," and that very few pairs are categorized as "partial overlap" or "segregation." When enlarging the neighborhood scale to 20 m, we find that many species pairs are still categorized as "mixing," but now the number of pairs categorized as "partial overlap" and "segregation" increases substantially, relative to the 5 m neighborhood (Figure 4.28b).

Figure 4.28c demonstrates how the number of species pairs that are categorized as a given co-occurrence type changes with neighborhood radius r. In this case, the GoF test was evaluated over the distance interval of 1 to

50 m. It is notable that approximately 75% of the species pairs (4246) do not show a significant amount of patterning at short neighborhood distances. This outlines the effect of stochasticity within small neighborhoods and that individuals of a given focal species will be surrounded, in general, by different sets of species. With a neighborhood of 50 m, the number of cases showing no pattern was reduced to approximately 45%. The mixing type peaks at 6 m, whereas almost no segregation or partial overlap can be found in small neighborhoods. Instead, the segregation and partial overlap categories show a proportional increase with larger neighborhood sizes, at the expense of mixing. As predicted by the simple model of patch type heterogeneities (Figure 4.27d), the reduction in the proportion of species pairs with mixing at increasing neighborhood radii r is compensated for by an increase in the proportion of species pairs with partial overlap. It is also notable that no species pair occurs with type IV association (Figures 4.28c,d). In summary, the recruit community at the BCI forest is characterized by many species pairs that do not show co-distribution patterns differing from those that would may arise by pure chance, but pairs that do show patterning are characterized by small-scale mixing and larger-scale partial overlap or segregation.

We may now ask how much of the spatial patterning observed for the recruit community is caused by the univariate spatial structure of the second pattern. Figure 4.26 showed that strong clustering in the second species may act to enlarge the simulation envelopes if we use the pattern reconstruction model for independence rather than the CSR null model. Indeed, when we test for independence using pattern reconstruction, the number of significant cases declines substantially (Figure 4.28d), especially for segregation, which almost completely disappears, and also for partial overlap, which is strongly reduced. However, the strong mixing signal still persists (Figure 4.28d). This indicates, in particular, that the larger-scale segregation patterns in Figure 4.28c are primarily caused by univariate clustering and that spatial configurations that appear as segregation, using the CSR null model, are in fact quite likely to arise by pure chance, under the independence null model, where this clustering is considered. The same applies for partial overlap, but we hypothesize that the mixing patterns are, in most cases, due to species interactions or shared habitat preferences.

The current example clearly demonstrates the importance of the selection of the appropriate null model. When we want to assess whether species are independently placed, we need to use a null model that conserves the univariate structure of the component patterns. However, even in cases where we know that two clustered patterns can arise by independent superposition, an important piece of information is provided when we can show that two patterns can be placed in the segregation or partial overlap categories over a range of neighborhood scales, when assessed using the CSR null model. This tells us that most (or a large proportion of) individuals of species 1 have no direct contact with individuals of species 2.

4.2.3 Point Processes Assuming Interactions

In this section, we describe point processes that explicitly assume an interaction between the two types of points. We consider simple cluster processes that incorporate an explicit and positive association between the two component patterns. We also introduce bivariate Gibbs processes that characterize negative interactions in bivariate patterns.

4.2.3.1 Bivariate Thomas Process with Shared Parents

We can generalize the simple univariate Thomas process presented in Section 4.1.4.2 to yield a bivariate point process with an explicit mechanism for producing positive associations. In Section 4.1.4.4, we already came across a bivariate point process (the bivariate parent–offspring Thomas processes) that explicitly produces a positive relationship between parents and offspring. If the offspring are clustered around the parents, a simple bivariate parent–offspring Thomas process can be used to quantify this relationship. Fitting this null model to the data allowed us to infer the underlying dispersal kernel, which is given by a bivariate normal distribution that governs the placement of offspring relative to their parents, the parents acting as cluster centers. Unfortunately, the simple univariate Thomas process makes very specific assumptions about the relationship between the two patterns that render it unsuitable for characterizing positive (i.e., mixing) relationships between bivariate patterns in general.

The *bivariate Thomas process with shared parents* is a simple extension of the simple univariate Thomas process (Equation 4.7), which allows us to quantify the association between two univariate patterns that can be approximated by simple Thomas processes. The underling idea of this point process is simple. While the simple univariate Thomas process is constructed by using one parent pattern and an associated offspring pattern, the bivariate Thomas process with shared parents uses the same principle, but now two different offspring patterns (with possibly different parameters σ_1 and σ_2 determining the cluster size) are generated around the same parents. In the next chapter we will show that this point process can be extended to cases of partly shared parents to produce bivariate associations ranging from strong attraction (all parents are shared and values of the parameters σ_1 and σ_2 are small) to independence (no parents are shared).

The parameters of the bivariate Thomas process with shared parents are the intensity ρ of the shared parents, the intensity λ_1 of offspring pattern 1, the intensity λ_2 of offspring pattern 2, the variance σ_1^2 determining the distance between parents and pattern 1, and the variance σ_2^2 determining the distance between parents and pattern 2. To calculate the bivariate pair-correlation function of this point process, we first estimate the bivariate product density, which is the sum of two cases: type 1 and type 2 points stem from different parents, and type 1 and type 2 points stem from the same parent. First, type 1 and type 2 points are independent if their locations can be traced to different parents since the parents are CSR. In this case, the contribution to the product density is $\lambda_2\lambda_1$, since the probability of finding an offspring of type 1 within disk dx_1 at location x_1 is given by $\lambda_1 dx_1$ and finding an offspring of type 2 within disk dx_2 at location x_2 is given by $\lambda_2 dx_2$. However, if the two offspring were produced by the same parent, we have to multiply the intensity ρ of the shared parents by the expected number of ordered pairs of offspring ($= \mu_1\mu_2 = \lambda_1\lambda_2/\rho^2$) and by the probability $h_2(r, \sigma_1, \sigma_2)$ that a pair of type 1 and type 2 offspring of the same parent are distance r apart. For $f_k = 1$ (i.e., p_s is a Poisson distribution) this yields the bivariate pair-correlation function:

$$g_{12}(r,\sigma_1,\sigma_2,\rho) = 1 + \frac{1}{\rho}h_2(r,\sigma_1,\sigma_2) = 1 + \frac{1}{\rho}h(r,\sqrt{\sigma_1^2 + \sigma_2^2}) \qquad (4.41)$$

Note that Equation 4.41 collapses to Equation 4.7 if patterns 1 and 2 are identical (i.e., the univariate case).

4.2.3.2 Bivariate Thomas Process with Partly Shared Parents

We can generalize the bivariate Thomas process with shared parents, which was presented in the last section, by considering the possibility that not all parent points are shared. The resulting *bivariate Thomas process with partly shared parents* allows us to consider a full range of situations spanning from the highly constrained case described in Equation 4.41 at one extreme to an independent pattern at the other. This is done by allowing only a proportion p_s of parents to be shared. This situation contrasts with the situation produced by Equation 4.41, which assumes that each point of pattern 1 (or pattern 2) has the same probability of being an offspring of any given parent. In this case, we may find that, by accident, a parent may have produced only offspring of type 1 or type 2, especially if the mean number of offspring per cluster (μ_1 or μ_2) is small. However, a broader range of conditions can be modeled using the bivariate Thomas process with partly shared parents, where the intensities of parents that produce only type 1 and type 2 points are ρ_1 and ρ_2, respectively, and the intensity of the parents producing both types of offspring is ρ_s, as defined above; thus, $\rho_2 + \rho_1 = \rho - \rho_s$. In this case, ρ is the combined intensity of both parental types. This model has the additional parameters λ_{1s} and λ_{2s}, which are the intensities of the component offspring patterns 1 and 2 stemming from the shared parents, respectively, and λ_1 and λ_2 are the intensities of patterns 1 and 2. Note that when only a proportion of the parents are shared, we obtain a pattern that is essentially an independent superposition of two bivariate patterns, one pattern exhibiting attraction comprising the offspring of the shared parents and one bivariate pattern where the two component patterns are independent, since the offspring are from nonshared parents.

We can extend Equation 3.131 for random superposition of a CSR and cluster pattern to the bivariate case to obtain the pair-correlation function for the bivariate Thomas process with partly shared parents. The only difference is that we have to replace the factor c^2 in Equation 3.131, where c was the proportion of points of the clustered pattern, by the factor $\lambda_{s1}\lambda_{s2}/\lambda_1\lambda_2$:

$$g_{12}(r,\sigma_1,\sigma_2,\lambda_{s1},\lambda_{s2},\lambda_1,\lambda_2,\rho_s) = 1 + \frac{\lambda_{s1}}{\lambda_1}\frac{\lambda_{s2}}{\lambda_2}\frac{1}{\rho_s}h_2(r,\sigma_1,\sigma_2) \qquad (4.42)$$

We can also derive this formula via the bivariate product density, which is the sum of two cases: (i) type 1 and type 2 points stem from the same parent and (ii) type 1 and type 2 points stem from different parents. The contribution of the first case yields $\lambda_{s1}\lambda_{s1}h_2(r,\sigma_1,\sigma_2)/\rho_s$ and the contribution of the second case yields $\lambda_1\lambda_2$, recalling that we divide the pair-correlation function by $\lambda_1\lambda_2$ to obtain the product density.

Under the reasonable assumption that nonshared parents have the same expected number of offspring as shared parents we find

$$g_{12}(r,\sigma_1,\sigma_2,\rho_1,\rho_2,\rho_s) = 1 + \frac{\rho_s\rho_s}{\rho_1\rho_2}\frac{1}{\rho_s}h(r,\sqrt{\sigma_1^2+\sigma_2^2}) \qquad (4.43)$$

In fact, if all parents are shared (i.e., $\rho = \rho_1 = \rho_2 = \rho_s$), Equation 4.43 collapses back to Equation 4.41 for a bivariate cluster process with shared parents. If there are no shared parents (i.e., $\rho_s = 0$), Equation 4.43 yields $g_{12}(r) = 1$, as expected for two independent patterns. The g-function in Equation 4.43 has the same functional form as the bivariate cluster process with shared parents (Equation 4.41):

$$g_{12}(r,\sigma_{12}^*,\sigma_2,\rho^*) = 1 + \frac{1}{\rho^*}h(r,\sigma_{12}^*) \qquad (4.44)$$

but with a modified "effective" intensity of parents ρ^*. Thus, we can fit Equation 4.41 to the bivariate pattern to estimate the "effective" intensity of parents ρ^* and, when using independent estimates for ρ_1 and ρ_2 derived from the univariate analyses of the two component patterns, we can estimate the intensity of the shared parents as

$$\rho_s = \frac{\rho_1\rho_2}{\rho^*} \qquad (4.45)$$

The "effective" intensity of parents ρ^* provides a means for distinguishing between different forms of association. The strongest association occurs if all parents of patterns 1 or 2 are shared (e.g., $\rho_1 = \rho_s$). In this case, we find that $\rho^* = \rho_2$. However, large values of ρ^* indicate that patterns 1 and 2 are independent, because Equation 4.45 will yield a very low intensity of shared parents.

Comparing the parameters fit to the two univariate, component patterns with those of the bivariate pattern allows for a diagnosis as to whether the point process yields consistent parameters. First, the fitted value for σ_{12}^* can be compared with the expected estimate for this point process, which is given by $\sigma_{12} = (\sigma_1^2 + \sigma_2^2)^{1/2}$. In this case, σ_1 and σ_2 are the fits for the corresponding univariate Thomas processes of the two component patterns. Second, one should also verify that $\rho_1 > \rho_s$ and $\rho_2 > \rho_s$, because the number of shared parents cannot be greater than the number of parents of pattern 1 or 2.

It is quite straightforward to simulate the bivariate Thomas process with partly shared parents. The total number of parents is given by $A(\rho_1 + \rho_2 - \rho_s)$. The parents can be arranged in random order, and numbered in such a way that $\rho_1 \geq \rho_2$. For pattern 1, we select parents randomly from among the parents numbered 1 to $A\rho_1$ and the parents for pattern 2 are selected at random from among parents assigned numbers ranging from $A(\rho_1 - \rho_s)$ to $A(\rho_1 + \rho_2 - \rho_s)$, which insures that ρ_s parents produce both types of offspring. The parents are distributed at random and the offspring are then distributed around the assigned parents following their respective bivariate normal distribution.

4.2.3.3 Application of Bivariate Thomas Process

As an example for the application of the bivariate Thomas process, contrasting versions with shared and nonshared parents, we can consider recruits of the typical light demanding tree species *C. insignis* and *M. argentea* (Figure 4.25a), which show a pattern of attraction at distances up to 20 m (Figure 4.25b). In a first step, we analyze the larger-scale clustering of the two patterns using the simple Thomas process. We fit the Thomas process for distances greater than 6 m, because both patterns exhibit a component of strong small-scale clustering (Figure 4.29a,b). The Thomas process does a good job of characterizing the larger-scale clustering. For *C. insignis* we obtain a parameter estimate for the number of parent points of $A\rho_1 = 14.9$ and an estimate of $\sigma_1 = 6.6$ m for the offsprings' scale of spread. The equivalent estimates for *M. argentea* yield values of $\sigma_2 = 4.4$ m and $A\rho_2 = 64.8$. Other than missing the small-scale clustering for *C. insignis*, the distribution function $D(r)$ of the distances to the nearest neighbor shows that the Thomas process characterizes the cluster structure of this species quite well; there is no indication of isolated points (inset Figure 4.29a). However, the distribution of *M. argentea* contains a substantial proportion of isolated recruits, with nearest-neighbor distances much greater than expected for the simple Thomas process (inset Figure 4.29b).

In contrast to the simple Thomas process for the univariate component patterns, the bivariate cluster process with shared and nonshared parents yields a surprisingly good fit of the observed bivariate pattern, with the exception

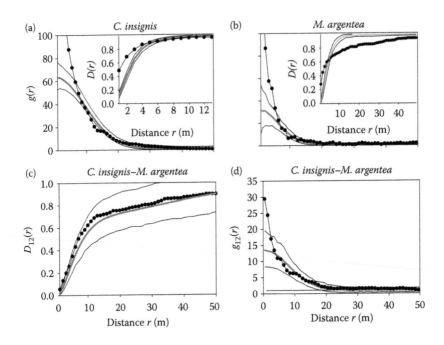

FIGURE 4.29
Application of the bivariate cluster process with shared and nonshared parents to an observed pattern for the recruits of the species *C. insignis* and *M. argentea* (Figure 4.25a), which exhibit a pattern of attraction for distances up to 20 m (Figure 4.25b). (a) Univariate pair-correlation analysis of the pattern of *C. insignis* with a simple Thomas process, with a small inset showing the corresponding results for the nearest-neighbor distribution function $D(r)$. (b) Same as panel (a), but for the species *M. argentea*. (c) Analysis of the bivariate nearest-neighbor distribution function $D_{12}(r)$, using the bivariate cluster process with shared and nonshared parents, fit with parameters from the univariate analyses. (d) Same as panel (c), but for the bivariate, pair-correlation function $g_{12}(r)$.

of some additional small-scale attraction at distances less than 3 m that is not captured by the point process (Figure 4.29d). To our surprise, the bivariate distribution function $D_{12}(r)$ of the distances from a *C. insignis* recruit to the nearest *M. argentea* neighbor lies fully within the simulation envelopes. The fitted parameters of the bivariate process yield $A\rho^* = 10.3$ and $\sigma_{12}^* = 6.52$ and we find that $(\sigma_1^2 + \sigma_2^2)^{1/2} = 5.65$ is reasonably close to the fitted value of σ_{12}^*. The total number of parents yield $A(\rho_2 + \rho_1 - \rho_s) = 14.9 + 64.8 - 10.3 = 69.3$ and the proportion of shared parents is 14.9%. Thus, only a few of the clusters of *C. insignis* are not shared by *M. argentea* [$A(\rho_1 - \rho_s) = 4.6$], but *M. argentea* has many clusters not shared by *C. insignis* [$A(\rho_2 - \rho_s) = 54.5$]. The latter diagnosis is supported by the visual inspection of the pattern (Figure 4.25a). However, note that the species *M. argentea* had a notable number of isolated recruits, which somewhat "inflates" the estimate of the number of clusters.

In summary, we find that the bivariate pattern of the two light-demanding species *C. insignis* and *M. argentea* show a spatial association structure than can be approximated by the bivariate cluster process with shared and nonshared parents. The lack of detail in the representation of the univariate patterns by the simple Thomas process (especially in missing additional fine-scale clustering) did not reduce the ability of the bivariate process to approximate the bivariate association pattern. The two species are found together in clusters with an approximate radius of 11 m and some of the individuals also share smaller-scale clusters. As in the earlier analysis of independence for this species pair, we hypothesize that they share canopy gaps, but *M. argentea* was able to use a wider range of sites than *C. insignis*.

4.2.3.4 Gibbs Processes to Model Interactions between Species

In the section on univariate patterns (Section 4.1.6.2), we briefly introduced Gibbs processes, which can be used to describe (mostly antagonistic) interactions among individuals. A logical extension of using Gibbs processes for univariate patterns is to use them to characterize bivariate or multivariate point processes. Here, we provide one example of a bivariate inhibition process that is modeled using pairwise-interaction Gibbs processes. At least four interaction parameters γ_{ij}, characterizing the effects of species j on species i, are required to model bivariate point patterns using the Gibbs process model, that is, γ_{11}, γ_{21}, γ_{21}, and γ_{22}. In general, the goal is to characterize the impact of negative interactions due to competition between species i and j. (For symmetric interactions, we have $\gamma_{21} = \gamma_{21}$.) The interaction parameters can range between zero and one: for $\gamma_{ij} = 0$ complete inhibition (i.e., a hardcore process) is produced and for $\gamma_{ij} = 1$ there are no interactions. The coordinates of the n_1 points of the first pattern and the n_2 points of the second pattern are given by x_{1o} and x_{2p} where $o = 1, \ldots n_1$ and $p = 1, \ldots, n_2$.

We can extend Equation 4.29 for a univariate Gibbs process, following the methods of Högmander and Särkkä (1999), so that the local density function of a bivariate point process is given by

$$f(x_{11}, \ldots, x_{n1}, x_{21}, \ldots, x_{2n2})$$

$$= \frac{1}{Z_{12}} \underbrace{\prod_{\substack{o,m \\ o<m}} h_{11}\left(\|x_{1o} - x_{1m}\|\right)}_{\substack{\text{Interactions among} \\ \text{points of pattern 1}}} \times \underbrace{\prod_{\substack{p,q \\ p<q}} h_{22}\left(\|x_{2p} - x_{2q}\|\right)}_{\substack{\text{Interactions among} \\ \text{points of pattern 2}}} \times \underbrace{\prod_{o=1}^{n_1} \prod_{p=1}^{n_2} h_{12}\left(\|x_{1o} - x_{2p}\|\right)}_{\substack{\text{Interactions between} \\ \text{points of pattern 1 and 2}}}$$

$$(4.46)$$

where $h_{11}(r)$ and $h_{22}(r)$ are the intraspecific interaction functions of patterns 1 and 2, respectively, $h_{12}(r)$ is the interspecific interaction function between individuals of patterns 1 and 2, and Z_{12} is a normalization constant that ensures that the local density function is a probability density function.

Similar to Illian et al. (2009), we can choose the pairwise interaction function $h_{ij}(r)$ to be

$$h_{ij}(r) = \begin{cases} \exp(-\beta_{ij}[1-(r/r_0)^2]^2) & 0 < r \le r_0 \\ 1 & r > r_0 \end{cases} \tag{4.47}$$

where the $\beta_{ij} = -\log(\gamma_{ij})$ are the competition coefficients. The interaction strength becomes maximal as the distance r between two points approaches zero, yielding $h_{ij}(r) = \exp(-\beta_{ij}) = \gamma_{ij}$. In contrast, if the distance between two points is equal to or greater than r_0, the interaction strength is minimal and yields $h_{ij}(r) = 1$. Thus, two points that are located at distance r interact only if they are closer than distance r_0 (i.e., the zone of influence). The strength of the interaction decreases with increasing distance up to r_0, at which point an asymptote is reached. With $\psi(r) = [1-(r/r_0)^2]^2$ being the corresponding pair potential function, we can rewrite Equation 4.47 as

$$h_{ij}(r) = \exp(\log(\gamma_{ij})\psi(r)) = \exp(\log(\gamma_{ij}^{\psi(r)})) = \gamma_{ij}^{\psi(r)} \tag{4.48}$$

and obtain the location density function of this point-process model, which is

$$f(x_{11}, \ldots, x_{n1}, x_{21}, \ldots, x_{2n2})$$

$$= \frac{1}{Z_{12}} \underbrace{\prod_{\substack{o,m \\ o<m}} \gamma_{11}^{\psi(\|x_{1o}-x_{1m}\|)}}_{\substack{\text{Interactions} \\ \text{pattern 1}}} \times \underbrace{\prod_{\substack{p,q \\ p<q}} \gamma_{22}^{\psi(\|x_{2p}-x_{2q}\|)}}_{\substack{\text{Interactions} \\ \text{pattern 2}}} \times \underbrace{\prod_{o=1}^{n_1}\prod_{p=1}^{n_2} \gamma_{12}^{\psi(\|x_{1o}-x_{2p}\|)}}_{\substack{\text{Interactions} \\ \text{pattern 1 and 2}}} \tag{4.49}$$

where $\|x_{1o} - x_{1m}\|$, $\|x_{2p} - x_{2q}\|$, and $\|x_{1o} - x_{2p}\|$ represent the distance r between the point pairs. To understand how this equation governs the Gibbs process, consider two simple examples. First, if pattern 1 (i.e., x_{11}, ... x_{1n1}) is a hard-core pattern with hard core distance r_0, we find that $\psi(r) = [1-(r/r_0)^2]^2 = 0$ for all pairs of points and the corresponding term $\gamma_{11}^{\psi(r)}$ yields the highest possible value (= 1) for all pairs of points because $r > r_0$ for all point pairs. The hard-core pattern for pattern 1 is, therefore, likely under the location density function 4.50, because this function is a measure of the likelihood of a particular spatial configuration of points of pattern 1 at locations $x_{11}, \ldots x_{1n1}$ under the Gibbs process. Second, if all points of pattern 1 are located within a small disk with radius $r_s \ll r_0$ the function $\psi(r) = [1-(r/r_0)^2]^2$ would be approximated by $\psi(r) = 1$ because in this case the distances r among all points r is close to zero and the term $\gamma_{11}^{\psi(r)}$ would yield its maximal value γ_{11} for all point pairs. As a consequence, the location density function would yield a very small value. Thus, this pattern would be very unlikely to occur. The same rationale works for the other two interaction terms in Equation 4.49.

Although parameterization of univariate Gibbs processes is already quite complex, the difficulties increase when bivariate and multivariate interactions need to be modeled. While bivariate or multivariate Gibbs processes are theoretically appealing, they have been rarely used in ecological applications. One exception is provided by the study of Grabarnik and Särkkä (2009), which introduces Gibbs processes and extends the univariate model to a multivariate hierarchical Gibbs process in which competition between types of points is not necessarily symmetric, but directed. For example, we can assume that large trees do not suffer much competition from small trees, but small trees suffer a great deal due to the influence of large trees. This assumption allows the formulation of the probability density of the hierarchical interaction model as a product of conditional intensities. Conditional intensities tell us basically how likely it is to add a point of type k to a pattern consisting of points of types 1 to k (Grabarnik and Särkkä 2009). This simplifies the parameter estimation procedure, especially if the number of types is large (see also Illian et al. 2009). We expect that the future development of point-process models incorporating Gibbs processes will provide more detailed and direct modeling approaches characterizing the complex interaction structures of bivariate and multivariate point patterns.

4.2.4 Accounting for Heterogeneity in Bivariate Patterns

As shown in Section 4.2.2.2, first-order heterogeneity of the univariate component patterns of a bivariate pattern can produce complex spatial relationships between the two patterns. The first-order heterogeneity of the two patterns can be quantified by their intensity functions $\lambda_1(x)$ and $\lambda_2(x)$. The heterogeneous distribution pattern may be a consequence of environmental dependency or the internal processes of population dynamics (e.g., dispersal limitation), which leave larger, but potentially suitable areas within the observation window unoccupied. If the two intensity functions $\lambda_1(x)$ and $\lambda_2(x)$ are tightly linked through a positive correlation structure we have *mixing*, but if they have a strong, negative correlation structure we have *segregation*. Additionally, if they are positively correlated in some areas of the observation window and negatively correlated in others, we have *partial overlap*.

Two species may show significant effects that can be classified as mixing, segregation, or partial overlap, even if they do not show direct interactions. We can explore these patterns to test for independence, using either the CSR null model or the pattern reconstruction null model (as presented in example Figure 4.26). This approach was illustrated for the simple patch-type heterogeneity (Figure 4.27), where the two species occupy only patches A_1 and A_2 of the study plot. In cases where the two species are independently placed within different patches, with little overlap (Figure 4.30a), the analysis using the CSR null model indicates strong segregation, because the two types of

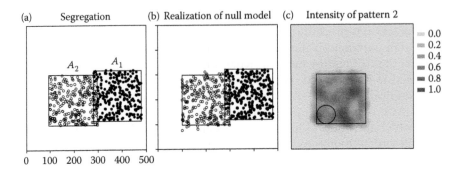

FIGURE 4.30

Heterogeneous Poisson null model for bivariate patterns. (a) Map of a pattern with patch type segregation where patterns 1 and 2 are CSR patterns inside their respective patches A_1 and A_2. The two patterns are distributed independently of one another. (b) Realization of HPP using the nonparametric intensity estimate of pattern 2 with displacement radius $R = 30$ m shown in panel (c). Note that a few points of the null model pattern fall outside of the original patch A_2.

points are more closely associated in the data than they are when the points of species 2 are relocated to random positions within the landscape. Note that under both the CSR null model (where species 1 is unchanged and species 2 follows CSR) and an implementation of the independence null model, the probability that a given location of the observation window receives a point of pattern 2 is constant (i.e., λ_2).

Although we can explore first-order effects, we may be more interested in determining whether *the individuals of the two species interact if they are located close to each other*, conditioned on the intensity functions $\lambda_1(x)$ and $\lambda_2(x)$. We can explore this by employing a null model that relocates the individuals of pattern 2 only locally. An example of a realization of such a null model is shown in Figure 4.30b, where pattern 1 is unchanged, but the points of pattern 2 are randomly displaced within their $R = 30$ m neighborhood (also see Section 4.1.2.1 which considers a similar null model for univariate patterns). The resulting large-scale pattern looks very similar to the observed large-scale pattern (cf. Figure 4.30a and b), and it is unlikely that a significant effect will be detected at larger scales. The reason for this is that the displacement of the points within the small neighborhood R will move the points to locations with a similar intensity function [i.e., $\lambda_2(x) \approx \lambda_2(x + eR)$], where e is a vector of length one. Because of this, the bivariate product density of the null model at distances $r > R$ will approximate those of the observed pattern. In this way, the null model factors out the first-order effects imposed by the nonconstant intensity function, at least approximately, and because of the separation of scales allows us to selectively study the small-scale interactions (see Section 4.1.2.2).

For bivariate patterns, the approach just described offers several possibilities for constructing null models. First, we can use the HPP with a nonparametric intensity estimate (Figure 4.30c) to randomize pattern 2 (Figure 4.30b), allowing for an approximate assessment of species interactions. However, if there is strong clustering in pattern 2, this model will not fully retain the observed clustering as it will partially "smear" the pattern. Instead, we must add an additional constraint to the small-scale structure of the pattern by using pattern reconstruction that conditions on $\lambda_2(x)$ of pattern 2 (see Sections 3.4.3.5 and 3.4.3.7). This will allow us to test independence, conditioned on the observed intensity function $\lambda_2(x)$. The implementation of the HPP for pattern 2 is the same as described for univariate patterns in Section 4.1.2.1.

Figure 4.31 provides three examples of a patch type heterogeneity, where patterns 1 and 2 follow CSR inside their respective patches, but both are

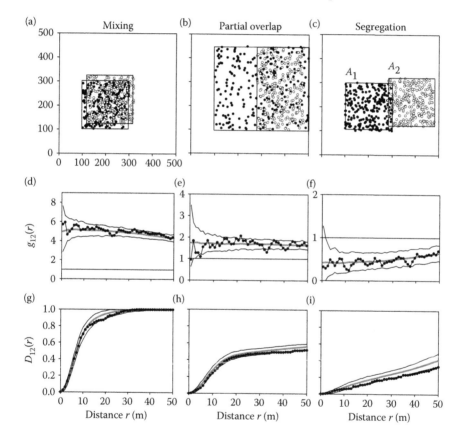

FIGURE 4.31
Factoring out first-order heterogeneity in a bivariate analysis by utilizing an HPP with nonparametric intensity estimate for pattern 2. (a) Mixing pattern comprising univariate patterns of 250 points within a 500×500 m observation window. The focal pattern is indicated by closed disks and the second pattern by open disks. (b) Same as panel (a), but for a bivariate pattern showing partial overlap. (c) Same as panel (a), but for a bivariate pattern showing segregation. (d)–(f) Results of the analysis with a bivariate pair-correlation function, using an HPP with nonparametric estimate of the intensity of pattern 2 with a bandwidth of $R = 30$ m. (g)–(i) Same as panels (d) through (f), but using a bivariate distribution function of nearest-neighbor distances. Analyses are conducted on the pattern located in the same column.

independent in the overlap area. Using an HPP that randomly displaces the points of pattern 2 within a neighborhood of $R = 30$ m reveals that the three bivariate patterns shown in Figure 4.31a–c are indeed independent, although the bivariate, pair-correlation function produces what would normally be a strong signal of attraction (i.e., $g_{12}(r) \gg 1$; Figure 4.31d,e) and segregation (i.e., $g_{12}(r) \ll 1$; Figure 4.31f). The only statistic with a signal indicating a significant departure from random is $D_{12}(r)$, which shows a slight tendency toward repulsion in Figure 4.31i for the segregated pattern. These results show that the HPP is able to factor out the effect of the large-scale heterogeneities shown in Figure 4.31a–c.

While the analysis can be conducted for any displacement distance R, it is desirable to use a distance that is likely to separate biological effects. For example, trees are expected to have a limited distance over which significant interactions take place. Hubbell et al. (2001) found that the conspecific, neighborhood density effects impacting survival disappeared within approximately 12–15 m of the focal plant. Several other studies using individual-based analyses of local neighborhood effects on tree growth and survival have confirmed this result (e.g., Stoll and Newbery 2005; Uriarte et al. 2005). What this suggests is that direct plant–plant interactions in forests may fade away at greater distances. As a consequence, the displacement distance R should be greater than the scale over which these interactions take place.

Another interesting question to consider in the current context is whether *separation of scales* occurs; that is, are the scales over which species interactions operate shorter than the scales of habitat heterogeneity. This can be tested in a simple way, using methods similar to those used in the univariate case as explained in Section 4.1.2.2. Because the HPP conditions on spatial structures at scales greater than R, it can only indicate significant effects at distances less than R; noncumulative summary statistics should be within the simulation envelopes for distances $r > R$. This occurs because the null model conserves the observed density $\lambda_2(x)$ of the second pattern at distances $r > R$. A focal point of pattern 1, which is held fixed, will have at distance $r > R$ approximately the same neighborhood density of pattern 2 as the observed pattern. As a consequence, the pair-correlation functions for the data and the null model will coincide. If a clear separation of scales occurs, we will observe that a noncumulative summary statistic, such as the pair-correlation function, will only show departures from the null model for distances $r < r_1$ and will lie within the simulation envelopes for the distance interval (r_1, R). However, in cases without separation of scales, we will find that $r_1 \approx R$. Thus, the summary statistics will approach the simulation envelopes only at distance R.

4.2.4.1 Examples for the Heterogeneous Poisson Null Model

Figure 4.32a shows two patterns that are driven to a large extent by environmental heterogeneity. The patterns were produced by trees with a dbh

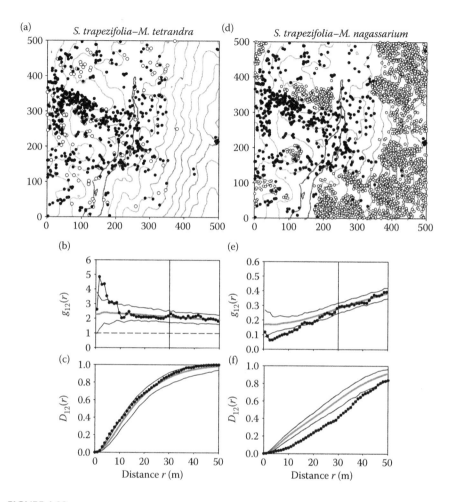

FIGURE 4.32
Separation of scales and bivariate analysis with a heterogeneous Poisson null model. (a) Map of large individuals (dbh l > 10 cm) of *S. trapezifolia* (closed disks; pattern 1) and *M. tetrandra* (open disks; pattern 2) within the tropical forest at Sinharaja, Sri Lanka (also analyzed in Figure 3.59). The map shows the contour lines of the plot, with the bold line indicating the lowest elevation. (b) Analysis of the bivariate pattern in panel (a), using a heterogeneous Poisson null model, with nonparametric estimate of the intensity function using a bandwidth of $R = 30$ m. (c) Distribution function of nearest-neighbor distances as in panel (b). (d) Map of *S. trapezifolia* (closed disks) and *M. nagassarium* (open disks). (e) Same as panel (b) for data shown in panel (d). (f) Same as panel (c) for data shown in panel (d).

greater than 10 cm for two species *Shorea trapezifolia* (pattern 1) and *Mastixia tetrandra* (pattern 2) in the tropical forest of Sinharaja, Sri Lanka. The map shows the contour lines of the plot (the bold line corresponds to the line with the lowest elevation) and the tree locations. The two species share the same topographically defined habitat; east facing slopes. Because of the shared habitat association, the two species are only distributed within one large patch on the left side of the observation window, leaving the right side almost empty. This species pair provides a good example for the "mixing" association pattern as shown in Figure 4.31a. The neighborhood density of species 2 around species 1 is roughly two times higher than expected for two independent homogeneous patterns [i.e., $g_{12}(r) = 1$], which would be the case if the two species were distributed over the entire study plot (Figure 4.32b). However, a question arises about the potential interactions between individuals of the two species, if they lie close together. Do they show positive or negative interactions? To answer this question, we need to selectively test for small-scale effects and therefore condition on the large-scale structure of the pattern. To accomplish this, we use the approach of the HPP with nonparametric intensity estimate as discussed above. We keep the individuals of the first pattern (i.e., *S. trapezifolia*; closed disks in Figure 4.32a) unchanged and randomly displace the individuals of the second species (i.e., *M. tetrandra*; open disks in Figure 4.32a) within a neighborhood of $R = 30$ m. We chose a 30 m neighborhood assuming that this was somewhat greater than the maximal interaction range for large trees.

The results of the analysis show that separation of scales occurred. The values of the pair-correlation function lie outside of the simulation envelopes for distances of 1–5 m, are located on the border for distances of 5–8 m, and lie completely inside the simulation envelopes for distances greater than 8 m (Figure 4.32b). The scale at which the pair-correlation function enters the envelopes is a much shorter distance than the 30 m that is guaranteed through the application of the HPP using a displacement radius of $R = 30$ m. This provides strong evidence that the process is able to accomplish a separation of scales. Thus, individuals of both species co-occur in the same habitat, but if they are close together they exhibit additional effects of small-scale attraction. The results for the distribution function of the distances to the nearest neighbor confirm this result; there is a weak and positive effect at small distances (Figure 4.32c). It is clear that using a null model that does not conserve the intensity of the two patterns will not be able to identify the subtle species interactions at shorter scales.

Figure 4.32d provides an example of a bivariate heterogeneous pattern with segregation, that is, the two species occupy largely disjunct areas of the study plot, with additional repulsion at small scales. The species *S. trapezifolia* (closed disks) is found predominantly on low-elevation, less-steep spurs, whereas the species *Mesua nagassarium* (open disks) is found predominantly on steep, upper-elevation spurs (Gunatilleke et al. 2006). The segregation of the two species is depicted quite well by the pair-correlation function; the neighborhood density of *M. nagassarium* around individuals of *S. trapezifolia*, at distances of 1–10 m, is 5 times lower than expected for an independent, homogeneous pattern [i.e., $g_{12}(r) \approx 0.2$]; at distances of 50 m, it is still more than half the expected value under independence. Using the heterogeneous

Poisson null model, with a displacement radius of $R = 30$ m, the overall segregation pattern remains unchanged, but the analysis allows us to infer what happens if individuals of the two species are accidently located close to each other. Figure 4.32e shows that there is a signal of small-scale repulsion at distances between 2 and 15 m. Again, using a null model that does not conserve the intensities of the two species will not allow us to detect the subtle effect of small-scale interactions that occur within the larger-scale pattern of segregation.

4.2.4.2 Pattern Reconstruction Using the Intensity Function

In the previous section, we used the HPP to selectively test for small-scale associations in heterogeneous bivariate patterns. However, this null model does not condition on the small-scale structures of the pattern, but instead randomizes the second pattern following its intensity function. A similar approach that selectively preserves the heterogeneous large-scale structures, as well as conserving the small-scale structure, is based on pattern reconstruction constrained by the intensity function (see Section 3.4.3.7). If a nonparametric kernel estimate is used, the reconstructed patterns show the same large- and small-scale properties as the original pattern, but the typical small-scale structures are randomly displaced. In contrast, the HPP somewhat "smears" the small-scale structures and makes them more diffuse.

Figure 4.33 provides an example of the application of pattern reconstruction as a null model for the pattern of the two species *S. trapezifolia* and *M. nagassarium* (Figure 4.32d), which were analyzed in the previous section using the HPP. The reconstruction of the pattern of species *M. nagassarium* used a nonparametric estimate of the intensity function with $R = 30$ m, and the summary statistics $g(r)$, $K(r)$, $H_s(r)$, and $D^k(r)$, with $k = 1, 2, 4, 6, 8, 12, 16, 20,$ 25, and 50.

Figure 4.33d shows one reconstruction, which does a good job of reassembling the observed large- and small-scale structures of the pattern. While the pattern reconstruction produces relatively abrupt boundaries of the species pattern, which can be seen in the observed pattern, a realization of the corresponding HPP will produce more diffuse, "smeared" boundaries. As a consequence, the analyses using pattern reconstruction produce different results when compared to the corresponding heterogeneous Poisson null model (Figure 4.32b,e). While the heterogeneous Poisson null model produces a clear signal of small-scale repulsion, this effect is weaker when using the corresponding pattern reconstruction null model. In particular, we observe that the nearest neighbor in the heterogeneous Poisson null model is much closer than observed (Figure 4.33c), whereas this is not the case using the pattern reconstruction approach (4.33f). The result for the heterogeneous Poisson null model is probably due to the "smearing" effect, which brings the nearest *M. nagassarium* neighbor closer to the individuals of

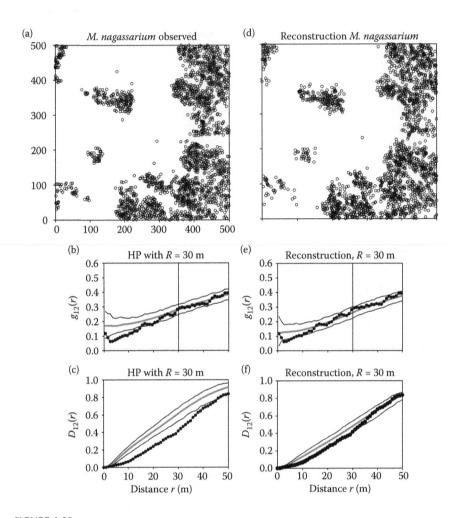

FIGURE 4.33
Comparison of results for null models based on pattern reconstruction and HPP for the pattern shown in (d). (a) Map of the observed pattern for species *M. tetrandra* (pattern 2). (b) Pair-correlation analysis of the bivariate pattern using the heterogeneous Poisson null model with nonparametric estimate of the intensity function with a bandwidth of $R = 30$ m. (c) Distribution function of nearest-neighbor distances as in panel (b). (d) An example of the reconstruction of the pattern shown in (a) based on the nonparametric estimation of the intensity function with bandwidth of $R = 30$ m. (e) Same as panel (b), but for the corresponding pattern reconstruction null model. (f) Same as panel (c), but for the corresponding pattern reconstruction null model.

S. trapezifolia than in the case of the more abrupt boundaries of the observed *M. nagassarium* pattern. This example illustrates the potential complexity of bivariate relationships quite well and emphasizes the need to employ the appropriate methods and null models in developing an understanding of the spatial structure of bivariate patterns.

4.3 Analysis of Multivariate Patterns

Classical methods of point-pattern analysis can only deal with univariate and bivariate patterns (and qualitatively marked patterns that have the same structure as bivariate patterns). However, in many applications in ecology more than two types of objects are of interest. For example, ecological communities usually contain more than two species and species-rich tropical forests may contain hundreds of species within one fully mapped 25 ha plot. Multivariate point patterns are composed of several univariate patterns that were created *a priori* by a different set of processes. Thus, a multivariate pattern comprises many different types of points. While bivariate analysis aims to characterize the spatial relationship between two types of points, *multivariate analysis is concerned with detecting nonrandom spatial structures in diversity*. As an extension of the bivariate case, we may be interested in how inter- and intraspecific interactions shape the observed spatial community pattern and if this pattern was influenced by environmental heterogeneity.

It is a largely unresolved question as to how we should characterize the spatial structure of multivariate point patterns. One of the principal problems in conducting multivariate analysis is grounded in *a lack of truly multivariate point pattern summary statistics* (see below) that summarize the spatial structure of multivariate point-pattern data in a simple and ecologically meaningful way. Clearly, we can use simple, straightforward, point-pattern extensions of the classical diversity indices, such as the Simpson index, the species richness index, or the Shannon index. However, multivariate point patterns are complex and each of these indices will only characterize a specific feature of a multivariate point pattern. An additional difficulty in multivariate point patterns is in formulating the fundamental division and suitable null models. Depending on the question and the summary statistic used, we can extend the fundamental division for bivariate patterns in two ways resulting in two forms of the fundamental division.

First, if our analysis aims to *reveal how an individual focal species is embedded within a multispecies community*, we can use multivariate summary statistics that view the community from the viewpoint of the focal species and quantify the diversity in the neighborhood of the individuals from that perspective. In this case, we ask if the characteristics of the diversity in the neighborhood of the individuals of a focal species differ from those encountered at random locations within the observation window. Similar to uni- and bivariate analyses, differences can arise from a variety of mechanisms. For example, the focal species may show an association with a species-rich habitat. In this case, the species richness in the neighborhood of the species will be greater than that expected at random locations. However, if the focal species is locally dominant and/or shows mostly negative interactions with conspecifics, it may have fewer species in its neighborhood than expected. Because the interest here is in an exploration of the placement of individuals of the focal species relative to the other species in the community, the fundamental division should condition on the patterns of all other species and only randomize the pattern of the focal species. Thus, the fundamental division in this case is the *independence of the focal species from all other species* and corresponds to that of the bivariate case, because only the focal species is randomized.

Second, if our interest is in revealing the *degree to which interspecific interactions shape the spatial diversity patterns within a community*, we need to generalize the fundamental division of bivariate independence to multivariate independence and generate *null communities* where species are pairwise independent from each other. We also need to insure that the univariate spatial structures are conserved. This null community approach is of special interest for questions of community assembly. Although theoretical studies have outlined the important role of species interactions (Chesson 2000b), several theories concerning the stochastic geometry of biodiversity, which assumes no species interactions, have been remarkably successful in predicting diversity patterns (e.g., Hubbell 2001; Morlon et al. 2008; McGill 2010). The *fundamental division of multivariate independence* can therefore be used to test the relative importance of species interactions in generating spatial patterns in diversity.

However, the null community approach can also be used to *reveal mechanisms underlying community assembly*. To this end, several null communities are formulated that work as classical null models. In each case, some aspects of the multivariate pattern are constrained, but others are randomized. For example, in the simplest form of null community, all species patterns are CSR and independent of each other. Such null communities assume lack of (i) univariate spatial structure, (ii) habitat association, and (iii) species interaction. Several stochastic realizations of the null communities, corresponding to the hypotheses on community assembly being considered, are generated and suitable multivariate summary statistics are estimated for the observed multivariate pattern and the multivariate patterns of the null communities. By progressively relaxing assumptions (i) and (ii), the relative importance of different mechanisms of community assembly can be studied (Shen et al. 2009). Note that relaxing (iii) and explicitly modeling species interactions is a very difficult task (Section 4.3.1.3).

In the following sections we discuss several approaches for summarizing and analyzing multivariate spatial patterns in ecology. For example, we present point-pattern extensions of classical diversity indices, we present the two fundamental divisions for multivariate patterns in more detail, and we discus several additional null models. All of these approaches involve either the application of bivariate analyses to all species pairs in a community or application of multivariate summary statistics.

4.3.1 Pairwise Analyses

A simple possibility for analyzing multivariate patterns is to look at pairs of component patterns and conduct a conventional bivariate analysis for each pair. For example, one may characterize the spatial relationship between different size classes of a given species by analyzing the pairwise relationships between the classes in a progressive sequence, for example, recruit–seedling, seedling–sapling, sapling–intermediate, and intermediate—adult (e.g., Wiegand et al. 2007c). Another important application of multivariate analyses in ecology is in the study of multispecies plant communities. Again, one may look at pairs of component patterns and conduct for each pair of species a conventional bivariate analysis. However, this becomes a quite laborious task for species-rich communities such as tropical forests, where the number of pairs can quickly exceed several thousand. Nearest-neighbor analysis of pairs of species in species-rich systems have been conducted by Lieberman and Lieberman (2007) for a tropical rainforest site in La Selva, Costa Rica, and Perry et al. (2009) in shrublands of the Eneabba sandplain, Australia. Point-pattern analyses with more complex null models, involving several hundred or thousand species pairs have been conducted by Wiegand et al. (2007a, 2012), and Wang et al. (2010) for the species-rich tropical forests in Sinharaja, BCI (Box 2.6) and the temperate Changbaishan (CBS) forest in China, respectively.

A potential problem with the pairwise approach is that species-rich communities are usually characterized by many species of low abundance. However, estimating pair-correlation functions or K-functions requires a relatively large sample size (such as say >70 per species; Wiegand et al. 2007a), which means that the analysis is limited to a consideration of the spatial patterns of the more abundant species and the results will be dominated by the behavior of those most abundant species. While it is technically not a problem to conduct thousands of analyses, it is somewhat more complicated to summarize the results of the numerous simulations involved. Another restriction of this approach is that only the pairwise effects can be detected; there is no possibility of incorporating the effects of the remaining species on the association of the species pairs under consideration. For example, we may imagine situations where the spatial patterns of two species are strongly influenced by a third species, but without information on the influence of the third species our ability to form inferences may be limited (e.g., Podani and Czárán 1997; Volkov et al. 2009).

4.3.1.1 Simple Methods for Summarizing Pairwise Analyses

One aspect of a multispecies community we can explore, using pairwise analysis, is a determination of how many species pairs show positive or negative patterns of association at different spatial scales. We can address this by determining the number of species pairs for which the bivariate pair-correlation function (or any other appropriate summary statistic) lies above or below the simulation envelopes at a given spatial scale r (e.g., Wiegand et al. 2007a, 2012; Wang et al. 2010). However, one problem with this approach is that we may overestimate significant effects, because of simultaneous inference (i.e., type I error, Loosmore and Ford 2006), if we simply tally up the number of species pairs with statistics lying outside of the simulation envelopes at each distance *r*. To avoid this problem, we might first conduct a GoF test for each species pair to determine whether the overall departure from the null model is significant for the distance intervals of interest. We then count departures from the null model as significant only for species pairs exhibiting a significant GoF test. In other words, we would consider species pairs having nonsignificant GoF tests as exhibiting independence at all scales.

Figure 4.34 shows the effects of removing cases from consideration that do not pass the GoF test, in assessing the significance of departures from the null model. The recruit community in the BCI forest (Box 2.6) is used as an example. This example was already analyzed in detail in Section 4.2.2.4 and Figure 4.28. We illustrate the differences arising in using the CSR null model (Figure 4.34a) versus the independence null model (Figure 4.34b). In both cases, species pairs were categorized according to co-distribution type for scales from 1 to 50 m and significance was assessed for the two summary statistics using the GoF test employing a 2.5% error rate. This yields an approximate 5% error rate overall, since we are testing two summary statistics simultaneously. Figure 4.34a,b provide a comparison of the categorization using the GoF test to one where the GoF was not used. This demonstrates the effects of over interpretation when assessing multiple scales within a single analysis, without accounting for multiple comparisons.

In examining the results for the CSR null model, we see only minor differences between the analysis that incorporates the GoF test and the one that does not (Figure 4.34a). Without the GoF test, there were somewhat fewer cases categorized as having no association type, with only a few more cases

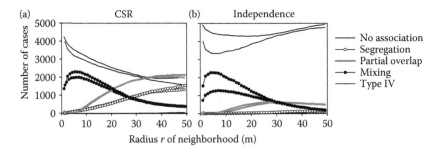

FIGURE 4.34
Effect of GoF test on interpretation of the results of the classification scheme for pairwise, species associations of the recruit community at BCI. (a) Dependence of classification into co-distribution types, with respect to the CSR null model, as a function of neighborhood radius *r*. Each curve is shown twice, once without taking the GoF test into account (upper curves for segregation, partial overlap, and mixing and lower curve for no association) and once incorporating information from GoF tests were counted. (b) Same as panel (a), but for the pattern reconstruction implementation of the independence null model.

being categorized as the mixing type. For the independence null model, when the GoF test was not used, we find basically the same proportion of species pairs categorized in the mixing type, as with the CSR model. In contrast, eliminating species pairs with nonsignificant GoF tests had a much greater effect, reducing the number of species pairs with mixing by almost half (Figure 4.34b). The effect of eliminating nonsignificant GoF tests for the independence model was negligible for the other co-distribution types. One reason for this may be that we conducted the GoF test for the interval of $r = 1$–50 m and mixing generally occurred within the smaller neighborhoods of this range. In contrast, segregation and partial overlap occurred within the larger neighborhoods. As a consequence, a significant effect within smaller neighborhoods could be overpowered by selecting too wide an interval in the GoF test.

The results in Figure 4.34a,b indicate that the signal of mixing detected by the CSR null model was also detected by the test for independence, but in the latter case the signal was somewhat weaker. However, there is a very noticeable difference between the CSR and independence null models in the consistent reduction of the number of cases categorized as segregation and partial overlap by the independence model. This is probably a result of heterogeneity and scale. Because the CSR null model produces homogeneous patterns, it cannot produce bivariate patterns with the appearance of larger-scale partial overlap or segregation and will more frequently indicate departures from the null model. However, the pattern reconstruction null model can produce heterogeneous patterns and will, therefore, agree with the null model more frequently than with CSR. However, the effects of (smaller-scale) mixing are probably caused by species interactions (or shared biotic habitat features such as canopy gaps) and are detected by both null models in a similar way.

4.3.1.2 Exploring the Interaction Structure of a Community

An interesting question to consider in the pairwise analysis of communities is whether or not there are key species that exhibit a higher (or lower) number of significant associations with other species than expected by chance (Wang et al. 2010; Wiegand et al. 2012). We can consider this question using a matrix-based approach, analyzing each of the $n \times (n - 1)$ possible bivariate patterns in a pairwise analysis of n species. This yields an $n \times n$ interaction matrix $M(k, l)$ containing values of 1 for species pairs (k, l) where the GoF test indicates a significant, positive departure from the null model, values of −1 for significantly negative departures, and values of 0 for cases where the GoF test does not detect a significant departure from the null model. The diagonal (i.e., $k = l$) of the matrix, which corresponds to univariate analyses, is excluded from consideration. Once the matrix is constructed, a randomization approach can be used to explore whether the significant interactions found among species pairs are randomly distributed among species, or

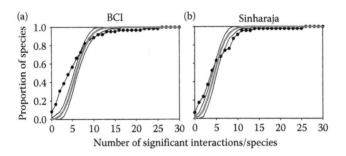

FIGURE 4.35

Exploring whether the interaction structure in communities of large individuals (dbh > 10 cm) is nonrandom. (a) The proportion $S(x)$ of species at the BCI forest (1995 census), which have x or fewer significant interactions with other species, including simulation envelopes for a null model where the observed significant interactions were distributed randomly (without replacement) over all species pairs in a given forest. (b) Same for the Sinharaja forest. (Adapted from Wiegand et al. 2012. *Proceedings B* 279: 3312–3320.)

whether some key species have a greater or lesser number of interactions with other species than expected by chance (Wiegand et al. 2012). The cumulative distribution function $S(x)$ for the $M(k, l)$ matrix can be used as a summary statistic for this purpose.

$S(x)$ gives the proportion of species that have fewer than x significant associations with other species. While $S(x)$ looks only at significant/nonsignificant associations, we can also use the two summary statistics $S^+(x)$ and $S^-(x)$, with $S^+(x) + S^-(x) = S(x)$, which specifically explore the occurrence of positive or negative associations. To assess whether $S(x)$ values are significant, a randomization test is performed. In doing this, the values of the matrix, excluding the diagonal, are randomly shuffled within the matrix and the $S(x)$ values are recalculated. The randomization procedure is repeated 199 times, assembling simulation envelopes from the results. These are then used to assess the significance of the observed $S(x)$ values.

In applying this approach for large trees (i.e., dbh > 10 cm) in the BCI forest (Box 2.6), Wiegand et al. (2012) found that small-scale species interactions were not randomly distributed among species (Figure 4.35a). There were some species with considerably more interactions than expected by chance and many species with fewer interactions than expected. Similar results were found for the community of large trees at the Sinharaja forest in Sri Lanka (Figure 4.35b).

4.3.1.3 Point-Process Modeling to Assess Species Interactions

In the previous section, we presented a simple way to assess whether or not there are species that exhibit an especially high or low number of significant associations with other species. Because this assessment was based on pairwise analyses, it cannot account for the effects of other species on the association of the focal pair of species being analyzed. There is clearly a need for more sophisticated methods that would allow for quantifying the inter-action structure among species co-occurring in a community. One promis-ing way for elucidating species interactions in more detail is the use of the recently developed approach of point-process modeling, such as the multi-variate Gibbs processes (e.g., Grabarnik and Särkkä 2009; Illian et al. 2009; Section 4.2.3.4). Such models assume parametric interaction structures and the fitted parameters contain information on the strength and direction of the interactions. Point-process models also allow adjustments for the effects of the remaining species on the association of species pairs (Illian et al. 2009). However, point-process models that consider all pairwise interactions quickly become intractable for species-rich communities. In some cases, sim-plifying assumptions can be made on the interaction structure to keep the method tractable. For example, Grabarnik and Särkkä (2009) and Illian et al. (2009) used hierarchical models to simplify the interaction structure based on biological arguments (e.g., small trees do not influence the pattern of large trees). This is a promising field for future research.

4.3.1.4 Maximum Entropy to Assess Species Interactions

An interesting approach for revealing species interactions (that however does not fully lie within the realm of point-pattern analysis) was proposed by Volkov et al. (2009). They divided the BCI tree community (Box 2.6) into equally sized 20×20 m^2 quadrats and reduced the detailed spatial informa-tion of the multivariate pattern into an abundance distribution $n = (n_1, n_2, n_3, n_4,...)$ of species within quadrats. The n_i are the abundances of species i in a quadrat. Next, they defined $P(n_1, n_2, n_3, n_4, ...)$ as the probability that a quadrat from the BCI plot contains the abundance distribution $n = (n_1, n_2, n_3, n_4,...)$. The size of the quadrat should be selected in such a way that it is large enough to contain a sufficient number of individuals, but small enough that there are sufficient quadrats for statistical averaging.

The underlying idea of the analysis of Volkov et al. (2009) is to maximize the entropy

$$H(P) = -\sum_n P(n)\ln(P(n)) \tag{4.50}$$

under a set of constraints given by information on the system that can be expressed as the distribution $P(n)$. The entropy has to be summed over all possible states n of the system. The first constraint is given by the condi-tion $\Sigma P(n) = 1$ (i.e., P is a probability distribution), the second constraint is given by the mean abundance of species i over all quadrats, and the third constraint is given by the observed covariance structure in the abundances of all pairs of species over the quadrats. Applying the principle of maximum entropy, Volkov et al. (2009) found that the elements of the inverted covariance matrix M_{ij} may be associated with the interaction strength between species i and j. It is clear from this that the matrix M_{ij} incorporates the joint effects of direct species interactions and co-distributions mediated by the environmental association of individual species or their history.

The results of the study by Volkov et al. (2009) showed that the interspecific interaction strengths were generally weak at the BCI forest, which is in agreement with the pairwise analyses by Wiegand et al. (2007a). Maximum entropy is linked to the amount of disorder in a system and stronger interspecific interactions would introduce more order into the system. Volkov et al. (2009) concluded that "there are many more ways of arranging the individuals within an ecological community when they are weakly interacting than when they are strongly interacting, p. 13857." Thus, it would be more likely to find a weak signal of species interactions in the spatial patterns of species-rich communities than a strong signal.

4.3.2 Multivariate Summary Statistics

4.3.2.1 Mean Compositional Information

Podani and Czárán (1997) made an early attempt at developing a truly multivariate summary statistic for characterizing departures from overall randomness in multivariate point patterns. They developed a method to obtain a single measure of the deviation from randomness that would be better suited for characterizing an overall multivariate pattern than summaries of pairwise measures. Their method utilized a nested set of circular plots around each individual of a community, spanning a range of increasing neighborhood distances r. Species composition was then determined within each neighborhood and further modified to only consider presence/absence information. Thus, their method scarifies detailed information on species abundances in the neighborhood of individuals, similar to the approach used in studying individual species–area relationships (ISARs) as discussed in Section 4.3.2.5.

Using the notation of Podani and Czárán (1997), species composition within distance r of an individual j was given by $F_j = (F_{j1}, \ldots, F_{jS})$, where $F_{ji} = 1$ if species i is present in the neighborhood of individual j and $F_{ji} = 0$ otherwise. Based on the neighborhoods of all individuals of a species, they estimated the probability of occurrence $P[F_j](r)$ of a particular pattern of species composition F_j. The compositional information for individual j was then defined as $I[F_j](r) = -\ln(P[F_j](r))$. If a particular compositional configuration F_j occurs only rarely in the community, its compositional information $I[F_j](r)$ is high, whereas very common configurations yield low values of $I[F_j](r)$. Finally, the *mean compositional information* $MCI(r)$ of the entire community is the average over the compositional information of all individuals j: $MCI(r) = (1/N(r))\Sigma_j I[F_j](r)$, where $N(r)$ is the total number of individuals used in the estimation. Note that $N(r)$ depends on the neighborhood radius r, if a buffer edge correction is used. If the neighborhood around individual j is not entirely inside the observation window, the species composition may not be correctly evaluated.

Podani and Czárán (1997) compared the observed $MCI(r)$ of the community with the expectation of a community where each species is distributed randomly and independently of the other species following CSR. The expectation $E[MCI(r)]_{CSR}$ under the null model yields the Shannon entropy function, based on the occurrence probabilities of all the possible species combinations under CSR: $E[MCI(r)]_{CSR} = -\Sigma_j P[F_j](r)\ln(P[F_j](r))$, where the average is taken over all 2^S possible species composition vectors F_j. The occurrence probability $P[F_j]$ of the composition F_j of a particular species j, under CSR, can be estimated based on the probability $\exp(n_i \pi r^2)$ that the neighborhood will not contain species i, if it has n_i individuals distributed in the observation window. The difference $\Delta MCI(r)$ between the observed and expected mean compositional information under CSR is an abstract measure of the "unlikeliness" of the observed species composition compared to the null model, but does not necessarily indicate the nature of the interactions between species. To test for the significance of departures from CSR, the individuals of all species are randomly and independently displaced to a random location within the observation window. This allows the estimation of simulation envelopes for $MCI(r)$. The mean compositional information can also be estimated for individual species. In this case, only focal individuals j of a given species are used. This allows the assessment of the role of individual species within the entire community.

Negative deviations from the CSR null model correspond roughly to a multivariate case of repulsion, with fewer species and a less diverse community around individuals than expected under randomness. Conversely, a positive deviation roughly corresponds to aggregation, where a wider range of unusual species combinations appears in the pattern. However, MCI has only rarely been used, probably because of its nonstraightforward interpretation (e.g., Bosdorf and Schurr 2000).

4.3.2.2 Nearest-Neighbor Matrices

Another possibility for analyzing the full array of species associations in fully mapped species-rich communities (with S species) is to use nearest-neighbor matrices. This is an S^2 square matrix containing the number of times that each species combination forms a nearest-neighbor pair (Perry et al. 2009). Because nearest-neighbor relationships are not necessarily symmetric, both, the i–j pairs and the j–i pairs are treated separately. There are S^2 possible nearest-neighbor pairs, of which S will be conspecific. Null models based on specific assumptions may be used to test the significance of the observed elements of the matrix. Higher dimensional nearest-neighbor contingency tables (Dixon 1994) can also be used to test for species associations in fully mapped species-rich communities.

4.3.2.3 Spatially Explicit Diversity Indices

Frequently used, nonspatial, diversity indices are the species richness index, the Simpson index (Simpson 1949), and Shannon diversity (Shannon 1948). Species richness characterizes diversity based on the number of species present within a plot, whereas the Simpson and Shannon indices are based on the number of individuals per species within the plot, incorporating information on richness and the degree of species evenness. High values of evenness occur when the abundances of all species are nearly alike, whereas low values occur when a few species are dominant and most other species are rare. These indices can be generalized within the framework of point-pattern analysis to capture diversity in a scale-dependent manner (e.g., Rajala and Illian 2012). They can then be used as summary statistics for fully mapped multivariate point patterns (Shimatani 2001; Shimatani and Kubota 2004).

The *species richness index* does not use species abundance information. It only considers the number of species present within the community inside an observation window. Shimatani and Kubota (2004) proposed a spatially-explicit point-pattern version of the species richness index. The *spatial species richness* $d(r)$ is defined as the expected number of species present within distance r of an arbitrarily chosen test point located in the observation window W. Clearly, this index is closely related to the the classical species-area relationship (SAR). While the Simpson index is based on the decomposition of community pattern into the sum of all univariate mark connection functions (see below), the spatial species richness index can be calculated by decomposition into the sum of all spherical contact distributions:

$$d(r) = \sum_{m=1}^{S} H_s^m(r) \tag{4.51}$$

$H_s^m(r)$ is the spherical contact distribution for species m and yields the probability that an individual of species m is located within distance r of an arbitrarily chosen test point in W. Decomposing the spatial species richness $d(r)$ into the "detectability" of each species allows the role of each species in the overall pattern of species diversity to be examined (Shimatani and Kubota 2004).

While the spatially-explicit, species richness index is location centered (i.e., the test points are arbitrary locations within the observation window), we can also define an individual-centered analog of this index, the *average individual-species area relationship* (ISAR) defined as the expected number of species (other than the focal species) present within distance r of an arbitrary individual of the community (see Equation 3.74 in Section 3.1.5.2):

$$\overline{\text{ISAR}}(r) = \frac{1}{\lambda} \sum_{f=1}^{S} \lambda_f \sum_{m=1,\neq}^{S} D_{fm}(r) \tag{4.52}$$

where $D_{fm}(r)$ is the probability that the nearest species m neighbor of an individual of the focal species f is located within distance r, λ_f/λ is the relative abundance of species f within the community and $\lambda = \Sigma\lambda_f$. Thus, this summary statistic is a decomposition based on bivariate distribution functions with respect to the nearest neighbor. The average ISAR, therefore, quantifies the average species richness in neighborhoods of the individuals of the community.

The classical *Simpson index D* (Simpson 1949) uses information on species abundances within an observation window (but not the information on spatial locations) and is defined as the probability that a randomly selected pair of individuals within a community belong to different species. Thus, $D = 1 - \Sigma p_i^2$ where $p_i = \lambda_i/\lambda$ is the relative abundance of species i within the community and $\lambda = \Sigma\lambda_i$. Shimatani (2001) extended the nonspatial Simpson index by introducing the additional condition that the pair of individuals should be located distance r apart. Thus, the *spatially explicit Simpson index* $\beta(r)$ is the conditional probability that a randomly selected pair of individuals separated by distance r belong to different species. The spatially explicit Simpson index can be expressed by using the mark connection functions $p_{ii}(r)$ of species i (Section 3.1.5.1):

$$\beta(r) = 1 - \sum_{i=1}^{S} p_{ii}(r) \tag{4.53}$$

This is reasonable because $p_{ii}(r)$ is the probability that an individual located distance r from a species i individual is a conspecific. Equation 4.53, therefore, first estimates the probability that two individuals located distance r apart are conspecifics, sums this up for all species i in the community and then subtracts the total from one, giving the probability that two randomly selected individuals distance r apart are not the same. If all species are CSR, we find that $p_{ii}(r) = p_i^2 = (\lambda_i/\lambda)^2$ and the classical Simpson index $D = 1 - \Sigma p_i^2$ is recovered. We can also consider the function $F(r) = 1 - \beta(r)$, which characterizes distance decay in community composition, since it represents the rate at which the likelihood of encountering a conspecific declines with distance r (Chave 2002; Condit et al. 2002). The cumulative version of the $\beta(r)$ index, which expresses the probability that two individuals separated by a distance of no more than r are heterospecific, is given by Shimatani (2001):

$$\alpha(r) = 1 - \sum_{j=1}^{S} \frac{\lambda_j^2 K_{jj}(r)}{\lambda^2 K(r)} \tag{4.54}$$

where the $K_{jj}(r)$ are the partial K-functions for species j, and $K(r)$ is the K-function for the entire community.

Taking a different approach, the spatially explicit Simpson index can also be formulated from the viewpoint of individual species. In this case, we can define $\beta_f(r)$ as the conditional probability that a randomly selected individual of the community, which is at distance r from the typical individual of the focal species f, is heterospecific. We can develop this idea by noting that the index $1 - \beta_f(r)$ measures the *local dominance of the focal species* at distance r away from the typical individual of the focal species. We begin constructing the local dominance index by first calculating the neighborhood density $O_{ff}(r) = \lambda_f g_{ff}(r)$ of conspecifics at distance r for the focal species f. Next, we sum up all the bivariate neighborhood densities $\Sigma_j O_{fj}(r) = \Sigma_j \lambda_j g_{fj}(r)$, which gives us the mean density of individuals of all heterospecific species at distance r centered on individuals of the focal species. Because the neighborhood density multiplied by $2\pi r dr$ yields the average number of individuals lying in a ring of radius r and width dr, we find that $1 - \beta_f(r) = O_{ff}(r)/\Sigma_j O_{fj}(r) = \lambda_f g_{ff}(r)/\Sigma_j \lambda_j g_{fj}(r)$. If all species are independent of the focal species [i.e., $g_{fj}(r) = 1$] and the focal species follows CSR [i.e., $g_{ff}(r) = 1$], we find that $\beta_f(r) = 1 - \lambda_f/\Sigma_j \lambda_j = 1 - \lambda_f/\lambda$ with $\lambda = \Sigma_j \lambda_j$. Thus, the nonspatial expectation of the individual Simpson index yields $\beta_f = 1 - \lambda_f/\lambda$. The cumulative, individual spatially-explicit Simpson index $\alpha_f(r)$ can also be used as a measure of the *local dominance of the focal species*, since $L_f(r) = 1 - \alpha_f(r)$ yields the mean proportion of conspecific neighbors within a neighborhood of radius r around the individuals of the focal species (Wiegand et al. 2007b).

Finally, the *Shannon diversity index* is defined as $H = -\Sigma p_i \ln(p_i)$, where $p_i = \lambda_i/\lambda$ is the relative abundance of species i within the community. A point-pattern extension of this index requires the conditional probability $p_i(r)$ that a point at distance r of an arbitrary individual of the community belongs to species i:

$$p_i(r) = \sum_{j=1}^{S} p_{ij}(r) \tag{4.55}$$

If all species are pairwise independent we find that $p_{ij} = \lambda_i \lambda_j/\lambda^2$, as expected, and $\Sigma_j p_{ij} = \lambda_i/\lambda$. The spatially explicit Shannon diversity index would yield

$$H(r) = -\Sigma p_i(r) \ln(p_i(r)) \tag{4.56}$$

It should be noted that the low abundance of rare species in small neighborhoods will lead to estimation problems. To our knowledge the Shannon index has not been used in a spatially-explicit point-pattern context (but see Reardon and O'Sullivan 2004 for its use in a sociological context). The spatially explicit point-pattern extensions of the classical diversity indices provide basic information on multispecies spatial patterns and spatial variation of species diversity. They consider all pairs of points at a given distance and measure the overall local diversity within a distance r in a point centric way. They also characterize the overall ecological dissimilarity at distance r. In Section 3.1.5 we provided examples for the Simpson index.

4.3.2.4 Species Mingling Index

The *mingling index* $\overline{M}_k$ is the k-nearest-neighbor analog to the cumulative Simpson index $\alpha(r)$ and is defined as the mean fraction among the k-nearest neighbors of the typical point that belong to a different species than the typical point. Thus, while the cumulative Simpson index $\alpha(r)$ evaluates all individuals of the community within distance r of the typical point and determines the proportion of heterospecifics, the mingling index uses the neighbor rank k as a distance measure and evaluates the first k neighbors of the typical point. To determine the mingling index, we first estimate for each individual x_n of the multivariate pattern the constructed mark $\overline{M}_k(x_n)$, which is the fraction of the k-nearest neighbors of x_n that belong to a different species than individual x_n (Illian et al. 2008, p. 314). In a second step, the average over all constructed marks of the points x_n is taken:

$$\overline{M}_k = \frac{1}{N} \sum_{i=1}^{N} \overline{M}_k(x_i) \tag{4.57}$$

An alternative method is to first estimate for each individual x_n the number of neighbors among the k-nearest neighbors that belong to a different species and estimate the corresponding *"mingling distribution"* $M_{i:k}$. This yields the probability that the typical point has i heterospecific neighbors among its k-nearest neighbors. The mingling index is then the mean of the mingling distribution (Illian et al. 2008, p. 321):

$$\overline{M}_k = \frac{1}{k} \sum_{i=0}^{k} i M_{i:k} \tag{4.58}$$

If the neighborhood rank k is interpreted as a distance measure, the mingling index can be plotted over neighborhood rank and an index similar to the cumulative Simpson index is obtained. For neighborhood rank $k = 1$ and a null model of independence among all species, we can estimate the expectation of the mingling index. The probability that the focal point is of species s yields λ_s/λ and the probability that the nearest neighbor is not of species s yields $(\lambda - \lambda_s)/\lambda$, thus the expected mingling index for complete independence among species yields the classical Simpson index:

$$\overline{M} = \sum_{s=1}^{S} \frac{\lambda_s}{\lambda} \frac{\lambda - \lambda_s}{\lambda} = 1 - \sum_{s=1}^{S} \frac{\lambda_s}{\lambda} \frac{\lambda_s}{\lambda} = D \tag{4.59}$$

when considering that the sum of all λ_i yields λ.

Pommerening et al. (2011) introduced a *mark mingling* function $v(r)$, which extends the idea of the mingling index by incorporating mark-correlation functions in the calculation. The test function $t(m_i, m_j)$ estimated for the two points x_i and x_j, with marks m_i and m_j, is given by

$$t_v(m_i, m_j) = \mathbf{1}(m_i \neq m_j) \tag{4.60}$$

The mark mingling function $v(r)$ should be closely related with the spatially explicit Simpson index (Equation 3.70), if the marks represent species.

4.3.2.5 Individual Species Area Relationships

Wiegand et al. (2007b) derived a truly multivariate summary statistic in an ecological context that generalizes the popular K-function for multivariate point patterns. While the bivariate K-function is based on the expected number of type 2 points within neighborhoods with radius r around a typical type 1 point, the analogous multivariate summary statistic determines the expected number of species within neighborhoods with radius r around a typical point of the focal pattern. Thus, similar to the mean compositional information statistic MCI, abundance information of different species within the neighborhoods of individual points is reduced to presence/absence information. Because of the analogy to SAR, this summary statistics is called the "individual-species area relationship" (ISAR).

The ISAR for a particular species is estimated by calculating the mean number of species within circular areas of radius r around individuals of the focal species f. This is done through the bivariate cumulative distribution functions $D_{fj}(r)$ of the distances from focal individuals to the nearest neighbor of species j, calculated for all species $j \neq f$. The $D_{fj}(r)$ are then summed up for all species $j \neq f$ present in the plot:

$$\text{ISAR}_f(r) = \sum_{j=1, j \neq f}^{S} D_{fj}(r) \tag{4.61}$$

In this case, S is the total number of species in the plot. Using $a = \pi r^2$ one can express the ISAR also in units of area to best resemble the common SAR.

ISAR is a monotonically increasing function of neighborhood radius r, with a maximum value of $S - 1$. For small distances r, it captures the species richness of the first few neighbors. Because the ISAR function contains explicit information about the spacing of the individuals of all species j relative to the position of the focal species f, it should reveal, similar to the K-function at the uni- or bivariate level, small-scale species interactions at the community level (Wiegand et al. 2007b). This is because the ISAR function provides a high-level summary of the net effect of the interactions of all $S - 1$ possible pairings of focal species f. However, at distances outside the range of direct species interactions (say neighborhoods of $r > 20$–30 m for trees), the ISAR should predominantly capture the effects of environmental heterogeneity, if present, where some areas of the plot have a higher (or lower) overall diversity than expected.

Equation 4.61 is one possibility for estimating the ISAR function. However, there is an alternative way of deriving the ISAR function that is closely related with the *landscape of local species richness*. We define the landscape of local species richness $LSR_f(r, x)$, for a given neighborhood radius r for each location x within the observation window, as the number of species within a circle with radius r centered on x (e.g., Figure 3.23). The $ISAR_f(r)$ is the mean value of $LSR_f(r, x)$ averaged over the locations x_{fi} of individuals i of focal species f. Thus, what the ISAR function is basically doing for a given neighborhood radius r is to explore whether the locations x_{fi} of individuals i of focal species f are positively or negatively associated with the "covariate" $LSR_f(r, x)$. Or in other words, we conduct a Berman test (Section 2.6.2.8) in which the test statistic is based on the average of the covariate values $LSR_f(r, x)$ at the points x_{fi} of the pattern.

The ISAR function can be used in different ways to explore how the individuals of a focal species are located within the "landscape" of local species richness and if they influence local species richness in their neighborhood. How this can be accomplished is explained in the following sections presenting null models for the ISAR function.

4.3.3 Null Models for "Individual" Multivariate Summary Statistics

In the following, we discuss several null models that can be used in conjunction with the ISAR (or other "individual" summary statistics) to explore how the individuals of a focal species are located within the "landscape" of local species richness. This approach can be used to determine whether a species is surrounded by more or less species than expected by a given null model. Because individual summary statistics, such as the ISAR function or the individual Simpson index $\beta_f(r)$, view the community from the perspective of the individuals of the focal species, inference is conditioned on the spatial pattern of all the other species in the community. Only the locations of the individuals of the focal species are randomized in the null model. Thus, we test for *independence of the focal species from the patterns of all other species*. This approach presents one possible generalization of the bivariate null model of independence, where only the focal species is randomized. For a given neighborhood radius r, we compare $ISAR_f(r)$, which is the same as the mean value of the landscape of local species richness $LSR_f(r, x)$ over observed locations x_{fi} of the focal species f, to the mean value of the locations generated by a null model. In this way, we test for positive or negative associations of the focal species with the landscape of local species richness under a given null model. The null models used to randomize the locations of the focal species are the same as for the bivariate case, but the summary statistic is different.

4.3.3.1 Focal Species CSR

Use of the CSR null model allows us to determine *whether a focal species is located in areas of relatively low or high species richness* given neighborhoods of different size. This can be tested by contrasting the observed ISAR curve to that of a null model that repeatedly relocates the individuals of the focal species to random locations within the entire plot (i.e., CSR; Wiegand et al. 2007b). This null model is the point-pattern analog of the common SAR (Shimantani and Kubota 2004) and links the ISAR concept with the SAR concept. If a focal species is mostly located in areas of low local species richness, the test with the CSR null model will yield a negative departure from the simulation envelopes. Conversely, if a focal species is mostly located in areas of high local species richness, the test with the CSR null model will yield a positive departure from the simulation envelopes.

Departures from the SAR null model may be produced by several mechanisms. First, a species may have a tendency toward strong clustering and high local dominance. As a consequence, fewer individuals of other species will be found in the neighborhood of this species and it is likely that the local species richness around this species will also be low. Second, a focal species may show an association to a habitat type that hosts more species and, consequently, the species richness in the neighborhood of the species will be higher than at random positions within the observation window (i.e., the SAR null model). The effect of a heterogeneous environment should persist also to larger neighborhoods *r*. Third, the density of individuals may vary over the observation window. For example, more frequent disturbances or a certain soil type may favor higher densities, but smaller individuals, within some areas of the plot. Because more individuals may also mean more species, areas of higher density may produce positive effects in the ISAR function. Finally, net positive or negative species interactions of the focal species with other species may yield positive or negative departures in the ISAR values as compared to the SAR null model. However, we expect such departures to occur only within relatively small neighborhoods *r*. It should be noted that the overall pattern may be overpowered by environmental effects. We need to factor out, to some extent, the effects of heterogeneity in the distributional pattern of the focal species to reveal effects of species interaction. This is explained in the next section.

4.3.3.2 Focal Species Heterogeneous Poisson

A subtle assessment of the *cumulative effects of the interactions of a focal species with all other species* can be provided by the ISAR statistic. We do this by determining the average observed species richness surrounding a focal species, which can be used to categorize nonrandom relationships into two basic categories. *"Diversity accumulators"* represent situations where positive, facilitative interactions with other species predominate. In this case, the focal species is surrounded by an overrepresentative amount of diversity. *"Diversity repellers"* represent situations where negative interactions predominate, and will occur for species that are surrounded by an underrepresentative proportion of species within their neighborhood. In contrast to these two cases, species richness in the neighborhood of the focal species would not differ from the expectation of no interactions, if positive and negative interactions are weak or cancel out. One complication in the assessment of these relationships is that a species may appear to be a diversity accumulator or repeller due to habitat association, rather than species interactions, if species richness differs within different habitats. In this case, the focal species will depart from the expectations under the SAR null model. This is analogous to the problem of heterogeneous patterns in bivariate point-pattern analysis. Specific methods are required to factor out the effects of habitat association to reveal the more subtle effects of species interactions.

One may use the methods already developed for bivariate patterns to account for the effects of habitat association and use a heterogeneous Poisson null model to randomize the focal species in accordance with the patterns of its intensity. The result is a habitat model for the target species that may be used for this purpose. A nonparametric kernel intensity estimate of the intensity function may also be used. However, note that using a habitat model to characterize the heterogeneity of the focal species will only capture the effects of habitat association, but cannot account for effects of dispersal limitation, which may also generate heterogeneous patterns. The HPP based on the parametric intensity estimate basically displaces each individual within a certain neighborhood, defined by the bandwidth R of a kernel function. The latter approach is suitable if "separation of scales" is likely. This will be the case when direct interactions among individuals operate at shorter scales than the scales of environmental heterogeneity. This is a likely situation when environmental heterogeneity is related to broad topographic features. When environmental heterogeneity occurs at scales similar to the scales of species interactions, such as would be the case with small-scale heterogeneity in edaphic factors, separation of scales will not be possible.

4.3.3.3 Focal Species Pattern Reconstruction

We may need to condition on the univariate pattern of the focal species in studying the relationship between a focal species and its surrounding community. Under these circumstances, we can use a pattern reconstruction approach, with or without a heterogeneous intensity function. This is similar to the approach used in testing for independence in bivariate patterns and may be essential, since certain spatial configurations may be unlikely under CSR models. As noted before, this is especially true for cases in which there is strong clustering by the focal species. Under these circumstances, realizations of the independence null model may frequently produce spatial patterns where clusters of individuals are mostly located in areas with local species richness below (or above) the values expected. In contrast, randomization with CSR will produce patterns where individuals are frequently located in areas below or above the expectation for local species richness, since local heterogeneity is not incorporated into the model. As a consequence, the CSR null model will yield simulation envelopes for the ISAR function that are too narrow, as it will underestimate the range of spatial configurations that may potentially occur using the independence model, as we have already seen in the bivariate case. The same situation may apply for the heterogeneous Poisson null model, but only if there are strong effects of small-scale clustering present that are not captured well by a heterogeneous Poisson null model that has been conditioned on the observed intensity of the focal species at a broader scale.

4.3.3.4 Examples of the Use of ISAR Analysis

Figure 4.36 provides an example of the application of the individual species–area relationship ISAR(r). We considered all large individuals (i.e., dbh > 10 cm) of the two canopy species *Trichilia tuberculata* and *Virola surinamensis* found in the BCI forest (Box 2.6). Our first goal was aimed at revealing how the individuals of the two species were located within the landscape of local species richness (Figure 3.23). To this end, we used a CSR null model to randomly displace the individuals of the focal species over the entire observation window. This null model produces the point-pattern equivalent of the classical SAR (Shimantani and Kubota 2004). Analysis with the ISAR function clearly shows that *T. tuberculata* (Figure 4.36a) is located within areas of lower than expected local species richness (Figure 4.36b). For example, within a radius of 20 m there are on average two species less than expected by the SAR. To better understand this result, we also estimated the average number of heterospecific individuals within neighborhoods of radius r around the individuals of the focal species *T. tuberculata,* using both the observed data and the CSR simulation data (Figure 4.36c). The reduced values for local species richness correlate very well with the reduced number of heterospecific individuals in the neighborhood of *T. tuberculata* individuals (cf. Figure 4.36b,c). For example, we find, on average, four fewer heterospecific individuals within 20 m of a *T. tuberculata* plant in the observed pattern than we do for random locations within the observation window. The expected number of heterospecific neighbors within this neighborhood for the null model was 47.6, whereas the observed number was 44.8. There were 6.1 conspecifics within this neighborhood, yielding a local dominance index of 6.1/51 = 0.12.

We obtain a different result when we repeat the same analyses for *V. surinamensis* (Figure 4.36d). Similar to the pattern observed for *T. tuberculata*, *V. surinamensis* exhibits a negative departure of the ISAR function from the null model within smaller neighborhoods, indicating that it is surrounded locally

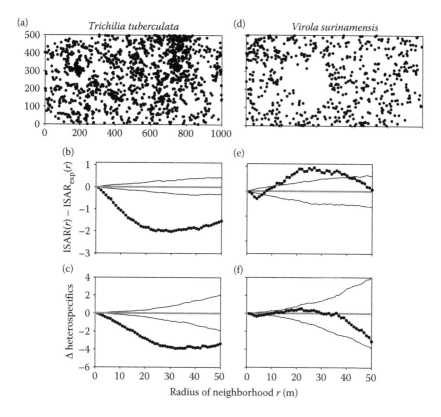

FIGURE 4.36

Examples of ISAR. (a) Map of all large individuals (dbh > 10 cm) of the species *Trichilia tubercu-lata* from the 2005 BCI census. (b) Results of the ISAR analysis using the CSR null model that provides the expectation of the SAR. The species shows negative departures from the SAR and is mostly located in areas of below-expected local species richness. (c) Same as panel (b), but for the number of heterospecific neighbors within distance *r* of the individuals of *T. tubercu-lata*, indicating that it is surrounded by a below-expected number of heterospecific neighbors. (d)–(f) Same as panels (a) through (c), but for the species *Virola surinamensis*.

by fewer species than expected for the SAR. In contrast, within broader scale neighborhoods of 10–40 m, *V. surinamensis* is located in areas containing more than expected local species richness. The analysis of the heterospecific neighbors shows that the negative, small-scale departure of the ISAR function is a consequence of fewer than expected heterospecifics, but the association with areas of higher species richness is not a consequence of a higher number of heterospecific neighbors (Figure 4.36f). Obviously, *V. surinamensis* is located in areas that show slightly higher species richness, but not higher tree densities.

To obtain an overview of the collective behavior of species at the BCI forest we applied the same analysis for the 49 most abundant species at the site, that is, those species that consisted of more than 90 individuals with a dbh greater than 10 cm. As a simple summary, we tallied up the number of species that were located in areas of below or above average species richness within a range of neighborhoods sizes (Wiegand et al. 2007b). Within a neighborhood radius of 4 m, almost all species exhibited a negative departure from the SAR null model when assessed by the ISAR function (Figure 4.37a). The number of species with negative departures from the expectation declined to 15 for neighborhoods of 20 m and larger (Figure 4.37a). At shorter distances (i.e., below 10 m) there were almost no positive departures from the null model, but for neighborhoods greater than 20 m there were approximately 15 species (range 10–17) exhibiting positive departures from the null model. Thus, approximately the same number of species were in areas of lower versus higher local species richness for neighborhoods greater than 20 m. Because almost all species showed negative departures for small neighborhoods r, some species must exhibit negative departures for small neighborhoods and positive departures for larger neighborhoods. For example, the species *V. surinamensis* exhibited negative departures within small neighborhoods, but positive departures within larger ones (Figure 4.36e).

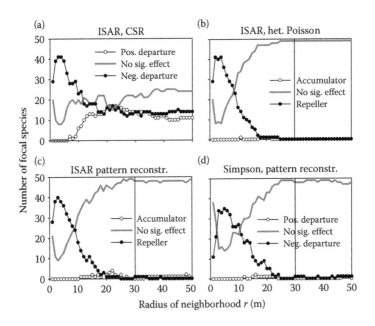

FIGURE 4.37
Summary of community wide analysis of the ISAR for large trees at the BCI forest. (a) CSR null model showing the number of species with positive (open disks) and negative (closed disks) departures from the null model as indicated by the ISAR function. (b) Same as panel (a), but for the heterogeneous Poisson null model with nonparametric intensity estimate with a bandwidth of $R = 30$ m. (c) Same as panel (b), but for the corresponding pattern reconstruction null model. (d) Same as panel (c), but for departures from the cumulative, individual, spatially explicit Simpson index $\alpha(r)$.

For the next step in the analysis of these patterns, we wanted to remove the effects of heterogeneity in the patterns of the focal species and reveal the potential effects of species interactions. To this end, we used the heterogeneous Poisson null model with a parametric intensity estimate. This null model randomly displaced individuals of the focal species within neighborhoods of $R = 30$ m. Because individuals are only relocated locally by this model, we can detect the smaller-scale effects of species interactions, while controlling for broader scale heterogeneity. The results of this analysis show that the peak in negative associations remains and that the larger-scale effects completely disappear (Figure 4.37b). The abrupt disappearance of the larger-scale effects at neighborhoods of 18 m indicates that we have been successful in separating scales, since the null model should allow us to examine effects up to 30 m. Thus, we find that large trees within the BCI forest show only negative small-scale interactions and do not interact at broader scales.

Although the heterogeneous Poisson null model controls for the observed broader scale heterogeneity of the focal species, it does not conserve the characteristics of the small-scale patterning (such as aggregation) and is, therefore, not suitable for testing independence under heterogeneity. Therefore, for this purpose we use a pattern reconstruction null model that incorporates the same intensity estimate as the heterogeneous Poisson null model. This allows us to condition on both the observed intensity function of the focal species and its small-scale patterning. Before turning to the entire community, we first continue the analysis of the two example species *T. tuberculata* and *V. surinamensis*. From this we find that the species *T. tuberculata* (Figure 4.36a) can be classified as a repeller species within neighborhoods smaller than 20 m, because local species richness within this size of neighborhood is lower than expected by the null model (Figure 4.38b). We used the index of local dominance $L_f(r) = 1 - \alpha_f(r)$, which is related to the cumulative individual Simpson index $\alpha_f(r)$, to better understand why *T. tuberculata* acts as a repeller species. First, we observe a maximum in the local dominance $L_f(r)$ of 14% at a neighborhood radius r of approximately 7 m (Figure 4.38c). The steep increase in the local dominance from a distance of $r = 1$ to $r = 7$ occurs because *T. tuberculata* exhibits a pattern of clustering superimposed on a soft-core pattern at short distances. This results in a peak of the neighborhood density at distances of $r = 7$ m (Figure 4.38a). We also observe that the local dominance of this species is larger than the nonspatial expectation at all investigated neighborhood scales, since the proportion of *T. tuberculata* within the community is approximately 6.9%. Thus, at its peak at $r = 7$ m, the local dominance is double the overall expectation.

Use of the pattern reconstruction null model, with nonparametric intensity estimate at a bandwidth of $R = 30$ m, indicates that the observed local dominance $L_f(r)$ of *T. tuberculata* is greater than predicted by the pattern reconstruction null model (Figure 4.38c). This also means that the proportion of heterospecifics around *T. tuberculata* is lower than expected for

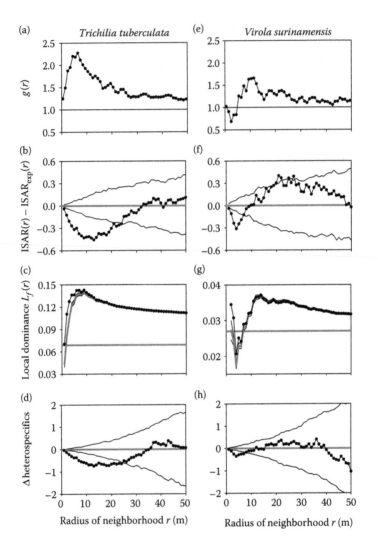

FIGURE 4.38
Various community level analyses for the species *Trichilia tuberculata* (a–d) and *Virola surinamensis* (e–h). (a) and (e) Pair-correlation function. (b) and (f) ISAR analysis given by subtracting expected values from observed values, using the pattern reconstruction null model with a nonparametric intensity estimate of bandwidth $R = 30$ m. *T. tuberculata* exibits a negative departure from the SAR and is mostly located in areas of below-expected local species richness. (c) and (g) Cumulative and local dominance index $L_f(r) = 1 - \alpha(r)$. (d) and (h) Number of heterospecific neighbors within distance r of focal individuals.

neighborhoods up to 10 m, because the pattern reconstruction null model conditions on the number of conspecific neighbors. This result indicates that the negative departure of the cumulative Simpson index from the null model is due to a lower than expected number of heterospecific individuals in the neighborhood of individuals, as shown in Figure 4.38d. Thus, heterospecific neighbors are, in general, repulsed by *T. tuberculata,* which may be a consequence of competition resulting in an overall lower number of heterospecific neighbors.

V. surinamensis (Figure 4.36d), like *T. tuberculata,* shows a repeller effect within neighborhoods up to 6 m (Figure 4.38f). This is a pattern that was already detected using the CSR null model (Figure 4.36f). Interestingly, the magnitude of the departure from the expectation using the CSR and pattern reconstruction null models is approximately the same, which indicates that the effects of larger-scale heterogeneity are minimal. We found the same result for other focal species, and the invariance in the peak of negative effects within small neighborhoods, when comparing the CSR and the heterogeneous Poisson null models (cf. Figure 4.37a,b) confirms this result. Thus, an interesting feature of the ISAR analysis is that small-scale interactions do not interact with larger-scale heterogeneity, as found in bivariate analyses using the pair-correlation or *K*-functions. A reason for this may be that the compound effects of heterogeneity revealed by the pair-correlation function depend strongly on locally elevated or reduced species densities, whereas the ISAR function is based on the nearest-neighbor distribution functions, which are much less prone to the effects of heterogeneity.

Within larger neighborhoods, *V. surinamensis* still has a weak effect on the surrounding community, as indicated by the positive departures at distances of 20–30 m seen using the pattern reconstruction null model (Figure 4.38f). Again, we find a positive departure in the local dominance index $L_f(r) = 1 - \alpha_f(r)$ within small neighborhoods (Figure 4.38g), indicating that the number of heterospecific individuals in the neighborhood of *V. surinamensis* is less than expected by the null model (Figure 4.38h). Given that *V. surinamensis* is a canopy tree, we suspect that the repeller effect of this species is a consequence of competitive interactions with neighbors, which produces a slight reduction in the number of heterospecific neighbors.

To obtain an overview of the collective interaction behavior of species at BCI, we can determine the number of significant repeller and accumulator cases within each neighborhood radius *r*. This produces almost identical results to those obtained with the heterogeneous Poisson null model (cf. Figure 4.37b,c), indicating that the detailed small-scale pattern of the focal species has basically no impact on the results of the ISAR analysis. Finally, we can also determine the number of positive and negative departures from the cumulative individual, spatially-explicit Simpson index $\alpha(r)$ for different neighborhoods *r*. This produces a very similar result to the ISAR function (cf. Figure 4.37c,d). Note that estimation of the Simpson index only makes sense if we use a pattern reconstruction null model that constrains the number of conspecific individuals within neighborhoods around the focal species. In this case, the variability of the Simpson index is only due to a different number of heterospecifics being encountered in the data as compared to the null model simulations. The peak of the negative departures of the Simpson index from the expectation (Figure 4.37d) is lower, but somewhat broader than that of the ISAR function (Figure 4.37c); otherwise the same qualitative results, as found before, indicate that the repeller effects within the BCI forest are mostly caused by the occurrence of a low number of heterospecifics within the neighborhood of the focal individuals.

4.3.4 Null Communities

In the previous Section 4.3.3, we presented null models for "individual" multivariate summary statistics, such as the ISAR or the individual Simpson index. These metrics view the community from the standpoint of focal species. The ISAR function allows us to assess how focal species are situated within a variable landscape of local species richness. In contrast, the Simpson index allows us to consider the proportion of heterospecifics in the neighborhood of a focal species and assess whether the values are lower or higher than expected using different null models. In both cases, the fundamental division for species-centered analysis is independence of the focal species from the pattern of all other species. However, we can take another approach. Multivariate patterns also allow us to consider a second form of fundamental division. In what we have considered so far, we have examined the placement of one species versus all other species whose distributions are kept fixed in space. We can also consider a second situation that aims *at identifying the underling assembly rules of the observed community*. In this instance, we need to randomize all of the species patterns following certain null model rules. We then compare the summary statistics of the observed community with those of the *null communities* assembled using the given null model (Shen et al. 2009). The corresponding fundamental division is the pairwise independence of all species.

Assembly of null communities is done by fitting a specific point-process model to the spatial pattern of all species present in the community, using rules corresponding to a given ecological assembly hypothesis (Waagepetersen 2007, Shen et al. 2009). Following the hypothesis, the point-process model keeps certain aspects of the data unchanged, but randomizes other aspects of the data. In the following sections we describe four such assembly hypotheses, the random placement hypothesis, the habitat-filtering hypothesis, the dispersal limitation hypothesis, and the combined model habitat filtering and dispersal limitation hypothesis.

The approach using null communities works as follows (Figure 4.39): First, one stochastic realization of the fitted point process is generated for each species, and the simulated patterns of all species are superimposed to yield one "null community." Because the simulated patterns of all species are independently superimposed, species interactions are not considered in the assembly of the null communities. Second, one generates a set of

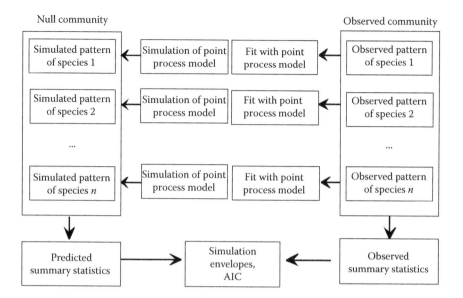

FIGURE 4.39

Assembly of null communities. First, the point-process model that represents a given hypothesis is fit to each observed species distribution pattern. Second, a stochastic realization of the fitted point process is generated. Third, the simulated patterns of all species are superimposed to yield one null community. One may generate for each hypothesis a total of 199 null communities and select from among the competing hypotheses the most parsimonious model by using Akaike's information criterion (AIC). Finally, one compares the observed and predicted summary statistics using simulation envelopes. (Adapted from Wang, X. et al. 2013. *Ecography* 36: 883–893.)

null communities, commonly 199, and calculates averages for the multivariate summary statistics to yield expectations for the hypothesis (Figure 4.39) and the corresponding simulation envelopes. Finally, this procedure is repeated for several competing hypothesis of community assembly, and the observed summary statistics are compared with the summary statistics predicted by the competing hypothesis. To evaluate the fit of a given hypothesis, we can use simulation envelopes of the summary statistics from the replicated null communities and apply a GoF test. We can then rank competing hypotheses using Akaike's information criterion (AIC) based on the sum of residuals and the number of parameters used in the different models (see Section 4.3.4.5). Applications of the null community approach can be found in Shen et al. (2009) and Wang et al. (2011, 2013).

Shen et al. (2009) used four competing models of community assembly: (i) the random placement hypothesis, where species show no habitat specificity, no mechanism of intraspecific aggregation (such as dispersal limitation), and no intraspecific interactions; (ii) the habitat filtering hypothesis, where the distribution pattern of species in the community is only determined by their observed habitat association, but there are no mechanisms of intraspecific aggregation or interspecific interactions; (iii) the dispersal limitation hypothesis, where there is only intraspecific aggregation but no habitat association or interspecific interactions; and (iv) the combined model, incorporating both the habitat filtering and dispersal limitation hypotheses. In the following, we present these four hypotheses in more detail.

4.3.4.1 Random Placement Hypothesis

The homogeneous Poisson process (CSR) is used to model the random placement hypothesis. This point-process model has no parameters to be fit, other than the number of individuals within the observational window as determined by the observed data. A completely random pattern (i.e., random placement) is generated for each species; the observed habitat associations, clustering or species interactions are not conserved. Null communities generated by the random placement hypothesis represent a complete randomization of the data and are a point of reference for evaluating more complex null models that incorporate specific, nonrandom mechanisms. Because plant species are spatially aggregated in most communities, the random placement hypothesis will generate null communities that are too well mixed. As a consequence, local species richness within a given area will usually be overestimated. Random placement communities will also show no distance decay, resulting in a Simpson index that is independent of distance.

4.3.4.2 Habitat Filtering Hypothesis

The habitat-filtering hypothesis assumes that the placement of individuals is only influenced by their species-specific, habitat suitability. Locations with higher habitat suitability will host more individuals of a given species than areas of lower habitat suitability. Given these considerations, the heterogeneous Poisson process (HPP) is used to represent the habitat-filtering hypothesis. Shen et al. (2009) used environmental variables related to topography and soil properties for parametric estimation of the intensity function $\lambda(x)$ of the HPP. The techniques they used, as presented in Waagepetersen and Guan (2009), were based on log-linear regression models that were designed to estimate the intensity function $\lambda(x)$ using environmental covariates (see Section 4.1.5.1; Equation 4.25). The resulting point-process model generates a pattern for each species, conserving the observed habitat associations, but not the observed clustering or species interactions. The HPP model is also able to indirectly generate (larger-scale) species associations, if two species share the same habitat or have opposing habitat requirements. Null communities based on the habitat-filtering hypothesis will produce distance decay, if species patterns are typically heterogeneous and are limited to restricted patches of the observation window. The strength of this effect can be measured with the Simpson index $\beta(r)$, which should increase in magnitude with increasing distance. However, if the species in the community show strong patterns of clustering, independent of habitat, the habitat filtering null model will underestimate the rate of distance decay and will overestimate the SAR, since the species will be too well mixed.

4.3.4.3 Dispersal Limitation Hypothesis

The dispersal limitation hypothesis posits patterns that arise from internal mechanisms, rather than external forces, the latter being considered by the habitat-filtering hypothesis. In the current context, we consider the concept of "dispersal limitation" to be quite general, referring to any internal mechanism of a population that can generate spatial structures, due to less than complete spatial mixing in the system. The most prominent mechanism for this, of course, is dispersal itself, which often yields clustered distributions of offspring around adults. The resulting pattern may show considerable heterogeneity, particularly if not all suitable sites can be easily reached.

Shen et al. (2009) used the simple homogeneous Thomas process (Section 4.1.4.2) to approximate aggregated patterns of individual species produced through "dispersal limitation." This is the simplest point-process model that can be used to create clustered patterns. It is also advantageous in having an analytical formula for the pair-correlation function $g(r)$, which facilitates fitting parameters to the model. However, the simple Thomas process only captures the second-order characteristics of the pattern in an approximate fashion, since it is based on one critical scale of clustering. Several of the potentially more detailed features of a univariate pattern, for example, double cluster structures, superposition patterns, and so on, cannot be represented by this simple point-process model (see Section 4.1.4). As a result, the Thomas process will often produce only an incomplete reconstruction of the observed spatial structures. Ultimately, it is not clear how imperfect the representation of the actual process will be using this approach, or how much this approach will affect the ability to produce suitable inferences regarding the underlying mechanisms driving the pattern. Therefore, a better representation of the dispersal limitation hypothesis might be obtained by using the nonparametric technique of pattern reconstruction. The latter approach is capable of reproducing more of the subtle aspects of the observed univariate pattern and, as a consequence, provides an appropriate methodology for modeling null communities. It can also produce the appropriate fundamental division for multivariate patterns, as it independently superimposes the patterns of all species based on their underlying univariate structures.

The homogeneous Thomas process, or pattern reconstruction using a homogeneous intensity function, can generate a pattern for each species that conserves the observed pattern of clustering, but does not incorporate the potential effects of species interactions or habitat association. Because the patterns are not constrained by the observed habitat associations, the intermediate scale co-occurrence patterns of the species are not well represented. What this means is that species will overlap in the null community more than observed, if there is habitat filtering among species. Species will overlap less than expected in the null community, if they share habitat associations in the observed pattern. The specific balance of habitat preferences among species and their relative abundances will determine if the modeled community is more or less mixed than the observed community and, therefore, will determine if the SAR and distance decay is over- or underestimated. It is clear that the "dispersal limitation" null hypothesis cannot adequately represent communities with clear "diversity hotspots," that is, patches with unusually high local species richness, as one might expect to occur, for example, in a disturbed area that hosts high densities of species.

4.3.4.4 Combined Habitat Filtering and Dispersal Limitation Hypotheses

The most complex null community model considered by Shen et al. (2009) accounted for the simultaneous effects of habitat association and heterogeneity generated by internal mechanisms due to "dispersal limitation." Thus, the model requires constraints on both habitat association and the univariate spatial structure of each species; only the potential effects of interspecific species interactions are removed. The null model, therefore, represents a test of independence under habitat association. Shen et al. (2009) used the heterogeneous Thomas process for this purpose (Sections 2.6.3.5, 4.1.5.1; Equation 4.26). The heterogeneous Thomas process results from "thinning" a homogeneous Thomas process with an intensity function $\lambda(x)$ (Waagepetersen 2007). The thinning operation deletes points, and the probability of a given point to be retained is given by $\lambda(x)/\lambda^*$, where λ^* is the maximal value of $\lambda(x)$ inside the plot. Biologically, this means that dispersal limitation creates a pattern that can be characterized by a homogeneous Thomas process, but habitat filtering removes points from areas of low habitat suitability. If $\lambda(x)$ is known, the parameters of the corresponding (pre-thinning) homogeneous Thomas process can be fit to the observed pattern of a given species using the inhomogeneous K-function (Waagepetersen 2007). The intensity function will be the same as the one generated by an analysis of the habitat filtering hypothesis. The heterogeneous Thomas process model is, therefore, a simple phenomenological description of aggregation that also accounts for the effect of environmental heterogeneity. It can, therefore, capture the larger-scale spatial dependencies of the species in the community.

As before, the best alternative for generation of the null communities for the combined habitat filtering and dispersal limitation hypotheses is to use pattern reconstruction, in this case incorporating the heterogeneous intensity function for the reconstruction. This allows the generation of null communities that conserve the observed habitat associations and univariate patterns of the component species. The null communities are, therefore, constrained by all of the observed characteristics of the individual species patterns at the broader scale, allowing an analysis of the small-scale association patterns that produces inferences regarding the importance of species interactions in producing community-scale distribution patterns.

An interesting issue to consider in using this approach is how much detail of the spatial pattern of the different species is really necessary to adequately characterize the different community level summary statistics. This can be addressed by comparing the results of the Thomas process versus pattern reconstruction in implementing the dispersal limitation and combined dispersal limitation/habitat filtering hypotheses. If the details of the spatial patterns do not matter (i.e., the coarse description of the Thomas process is already sufficient), we may also expect that species interactions (which determine the smaller-scale placement of individuals within the community) may be less important.

4.3.4.5 Evaluating the Fit of the Different Hypotheses

We need to use summary statistics that operate at the community scale, if we are to evaluate the ability of a null community model to appropriately depict the patterns exhibited by the observed community. Shen et al. (2009) and Wang et al. (2011), for example, used the classical SAR for this purpose. However, it is clear that a single summary statistic will only be able to characterize one particular aspect of the complex multivariate spatial pattern; therefore, use of several multivariate summary statistics is required. Wang et al. (2011) used both the SAR and the distance decay function, but they calculated them within a quadrat-based framework, where the spatially explicit information on the location of individuals within the quadrats is lost. It is, therefore, recommended to use multivariate point-pattern summary statistics that use all of the information available regarding placement of individuals within the community. All the multivariate point-pattern summary statistics presented in Section 4.3.2 can be used for this purpose.

We can use additional summary statistics that evaluate functional diversity relative to species richness, if additional information on species traits or pairwise phylogenetic distances for all the species of the community are available. For example, Wang et al. (2013) used the relationships coupling phylogenetic and functional diversity with area (termed PDAR and FDAR), which were formulated in a manner analogous to the SAR. Wang et al. (2013) estimated the SAR, PDARs, and FDARs by randomly selecting quadrats of increasing size within the observation window. The phylogenetic and functional diversity in each quadrat were calculated using the PD (Faith 1992) and FD (Petchey and Gaston 2006) metrics, which are identical, except that PD utilizes a phylogeny versus FD, which derives a dendrogram from trait data. Alternatively, one may also use phylogenetically based point-pattern summary statistics, such as the phylogenetic mark-correlation function, which generalizes the Simpson index $\beta(r)$ so that it includes a continuous measure of species distance (Section 3.1.7.6), and the rISAR function, which generalizes the ISAR to include a continuous measure of species distance (Section 3.1.5.3).

Two methods can be used to rank competing hypotheses, which are represented by different assembly rules for the null communities. First, we can estimate the expectation and the simulation envelopes of the summary statistics from the replicated null communities. A hypothesis is considered adequate if the observed summary statistic falls within the 95% simulation envelope of the null communities, otherwise, the model is rejected. A GoF test can then be used to test for the overall acceptance of the hypothesis. A second method is to compare competing models based on the sum of residuals and the number of parameters in different models, using Akaike's information criterion (AIC) (Webster and McBratney 1989; Shen et al. 2009; Wang et al. 2011). The latter method allows for a consideration of model complexity in evaluating model fit. Because AIC is calculated on the community level (i.e., by superposing realizations of the point-process models for individual species), the number of parameters in a given point-process model is the number of covariates that were used in the hypothesis at least once.

4.4 Analysis of Qualitatively Marked Patterns

Qualitatively marked patterns usually comprise two types of points, where *a posteriori qualitative marks distinguish between the two types within a given univariate pattern.* The *a posteriori* qualitative marks are essentially a binary property of an ecological object, such as surviving versus dead, burned versus unburned, infected versus noninfected, and so on. It is important to distinguish between qualitatively marked patterns and bivariate patterns that have a similar structure (i.e., two types of points). In qualitatively marked patterns a given univariate pattern is marked *a posteriori* by a process that distinguishes between two types of points (e.g., surviving vs. dead plant), whereas in bivariate patterns two different *a priori* sets of processes generate the two component patterns (Goreaud and Pelissier 2003).

Analysis of qualitatively marked patterns is concerned with the characterization of the process that created the qualitative mark over the existing univariate pattern. Therefore, the analysis aims to reveal the spatial correlation structure of the marking, conditional on the given univariate pattern. The fundamental division of qualitatively marked patterns is given by *random labeling,* where the process that creates the marks shows no spatial dependency and allocates the mark randomly over the points of the univariate pattern. Thus, the fundamental difference between bivariate patterns and qualitatively marked patterns is also manifested in the way the null models work: while bivariate analysis requires null models where the points of the second pattern are displaced to random locations within the observation window, the null models for qualitatively marked patterns distribute the mark of the first (or second) type randomly over the existing locations of the underlying univariate pattern.

All bivariate summary statistics, such as the pair-correlation function or Ripley's K-function, can be used to characterize the spatial structure contained in the marks; however, they must be interpreted with respect to the fixed spatial structure of the underlying univariate pattern (e.g., both dead and surviving trees). However, because of the fixed underlying univariate pattern, better adapted test functions can be derived that are able to test for specific aspects of interest that elucidate how a pattern may depart from the random labeling null model.

Because the pattern of individuals generated by random labeling is basically a random thinning of the underlying univariate pattern of all individuals we can easily estimate the expectation of several summary statistics under random labeling. The expectation under random labeling of all "partial" pair-correlation functions $g_{11}(r)$, $g_{22}(r)$, $g_{12}(r)$, and $g_{21}(r)$ is the pair-correlation function $g_{1+2,\,1+2}(r)$ of the univariate pattern (indicated by the $1+2$ subscript): that is, under random labeling we find

$$g_{11}(r) = g_{22}(r) = g_{12}(r) = g_{21}(r) = g_{1,1+2}(r) = g_{2,1+2}(r) = g_{1+2,\,1+2}(r) \qquad (4.62)$$

The same is true for the expectation of the partial K-functions. The identities given in Equation 4.62 are the key for the development of specific test statistics for random labeling.

4.4.1 Complex Departures from Random Labeling

A good example for illustrating the random labeling null model, and the different test statistics used in diagnosing departures from random labeling, is the mortality of saplings in forest communities. In the following, we illustrate the different perspectives of testing the random labeling hypothesis with an example of surviving and dead recruits of the shrub species *Palicourea guianensis,* as found in a census for the tropical forest at BCI, Panama (Box 2.6) (Figure 4.40a). The marks for the recruits are dead (filled circles; pattern 1) and living (open circles; pattern 2).

Potential mechanisms that may cause the mortality of saplings in tropical forests are often associated with negative density dependence, which occurs when nearby conspecifics impair performance. One outcome of negative, density dependence is self-thinning (Kenkel 1988), where intraspecific competition is most intense in areas of high sapling density, resulting in higher mortality. Sapling mortality may also be influenced by interspecific competition with nearby individuals of other species, or by the Janzen-Connell effect, where host-specific pests or pathogens produce higher mortality of recruits near conspecific adults and/or in areas of high sapling density (Janzen 1970; Connell 1971). Finally, mortality may depend on the local light regime or environmental covariates, independent of, or in conjunction with inter- or intraspecific

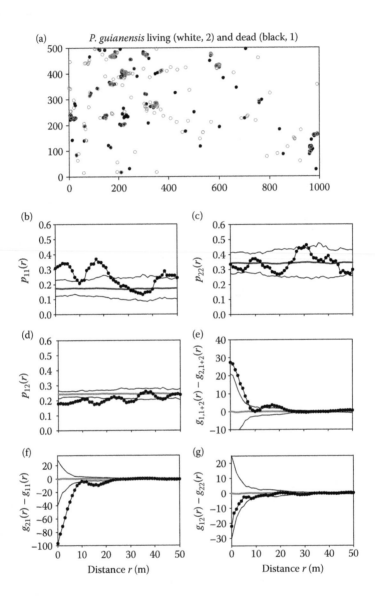

FIGURE 4.40
Typical examples of the analysis of a point pattern with binary marks. (a) Map of the point pattern given by the locations of recruits of the shrub species *Palicourea guianensis* from the second census of a 50 ha plot in the tropical forest at BCI, Panama. The marks are dead (filled circles) and living (open circles). (b)–(g) Test statistics revealing different aspects of departures from the random labeling null model.

interactions. Note that scramble competition and the density-dependent Janzen Connell effect produce the same type of negative density-dependent mortality pattern, where mortality is greatest in areas with a high density of individuals. These considerations indicate that situations that may lead to departures from random labeling are more complex than for uni- and bivariate patterns, where the fundamental division is related to one clear dividing hypothesis (i.e., complete spatial randomness separates clustering vs. regularity, or independence separates attraction vs. repulsion). Summary statistics tailored specifically to the situation are required, if we want to adequately characterize the possible mechanisms producing departures from random labeling (Dixon 2002).

4.4.1.1 Univariate Perspective

From a univariate perspective, we ask if the mark that characterizes pattern 1 (dead in our case) shows more or less aggregation (or regularity) than the unmarked pattern. We are also interested in determining at what distances this occurs. From this perspective, departures from random labeling are produced by a process that promotes aggregation or hyperdispersion of the mark within the univariate pattern. Appropriate summary statistics for the univariate perspective are the univariate pair-correlation function $g_{11}(r)$ and the univariate mark connection function $p_{11}(r)$, which factor out the signal of the spatial structure of the pattern for the unmarked pattern. Older studies mostly used $K_{11}(r)$, the univariate K-function, instead of the pair-correlation function, as the primary summary statistic for random labeling (e.g., Kenkel 1988; Goreaud and Pelissier 2003). However, because of the cumulative nature of the K-function, it is very difficult to identify the specific scales at which the effects are produced. We, therefore, recommend use of the noncumulative summary statistics.

Under scramble competition (or density-dependent Janzen Connell effects), we expect that clusters of interacting recruits should die together, thus dead recruits should be spatially aggregated (e.g., Kenkel 1988). The dead recruits of the species *P. guianensis* tend to be aggregated (Figure 4.40b). There is a strong departure from expectation by the univariate mark connection function of dead recruits up to distances of 8 m, which corresponds with the size of dense clumps of dead recruits that can be observed in the data. There is also a second peak in the mark connection function at a distance of 17 m, which depicts a typical pattern of spacing between clumps of dead individuals. Interestingly, the surviving recruits do not show any departure from the random labeling null model (Figure 4.40c). This indicates that mortality acted in nonrandom way over the recruits and individuals growing in clumps were more susceptible.

4.4.1.2 Bivariate Perspective

From a bivariate perspective, we ask if two marks (e.g., dead and living) show segregation or attraction, and at what spatial scale. Appropriate summary statistics for the bivariate perspective are the bivariate pair-correlation function $g_{12}(r)$ or the bivariate mark connection function $p_{12}(r)$, which factor out the signal of the spatial structure of the unmarked pattern. If competition produces a pattern of dead vs. living, the dead recruits may be segregated from or attracted by living recruits, conditioned on their pre-mortality locations. If the pattern is produced by scramble competition, recruits will likely exhibit a clumped mortality pattern; there should be segregation between surviving and dead recruits, since surviving recruits should be located in areas of lower density. However, in the case of one-sided competition (i.e., asymmetric contest competition), the bivariate analysis should show attraction, because the suppressed (dead) recruits (the "losers") will be located near superior rivals.

Indeed, dead and surviving recruits of the species *P. guianensis* are segregated up to distances of 30 m, within the pattern of all recruits (Figure 4.40d). Thus, clumps of dead individuals and more isolated surviving individuals are located within somewhat different areas of the plot. This points again to negative density-dependent mortality and supports the hypothesis of scramble competition or density-dependent Janzen Connell effects.

4.4.1.3 Density-Dependent Perspective

From a density-dependent perspective, we can ask if one type of mark (e.g., dead individuals) is surrounded, on average, by more individuals with the same mark or another mark (e.g., surviving individuals). Again, we are interested in this as a function of distance. An appropriate summary statistic for an initial test of density-dependent effects is given by $g_{1,1+2}(r) - g_{2,1+2}(r)$. This statistic is especially well suited for detecting density-dependent effects in mortality, since it compares the density of a neighborhood composed of both dead and surviving recruits (i.e., as indicated by the subscript $1 + 2$) surrounding dead recruits (i.e., pattern 1) with the neighborhood surrounding surviving recruits (i.e., pattern 2). The expected value of $g_{1,1+2}(r) - g_{2,1+2}(r)$ is zero under random labeling (Equation 4.62). In contrast, under density-dependent mortality dead recruits would be more likely to occur in areas with high premortality densities, that is, $g_{1,1+2}(r) > g_{2,1+2}(r)$. Conversely, under facilitation, surviving recruits would be more likely to be surrounded by more neighbors, that is, $g_{1,1+2}(r) < g_{2,1+2}(r)$. The use of $g_{1,1+2}(r) - g_{2,1+2}(r)$ as a summary statistic was introduced by Yu et al. (2009) to test for density-dependent mortality in a Mongolian pine forest. It has also been employed by Jacquemyn et al. (2010) to analyze mortality of recruits of *Primula vulgaris*, and by Raventos et al. (2010) to investigate mortality in a Mediterranean gorse shrubland.

In the example we have been using for this section, the recruits of *P. guianensis* show clear indications of negative density-dependent mortality. The test statistic indicates that $g_{1,1+2}(r) > g_{2,1+2}(r)$ for distances up to 7 m, as $g_{1,1+2}(r) - g_{2,1+2}(r)$ lies above the simulation envelope for the random labeling null model at those scales (Figure 4.40e). The test statistic also contains a weak indication for a second peak of density-dependent mortality at distances of 17 m, which corresponds to the typical distance separating clumps of dead individuals. This direct test of density-dependent effects confirms a diagnosis made using the uni- and bivariate mark connection functions (Figure 4.40b–d).

4.4.1.4 Cluster Perspective

In some studies, especially in an epidemiological context (Diggle and Chetwynd 1991), the test statistics $K_{11}(r) - K_{22}(r)$ [or $g_{11}(r) - g_{22}(r)$] are used to examine the degree of clustering in the two component patterns. In the case of negative density-dependent mortality, we would expect that the dead recruits would be more clustered than the surviving recruits, so $g_{11}(r) - g_{22}(r) < 0$. This test statistic is important for epidemiological studies under a case control design (Section 4.4.1.7), where pattern 2 serves as a control for the underlying heterogeneity (e.g., to describe the distribution of the population at risk) and pattern 1 represents the cases (e.g., infections). The special interest here is to determine whether the cases are more clustered than the underlying population at risk, which is tested typically by using the test statistics $K_{11}(r) - K_{22}(r)$ [or $g_{11}(r) - g_{22}(r)$].

4.4.1.5 Additional Patterns in the Mark

To test whether the process that distributes the marks over the locations of the pattern affected only one type of point (e.g., dead individual), but not the other (e.g., surviving individual), one can compare the results of the two summary statistics $g_{12}(r) - g_{22}(r)$ and $g_{21}(r) - g_{11}(r)$ (Getzin et al. 2006). This test can, for example, confirm the presence of negative density-dependent mortality, when the dead individuals are clustered, but not the surviving individuals. Thus, the pattern of dead individuals shows an additional effect not present in the surviving individuals. A value of $g_{12}(r) - g_{22}(r) \approx 0$ indicates that the density of surviving recruits (pattern 2) around dead recruits (pattern 1) at scale r is approximately the same as the density of surviving recruits around surviving recruits. However, if additional clustering exists within the dead recruits, which is independent of the pattern of surviving recruits (e.g., caused by competition from adults or environmental conditions), this would not be identified by the summary statistic $g_{12}(r) - g_{22}(r)$. Instead, we would expect $g_{21}(r) - g_{11}(r) \ll 0$. Thus, $g_{12}(r) - g_{22}(r)$ can be used to reveal if surviving and dead recruits follow the same overall pattern and $g_{21}(r) - g_{11}(r)$ can be used to determine if there is an additional pattern within dead recruits that is independent of the location of the surviving recruits (Getzin et al. 2006, 2008; Watson et al. 2007; Jacquemyn et al. 2010).

The test statistic $g_{21}(r) - g_{11}(r)$ exhibits a very strong and negative departure from the null model in our example using *P. guianensis*. This is caused by the aggregation of dead individuals (Figure 4.40f), whereas $g_{12}(r) - g_{22}(r)$ does not exhibit any departure from the random labeling null model (Figure 4.40 g). Thus, there is an additional pattern within dead recruits that is independent of the location of the surviving recruits. This pattern is caused by negative density-dependent mortality.

4.4.1.6 Trivariate Perspective

From a trivariate perspective, we ask if a mark is influenced by a third pattern, and at what spatial scales (De la Cruz et al. 2008; Biganzoli et al. 2009; Jacquemyn et al. 2010; Raventos et al. 2010). A good example for this is the Janzen–Connell or competition effect where the proximity of adult conspecific trees should increase the risk of mortality among young recruits. This has been tested by Jacquemyn et al. (2010). Other examples for a third pattern impacting marks within a second pattern are competing species. This perspective is also appropriate in cases where fire mortality of nonserotonous plants might be related to proximity to serotonous plants (i.e., a kill the neighbor strategy; Biganzoli et al. 2009).

The appropriate null model for this question is again random labeling, but the summary statistic considers the impact of plants of a third pattern on the distribution of the mark. In this case, we have adult plants (subscript a), dead recruits (subscript 1) and surviving recruits (subscript 2). The summary statistic estimates the probability of survival by recruits as a function of distance r from adults

$$p_{a,2}(r) = \frac{\lambda_2}{(\lambda_1 + \lambda_2)} \frac{g_{a,2}(r)}{g_{a,1+2}(r)} \tag{4.63}$$

The quantities λ_1 and λ_2 are the intensities of the dead recruits and surviving recruits, respectively, and $g_{a,1+2}(r)$ and $g_{a,2}(r)$ are the bivariate pair-correlation functions measuring the intensity normalized neighborhood density around adults (a) of the combined pattern of surviving and dead recruits $(1 + 2)$ and surviving recruits (2), respectively.

The expectation of this summary statistic under random labeling is the overall probability of survival, that is, number of surviving recruits divided by number of recruits. If negative interactions are exerted by adults at distance r (i.e., competition), we expect a lower probability of survival, that is, $p_{a,2}(r) < \lambda_2/(\lambda_1 + \lambda_2)$, whereas positive interactions would be indicated by a higher probability of survival, that is, $p_{a,2}(r) > \lambda_2/(\lambda_1 + \lambda_2)$. Examples for the application of trivariate random labeling are given in Section 3.1.6.2.

Note that this summary statistic is formally analogous to mark-connection functions (Getzin et al. 2008; Illian et al. 2008), because we normalize with the expectation under random labeling, thereby removing the effect of the spatial structure of the premortality pattern from the summary statistic. The summary statistic $p_{a,i}$ therefore, shows the effects we are interested in (i.e., influence of adults) much more clearly than simply using the pair-correlation function $g_{a,2}(r)$ or the L-function.

4.4.1.7 Case–Control Design

The case–control design is frequently used in the context of epidemiology to detect clustering of disease cases relative to the natural variation in the background population (Diggle 2003). The spatial pattern of cases alone is not very informative, unless it is assessed relative to a "control" sample of the pattern of susceptible individuals (i.e., the population at risk), which may itself show spatial clustering. The control pattern represents, from the point of view of the cases, the environmental heterogeneity that exists in the background, since the disease acts over the given locations of the population at risk. Thus, all n_1 cases of the disease (type 1), which occurred over a defined time period, need to be mapped within the observation window. The case–control design also requires that n_2 control points (type 2) from the underlying population at risk need to be sampled within the observation window. This is used to represent preexisting environmental heterogeneity. It is clear that the number of control points should be substantially greater than that of the cases, but it is not necessary to sample all control points. The only condition is that the control points used are a random sample of all control points. If there were an absence of clustering among the cases, relative to the control, we would expect that the cases represent a random sample of the combined pattern of cases and controls. This would indicate that the pattern follows a pattern of random labeling. The typical test statistic for random labeling in the epidemiological context is $K_{11}(r) - K_{22}(r)$, which compares the clustering of the cases with that of the controls.

For some working hypotheses, we can also apply the case-control design for bivariate patterns (e.g., Diggle et al. 2007). For example, Getzin et al. (2008) used the case–control design to assess a decline in clustering of western hemlock trees (*Tsuga heterophylla*) with increasing size class (i.e., self-thinning). The working hypothesis was that the large-scale pattern of mature trees reflects patterns of environmentally driven habitat quality, for example, caused by rock outcrops or wet drainage sites, because mature trees have undergone excessive thinning and are expected to have exploited all available sites (Getzin et al 2008). In contrast, individuals categorized as seedlings, small saplings, or large saplings in their study, exhibited substantial small-scale clustering on top of the pattern of variation due to larger-scale patterns of environmental suitability. The observed pattern was probably caused by internal clustering mechanisms or small microsites for establishment. The question was to determine how the initial pattern of strong clustering of small western hemlock trees disappeared with increasing size class. Thus, Getzin et al. (2008) used large western hemlock trees as a control and the smaller western hemlock trees as cases. A decline in clustering with increasing size reflects self-thinning and the strength of this decline relative to the "control" pattern reflects the strength of density-dependent thinning. They conducted random labeling, where the control was type 1 (i.e., adults) and the case was type 2 (i.e., seedlings, small saplings, or large saplings). They used mark-connection functions and $g_{21}(r) - g_{22}(r)$ versus $g_{12}(r) - g_{11}(r)$ as test statistics. Note that this approach assumes that the pattern of large-scale habitat suitability does not change during the transition from seedlings to saplings to adults.

Diggle et al. (2007) applied the case–control design to an analysis of juvenile and adult trees in the Sinharaja forest, Sri Lanka. Their working hypothesis was opposite that of Getzin et al. (2008). They assumed that the spatial pattern of juveniles reflects the underlying spatial variation in the environment, while the distribution of adults reflected the effects of environmental variation combined with spatially structured effects of survival to adulthood. This assumption is reasonable in cases where juveniles are very abundant, but not strongly clustered, occupying more or less all of the suitable sites available; whereas adults exhibit a sparser and more clustered distribution. Such situations may arise if survival of young trees in unsuitable habitats is reduced and causes adults to be more strongly associated with suitable habitat than juveniles (Bagchi et al. 2011). In this case, the intensity function $\lambda_A(x)$ of adult trees can be given by $\lambda_A(x) = f(x)\,\lambda_J(x)$ where $\lambda_J(x)$ is the intensity of juveniles and the function $f(x)$ represents a possible spatial structure in the pattern of survival. If the function $f(x)$ is constant, the cases form an HPP with an intensity function proportional to that of the controls. If the number of controls is high, a random selection of cases from the combined pattern of cases and controls (i.e., random labeling) approximates an HPP. Diggle et al. (2007), however, were especially interested in factoring out the effect of a nonconstant function $f(x)$ (i.e., changes in habitat suitability with increasing size) to be able to assess the residual clustering of adult trees. As a consequence, they used a null model in which the probability of being a case was dependent on the function $f(x)$ (see Section 4.4.2.1).

4.4.2 Null Models Other than Random Labeling

The random labeling null model is the simplest null model used to test qualitatively marked patterns and serves the purpose of detecting nonrandom spatial structures in mortality. However, once nonrandom structures are detected, we may use more refined null models to explore the possible causes of the departures from random labeling in more detail. In the following, we present null models that consider heterogeneity in mortality. We also consider a refined case-control design that is able to consider differences in the intensity functions of cases and controls. This can be done, in a parametric fashion, by using a covariate that characterizes the probability of mortality and, in a nonparametric fashion, by conducting random labeling at a local scale. We also present a random labeling null model that can be used for community wide analyses.

4.4.2.1 The Covariate Perspective

Similar to the approach used in univariate and bivariate analysis, a qualitatively marked pattern may not be fully homogeneous and the probability of mortality may vary within the observation window. From a covariate perspective, we can ask whether nonrandom labeling was related to a (environmental) covariate, a spatially varying probability of mortality, or a constructed covariate that may represent the overall density of large trees. For example, the probability of mortality of saplings may be higher in areas of lower habitat suitability, produced, for example, by lower moisture or nutrient content. In the best case, if we can derive a statistical model that characterizes the probability of mortality dependent on environmental variables (a type of logistic regression model), we can use the resulting map of mortality probabilities to govern the distribution of the mark "dead" in the null model. In this case, the statistical model for the probability of mortality plays the role of the intensity function in the HPP, as applied to univariate and bivariate patterns. An appropriate null model for this case will thus modify the random labeling null model slightly. The n_1 "dead" marks are not randomly distributed over all points of the unmarked pattern, but a tentative "dead" mark of a random point from the univariate pattern is selected with a probability that is proportional to the value of the covariate.

Raventos et al. (2012) used this approach to investigate the potential role of the resprouting herb *Brachypodium retusum* on the observed density-dependent mortality of the dominant obligate, seed recruiting species *Ulex parviflorus*, *Cistus albidus*, *Helianthemum marifolium*, and *Ononis fruticosa* in a Mediterranean shrubland. They hypothesized that survival of the seed-recruiting species should be higher in places with higher biomass of *B. retusum* and lower in places with lower biomass. To implement this hypothesis they used a covariate that represented the biomass of *B. retusum*.

Note that the third pattern used in the trivarate perspective can also be represented as an environmental covariate. For example, the covariate could be the distance to the nearest individual (or density) of one species associated with each point of a marked pattern of a second species. In fact, "individual-based analyses" (e.g., Hubbell et al. 2001; Peters 2003; Uriarte et al. 2004), which are a form of likelihood-based regression model, take this approach to an extreme in analyzing the effects of spatial configuration, sizes, and species of neighboring trees on the probability of mortality. However, this approach is outside the scope of point-pattern analysis.

Diggle et al. (2007) and Henry and Brown (2009) proposed a refined form of case-control analyses that is able to take differences in the intensity functions of cases and controls into account. For example, in the context of forest communities, their methods would allow for (first-order) changes in the habitat suitability of trees with increasing size class. In developing their approach, Diggle et al. (2007) and Henry and Brown (2009) were mainly interested in second-order effects, such as negative, density-dependent mortality, but not in the first-order effects produced by different intensity functions for cases and controls. The latter have the potential to mask the effects of negative density-dependent mortality (Bagchi et al. 2011). In their studies, this was a problem because several of their tree species increased in clustering with increasing size class, which is contrary to the usual model of self-thinning, where tree species should become less clustered with increasing size (e.g., Condit et al. 2000).

In applying the case-control approach, Diggle et al. (2007) and Henry and Brown (2009) modeled the intensity of the cases $\lambda_1(x)$ relative to that of the controls $\lambda_2(x)$ as $\lambda_1(x) = f(x) \lambda_2(x)$ to consider changes in first-order habitat suitability as a function of size class. Clearly, if the function $f(x)$ is not constant, it may induce an additional effect into the spatial pattern of controls, which may confound the effects of pure second-order effects, such as negative density dependence, in a random labeling analysis. Diggle et al. (2007) assumed the function $f(x)$ to be log-linear with respect to the covariates $v_i(x)$, that is, $f(x) = \exp[\alpha + \Sigma \beta i \, v_i(x)]$. Essentially, the function $f(x)$ is the ratio of the case intensity to the control intensity and can be interpreted as an odds ratio. Thus, the probability that location x_i of the joint pattern of cases and controls is a case, given the covariate values $v_j(x_i)$ at x_i, is given by $P(\text{case}|v_j(x)) = f(x)/(1 + f(x))$ and can be estimated based on a logistic regression model:

$$P(\text{case}|v_i(x)) = \exp[\alpha + \Sigma \beta_j \, v_j(x)]/(1 + \exp[\alpha + \Sigma \beta i \, v_j(x)]) \qquad (4.64)$$

Diggle et al. (2007) then used a specific procedure for assessing additional clustering in the cases. First, they estimated the inhomogeneous K-function $K_1(r)$ of the cases based on the intensity function $\lambda_1(x) = f(x) \lambda_2(x)$. The intensity $\lambda_2(x)$ of the controls was estimated nonparametrically using a kernel estimate and the parameters of the function $f(x)$ were determined based on Equation 4.64. Thus, they used a hybrid method, where the intensity of the cases was the product of a nonparametric baseline intensity $\lambda_2(x)$ and a parametric function of spatially referenced covariates.

After the model was constructed, Diggle et al. (2007) simulated realizations of the underlying process by random relabeling the cases and controls. In this way, they produced null model patterns for the cases. They based the labeling on the spatially varying probabilities given by Equation 4.64. Thus, the probability that a location x was selected as a case was proportional to $f(x)/(1 + f(x))$. This procedure conditions on an additional (known) first-order effect of habitat in the relationship between cases and controls. This is in addition to the overall habitat effect characterized by the nonparametrically estimated, baseline intensity $\lambda_2(x)$ of the controls. To reveal additional clustering of the cases relative to the control and the effect of changes in habitat suitability with size, which was depicted by $f(x)$, they also estimated the inhomogeneous K-function of the cases of the null model simulation. Because $\lambda_1(x) = f(x)\lambda_2(x)$, they first needed to nonparametrically estimate, for each null model pattern, the intensity $\lambda_2(x)$ to estimate the intensity $\lambda_1(x)$ for the cases, as subsequently required for estimating the inhomogeneous K-function $K_1(r)$. Based on the summary statistic $K_1(r)$, they developed simulation envelopes and a form of GoF test based on a two-sided standard T-test as given in Equation 2.10.

Henry and Brown (2009) expanded the approach taken by Diggle et al. (2007). While Diggle et al. (2007) assumed that the cases in the null model simulations followed an HPP based on the control locations and the spatially varying probability given in Equation 4.64; they also allowed the control to have additional clustering (e.g., being a inhomogeneous Neyman Scott process). They controlled for this in Equation 4.64. To this end, they modeled both the cases and controls as an inhomogeneous Neyman Scott processes. Because cases and controls could show different patterns of clustering using this approach, they needed to select the cases in the null model simulations only from the control pattern. Application of this approach by Bagchi et al. (2011) to 139 tree species in four large (>25ha) plots of tropical forest showed that accounting for the influence of environmental gradients revealed a pattern of negative density dependence that was otherwise undetectable in 10 of the species considered. They found that forest plots with less pronounced topographic structuring had the largest proportion of species showing a spatial signal of negative density dependence after controlling for the effect of topography. This suggested that negative density dependence was stronger in the more homogeneous sites (or that other environmental gradients in the more heterogeneous forest plots were also more pronounced, or less correlated with topography) (Bagchi et al. 2011).

4.4.2.2 Local Random Labeling

If there is a larger-scale spatial dependence (i.e., the ratio of the occurrence of two marks is not constant within the observation window) in the probability of a mark (e.g., "dead"), we can also use a nonparametric test that factors out the effect of this heterogeneity in an approximate way. For qualitatively marked patterns, we can use a null model that is similar to the case of the HPP with nonparametric intensity estimate for univariate and bivariate patterns. While random labeling basically switches the marks of two points randomly selected from the entire observation window, we can restrict switching of marks to involve only point pairs that are located within a given distance R. We call this null model *local random labeling*. Switching labels of points that are located within distance R in qualitatively marked patterns is analogous to randomly displacing a point within distance R for univariate and bivariate patterns. For example, we can expect that the probability of mortality for two trees separated by a small distance will be similar, if the rate of mortality is due to environmental conditions (e.g., lower mortality occurs in wetter habitats). Permutation of the "dead" mark only within a local area assures that the overall environmentally driven probability of mortality is approximately the same within a small area, allowing us to reveal second-order effects of the smaller-scale placement of saplings on mortality (e.g., due to negative-density dependent mortality).

To illustrate this null model, we again use the example of surviving and dead recruits of the shrub species *Palicourea guianensi* (Figure 4.40a).

However, we now restrict the random labeling to recruits located within a separation distance of 100 m to factor out large-scale environmental dependence in the probability of mortality. Results of local random labeling differ from results of (global) random labeling (cf. Figures 4.40 and 4.41). The most obvious feature observed with local random labeling is that the expectations produced by the null model follow the underlying cluster structure closely. This is indicated by the expectations of the p_{11} and p_{22} test statistics (Figure 4.41a,b), which show a wave-like structure similar to the observed test statistics. Looking at the observed pattern (Figure 440a), we see an indication that clusters of high density *P. guianensis* recruits are somewhat isolated at the 100 m scale and local random labeling, therefore, mostly switches the marks inside the clusters. As a consequence, the signal of density-dependent mortality as depicted by the test statistic $g_{1,1+2}(r) - g_{2,1+2}(r)$ disappears (Figure 4.41d), but the signal for aggregation of dead recruits (Figure 4.41a) and segregation of surviving and dead recruits (Figure 4.41c) remains. Because the

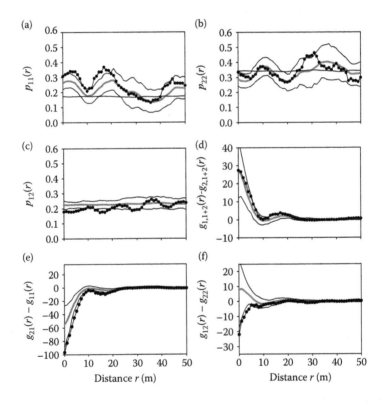

FIGURE 4.41
Localized random labeling, a nonparametric approach accounting for the effects of a potential environmental dependency in the probability of mortality. Same as Figure 4.40, but the marks of two individuals were only switched if the recruits were located within a distance of 100 m. Panels (a) and (b) show the expectation of random labeling as a horizontal line.

signal of negative density dependence disappears, the significance of the two, test statistics $g_{12}(r) - g_{22}(r)$ and $g_{21}(r) - g_{11}(r)$ also changes (Figure 4.41e,f); the signal of the additional pattern in the dead individuals is now much weaker (cf. Figures 4.41e and 4.40f).

4.4.2.3 Community Wide Random Labeling

The previous examples of random labeling involved only analysis of a qualitatively marked univariate pattern and allowed the analysis of a single species or groups of species. However, an interesting question is how much of the mortality patterns of individual species can be explained by mortality patterns of all species (or all species of a functional group) together. To explore this question we can contrast two random labeling null models, a null model of random labeling that ignores species (species-blind random labeling) and a null model of species-specific random labeling. The null model of *species-blind random labeling* randomizes the "dead" mark among all small saplings, regardless of species-specific mortality rates and, therefore, corresponds to standard random labeling of the marked pattern of all small saplings. However, the null model corresponding to *species-specific random labeling* randomizes the "dead" mark only within species, thereby conserving the species-specific mortality rates. Because the expectation of species-specific random labeling is constrained by the distribution pattern of individual species we may find substantial differences between the results of these two null models. For example, an interesting question is how much of the mortality pattern of an individual species can be explained by mortality patterns of all species (or all species of a functional group) together. For example, if species-blind, negative, density-dependent mortality is a key process that governs mortality of small saplings, we would expect a stronger signal of negative density dependence when analyzing all species together compared with the joint result of all individual species analyses (i.e., species-specific random labeling). A contrasting result should emerge if the signal of density-dependent mortality was species specific, as for example assumed by the Janzen Connell hypothesis, which imagines that host-specific pests and pathogens cause individuals located in clumps to have a higher risk of mortality. In this case, we would expect significant departures due to the species-specific random labeling null model.

We applied the species-specific versus species-blind random mortality null models to the surviving and dead saplings (1 cm $\leq$ dbh < 2 cm) of gap species found at the BCI forest (Box 2.6) (Figure 4.42a). Because saplings of gap species find especially favorable conditions in canopy gaps, we expect strong effects of species-blind, negative, density-dependent mortality. This could be caused either by mortality due to overall sapling crowding or because the gaps that were favorable for establishment have been closed by the growth of larger trees and are now more shaded, thus affording less favorable conditions for small saplings. Applying the species-blind, random labeling, null

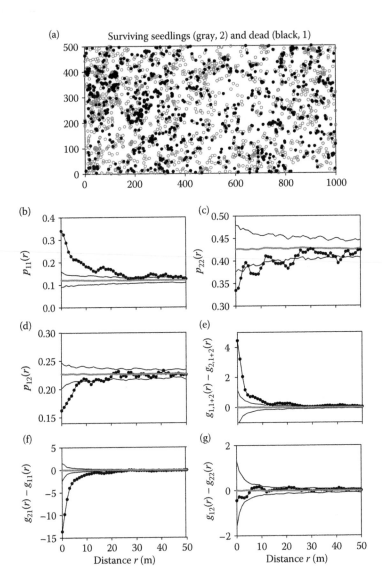

FIGURE 4.42
Community wide, species-blind, random labeling for small saplings of gap species at the BCI
forest. In this example, we used species-blind random labeling, which did not conserve species-specific mortality rates. (a) Map of the locations of small saplings of gap species from the
1985 census for the BCI forest. The marks are dead (filled disks) and living (open gray disks).
(b)–(g) Various test statistics assessed by randomizing the "dead" mark regardless of species.
The analyses reveal different aspects of departures in the observed pattern from the species-
blind, random labeling null model.

model we find that dead saplings are strongly aggregated within all small saplings (Figure 4.42b). The aggregation of dead individuals is similar to that observed for the species *P. guianensis* (Figure 4.40b), with strong aggregation at distances below 5 m and a second peak at 20 m, probably depicting the typical distance among clusters of the dead saplings of the gap species. Surviving individuals show hyperdispersion, at shorter distances when considering all small saplings together (Figure 4.42c). There is also strong segregation by small, surviving and dead saplings at distances below 10 m (Figure 4.42d). The test function $g_{1,1+2}(r) - g_{2,1+2}(r)$ indicates very strong effects of negative, density-dependent mortality, especially at distances below 5 m (Figure 4.42e), indicating that dead saplings cluster in multispecies groups. This additional pattern in dead recruits is also shown by the finding that the test statistic $g_{12}(r) - g_{22}(r)$ does not show significant departures from the null model (Figure 4.42 g), but $g_{21}(r) - g_{11}(r)$ does (Figure 4.42f).

To determine whether the strong signal of species-blind, negative density dependence was caused by larger-scale heterogeneity, we also applied the null model of local random labeling presented in the last chapter to the small sapling data combining all the gap species. The null expectation of local random labeling (using a neighborhood distance of 50 m) did not differ from the null expectation of standard random labeling (results not shown). This indicates that larger-scale habitat effects are unlikely to cause the observed departures from the standard random labeling null model.

The species-specific random labeling null model, which conserves the mortality rates of individual species, yields substantially different results. First, we find that most of the aggregation of dead saplings, and regularity of surviving saplings, is explained by the species-specific mortality rates (Figure 4.43a,b), whereas the segregation of surviving and dead saplings remains (Figure 4.43c). Notably, the signal of negative density dependence disappears; the data are fully within the simulation envelopes (Figure 4.43d). Thus, when conducting random labeling within species and combining the results of the individual analyses (this is what species-specific random labeling basically does) we find only weak departures from the random mortality hypothesis, but no negative density dependence.

In summary, we find that the species-blind, random labeling, null model indicates a strong effect of negative density dependence, but under the species-specific random labeling null model we find no signal of negative density dependence. This suggests that small saplings of gap species, which are preferentially located in gaps, have an overall risk of mortality that is higher, even if conspecific density does not influence the mortality pattern. Thus, species-blind, negative, density-dependent mortality could be a key process that governs mortality of small saplings of gap species at BCI. However, the lack of a signal for negative density-dependent mortality in the species-specific analysis suggests that Janzen Connell effects may be relatively unimportant for the mortality of small saplings of gap species during this observation period.

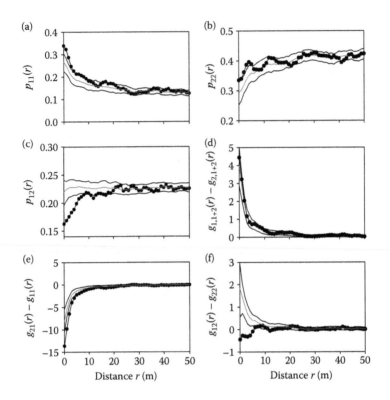

FIGURE 4.43
Community wide, species-specific, random labeling for small saplings of gap species at the BCI forest. In this example, we used species-specific random labeling, which conserved species-specific mortality rates. Same as Figure 4.42 but for species-specific random labeling.

In a second example, we contrast the results found for the gap species with those obtained for shade tolerant species. Because of the high abundance of small saplings of shade-tolerant species at BCI, we analyzed only a 4 ha subplot, yielding a total of 4576 shade-tolerant small saplings. We found that the expectations of the species-specific random labeling null model fit those of the species-blind null model remarkably well (cf. solid black lines with bold gray lines in Figure 4.44b–e). This is in strong contrast to the results found for the gap species and indicates that mortality among the shade-tolerant species works mostly in a species-blind manner. We also found that small, dead saplings of shade-tolerant species were strongly clustered at distances up to 10 m (Figure 4.44b), whereas surviving individuals were weakly clustered within all small saplings (Figure 4.44c). Additionally, there was segregation between surviving and dead individuals up to distances of 20 m (Figure 4.44d) and we find that a tendency toward positive density-dependent mortality was only weakly significant (Figure 4.44e). The strong contrast between the results obtained for gap species and shade tolerant species

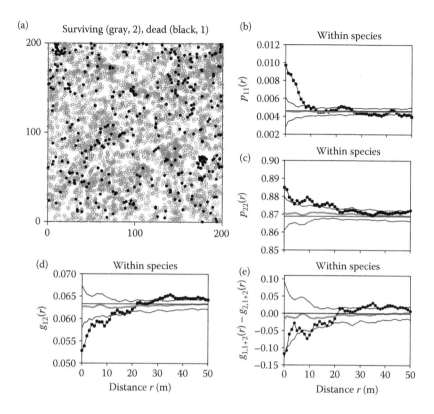

FIGURE 4.44
Community-wide random labeling for small saplings of shade-tolerant species. (a) Map of the locations of small saplings of shade-tolerant species found for the 1985 census within a 200 × 200 m subplot of the BCI forest. The marks are dead (filled disks) and living (open gray disks). (b)–(e) Various test statistics assessed with the species-specific random labeling null model, which randomizes the "dead" mark within species.

shows that there is a strong signal of species properties in the spatial patterns of mortality of small saplings at the BCI forest.

4.5 Analysis of Quantitatively Marked Patterns

Only few examples exist for the analysis of qualitatively marked patterns in ecology, and most of the methods presented are rather new. In Section 3.1.7, we provided a detailed treatment of the different types of data types and analyses available for qualitatively marked patterns, enriched with detailed examples. For this reason, we do not treat this data type in detail in Chapter 4.

4.6 Analysis of Objects with Finite Size

Similar to the case for quantitatively marked patterns, there are only few ecological examples of the analysis of objects with finite size, and Section 3.1.8 provided a detailed treatment of the available methods and examples. For this reason, we do not treat this data type in detail in Chapter 4.

5

A Course Outline Based on the Book

This chapter provides an outline for a course on spatial point-pattern analysis based on the course *"Patrones espaciales en ecología: modelos y análisis"* that the first author presented several times at the graduate school Alberto Soriano in the Agronomy department of the University of Buenos Aires, Argentina and at the Mediterranean Agronomic Institute of Zaragoza (IAMZ-CIHEAM), Spain. The course is designed to be covered during a 2-week period, lasting approximately 6 h/day. The point-pattern part of the course includes approximately 1 week of activities. Table 5.1 suggests a rough timetable for the course. In general, there should be 4 hours of lecture in the morning and 2 or more hours of supervised hands-on activities, during which the students apply the material presented and conduct their own analyses with the appropriate point-pattern software. The material is closely related to the contents of the book and we provide chapter numbers as guidance.

5.1 Introduction

The opening lecture should provide an answer to the basic question: what is point-pattern analysis and why is it used? This should be illustrated with examples representing the different data types and should introduce the typical questions asked in ecology associated with the data types. Based on simple examples of univariate patterns, the concepts of (i) summary statistics, (ii) null models, (iii) point-process models, (iv) comparison of models and data, and (v) homogeneous versus heterogeneous patterns should be informally introduced. Suitable background material for this lecture is provided in Chapter 1 and Section 2.1.

5.1.1 Data Types

After the opening lecture, the students gain an overall idea of the typical questions addressed and the steps taken in conducting a point-pattern analysis. At this point, the different data types can be introduced and summarized more formally. A definition of point patterns and different data types is first provided, together with examples illustrating each aspect of importance:

TABLE 5.1

Rough Outline for a 1-Week Course on the Basics of Spatial, Point-Pattern Analysis.
One Lesson Would Be Approximately 1 Hour

	Day 1	Day 2	Day 3	Day 4	Day 5
Lesson 1	Opening lecture: what is point-pattern analysis and why is it used	Univariate analysis summary statistics	Bivariate analysis summary statistics	Qualitatively marked patterns, random labeling	Quantitatively marked patterns, test functions, mark correlation functions
Short break					
Lesson 2	Data types, typical questions, mapping issues, summary statistics	Null models and point-process models for univariate analysis	Null models and point-process models for bivariate analysis	Test statistics for random labeling	Examples for mark correlation functions
Long break					
Lesson 3	Null models, point-process models	Heterogeneous patterns, classification, and testing	Null models and point-process models for bivariate analysis	Examples	Analysis of objects with finite size and real shape
Short break					
Lesson 4	Compare models and data: simulation envelopes, fitting models to the data	Methods to deal with heterogeneous patterns	Multivariate analysis	Trivariate random labeling	Examples

- A spatial point pattern consists of a set of ecological objects that can be characterized by their locations in space. Two types of additional information may be available: *marks* that provide additional information characterizing the objects (e.g., size or status), and mapped *covariates* that describe properties of the environment that may potentially influence the location of the objects
- Unmarked patterns
  - Univariate (one type of point)
  - Bivariate (two types of points, with *a priori* differences)
  - Multivariate (several types of points, with *a priori* differences)
- Qualitatively marked patterns (univariate pattern with an *a posteriori* mark, such as surviving vs. dead)

- Quantitatively marked patterns (the mark is given by a property of the points, such as size)
- Objects of finite size and real shape

Suitable background material for this lecture is provided in Section 2.2. This lecture can also provide some background on the issues of mapping, optimal plot size and shape, and the issue of one large versus several small plots.

5.1.2 Summary Statistics

Once the students understand the diversity of data types that can be analyzed with point-pattern analysis, the topic of how to quantify the statistical properties of a point pattern is broached. This introduces the ideas underlying the notion of summary statistics. Section 2.3 introduces the different summary statistics. The key points are as follows:

- Summary statistics quantify the statistical properties of spatial point patterns and provide a brief description of point patterns using numbers, functions, or diagrams
- Each data type is associated with specific summary statistics and the different types of questions that are appropriate to address with each data type are presented
- Presentation of indices versus functional summary statistics
- Concepts of location-related versus point-related summary statistics
- Concept of the typical point for homogeneous patterns
- Heterogeneous patterns

5.1.3 Null Models and Point-Process Models

Once the students know that summary statistics are used to quantify the statistical properties of a point pattern, the next issue to address is whether these properties are compatible with stochastic effects of chance events or whether there is a "pattern" in the point-pattern data. This leads naturally to the use of null models. Suitable background material for this lecture is provided in Section 2.4. The key points are as follows:

- A null model is a pattern-generating model that randomizes certain aspects of the data, while holding other factors constant. The task is to find out whether the data contain spatial structure that is not contained in the null model. A good starting example is CSR as a null model for several examples using point-pattern analysis software.
- Each data type requires its own appropriate null models.

- Null models are often too simple to address questions for real-world patterns. Therefore, more complex point-process models are required. This idea can be introduced by using an example involving a clustered pattern demonstrated with point-pattern software.
- *Exploratory context*: point-process models are used to provide a detailed description of the spatial structure of a pattern. In this case, a point-process model is presented as a parametric model that is fit to the data to provide a succinct description of the spatial structure.
- *Confirmatory context*: point-process models are explicitly constructed to represent a hypothesized process. In this case, the task is to confirm that the observed pattern cannot be distinguished from the patterns produced by the point process.

5.1.4 Techniques to Compare Models and Data

Once the concepts of null models and point-process models are understood, the next task is to compare the patterns generated by the models with observed data. Here, the summary statistics are applied to both simulated and observed point-pattern data and several methods to compare the summary statistics of simulated and observed patterns are presented. Section 2.5 provides background material for this lecture. The key points are as follows:

- Simulation envelopes
  - Interpretation and construction
  - Type I error and GoF tests
- Fit of parametric models to the data
  - Minimum contrast
  - (If there is time: Maximum likelihood and pseudolikelihood methods)

The different methods should be directly illustrated with point-pattern software for simple univariate example patterns.

5.2 Analysis of Univariate Patterns

The concepts presented in the introductory material are now presented in more detail for univariate patterns. Some examples of univariate patterns are given and the objective of the analysis of univariate patterns is defined (i.e., to describe spatial structure of univariate pattern).

5.2.1 Summary Statistics for Univariate Patterns

Because the ideas underlying the univariate summary statistics reappear for the other data types, they should be presented in greater detail for univariate summary statistics than for other data types. After their definition, the different summary statistics should be illustrated with several examples directly with point-pattern analysis software. The different univariate summary statistics are explained in Sections 2.3.1 through 2.3.5, and interested students can find technical detail on estimators in Sections 3.1.1 through 3.1.3. The key points are as follows:

- Intensity λ and intensity function $\lambda(x)$
- K-function $K(r)$
  - Idea of the estimator and methods of edge correction (Ripley, Ohser)
  - L-function and its interpretation
- Pair correlation function $g(r)$
  - Idea behind $g(r)$ and its interpretation
  - Relationship between $K(r)$ and $g(r)$
  - "Memory" of the K-function
  - Estimation, edge correction, and bandwidth
- Distribution function $D^k(r)$ of distances to kth neighbor
  - Idea behind $D^1(r)$ and its interpretation
  - Clarification of contrasting nature of $D^1(r)$ compared to $g(r)$ and $K(r)$
  - Estimator and edge correction methods
  - Generalization to $D^k(r)$
- Mean distance to kth neighbor
- Spherical contact distribution $H_s(r)$ and its interpretation and estimation.

5.2.2 Null Models and Point-Process Models for Univariate Patterns

This lesson presents several useful null models and point-process models for univariate analysis. Several examples should be analyzed directly with point-pattern software. It is important to show the results of the analyses for several summary statistics. Section 4.1 provides a systematic presentation of different relevant null models and examples. The key points are the following:

- Homogeneous Poisson process
  - Interpretation of departures from CSR
  - Interactions among points versus environmental heterogeneity
  - Separation of scales

- Heterogeneous Poisson process
  - Intensity function
  - Kernel estimate
  - Parametric fits of intensity function with environmental variables
- Thomas process
  - Construction and interpretation
- Soft and hard core processes
  - Construction and interpretation

5.2.3 Heterogeneous Patterns

This lesson should create awareness of potential problems in the analysis of heterogeneous point patterns and provide some methods for diagnosing and dealing with heterogeneity. Section 2.6 provides background information for this lecture. The key points are as listed below:

- The concept of the typical point does not hold for heterogeneous patterns
- Classification of heterogeneous patterns (Section 2.6.1)
  - Heterogeneity in intensity
  - Heterogeneity in point–point interactions
- Testing for heterogeneity (Section 2.6.2)
  - Habitat model for intensity function (parametric estimate)
  - Inspect the pair-correlation function and the L-function
  - Analyze patterns in subwindows
- Approaches for dealing with heterogeneity (Section 2.6.3)
  - Ignore heterogeneity
  - Identify and analyze homogeneous areas (irregular observation window)
  - Heterogeneous Poisson process with nonparametric intensity estimate
  - Heterogeneous Poisson process with parametric intensity estimate
  - Inhomogeneous summary statistics

5.3 Analysis of Bivariate Patterns

Typical examples for bivariate patterns are provided and the objective of the analysis of bivariate patterns is defined (i.e., to describe the spatial relationship between two univariate patterns).

5.3.1 Summary Statistics for Bivariate Patterns

The summary statistics for bivariate patterns can be presented very briefly because they represent a direct generalization of issues already addressed for univariate summary statistics. Interested students can find technical detail on estimators of the bivariate summary statistics in Section 3.1.4.

5.3.2 Null Models and Point-Process Models for Bivariate Patterns

Presentation of null models and point-process models that are useful for the analysis of bivariate patterns. It is important to clarify that the basic null model for bivariate patterns (i.e., independence) must remove the interdependence of the two univariate component patterns, but conserve the univariate structures of the component patterns. The null models here must be clearly separated from the random labeling null model required for the superficially similar data type of qualitatively marked patterns. The null models should be directly demonstrated with examples using point-pattern analysis software. Section 4.2 provides a systematic presentation of different relevant null models and examples. The key points are the following:

- Basic null model: independence
  - Toroidal shift or parametric point-process models for independence
  - Effects of interactions versus environmental heterogeneity
- Heterogeneous Poisson process due to changes in the intensity of the focal pattern
  - Approximate separation of species interactions from environmental heterogeneity
  - Separation of scales
- Heterogeneous Poisson process due to changes in the intensity of the nonfocal pattern
  - Gaussian (or other) kernel function used for the model, for example, recruits around adult plants
- Perhaps introduce a discussion of a simple Thomas process for bivariate patterns

5.3.3 Analysis of Multivariate Patterns

This section may be skipped if time is limited. Typical examples and questions for multivariate patterns are presented and the objective of multivariate analysis is defined (i.e., characterize diversity expressed through spatial structures). In general, the complexity of multivariate patterns poses difficulties in analyzing them, both with respect to summary statistics and null models. Classical plot-based methods of ecology are sometimes used, but they disregard a great deal of information that would be contained in fully mapped patterns. Sections 3.1.5 and 4.3 present different methods to analyze multivariate patterns. Some approaches to introduce include the following:

- Conducting many pairwise bivariate analyses. This requires methods for summarizing results of pairwise analyses.
- Using plot-based methods of ecology based on species co-occurrence matrices.
- Using parametric point-process models.

Approaches that "truly" use multivariate point-pattern analysis are the following:

- Generalization of commonly used diversity indices to point-pattern analysis, such as the point-pattern species–area relationship, the spatially explicit Simpson index (work by Shimatani), or the individual species–area relationship (ISAR).

5.4 Analysis of Qualitatively Marked Patterns

The lesson may be introduced with the motivating example of the analysis of mortality of trees. The objective of the analysis of quantitatively marked point patterns is to describe statistical properties of processes that distribute labels to points of an already existing univariate pattern (i.e., determine spatial structure of marks within given points).

5.4.1 Null Model of Random Labeling

The consideration of qualitatively marked point patterns leads directly to the null model employing random labeling. This null model must be clearly distinguished from the analysis of bivariate patterns, although summary statistics for bivariate patterns can be used for the analysis. Background material for this lecture is provided in Sections 2.2.2, 3.1.6, and 4.4. A key point is that

- The interest in this type of analysis is focused on the spatial structure of the marks (e.g., "surviving" versus "dying"), which is conditioned on the locations of the underlying univariate pattern. In contrast, the null model of independence, which was the topic of the previous Section 5.2, considers potential interdependence between two univariate patterns, conditioned on their univariate structure. Thus, the locations of the two univariate component patterns are not fixed.

5.4.2 Summary Statistics to Detect Departures from Random Labeling

Because the locations are fixed, several test statistics composed of pair-correlation functions can be used to test for departures from random labeling. Each of these captures a slightly different biological aspect. The key points are as follows:

- Present and interpret the different test statistics based on pair-correlation functions
  - $g_{11}, g_{12}, g_{11} - g_{22}, g_{12} - g_{11}, g_{1,1+2} - g_{2,1+2}$
- Summary statistics adapted to data type: mark connection functions
  - Definition and estimation of mark connection functions
  - Factor out the signal of the univariate structure
- Trivariate random labeling (independent labeling)
  - Motivating example
  - Summary statistics

5.5 Analysis of Quantitatively Marked Patterns

The lesson may be introduced with a motivating example of the analysis, such as a mark based on the "size" of trees. The objective of the analysis of quantitatively marked point patterns is to determine possible spatial correlations in quantitative marks, conditioned on a given univariate pattern. Background material for this data type is provided in Sections 2.2.3 and 3.1.7.

5.5.1 Summary Statistics and Test Functions

Mark correlation functions, which are appropriate for dealing with quantitatively marked patterns, can be motivated as a generalization of the mark connection functions. First, different test functions are introduced:

- $m_1 * m_2, m_1$ or $m_2, (m_1 - m_2)^2/2, (m_1 - m)(m_2 - m)$

and the corresponding (nonnormalized and normalized) mark correlation functions are defined. The key points are as listed:

- (Nonnormalized) mark correlation functions yield the average of the test function over all pairs of points at distance r
- The conditional nature factors out the signal of the univariate structure and focuses on distance-dependent correlations among marks
- Interpretation and estimation of mark correlation functions

- Null model (independent marking)
- Differences with geostatistics
- Generalization to more marks

5.6 Analysis of Objects of Finite Size and Real Shape

This section may be skipped if time is limited. The lesson may be begun with some typical examples (e.g., shrublands) and points to limitations of the usual point approximation. The objective of the analysis of objects of finite size and real shape is to describe spatial structures in patterns of objects analogous to uni- and bivariate analysis. Background material for this lecture is provided in Section 3.1.8.

5.6.1 Summary Statistics

Present the underlying idea of this type of analysis, how objects are represented with the help of an underlying grid, and how a point pattern is constructed from the grid data. Discuss how this approach allows the basic application of the summary statistics typically used for uni- and bivariate patterns. However, there are additional features to be considered when analyzing objects such as the following:

- Overlap among objects
- Count or do not count cell pairs of the focal object
- Use all pairs of points between objects or only the shortest distance between pairs of objects
- Edge correction

5.6.2 Null Models

Although the null models used in studying objects of finite size and real shape are analogous to those of uni- and bivariate analyses, there are additional features to be considered:

- How to randomize entire objects
- Masking (space limitation)
- Treatment of edge

Once these differences are presented, examples of uni- and bivariate analyses are given.

Frequently Used Symbols

Numbers under the column "Key References" that are not preceded by "Section" or "Sections" are equation numbers.

Symbol	Explanation	Key References
$\mathbf{1}(cond)$	Indicator function with value 1 if *cond* is true and 0 otherwise	3.2, 3.33, 3.43, 3.102, 4.60
$\mathbf{1}(\|x_i - x_j\|, r)$	Count (or indicator) function, yields 1 if $\|x_i - x_j\| \leq r$ and 0 otherwise. Used for estimation of the K-function	2.19, 3.32, 3.93
A	Area of observation window W	
A_m	Area of a subwindow W_m	3.108
α	Approximate error rate of simulation envelopes	Section 2.5.11
$\alpha(r)$	Cumulative and spatially explicit Simpson index	3.71, 4.54
$A_A(r)$	Morphological function, being equal to $H_s(r)$	Sections 2.3.5, 3.1.3.3
Area[X]	Operator estimating area of X within observation window W	3.103, 3.104, 3.118, 3.120, 3.136
b	Width of buffer, used for estimators of $D_k(r)$	3.44, 3.47, 3.66
BMW	Used as superscript to refer to the estimator of inhomogeneous summary statistics of the Baddeley, Møller, and Waagepetersen (2000) paper	2.20, 3.25, 3.38, 3.55, 3.62
β_i	Distance from point i to nearest border of W, used for estimators of $D^k(r)$	3.44, 3.45, 3.66, 3.67
$\beta(r)$	Spatially explicit Simpson index	3.69, 3.70, 4.53
$\beta_{\text{phy}}(r)$	Phylogenetic Simpson index	3.100, 3.101,
β^*	Nonspatial expectation of $\beta(r)$, the classical Simpson index	3.101
β^*_{phy}	Mean pairwise phylogenetic distance over all pairs of individuals present in the observation window; expectation of the phylogenetic Simpson index $\beta_{\text{phy}}(r)$	3.101

Symbol	Explanation	Key References
$b(x, R)$	Disk of radius R centered at location x	3.17
$\partial b(x_i, r)$	Circumference of $b(x_i, r)$	3.17
c	Tuning constant used in minimal contrast method	2.11, Section 2.5.2.1
$C_i(r)$	Circle with radius r centered on point i	3.104
$C_m(x_i)$	Indicator function which yields a value of 1 if point x_i is of type m and 0 otherwise	3.53, 3.60, 3.65, 3.83
$C_{lm}(x_i, x_j)$	Indicator function which yields a value of 1, if point i (with coordinates x_i) is of type l and point j (with coordinates x_j) is of type m and 0 otherwise. Used in estimators of partial (bivariate) summary statistics	3.51, 3.65, 3.80, 3.97
CE	Clark–Evans index	Sections 2.3, 3.1.3.2
$c_t(r)$	(Nonnormalized) mark-correlation function based on test function t	3.84
$c_{lm,t}(r)$	(Nonnormalized) mark-correlation function for a qualitatively marked pattern (with type l and m points) and one quantitative mark based on test function t	3.97
c_t	Normalization constant of mark-correlation function based on test function t	3.87
CSR	Complete spatial randomness	Sections 2.6.1.1, 4.1.1
dbh	Diameter at breast height, a standard measure of the size of trees	
$d(r)$	Probability that the nearest neighbor is located at distance r from a typical point	Section 2.3.4.1
$d(r)$	Spatial species richness introduced by Shimatani and Kubota (2004), point-pattern analog of species–area relationship	3.72, 4.51
$d(f, m)$	(Phylogenetic) dissimilarity between species f and m	3.75, 3.76, 3.100, 3.102,
d_i	Distance from point i to the nearest neighbor, see also d_{ik}	3.2
d_{ik}	Distance from point i to kth neighbor	3.43–3.45, 3.47, 3.65–3.67, 3.111

Symbol	Explanation	Key References
$\mathbf{d}x_1$	Area of small disk at location x_1, used in relation with product densities	Sections 2.3.2, 3.1.2.3, 3.3.2, 4.1.4.2,
$D(r)$	Nearest-neighbor distribution function, the univariate and cumulative distribution function of the distance r to the nearest neighbor, see also $D^k(r)$	3.2, Sections 2.3.4.1, 3.1.3.1
$D_{12}(r)$	Bivariate distribution function of the distance r from a point of type 1 to the nearest point of type 2, see also $D^k_{lm}(r)$	4.36, 4.39
$D^k(r)$	Cumulative distribution function of the distances r to the kth neighbor. Sometimes the subscript k is used instead of the superscript	3.43–3.45, 3.47
$D^k_{lm}(r)$	Cumulative partial distribution function of the distances r from type l points to their kth type m neighbor. With superscript k,b: buffer edge correction, and with superscript k,H: Hanisch edge correction	3.65–3.67, 3.73, 3.75
$D^{k,b}(r)$	Buffer estimator of $D^k(r)$. With subscript k,H: Hanisch estimator	3.44, 3.45, 3.47
$\Delta(\theta)$	Minimal contrast for parameterization θ	2.11
Δ^p_f	Expectation of $r\mathrm{ISAR}_f(r)$ for large neighborhoods r	3.77
$\Delta^p_f(x, r)$	Landscape of phylogenetic neighborhood dissimilarity, estimated for each location x and focal species f as the mean pairwise phylogenetic distance between the focal species f and all other species located within a given distance r	Section 3.1.5.3
$E^\varphi_i(\psi)$	Partial energy of reconstructed pattern ψ with respect to the ith summary statistic of the observed pattern φ. Used in pattern reconstruction	3.137
$E^\varphi_{\text{total}}(\psi)$	Total energy of reconstructed pattern ψ. Used in pattern reconstruction	3.138
$f^\psi_i(x)$	Functional summary statistic i estimated for pattern ψ. The x can be, for example, distance r or neighborhood rank k	3.137

Symbol	Explanation	Key References
f_k	$f_k = (1 + k)/k$, in the formula of the pair-correlation function of a Thomas process, accounts for a clustered distribution of the number of points over the clusters, as described by a negative binominal distribution with parameter k	4.7, 4.9, 4.10, 4.11, Section 4.1.4
f_{k1}, f_{k2}	In the formula of the pair-correlation function of nested double-cluster Thomas processes, accounts for clustered distributions of the number of small clusters over the large clusters and the number of points of the pattern over the small clusters, respectively	4.14, 4.15, 4.17, 4.18
$F(x)$	Some studies use this symbol for $H_s(r)$	
$G(y)$	Some studies use this symbol for $D(r)$	
$f(x_i, r), g(x_i, r)$	Functions which depend on the focal point i and other points within or at distance r from the focal point. Used for combining replicates	3.106
$g(r)$	Pair-correlation function, may have superscripts O, WM, R, S, and BMW indicating the estimator used	Sections 2.3.2.1, 3.1.2.4
$g_{12}(r)$	Bivariate (or partial) pair-correlation function	Section 3.1.4.2
$\hat{g}_{lm}^{WM}(r)$	Estimator of partial pair-correlation function for type l and m points based on the WM estimator. Substitution of WM by superscripts R, S, and O indicate the Ripley, Stoyan, and Ohser estimators, respectively	3.54
$\hat{g}_{lm}^{BMW}(r, \lambda_l(x), \lambda_m(x))$	Estimator of inhomogeneous partial pair-correlation function for type l and m points based on the BMW estimator and the two partial intensity functions $\lambda_l(x)$ and $\lambda_m(x)$. Substitution of BMW by superscripts O and WM indicates the alternative estimators based on the Ohser and WM method	3.55, 3.56, 3.58
$\hat{g}_{inhom}^{O}(r, \lambda(x))$	Estimator of the inhomogeneous pair-correlation function based on the intensity function $\lambda(x)$ and an extension of the Ohser edge correction	3.29

Symbol	Explanation	Key References
$\hat{g}_{\text{inhom}}^{\text{WM}}(r, \lambda(x))$	Estimator of the inhomogeneous pair-correlation function based on the intensity function $\lambda(x)$ and an extension of the WM edge correction	3.31
$g_{mm}(r)$	Multiplicatively weighted, pair-correlation function	3.91
$h(r)$ or $h(r, \sigma)$	Bivariate probability density function, such as the normal distribution, to characterize the position of points in a cluster, relative to the cluster center	4.1, 4.9, 4.15, 4.19
$h_2(r)$ or $h_2(r, \sigma)$	Probability that two points of the same cluster are distance r apart. Quantity used to derive analytical expressions for pair-correlation function of Thomas processes	4.1, 4.2, 4.6, 4.9
$\gamma_W(\mathbf{r})$	Set covariance, the area of the intersection of rectangle W with W_r, where W_r is the rectangle W shifted by a vector $\mathbf{r}$	3.8
$\bar{\gamma}_W(r)$	Isotropized set covariance, used for Ohser edge correction for second-order summary statistics	3.6, 3.7
$\bar{\gamma}_W(r, \lambda(x))$	Generalized isotropized set covariance for estimation of the inhomogeneous pair-correlation function based on intensity function $\lambda(x)$	3.30
$\gamma_m(r)$, $\gamma_{m1m2}(r)$	Univariate and bivariate mark vario-grams, respectively; mark-correlation functions resulting from test function t_4	Sections 3.1.7.1, 3.1.7.3, 3.1.7.4
h	Bandwidth of kernel function	Box 2.4
$H_s(r)$	Spherical contact distribution	3.49, Sections 2.3.5.1, 3.1.3.3, 3.49
$H_s^m(r)$	Partial spherical contact distribution of type m points	3.72
$I_m(r)$, $I_{m1m2}(r)$	Univariate and bivariate mark-correlation functions, respectively, resulting from test function t_6; yield a spatial variant of the classical Pearson correlation coefficient	Sections 3.1.7.1, 3.1.7.3, 3.1.7.4
$\text{ISAR}_f(r)$	Individual species–area relationship for focal species f	3.73, 3.74, 3.75

Symbol	Explanation	Key References
$\overline{\mathrm{ISAR}}(r)$	$\mathrm{ISAR}_f(r)$, averaged over all species f and weighted with their relative abundances. Yields the expected number of species around the typical individual of the community	3.74
φ_{ij}	Angle used for calculating the Ripley edge correction weights for two points i and j	3.5
φ_h	Symbolizes a homogenous pattern in the development of inhomogeneous second-order summary statistics	Sections 3.1.2.5, 3.1.2.6
φ_o	Observed inhomogeneous pattern that results from thinning a homogeneous pattern φ_h with thinning surface $p(x)$. Used for inhomogeneous second-order statistics	Sections 3.1.2.5, 3.1.2.6
k, k_1, k_2	Clumping parameters for the negative binominal distribution, used in Thomas cluster processes	Sections 4.14, 4.1.4.2, 4.1.4.4, 4.1.4.5
k	Neighbor rank	
$k_d(r)$	Phylogenetic mark-correlation function	
$k_R^B(z)$	(Two-dimensional) box kernel function, where R is the bandwidth and z the argument, usually being the distance between two points	3.15, Box 2.4
$k_R^E(z)$	(Two-dimensional) Epanechnikov kernel, where R is the bandwidth and z the argument, usually being the distance between two points	3.16, Box 2.4
$k_t(r)$	Normalized mark-correlation function based on test function t	3.87
$k_t^{\mathrm{cum}}(r)$	Cumulative and normalized mark-correlation function based on test function t	3.95
$\hat{\kappa}(r)$	Quantity analogous to the product density, used for estimation of the K-function, estimates $\lambda^2 K(r)$. May have superscripts O, WM, R, S, and BMW indicating the estimator used	3.32, 3.34, 3.37
$K(r)$	Ripley's K-function, may have superscripts O, WM, R, S, and BMW indicating the estimator used	Sections 2.3.2.2, 3.1.2.7

Symbol	Explanation	Key References
$\hat{K}^{BMW}(r, \lambda(x))$	Estimator of inhomogeneous K-function using the BMW method of Baddeley et al. (2000). With subscript WM or O: alternative estimator of inhomogeneous K-function based on Ohser edge correction	2.20, 3.38, 3.39, 3.42
$\hat{K}^{WM}_{lm}(r)$	Estimator of partial K-function for type l and m points based on the WM estimator. Substituting superscripts R, S, and O for WM indicates the Ripley, Stoyan, and Ohser estimators, respectively. The inhomogeneous estimators are indicated with superscripts "BMW," "O, inhom," and "WM, inhom"	3.61
l_x, l_y	x- and y side length of rectangle (usually used in reference to observation window W)	3.46, 3.68
$K_{12}(r)$	Bivariate K-function	
$K2(r)$	The $K2$ function, the derivative of the K-function introduced in Schiffers et al. (2008). The $K2$ function has the interesting property that it can diagnose aggregation or hyperdispersion despite the occurrence of heterogeneity in the intensity function	2.5
$K_t(r)$	Cumulative mark product function based on test function t	3.93
$L(r)$	Transformation $L(r) = [K(r)/\pi]^{0.5} - r$ of K-function to stabilize the variance and yield an expectation of zero over all distances r under CSR	Section 2.3.2.2
$L_1(r), L_2(r)$	Two different transformations of the K-function, where $L_1(r) = [K(r)/\pi]^{0.5}$ and $L_2(r) = L(r) = L_1(r) - r$. $L_1(r)$ is preferred by statisticians while $L_2(r)$ is usually used in ecological applications	2.3, 2.4
λ	Intensity of a homogeneous pattern, the mean number of points per unit area	Section 2.3.1.1
$\lambda(x)$	Intensity function depending on location x	Section 2.3.1.1
$\lambda_m(x)$	"Partial" intensity function of type m points of a bi- or multivariate pattern	

Symbol	Explanation	Key References
$\hat{\lambda}_n$	Natural estimator of the intensity, number of points n divided by area A	3.19, 3.20, 3.23, 3.29, 3.37
$\hat{\lambda}_S(r), \hat{\lambda}_V(r)$	Adapted estimators of the intensity used for estimation of the pair-correlation function and the K-function, respectively	3.22, 3.35
μ	In Thomas process: mean number of points per cluster	
μ	In mark-correlation functions: the mean mark taken over the entire quantitative pattern	3.89
μ_1, μ_2	In mark-correlation functions: mean of first and second mark, respectively	
$\mu(r)$	In mark-correlation functions: mean of the first mark of two points separated by distance r	3.88
$\mu_1(r), \mu_2(r)$	In bivariate mark-correlation functions: mean of the first and second marks of two points separated by distance r, respectively	
$\Lambda(x)$	Stochastic process that generates an intensity function $\lambda(x)$ as a single realization. Used in Cox processes	4.22, 4.24, 4.26
m_i	Mark of point i	
$m_{i,1}, m_{i,2}$	First and second mark of point i, respectively	
m_D	The mean distance to the nearest neighbor	
n	Number of points of a univariate pattern within an observation window	
n_m	Number of points of a univariate pattern within subwindow W_m	3.108, 3.109
$n_i^c(r), n_i^R(r)$	The number of additional points within (or at) distance r from point i, respectively. The superscripts c and R refer to circle and ring, respectively	3.1, 3.10, 3.14
$\bar{n}^R(r)$	Mean number of points within a ring of radius r (and width dr) around the points of a pattern, part of the naive estimator of $O(r) = \lambda g(r)$. The superscript R refers to ring	3.10

Symbol	Explanation	Key References
$\bar{n}^c(r)$	Mean number of additional points in disks with radius r around the points of a pattern, naive estimator of $\lambda K(r)$. The superscript c refers to circles	3.1
$\bar{n}^c_{\ominus r}(r)$	Same as $\bar{n}^c(r)$, but using only points i that are further away from the border than distance r. The superscript c refers to circles	3.3
$\hat{N}^g_n(r)$	Naive estimator of the expected number of points within rings with radius r and width dr that does not consider edge correction	
$n^R_i(r)$	The number of additional points within a ring with radius r and width dr around point i of the pattern. The superscript R refers to ring	3.10
$n_{\ominus r}$	Number of points in the reduced observation window $W_{\ominus r}$ where the area closer than distance r from the border of W is excluded	
$nn(k)$	Mean distance to the kth neighbor	3.48
$O(r)$	O-ring statistic, satisfies $O(r) = \lambda g(r)$	Sections 2.3.2.1, 3.1.2.2
$\hat{O}^n(r)$	Naive estimator of $O(r)$ without edge correction	3.10
$O_{12}(r)$	Bivariate O-ring statistic	
$p(k, r)$	Proportion of points that have k neighbors at distance r	
$P(k, r)$	Proportion of points that have k neighbors within distance r. Cumulative version of $p(k, r)$	
$p_{lm}(r)$	Mark-connection functions with $l, m = 1$ or 2. A mark-connection function can be interpreted as the conditional probability that the first point i is of type l and the second point j is of type m, given that there is a point of the process at a location x_i and another point of the process at a location x_j separated by a distance $\|x_i - x_j\| = r$. Can also be generalized for more than two types of points	2.6, 3.78

Symbol	Explanation	Key References
$p_{fl}(r)$	Summary statistic used for trivariate random labeling; probability that a point of a second pattern (which is marked with qualitative marks l and m) at distance r from a typical point of the focal pattern f is of type l. Can be interpreted as a mark-connection function	3.81
$p_S(\mu)$	Poisson distribution with mean μ, used for cluster processes to describe the distribution of S, the number of points being part of a given cluster	Box 2.5
p_m	Proportion of points being of type m, used to estimate expectation for mark-connection functions	2.6, 3.78
$p(x)$	Thinning function $p(x) = \lambda(x)/\lambda^*$, being the probability of accepting a tentative point when simulating a heterogeneous Poisson process (λ^* is the maximal value of the intensity function within W)	Sections 2.6.3.5, 3.1.2.5
$P(r)$, $M(r)$	Two axes of a classification scheme of bivariate co-occurrence patterns based on the K-function and the nearest-neighbor distribution function, respectively	4.36, 4.38, 4.40
$PISAR_f(r)$	Average phylogenetic dissimilarity $d(f, m)$ between all species m located within neighborhoods of radius r around the typical individual of the focal species f	3.75, 3.76
$Points_m[X]$	Operator that counts number of type m points within subarea X of observation window	3.103, 3.104, 3.118, 3.120, 3.136
r	Variable indicating distance between two points, often the distance from the typical point	
$\mathbf{r}$	Vector distance between two points	
R	Radius of a disk or bandwidth of kernel estimate of intensity $\lambda(x)$	
r_C	Cluster size of a Thomas process	Section 4.1.4.2

Symbol	Explanation	Key References
r_C	Parameter used in the cluster detection algorithm, describes the minimal distance between two clusters. If the distance between two clusters is less than r_C, the algorithm merges these clusters	Section 3.4.1
$rISAR_f(r)$	$= PISAR_f(r)/ISAR_f(r)$: yields the mean phylogenetic dissimilarity between the typical individual of a focal species f and the species present within distance r of the typical individual	3.76
(r_0, r_{max}), (r_{min}, r_{max})	Distance intervals for minimal contrast method or GoF test, respectively	Sections 2.5.1.2, 2.5.2.1, 4.1.4.3
$\rho^{(2)}(x_1, x_2)$	Second-order product density depending on two locations x_1 and x_2	Section 3.1.2.3
$\rho(r)$	Product density for homogeneous patterns that depends only on the distance r between points, may have superscripts O, WM, R, and S indicating the estimator used	
$\rho_t(r)$	Product density for quantitatively marked patterns and test function t	3.85
$\rho_{lm}(r)$	"Partial" product densities for bi- or multivariate patterns with type l and m points. Yields a univariate partial product density if $l = m$	2.6
ρ	In Thomas process the intensity of parent points	4.7
σ	Parameter of Thomas process, proportional to the cluster size	
σ_μ^2	Variance of marks of a univariate, quantitatively marked pattern	3.89
$T(r)$	T-function, a third-order summary statistic	
T_i	Two-sided test statistic (based on the standard T-test) constructed from an observed summary statistic and summary statistics from simulations of the null model	2.10
$t(m_i, m_j)$	Test function used for mark-correlation function with marks m_i and m_j	3.84, 3.85, 3.88, 3.93, 3.95, 4.60

Symbol	Explanation	Key References
$t(m_{il}, m_{jm})$	Test function used for mark-correlation function with quantitative marks m_{il} and m_{jm} taken from a pair of points where the first point i is of type l and the second point j of type m	3.96, 3.97, 3.98
u_i	Test statistic for goodness-of-fit test that represents the accumulated deviation of the observed summary statistic from the expected test statistic under the null model, summed up over an appropriate distance interval (r_{min}, r_{max})	2.7
v, v_i	Area of observation window W or subwindows W_i, used in context of estimators	3.11, 3.12
$\bar{v}(r)$	Mean area of (the potentially incomplete) rings with radius r in W around the points i of the pattern	3.13
$v(X)$	Area operator that returns the area of a geometric object X	3.35, 3.36
$v_{d-1}(X)$	Area of the border of a d-dimensional geometric object X	3.17, 3.21, 3.22, 3.23
$v(x), v_i(x)$	Covariates that depend on location x	
W	Observation window where the coordinates of the point pattern are recorded	
$W \cap \partial b(x_i, r)$	Circumference of disk $b(x_i, r)$ with center x_i and radius r that is located in W	3.17, 3.21, 3.22, 3.22
$W_1, .. , W_k$	Subwindows	
$w(d_{ik})$	Edge correction weight for point i in estimator of $D^k(r)$, dependent on distance d_{ik} from point i to kth neighbor	
w_{ij}	Weight for edge correction in estimators of second-order statistics, which compensates for incomplete rings or circles around a point at location x_i	3.5
$w_{i,j}^R(r)$	Weight for edge correction in the Ripley estimator for points i and j	3.5, 3.9
$w^O(r)$	Weight for edge correction in the Ohser estimator, only depends on distance r, not on the specific point pair	3.6, 3.9
$w^{WM}(r)$	Weight for edge correction in the WM estimator, only depends on distance r, not on the specific point pair	3.9

Symbol	Explanation	Key References
$w^S_{i,j}(r)$	Weight for edge correction in the Stoyan estimator for points i and j	3.8, 3.9
$x = (x, y)$	Location in two-dimensional space with coordinates x, y	2.1
$x_i = (x_i, y_i)$	Location of point i in two-dimensional space with coordinates x_i, y_i	
Z_1	Test statistic of Berman test	Section 2.6.2.8
$\hat{Z}^g_n(r)$	Naive estimator of the mean number of points at distance r from the points of the pattern	

Glossary

Adapted estimators: Estimators of summary statistics that are given as quotients where the same estimation rule is used in the numerator and denominator. Ensures that numerator and denominator have similar fluctuations, which partially cancel out through division.

Aggregation: Pattern shows elevated neighborhood density relative to a completely random pattern.

Aggregation formulas: Allows the estimation of a single summary statistic based on data from several *replicate patterns*.

Attraction: Points of two patterns are closer together than expected under independence. See also *repulsion*.

Average point of the pattern: Several summary statistics are based on neighborhood characteristics of points (e.g., number of points within distance r), which are estimated, in practice, as an average over all points of the pattern. If the pattern is homogeneous, the *typical point* is the same as the average point. However, for heterogeneous patterns no *typical point* exists, but the average point can be used to describe the average neighborhood characteristics of the pattern.

Berman test: Can be used to test if a point pattern shows a significant association with a continuous environmental variable. Compares the mean value of the covariate at the data points with that expected under a given null model (CSR, toroidal shift or, better yet, the pattern reconstruction null model).

Bivariate point pattern: Pattern consisting of two *univariate patterns* with *a priori* different types of points (e.g., two different species). Also called an *a priori qualitatively marked pattern*. Care must be taken in distinguishing this type of pattern from (*a posteriori*) *qualitatively marked patterns* (e.g., the pattern of dead and surviving trees of a given species).

Buffer edge correction: Mostly used for nearest-neighbor summary statistics. Uses only focal points i located within an "inner window," surrounded by a buffer of width b, to safely estimate point-pattern statistics for distances $r \le b$. Loses some information by not utilizing some points in making estimates.

Case control design: Frequently used in the context of epidemiology to detect clustering of disease cases relative to the natural variation in the background population. Uses random labeling as a null model.

Categorical habitat types: Required for *torus translation test*. Either determined statistically using *regression tree analysis* or based on *ad hoc* assumptions.

Clumping: Terminology used in the book in the context of negative binominal distributions and Thomas cluster processes; refers to the way the number of points per cluster are distributed over the clusters. This can be described with the clumping index k.

Cluster process: Point process that generates clustered patterns, for example, a Thomas process.

Clustered pattern: *Aggregated pattern*, having an elevated neighborhood density so that objects tend to lie closer to one another than would be expected with a CSR distribution. See also *hyperdispersion*.

Co-distribution patterns: Quantifies, in the context of point patterns (as opposed to quadrat-based approaches), how a second species is located in neighborhoods around individuals of a given focal species. The different types of co-distribution patterns include mixing, segregation, partial overlap, and no association. For determination of the co-distribution type of a bivariate pattern the two bivariate summary statistics $K_{12}(r)$ and $D_{12}(r)$ are used.

Confirmatory point pattern analysis: A *point process* that is used as an implementation of an ecological hypothesis about the expected stochastic properties of the observed point pattern. One can also formulate several point-process models representing alternative hypotheses. Because statistical significance is important for this approach, more simulations of the point-process model are conducted as compared to the *exploratory* approach and specific methods are required to minimize the error rates of the statistical tests.

Covariates: Properties of the environment within an *observation window* (such as elevation, slope, or wetness) that may influence the location of ecological objects. The covariate is usually measured (or interpolated) and characterized as a distribution pattern distributed as cells within a grid.

Cox processes: Broad class of point-process models to generate and describe clustered point patterns. Can accommodate clustering and environmental heterogeneity at the same time.

CSR: Complete spatial randomness, an important null model in which points are independently distributed under constant intensity. The *fundamental division* of univariate patterns. More formally it is referred to as a *homogeneous Poisson process*.

Dispersal kernel: The probability density function $f(r)$ of objects at distance r from the location of a parent object. Can be estimated non parametrically using point-pattern analysis.

Edge correction: May be required if the estimation of *summary statistics* is sensitive to knowledge of points outside the observation window W. Edge correction methods may use only focal points that do not require knowledge of points outside W, may be used to reconstruct unknown points outside W, may be accomplished by introducing weights that appropriately compensate for the lack of knowledge regarding outside points, or may directly estimate the bias caused by missing points outside W.

Exploratory point pattern analysis: The form of analysis employed when the researcher's goal is to characterize the properties of an observed point pattern as closely as possible. In this case, a *point process* with known properties is used that constitutes a "point of reference" or benchmark. Stochastic realizations of a point-process model are generated and compared with the observed data. Parametric point processes can be fit to the data and, if the fit is satisfying, the parameters can be considered to be a succinct summary of the properties of the point pattern.

Functional summary statistics: Characterizes a point pattern as a function of distance r, or neighborhood range k, or any other variable of interest. Examples include almost all *summary statistics* presented here.

Fundamental division: The simplest possible randomization of a given data type. The *fundamental division* is used to determine if the data contain a signal that can be distinguished from pure stochastic effects that arise for the given data structure with the observed sample size. For example, CSR is the fundamental division for univariate patterns, independence for bivariate patterns, and random labeling for qualitatively marked patterns.

Gibbs processes: General class of point processes that focuses on interactions among points. Motivated by techniques in physics and requires complex simulation methods.

Goodness-of-fit test: Allows assessment of the fit of a point-process model to the summary statistics of the data using a single number (i.e., test statistic) by reducing the differences between the values of the observed *functional summary statistic* and those of the replicate simulations of a null model. Avoids type I error inflation and should be used in conjunction with *simulation envelopes*.

Gradient heterogeneity: One (simple) extreme case of heterogeneity where the intensity function is a (linear) gradient, used to demonstrate effects of heterogeneity on summary statistics.

Habitat model: Parametric model to estimate the intensity function of a point pattern based on environmental variables, see also *species distribution models.*

Hanisch edge correction: Edge correction method for nearest-neighbor summary statistics that uses all focal points for which the distance to the kth neighbor can be correctly estimated.

Hard core pattern: A *hyperdispersed* pattern where the points are restricted to be separated by at least a minimal distance (i.e., the hard core distance). This pattern can be produced by randomly placing non-overlapping disks with radius R within the observation window and treating the centers of the disks as the pattern.

Heterogeneous pattern: The intensity of the pattern depends on location (i.e., a spatial trend) and/or the statistical properties of the interactions among neighboring points of the pattern change with location. No *typical point* exists.

Heterogeneous Poisson process: A point-process model that is completely determined by its intensity function $\lambda(x)$. See Box 2.3.

Homogeneous pattern: The properties of the pattern are the same in all directions and at all locations of the observation window. The *intensity* λ of the pattern is constant. The properties of the pattern can be characterized as properties of a *typical point* of the pattern (which is in practice determined by averaging over all points), exhibiting only stochastic deviations from the properties of the *typical point*.

Homogeneous Poisson process: see CSR.

Hyperdispersion: Occurs when neighborhood densities are reduced, i.e., objects tend to be farther apart than expected with a CSR distribution. See also *aggregation*.

Independence null model: Fundamental division for bivariate patterns. Must maintain the observed spatial structure of the two component patterns, but break their potential interdependence. Can be achieved parametrically by fitting a point-process model to one of the component patterns or nonparametrically using *pattern reconstruction*. An older, imperfect solution was the *toroidal shift* proposed by Lotwick and Silverman (1982).

Independent marking: Fundamental division for quantitatively marked patterns. The marks (e.g., size of trees) are randomly shuffled over the points.

Individual species area relationship ISAR$_f$(r): The point-related analogue of spatial species richness $d(r)$, which can be decomposed into the sum of the partial nearest-neighbor distribution functions of all species present in W. It is defined as the expected number of species present within distance r of the typical point of a focal species f. Is intermediate between the K-function (which operates at the species level) and the species–area relationship (SAR; which operates at the community level). Can be used to explore how a given species is located within the *landscape of local species richness*. Can be generalized to incorporate phylogenetic or functional dissimilarity between species.

Inhomogeneous summary statistics: An approach to adapt second-order statistics to patterns that show a nonconstant intensity function. The inhomogeneous summary statistics basically factor out the (first-order) heterogeneity and reveal the "pure" homogeneous characteristics of the pattern. Is especially useful if a parametric point process is fit to the data. Requires second-order, *intensity-reweighted stationarity* of the pattern; see *thinning*.

Intensity λ and intensity function $\lambda(x)$: A fundamental location-related summary statistic of a univariate point pattern, it is the mean number of points per unit area. In general, the intensity will depend on the location x because the local point density may vary with location. Homogeneous patterns have a constant intensity. The intensity function can be estimated parametrically and nonparametrically.

Irregular observation windows: Most applications of point-pattern analysis use rectangular observation windows. However, in some cases observation windows with irregular boundaries must be selected, for example, to exclude heterogeneities. Estimation of summary statistics within irregular observation windows is difficult due to edge correction.

Isotropic point patterns: Have properties that are invariant under rotations around the origin.

Isotropized set covariance $\bar{\gamma}_W(r)$: A quantity developed in set theory that is used in constructing *Ohser edge correction* weights. $\bar{\gamma}_W(r)/A$ is the expected length of the circumference of a randomly placed circle with radius r lying inside W, divided by the circumference of the corresponding full circle. Can be calculated analytically for rectangular observation windows and generalized for heterogeneous patterns with known intensity $\lambda(x)$ to derive estimators of inhomogeneous second-order statistics.

K-function: Important summary statistic used predominantly in older point-pattern studies. It is proportional to the average number of additional points within distance r of the typical point. It is the cumulative version of the pair-correlation function. Often called Ripley's K-function.

K2 function: Introduced in Schiffers et al. (2008) to diagnose aggregation or hyperdispersion of univariate patterns despite heterogeneity in the intensity function. The $K2$ function is the derivative of the pair-correlation function $g(r)$ and describes how quickly the neighborhood density declines with distance. It is insensitive to heterogeneity because heterogeneity usually introduces a slow change in $g(r)$ as r changes, whereas second-order effects cause a rapid change in $g(r)$.

Kernel function $k(x)$: Used for estimation of summary statistics based on the pair-correlation function (determines whether or not two points i and j are located distance r apart) or for *nonparametric intensity estimation*.

L-function: Transformation of the K-function to yield an expectation of zero under CSR, mostly used instead of the K-function.

Landscape of local species richness: Measures, at each location x of the observation window, the mean number of species within distance r. Related to the ISAR function.

Landscape of phylogenetic neighborhood dissimilarity: Measures, at each location x of the observation window, the mean pairwise phylogenetic distance $\Delta_f^p(x, r)$ between the focal species f and all species located within a given distance r. Collapses to the landscape of local species richness if a binary distance measure is used (i.e., distance 0 for conspecifics and distance 1 for heterospecifics).

Location-related summary statistics: Statistics evaluated from test locations that are placed within the observation window W independently of the points in the pattern (e.g., species–area relationship, intensity, spherical contact distribution, ...).

Mark: Additional information characterizing ecological objects. Marks can be (i) *a priori qualitative marks* defining different types of points created by *a priori* processes (e.g., different species; treated here as *bivariate or multivariate* patterns), (ii) *a posteriori qualitative marks* characterizing points created by a process acting *a posteriori* (e.g., dead vs. surviving individuals of the same species, trees occupied vs. non-occupied by a epiphyte) and (iii) *quantitative marks* characterizing points (e.g., size of trees, number of fruits).

Mark-connection functions: Adapted summary statistics for *qualitatively marked patterns*. For two points l and m of a pattern separated by distance r, the mark-connection function $p_{lm}(r)$ gives the probability that the first point is of type l and the second of type m. Related to the *distance decay F(r)* and *Simpson index β(r)* in the context of multivariate patterns.

Mark-correlation function: One of the mark-correlation functions, based on the test function $t(m_i, m_j) = m_i\, m_j$.

Mark-correlation functions: Adapted summary statistics for *quantitatively marked patterns*. A (nonnormalized) mark-correlation function $c_t(r)$ yields the conditional mean of a test function $t(m_i, m_j)$ calculated from the marks m_i and m_j of two points i and j, which are separated by distance r. Usually normalized with the nonspatial expectation of the test function taken over all pairs of points. Several different test functions are used, depending upon the questions being addressed, e.g., the r-mark-correlation function, the mark variogram $\gamma_m(r)$, or the Moran's I type mark-correlation function $I_m(r)$.

Mark variogram $\gamma_m(r)$: Mark-correlation function with similar structure to a variogram based on the test function $t(m_i, m_j) = (m_i - m_j)^2/2$. Small values in the mark variogram indicate similarity in the marks between points separated by distance r, whereas large values indicate dissimilarity.

Mean distance to kth neighbor: Summary statistic $nn(k)$ that yields the mean distance to the kth neighbor, as a function of the neighborhood rank k. Can be approximated by a power law.

Minimum contrast methods: Method used to fit a point-process model to the data where the summary statistic of the point-process model must be known analytically. Allows for efficient parameter estimation.

Minus-sampling: Edge correction method where the estimation of the summary statistics is only based on focal points that need no edge correction. It is an exact method that, however, disregards information.

Moran's *I* type mark-correlation function $I_m(r)$: Mark-correlation function that investigates how the marks of two points separated by distance r differ from their conditional mean value $\mu(r)$. Is a spatial variant of the classical Pearson correlation coefficient and is based on the test function $t(m_i, m_j) = [\mu(r) - m_i][\mu(r) - m_j]$.

Multiplicatively weighted, pair-correlation function $g_{mm}(r)$: Variant of a mark-correlation function based on test function $t(m_i, m_j) = m_i\, m_j$, which is not normalized with the pair-correlation function of the underlying univariate pattern. If the marks show no spatial correlation, $g_{mm}(r)$ collapses to the pair-correlation function, and if the univariate pattern is a random pattern, $g_{mm}(r)$ collapses to the mark-correlation function. It characterizes the spatial distribution of the "mark mass" rather than the point distribution.

Multivariate pattern: A point pattern consisting of the union of several *univariate patterns* with *a priori* different types of points (e.g., different species).

Naive estimators: Estimators of summary statistics that do not consider edge correction.

Nearest-neighbor distribution functions $D^k(r)$: The cumulative distribution functions of the distances r to the kth neighbor. Sometimes we use a subscript k instead of the superscript. Important summary statistics that quantify the clustering of a pattern. Is complemented by the *spherical contact distribution function* that quantifies the gaps in a pattern

Neyman–Scott process: General class of cluster processes.

Nonparametric intensity estimate: Basically a smoothing technique based on a kernel estimate of the intensity function. Usually an Epanechnikov kernel with bandwidth R is used. An important application together with *heterogeneous Poisson processes* yields a null model that basically produces a local displacement of the points within a neighborhood with radius R (in contrast to a global displacement done by CSR).

Null community: Null communities are constructed by independently superposing one realization of a given point-process model for each species present in the community. They are usually constructed based on point-process models that conserve certain aspects of the observed univariate spatial structure of each species, which is then attributed to the action of certain processes (such as habitat association or dispersal limitation).

Null model: Null models in ecology are usually viewed as pattern-generating models that randomize certain aspects of the data, while other aspects are held constant. A null model in this role functions as a standard, statistical, null hypothesis for detecting patterns. The objective is to find out if there is structure in the data that does not exist in the null model. The *fundamental division* is the simplest null model for a given data structure that completely randomizes the data conditional on its type.

Numerical summary statistics: Indices that characterize spatial pattern with a single value. Examples include the intensity λ or the mean distance m_D to the nearest neighbor. Frequently used in forestry.

Objects of finite size and real shape: In some cases, the assumption does not hold that an ecological object can be approximated as a point. Extensions of point-pattern analysis can be used to allow analysis of objects of finite size and real shape. In this approach, objects are generally approximated by an underlying grid, where the center of each cell of the object is represented as a point, and then estimators of summary statistics analogous to those used in standard point-pattern analysis are applied to this constructed pattern. Null models using this approach require additional rules to define objects, potential overlap among objects, and randomization of objects.

Observation window W: A specific subset of space within which all locations of the ecological objects of interest are recorded. Is usually a rectangle, but can also have *irregular shape*.

Ohser edge correction: Edge correction method for second-order summary statistics where the bias caused in the estimator by unobserved points is corrected, not for individual point pairs (as in Ripley edge correction), but by a single function depending only on distance r, which characterizes the expected bias.

O-ring statistic $O(r)$: A summary statistic closely related to the pair-correlation function: $O(r) = \lambda\, g(r)$. It is the average density of points at distance r away from the typical point. Sometimes called the palm intensity function or neighborhood density function.

Pair-correlation function $g(r)$: One of the most important summary statistics. It is proportional to the average density of points at distance r away from the typical point.

Parametric intensity estimate: Statistical methods to relate presence (or presence/absence, abundance) data to environmental covariates in estimating the intensity function. Closely related to habitat and species distribution modeling.

Partial summary statistics: Summary statistics for bivariate and multivariate patterns that selectively evaluate one or two types of points, resulting in partial uni- and bivariate summary statistics, respectively.

Patch heterogeneity: One (simple) extreme case of heterogeneity where a pattern is confined to one patch and the points of the pattern are distributed as CSR inside the patch. Used to demonstrate effects of heterogeneity on summary statistics.

Pattern reconstruction: Nonparametric technique based on simulated annealing, which allows the generation of point patterns exhibiting the same summary statistics as an observed pattern. Important for developing null models, especially those exhibiting independence. Can also be conditioned by an intensity function. Can also be used to assess "information content" of different summary statistics.

Phylogenetic individual species–area relationship PISAR$_f(r)$: Weights each species m in the estimation of the ISAR function with an index of phylogenetic (or functional) dissimilarity $d(f, m)$ with respect to the focal species f, which has a value of zero if $f = m$ and one if the two species show maximal dissimilarity. Can be normalized by the ISAR function to investigate the "pure" spatial phylogenetic signal, independent of species richness within the neighborhood r.

Phylogenetic mark-correlation functions $k_d(r)$: A mark-correlation function, where the test function is the phylogenetic (or functional) dissimilarity between two individuals located at distance r. Closely related with the Simpson indices, that is, $k_d(r) = (\beta^*/\beta^*_{phy}) (\beta_{phy}(r)/\beta(r))$. It is able to selectively describe phylogenetic spatial structure in the smaller-scale placement of individuals, independent of the overall phylogenetic community structure.

Phylogenetic Simpson index $\beta_{phy}(r)$: Weights each point pair i and j in the estimation of the Simpson index with an index of phylogenetic (or functional) dissimilarity $d(sp_i, sp_j)$, where sp_i is the species identifier of point i. The $d(sp_i, sp_j)$ has a value of zero if $sp_i = sp_j$ and one if the two species show maximal dissimilarity. For large values of r, $\beta_{phy}(r)$ approaches the index β^*_{phy}, which is the mean pairwise phylogenetic (or functional) dissimilarity. If normalized with the Simpson index and the nonspatial Simpson indices β^*_{phy} and β^*, it yields the phylogenetic mark-correlation function.

Plus-sampling: Edge correction method where the estimator of the summary statistics reconstructs the missing points outside the observation window by enlarging the pattern using torus translations, mirroring, etc.

Point pattern: A spatial point pattern consists of a set of ecological objects that can be characterized by their locations in space. Two types of additional information may be available: *marks* that provide additional information characterizing the objects (e.g., size or status) and mapped *covariates* that describe properties of the environment that may potentially influence the location of the objects.

Point process: The points of a point pattern are assumed to be generated by some form of stochastic mechanism, and *point processes* are mathematical models characterizing those stochastic mechanisms (but not the dynamic processes that actually generate the spatial patterns in space and time, such as birth, death, dispersal). Point processes are mathematical constructs used to characterize observed, static, point patterns. In contrast to null models, more complex point-process models are often used to model the data and the task is to confirm that the observed pattern cannot be distinguished from the patterns produced by the point process.

Point-related summary statistics: Describe the spatial structure of the pattern from the viewpoint of the points of the pattern, often the typical point, and summarize properties of the neighborhood of the points. Examples include the pair-correlation function, the nearest-neighbor distribution functions and mark-correlation functions.

Product density: The kth-order product density $\rho^{(k)}$ characterizes the frequency of possible configurations of k points in space. Important product densities are the first-order product density, which yields the intensity function $\lambda(x)$, and the second-order product density, which is related to the pair-correlation function $g(r)$.

Qualitative mark: A nonnumerical property of points, such as surviving vs. dead or infected vs. noninfected. Note the difference to a *bivariate pattern* in which the property (e.g., species 1 and 2) distinguished among the points *a priori* whereas a qualitative mark was created *a posteriori* over an existing pattern.

Quantitative mark: A numerical property of a point, such as size of a tree or number of fruits produced.

Random labeling: Fundamental division for qualitatively marked patterns, the marks are randomly shuffled over the points of the pattern.

Regression tree analysis: A constrained cluster method. Can be used to define assemblages in multivariate patterns based on environmental variables and delineates subareas within the observation window with most similar assemblages that are constrained by the environmental variables.

Replicate patterns: Often several small plots are mapped to form statistical replicates, when sample sizes are small in each individual plot. Application of *aggregation formulas* allows one to combine the data from several replicate plots into one summary statistic; this is accomplished basically by estimating weighted averages of the summary statistics. Can also be used to implement null models that otherwise would require specific software.

Repulsion: When points of two patterns at short distances are further apart than expected under independence. See also *attraction*.

Ripley edge correction: Popular edge correction method for second-order statistics, where point pairs in the estimator are individually weighted, depending on how much of the circle centered on the first point and passing through the second point overlaps the observation window W.

rISAR function: $r\mathrm{ISAR}_f(r) = \mathrm{PISAR}_f(r)/\mathrm{ISAR}_f(r)$, which is the expected phylogenetic distance between the focal species and an arbitrarily chosen species from a neighborhood with radius r around the typical point of the focal species. Will asymptotically approach the mean phylogenetic distance Δ_f^p of the focal species to all other species in the observation window.

r-mark-correlation function: Important mark-correlation function, yields the mean value $\mu(r)$ of the marks of points, which have a neighbor at distance r, divided by the mean mark μ.

Second-order statistics: Based on the spatial relationships of pairs of points. Includes the pair-correlation function, the K-function and mark-correlation and mark-connection functions.

Segregation: The points of a bivariate pattern tend to occur within different areas of the observation window.

Simpson index β^*: Measure of alpha diversity, the probability that two randomly selected individuals of a multivariate pattern (e.g., a local community) are of different types (e.g., species). This index quantifies the degree of evenness. β^* is small if one (or a few) types dominate the multivariate pattern, and β^* reaches its maximum $1 - 1/S$ if the abundances of all types are equal.

Simulation envelopes: Essential tool in point-pattern analysis used to evaluate the observed summary statistics relative to the range of values expected under the null model or a point-process model. Typically the 5th lowest and highest values of the summary statistics estimated from 199 simulations of the null model are used as simulation envelopes. In this case, the test has, at each distance value r, when viewed in isolation, an error rate of $\alpha = 0.05$. Simulation envelopes should be accompanied by a goodness-of-fit test.

Soft core pattern: A *hyperdispersed* pattern that could be produced by randomly placing nonoverlapping disks with variable radii R within the observation window and treating the centers of the disks as the pattern.

Spatial species richness $d(r)$: The point-pattern analogue to the species–area relationship SAR. It is defined as the expected number of different species present within distance r of an arbitrarily chosen point in an observation window W. Can be decomposed into the sum of the partial spherical contact distributions of all species present in W. Is a location-related summary statistic.

Spatially explicit Simpson index $\beta(r)$: Generalization of the Simpson index by introducing the condition that the two points are distance r apart. $\beta(r)$ is the conditional probability that two points separated by distance r belong to different types (species). Is sensitive to aggregation of the different species. $\beta(r)$ can be expressed as the sum of the partial pair-correlation functions. Can be formulated also as a cumulative index $\alpha(r)$, expressed as a sum of partial K-functions.

Separation of scales: Assumption in the application of heterogeneous Poisson processes with parametric intensity estimate, needed to separate small-scale effects of species interactions from larger-scale effects of environmental heterogeneity.

Species distribution models: See *habitat models* and *parametric intensity estimate*.

Spherical contact distribution function $H_s(r)$: The cumulative distribution function of the distances r from test locations to the nearest point of the pattern. Complementary to the *nearest-neighbor distribution functions*, but quantifies the "gaps" in the pattern.

Stationary pattern: Homogeneous pattern, that is, statistical properties of the point patterns are the same everywhere within the observation window.

Stoyan edge correction: Edge correction method where point pairs in the estimator are individually weighted by the factor $(a_x\,b_x)/((a_x - r_x)(b_x - r_y))$, where $\mathbf{r} = (r_x,\,r_y)$ is the distance vector between the two points, and a_x and b_x are the side lengths of W.

Strauss process: A Gibbs process that makes specific assumptions regarding the functional form of the point interactions.

Summary statistics: Summary statistics provide a brief and concise description of point patterns using numbers, functions or diagrams. They quantify the statistical properties of spatial point patterns, often as a function of distance or neighbor rank. Different data structures may require different summary statistics. In general, we may need several summary statistics to characterize different features of complex point patterns.

Superposition of point processes: Formulas that allow an estimation of the resulting summary statistics of a point process that represent the independent superposition of two point processes. Estimation is based on the summary statistics of the two point processes. A useful tool to broaden the toolbox of point-process models, especially in cluster point processes.

Test function $t(m_i, m_j)$: The core concept of the *mark-correlation functions*. Depends on the marks m_i and m_j of two points i and j, which are distance r apart.

Thinning: Important concept for estimation of inhomogeneous summary statistics. It is assumed that the given pattern could be produced in principle by independent thinning of an (unknown) homogeneous pattern with a (known) intensity function $\lambda(x)$ (*second-order intensity reweighted stationary*). The probability of accepting a point of the homogeneous pattern is given by $p(x) = \lambda(x)/\lambda^*$, where λ^* is the maximal value of the intensity function within the observation window. This allows us to "reconstruct" the summary statistics of the underlying homogeneous pattern. Note that not all heterogeneous patterns fulfill the above condition.

Thomas process: Important point-process model that generates aggregated point patterns. Has the advantage that the pair-correlation function can be estimated analytically. Can be used as a basic building block of more complex cluster processes, for example, point processes with two nested scales of clustering.

Torus translation test: Method used to determine if the spatial pattern of a species shows significant association to *categorical habitat types*. The test explicitly accounts for the univariate spatial pattern of the species by creating a null distribution of the habitat map based on toroidal shifts.

Trivariate random labeling: Tests the impact of a focal pattern (e.g., adult trees) on the process that created the qualitative mark of a second pattern (e.g., surviving versus dead saplings). For example, the test function may estimate the probability of a seedling being dead as a function of distance from an adult tree.

Typical point of the pattern: If a pattern is homogeneous, differences in point configurations at specific locations result only from random fluctuations that follow the same laws at all locations. In this case one may define a typical point of the pattern and develop summary statistics that characterize the specific point configuration in the neighborhood of the typical point. For example, the K-function $K(r)$ is related to the number of points within distance r of the typical point. Mathematically, the concept of the typical point is defined by means of palm distribution theory.

Univariate pattern: Simplest type of point pattern where all points are of the same type and the only characteristics distinguishing among them are their locations.

Unmarked patterns: Univariate, bivariate, and multivariate patterns. Unmarked patterns may only contain information on the *a priori* type of a point (e.g., the species label) but not *a posteriori* information (such as the categories surviving vs. dead), which was added by a process that acted *a posteriori* over an existing pattern. Other authors call bi- and multivariate patterns *a priori* qualitatively marked patterns.

Virtual aggregation: Arises if large-scale heterogeneity is present in the pattern (e.g., the pattern comprises one large patch with surrounding areas void of points such as a *patch heterogeneity*), which imprints on the pair-correlation function and the L-function a spurious signal of aggregation. In this case, the pair-correlation function typically shows values greater than 1 at smaller to intermediate distances and the L-function increases linearly over the same scales, creating a signal that implies there is "aggregation at all scales." The $K2$ function is not affected by this problem and can be used in conjunction with the pair-correlation function to diagnose the "real" small-scale spatial structure in the data. Other methods to avoid virtual aggregation include using the heterogeneous Poisson process as a null model or excluding void areas from the observation window.

WM edge correction: Edge correction method for second-order statistics of the "Ohser" type, which corrects for the bias in the naive estimator with a factor that only depends on the distance r between points and additionally on the locations of all points of the pattern.

References

Adler, F.R. 1996. A model of self-thinning through local competition. *Proceedings of the National Academy of Sciences, USA* 93: 9980–9984.

Baddeley, A. 2010. Modelling strategies. In: *Handbook of Spatial Statistics*, eds. A.E. Gelfand, P.J. Diggle, P. Guttorp, and M. Fuentes, 339–369. CRC Press, Boca Raton, FL.

Baddeley, A., P. Gregori, J. Mateu, R. Stoica, and D. Stoyan (eds.) 2006. *Case Studies in Spatial Point Pattern Modelling.* Lecture Notes in Statistics 185, Springer-Verlag, New York.

Baddeley, A., Møller, J., and R. Waagepetersen. 2000. Non- and semi-parametric estimation of interaction in inhomogeneous point patterns. *Statistica Neerlandica* 54: 329–350.

Baddeley, A. and R. Turner. 2000. Practical maximum pseudolikelihood for spatial point patterns. *Australian and New Zealand Journal of Statistics* 42: 283–322.

Baddeley, A. and R. Turner. 2005. Spastat: An R package for analyzing spatial point patterns. *Journal of Statistical Software* 12: 1–42.

Baddeley, A. and R. Turner. 2006. Modelling spatial point patterns in R. In: *Case Studies in Spatial Point Pattern Modelling*, eds. A. Baddeley, P. Gregori, J. Mateu, R. Stoica, and D. Stoyan, 23–74. Lecture Notes in Statistics 185. Springer-Verlag, New York.

Baddeley, A., R. Turner, J. Møller, and M. Hazelton. 2005. Residual analysis for spatial point processes. *Journal of the Royal Statistical Society Series B—Statistical Methodology* 67: 617–666.

Bagchi, R., P.A. Henrys, P.E. Brown, D.F.R.P. Burslem, P.J. Diggle, C.V.S. Gunatilleke, I.A.U.N. Gunatilleke et al. 2011. Spatial patterns reveal negative density dependence and habitat associations in tropical trees. *Ecology* 92: 1723–1729.

Baraloto C., F. Morneau, D. Bonal, L. Blanc, and B. Ferry. 2007. Seasonal water stress tolerance and habitat associations within four neotropical tree genera. *Ecology* 88: 478–489.

Barot, S., J. Gignoux, and J.C. Menaut. 1999. Demography of a savanna palm tree: Predictions from comprehensive spatial pattern analyses. *Ecology* 80: 1987–2005.

Batista, J.L.F., Maguire D.A. 1998: Modeling the spatial structure of tropical forests. *Forest Ecology and Management* 110: 293–314.

Beale, C.M., J.J. Lennon, J.M. Yearsley, M.J. Brewer, and D.A. Elston. 2010. Regression analysis of spatial data. *Ecology Letters* 13: 246–264.

Berman, M. 1986. Testing for spatial association between a point process and another stochastic process. *Applied Statistics* 35: 54–62.

Bertalanffy, L. von. 1968. *General System Theory: Foundations, Development, Applications.* Braziller, New York.

Besag, J. 1975. Statistical analysis of non-lattice data. *The Statistician* 24: 179–195.

Besag, J. 1977. Contribution to the discussion of Dr. Ripley's paper. *Journal of the Royal Statistical Society*, Series B 39: 193–195.

Biganzoli, F., T. Wiegand, and W.B. Batista. 2009. Fire-mediated interactions between shrubs in a South American temperate savannah. *Oikos* 118: 1383–1395.

Blanco, P.D., C.M. Rostagno, H.F. del Valle, A.M. Beeskow, and T. Wiegand. 2008. Grazing impacts in vegetated dune fields: Predictions from spatial pattern analysis. *Rangeland Ecology and Management* 61: 194–203.

Bolker, B.M. and S.W. Pacala. 1999. Spatial moment equations for plant competition: Understanding spatial strategies and the advantages of short dispersal. *American Naturalist* 153: 575–602.

Bossdorf, O., F. Schurr, and J. Schumacher. 2000. Spatial patterns of plant association in grazed and ungrazed shrublands in the semi-arid Karoo, South Africa. *Journal of Vegetation Science* 11: 253–258.

Bourguignon, T., M. Leponce, and Y. Roisin. 2011. Are the spatio-temporal dynamics of soil-feeding termite colonies shaped by intraspecific competition? *Ecological Entomology* 36: 776–785.

Brix, A., R. Senoussi, P. Couteron, and J. Chadoeuf. 2001. Assessing goodness of fit of spatially inhomogeneous Poisson processes. *Biometrika* 88: 487–497.

Brodatzki, U. and K. Mecke. 2002. Simulating stochastic geometries: Morphology of overlapping grains. *Computer Physics Communication* 147: 218–221.

Brown, C., R. Law, J.B. Illian, and D.F.R.P. Burslem. 2011. Linking ecological processes with spatial and non-spatial patterns in plant communities. *Journal of Ecology* 99: 1402–1414.

Burnham, K.P. and D.R. Anderson. 2002. *Model Selection and Multi-model Inference: A Practical Information-Theoretical Approach*. Springer-Verlag, New York.

Calabrese, J.M., F. Vazquez, C. López, M. San Miguel, and V. Grimm. 2010. The independent and interactive effects of tree-tree establishment competition and fire on savanna dynamics and spatial structure. *American Naturalist* 175: E44–E65.

Cale, W.G., G.M. Henebry, and J.A. Yeakly. 1989. Inferring process from pattern in natural communities. *BioScience* 39: 600–606.

Cavender-Bares, J., A. Keen, and B. Miles. 2006. Phylogenetic structure of Floridian plant communities depends on taxonomic and spatial scale. *Ecology* 87: S109–S122.

Chave, J. and E.G. Leigh. 2002. A spatially explicit neutral model of beta-diversity in tropical forests. *Theoretical Population Biology* 62: 153–168.

Cheng, J., X. Mi, K. Nadrowski, H. Ren, J. Zhang, and K. Ma. 2012. Separating the effect of mechanisms shaping species-abundance distributions at multiple scales in a subtropical forest. *Oikos* 121: 236–244.

Chesson, P. 2000a. Mechanisms of maintenance of species diversity. *Annual Review of Ecology, Evolution, and Systematics* 31: 343–366.

Chesson, P. 2000b. General theory of competitive coexistence in spatially-varying environments. *Theoretical Population Biology* 58: 211–237.

Chuyong, G.B., D. Kenfack, K.E. Harms, D.W. Thomas, R. Condit, and L.S. Comita. 2011. Habitat specificity and diversity of tree species in an African wet tropical forest. *Plant Ecology* 212: 1363–1374.

Clark, J.S. M. Silman, R. Kern, E. Macklin, and J. HilleRisLambers. 1999. Seed dispersal near and far: Patterns across temperate and tropical forests. *Ecology* 80: 1475–1494.

Clark, P.J. and F.C. Evans. 1954. Distance to nearest neighbour as a measure of spatial relationships in populations. *Ecology* 35: 445–453.

Comas, C., P. Delicado, and J. Mateu. 2011. A second order approach to analyse spatial point patterns with functional marks. *TEST* 20: 503–523.

Comita, L.S., R. Condit, and S.P. Hubbell. 2007. Developmental changes in habitat associations of tropical trees. *Journal of Ecology* 95: 482–492.

Condit, R. 1998. Tropical forest census plots. Springer-Verlag, Berlin, and R. G. Landes Company, Georgetown, Texas.

Condit, R., P.S. Ashton, P. Baker, S. Bunyavejchewin, C.V.S. Gunatilleke, I.A.U.N. Gunatilleke, S.P. Hubbell et al. 2000. Spatial patterns in the distribution of tropical tree species. *Science* 288: 1414–1418.

Condit, R., N. Pitman, E.G. Leigh, J. Chave, J. Terborgh, R.B. Foster, P. Nuñez et al. 2002. Betadiversity in tropical forest trees. *Science* 295: 666–669.

Connell, J.H. 1971. On the role of natural enemies in preventing competitive exclusion in some marine animals and in rain forest trees. In *Dynamics of Numbers in Populations*, eds. P.J. den Boer and G.R. Gradwell, 298–312. PUDOC, Wageningen, the Netherlands.

Coomes, D.A., M. Rees, and L. Turnbull. 1999. Identifying aggregation and association in fully mapped spatial data. *Ecology* 80: 554–565.

Couteron, P., J. Seghieri, and J. Chadoeuf. 2003. A test for spatial relationships between neighbouring plants in plots of heterogeneous plant density. *Journal of Vegetation Science* 14: 163–172.

Cressie, N. 1993. *Statistics for Spatial Data, Revised Edition*. John Wiley & Sons, New York.

Cutler, N.A., L.R. Belyea, and A.J. Dugmore. 2008. The spatiotemporal dynamics of a primary succession. *Journal of Ecology* 96: 231–246.

Dale, M.R.T. 1999. *Spatial Pattern Analysis in Plant Ecology*. Cambridge University Press, Cambridge.

Daws, M.I., C.E. Mullins, D.F.R.P. Burslem, S.R. Paton, and J.W. Dalling. 2002. Topographic position affects the water regime in a semideciduous tropical forest in Panama. *Plant and Soil* 238: 79–90.

De'ath, G. 2002. Multivariate regression trees: A new technique for modeling species–environment relationships. *Ecology* 83: 1105–1117.

Degenhardt, A. 1999. Description of tree distribution and their development through marked Gibbs processes. *Biometrical Journal* 41: 457–470.

De la Cruz, M., R.L. Romao, A. Escudero, and F.T. Maestre. 2008. Where do seedlings go? A spatio-temporal analysis of seedling mortality in a semi-arid gypsophyte. *Ecography* 31: 1–11.

De Luis, M., J. Raventós, T. Wiegand, and J.C. González-Hidalgo. 2008. Temporal and spatial differentiation in seedling emergence may promote species coexistence in Mediterranean fire-prone ecosystems. *Ecography* 31: 620–629.

Detto, M. and H.C. Muller-Landau. 2013. Fitting ecological process models to spatial patterns using scale-wise variances and moment equations. *The American Naturalist* 181: E68–82.

Diggle, P.J. 1983. *Statistical Analysis of Spatial Point Patterns*. Academic Press, London.

Diggle, P.J. 2003. *Statistical Analysis of Spatial Point Patterns*. 2nd ed. Arnold, London.

Diggle, P.J. and A. Chetwynd 1991. Second-order analysis of spatial clustering for inhomogeneous populations. *Biometrics* 47: 1155–1163.

Diggle, P.J., V. Gómez-Rubio, P.E. Brown, A.G. Chetwynd, and S. Gooding. 2007. Second-order analysis of inhomogeneous spatial point processes using case–control data. *Biometrics* 63: 550–557.

Dixon, P.M. 1994. Testing spatial segregation using a nearest-neighbor contingency table. *Ecology* 75: 1940–1948.

Dixon, P.M. 2002. Ripley's K function. *Encyclopedia of Environmetrics* 3: 1796–1803.

Dormann, C.F., J.M. McPherson, M.B. Araujo, R. Bivand, J. Bolliger, G. Carl, R.G. Davies et al. 2007. Methods to account for spatial autocorrelation in the analysis of species distributional data: A review. *Ecography* 30: 609–628.

Dovčiak, M., L.E. Frelich, and P.B. Reich. 2001. Discordance in spatial patterns of white pine (*Pinus strobus*) size-classes in a patchy near-boreal forest. *Journal of Ecology* 89: 280–291.

Duboz, R., D. Versmisse, M. Travers, E. Ramat, and Y.J. Shin. 2010. Application of an evolutionary algorithm to the inverse parameter estimation of an individual-based model. *Ecological Modelling* 221: 840–849.

Elith, J. and J.R. Leathwick. 2009. Species distribution models: Ecological explanation and prediction across space and time. *Annual Review of Ecology, Evolution, and Systematics* 40: 677–697.

Faith, D.P. 1992. Conservation evaluation and phylogenetic diversity. *Biological Conservation* 61: 1–10.

Fang, W. 2005. Spatial analysis of an invasion front of *Acer platanoides*: Dynamic inferences from static data. *Ecography* 28: 283–294.

Felinks, B. and T. Wiegand. 2008. Analysis of spatial pattern in early stages of primary succession on former lignite mining sites. *Journal of Vegetation Science* 19: 267–276.

Galiano, E.F. 1982. Pattern detection in plant populations through the analysis of plant-to-all-plants distances. *Vegetatio* 49: 39–43.

Gelfand, A.E., P.J. Diggle, M. Fuentes, and P. Guttorp (eds.) 2010. *Handbook of Spatial Statistics*, CRC Press, Boca Raton, FL.

Getis, A. and J. Franklin. 1987. Second-order neighborhood analysis of mapped point patterns. *Ecology* 68: 474–477.

Getzin, S., C. Dean, F. He, T. Trofymow, K. Wiegand, and T. Wiegand. 2006. Spatial patterns and competition of tree species in a Douglas-fir chronosequence on Vancouver Island. *Ecography* 29: 671–682.

Getzin, S., T. Wiegand, K. Wiegand, and F. He. 2008. Heterogeneity influences spatial patterns and demographics in forest stands. *Journal of Ecology* 96: 807–820.

Gil A., A. Lobo, M. Abadi, L. Silva, and H. Calado. 2013. Mapping invasive woody plants in Azores protected areas by using very high-resolution multispectral imagery. *European Journal of Remote Sensing* 46: 289–304.

Goreaud, F. and R. Pelissier. 1999. On explicit formulas of edge effect correction for Ripley's K-function. *Journal of Vegetation Science* 10: 433–438.

Goreaud, F. and R. Pelissier. 2003. Avoiding misinterpretation of biotic interactions with the intertype K-12-function: Population independence vs. random labelling hypotheses. *Journal of Vegetation Science* 14: 681–692.

Gotelli, N.J. and A.M. Ellison. 2004. *A Primer of Ecological Statistics*. Sinauer Associates, Sunderland, MA.

Gotelli, N.J. and G.R. Graves. 1996. *Null Models in Ecology*. Smithsonian Institution Press, Washington, DC.

Grabarnik, P., M. Myllymäki, and D. Stoyan. 2011. Correct testing of mark independence for marked point patterns. *Ecological Modelling* 222: 3888–3894.

Grabarnik, P. and A. Särkkä. 2009. Modelling the spatial structure of forest stands by multivariate processes with hierarchical interactions. *Ecological Modelling* 220: 1232–1240.

Grimm, V. and S.F. Railsback. 2012. Pattern-oriented modelling: A "multiscope" for predictive systems ecology. *Philosophical Transactions Royal Society London B* 367: 298–310.

Grimm, V., E. Revilla, U. Berger, F. Jeltsch, W. Mooij, S.F. Railsback, H. Thulke, J. Weiner, T. Wiegand, and D.L. DeAngelis. 2005. *Individual-Based Modeling and Ecology*. Princeton University Press, Princeton, NJ.

Grimm, V., E. Revilla, U. Berger et al. 2005. Pattern-oriented modeling of agent-based complex Systems: Lessons from ecology. *Science* 310: 987–991.

Guan, Y. 2008. A KPSS test for stationarity for spatial point processes. *Biometrics* 64: 800–806.

Guisan, A. and W. Thuiller. 2005. Predicting species distribution: Offering more than simple habitat models. *Ecology Letters* 8: 993–1009.

Guisan, A. and N.E. Zimmermann. 2000. Predictive habitat distribution models in ecology. *Ecological Modelling* 135: 147–186.

Gunatilleke, C.V.S., I.A.U.N. Gunatilleke, S. Esufali, K.E. Harms, P.M.S. Ashton, D.F.R.P. Burslem, and P.S. Ashton. 2006. Species–habitat associations in a Sri Lankan dipterocarp forest. *Journal of Tropical Ecology* 22: 371–384.

Haase, P. 1995. Spatial pattern analysis in ecology based on Ripley's K-function: Introduction and methods of edge correction. *Journal of Vegetation Science* 6: 575–582.

Haase, P. 2001. Can isotropy vs. anisotropy in the spatial association of plant species reveal physical vs. biotic facilitation? *Journal of Vegetation Science* 12: 127–136.

Hahn, U., E.B.V. Jensen, M.N.M. van Lieshout, and L.S. Nielsen. 2003. Inhomogeneous spatial point processes by location-dependent scaling. *Advances in Applied Probability* 35: 319–336.

Hanisch, K.-H. 1984. Some remarks on estimators of the distribution function of nearest-neighbour distance in stationary spatial point patterns. *Statistics* 15: 409–412.

Hardy, O.J. and B. Senterre. 2007. Characterizing the phylogenetic structure of communities by an additive partitioning of phylogenetic diversity. *Journal of Ecology* 95: 493–506.

Harms, K.E., R. Condit, S.P. Hubbell, and R.B. Foster. 2001. Habitat associations of trees and shrubs in a 50-ha neotropical forest plot. *Journal of Ecology* 89: 947–959.

Harper, J.L. 1977. *Population Biology of Plants*. Academic Press, London

Hartig, F.J. Calabrese, B. Reineking, T. Wiegand, and A. Huth. 2011. Statistical inference for stochastic simulations models—Theory and application. *Ecology Letters* 14: 816–827.

He, F. and R.P. Duncan. 2000. Density-dependent effects on tree survival in an old-growth Douglas fir forest. *Journal of Ecology* 88: 676–688.

Henrys, P.A. and P.E. Brown. 2009. Inference for clustered inhomogeneous spatial point processes. *Biometrics* 65: 423–430.

Hilborn, R. and M. Mangel. 1997. *The Ecological Detective. Confronting Models with Data*. Princeton University Press, Princeton.

Högmander, H. and A. Särkkä, 1999. Multitype spatial point patterns with hierarchical interactions. *Biometrics* 55: 1051–1058.

Howe, H.F. 1989. Scatter- and clump-dispersal and seedling demography: Hypothesis and implications. *Oecologia* 79: 417–426.

Hubbell, S.P. 2001. *The Unified Neutral Theory of Biodiversity and Biogeography*. Princeton University Press, Princeton, NJ.

Hubbell, S.P., J.A. Ahumada, R. Condit, and R.B. Foster. 2001. Local neighborhood effects on long-term survival of individual trees in a neotropical forest. *Ecological Research* 16: S45–S61.

Hubbell, S.P. and R.B. Foster. 1983. Diversity of canopy trees in a neotropical forest and implications for conservation. In *Tropical Rain Forest: Ecology and Management*, eds. S.L. Sutton, T.C. Whitmore, and A.C. Chadwick, 25–41. Blackwell Scientific, Oxford.

Hubbell, S.P. and R.B. Foster. 1986. Commonness and rarity in a neotropical forest: Implications for tropical tree conservation. In *Conservation Biology: The Science of Scarcity and Diversity*, ed. M. Soule, 205–231. Sinauer Associates, Sunderland, MA.

Hubbell, S.P., F. He, R. Condit, L. Borda-de-Água, J. Kellner, and H. ter Steege. 2008. How many tree species are there in the Amazon and how many of them will go extinct? *Proceedings of the National Academy of Sciences, USA* 105S: 11498–11504.

Hurtt, G.C. and S.W. Pacala. 1995. The consequences of recruitment limitation: Reconciling chance, history and competitive differences between plants. *Journal of Theoretical Biology* 176: 1–12.

Hutchinson, G.E. 1957. Concluding remarks. *Cold Spring Harbor Symposium on Quantitative Biology* 22: 415–427.

Illian, J.B., J. Møller, and R. Waagepetersen. 2009. Hierarchical spatial point process analysis for a plant community with high biodiversity. *Environmental and Ecological Statistics* 16: 389–405.

Illian, J.B., A. Penttinen, H. Stoyan, and D. Stoyan. 2008. *Statistical Analysis and Modelling of Spatial Point Patterns*. John Wiley & Sons, Chichester, England.

Jacquemyn, H., R. Brys, O. Honnay, I. Roldán-Ruiz, B. Lievens, and T. Wiegand. 2012a. Non-random distribution of orchids reflects different mycorrhizal association patterns in a hybrid zone of three Orchis species. *New Phytologist* 193: 454–464.

Jacquemyn, H., R. Brys, B. Lievens, and T. Wiegand. 2012b. Spatial variation in belowground seed germination and divergent mycorrhizal associations correlate with spatial segregation of three co-occurring orchid species. *Journal of Ecology* 100: 1328–1337.

Jacquemyn, H., R. Brys, K. Vandepitte, O. Honnay, I. Roldán-Ruiz, and T. Wiegand. 2007. A spatially-explicit analysis of seedling recruitment in the terrestrial orchid *Orchis purpurea*. *New Phytologist* 176: 448–459.

Jacquemyn, H., P. Endels, O. Honnay, and T. Wiegand. 2010. Spatio-temporal analysis of seedling recruitment, mortality and persistence into later life stages in the rare *Primula vulgaris*. *Journal of Applied Ecology* 47: 431–440.

Janzen, D.H. 1970. Herbivores and the number of tree species in tropical forests. *The American Naturalist* 104: 501–528.

Jeltsch, F., K.A. Moloney, and S.J. Milton. 1999. Detecting process from snap-shot pattern: Lessons from tree spacing in the southern Kalahari. *Oikos* 85: 451–467.

John, R., J.W. Dalling, K.E. Harms, J.B. Yavitt, R.F. Stallard, M. Mirabello, S.P. Hubbell et al. 2007. Soil nutrients influence spatial distributions of tropical tree species. *Proceedings of the National Academy of Sciences, USA* 104: 864–869.

Journel, A.G. and C.J. Huijbregts, 1978. *Mining Geostatistics.* Academic Press, London.

Kanagaraj, R.T. Wiegand, L. Comita, and A. Huth. 2011b. Tropical tree species assemblages in topographic habitats change in time and with life stage. *Journal of Ecology* 99: 1441–1452.

Kanagaraj, R., T. Wiegand, S. Kramer-Schadt, M. Anwar, and S.P. Goyal. 2011a. Assessing habitat suitability for tiger in the fragmented Terai Arc Landscape of India and Nepal. *Ecography* 34: 970–981.

Kenkel, N.C. 1988. Pattern of self-thinning in jack pine: Testing the random mortality hypothesis. *Ecology* 69: 1017–1024.

Kenkel, N.C., M.L. Hendrie, and I.E. Bella. 1997. A long-term study of *Pinus banksiana* population dynamics. *Journal of Vegetation Science* 8: 241–254.

Kirkpatrick, S., C.D. Gelatt, and M.P. Vecchi. 1983. Optimization by simulated annealing. *Science* 220: 671–680.

Klaas, B.A., K.A. Moloney, and B.J. Danielson. 2000. The tempo and mode of gopher mound production in a tallgrass prairie remnant. *Ecography* 23: 246–256.

Komuro, R., E.D. Ford, and J.H. Reynolds. 2006. The use of multicriteria assessment in developing a process model. *Ecological Modelling* 197: 320–330.

Koukoulas, S. and A. Blackburn. 2005. Spatial relationships between tree species and gap characteristics in broad-leaved deciduous woodland. *Journal of Vegetation Science* 16: 587–596.

Kraft, N.J.B., W.K. Cornwell, C.O. Webb, and D.D. Ackerly. 2007. Trait evolution, community assembly, and the phylogenetic structure of ecological communities. *American Naturalist* 170: 271–283.

Krebs, C. 1978. *Ecology: The Experimental Analysis of Distribution and Abundance*. Harper & Row, New York.

Kubota, Y., H. Kubo, and K. Shimatani. 2007. Spatial pattern dynamics over 10 years in a conifer/broadleaved forest, northern Japan. *Plant Ecology* 190: 143–157.

Lai, J.S., X.C. Mi, H.B. Ren, and K.P. Ma. 2009. Species–habitat associations change in a subtropical forest of China. *Journal of Vegetation Science* 20: 415–423.

Law, R., J. Illian, D.F.R.P. Burslem, G. Gratzer, C.V.S. Gunatilleke, and I.A.U.N. Gunatilleke. 2009. Ecological information from spatial point patterns of plants, insights from point process theory. *Journal of Ecology* 97: 616–628.

Legendre, P., X. Mi, H. Ren, K. Ma, M. Yu, I-F. Sun, and F. He. 2009. Partitioning beta diversity in a subtropical broad-leaved forest of China. *Ecology* 90: 663–674.

Leigh, E.G. 1999. *Tropical Forest Ecology: A View from Barro Colorado Island*. Oxford University Press, New York.

Leigh, E.G.J., S. Lao, R. Condit, S.P. Hubbell, R.B. Foster, and R. Perez. 2004. Barro Colorado Island forest dynamics plot, Panama. In *Tropical Forest Diversity and Dynamism: Findings from a Large-Scale Plot Network*, eds. E. Losos and E.G.J. Leigh, pp. 451–463. University of Chicago Press, Chicago.

Lele, S.R. 2009. A new method for estimation of resource selection probability function. *Journal of Wildlife Management* 73: 122–127.

Lele, S.R. and J.L. Keim. 2006. Weighted distributions and estimation of resource selection probability functions. *Ecology* 87: 3021–3028.

Levin, S.A. 1992. The problem of pattern and scale in ecology. *Ecology* 73:1943–1967.

Lieberman, M. and D. Lieberman. 2007. Nearest-neighbor tree species combinations in tropical forest: The role of chance, and some consequences of high diversity. *Oikos* 116: 377–386.

Lieshout, M.N.M. van, and A.J. Baddeley. 1996. A nonparametric measure of spatial interaction in point patterns. *Statistica Neerlandica* 50: 344–361.

Lin Y.C., L.W. Chang, K.C. Yang, H.H. Wang, and I.F. Sun. 2011. Point patterns of tree distribution determined by habitat heterogeneity and dispersal limitation. *Oecologia* 165: 175–84.

Loosmore, N.B. and E.D. Ford. 2006. Statistical inference using the *G* or *K* point pattern spatial statistics. *Ecology* 87: 1925–1931.

Lotwick, H.W. and B.W Silverman. 1982. Methods for analysing spatial processes of several types of points. *Journal of the Royal Statistical Society B* 44: 406–413.

Maheu-Giroux, M. and S. de Blois. 2007. Landscape ecology of *Phragmites australis* invasion in a network of linear wetlands. *Landscape Ecology* 22: 285–301.

Manly, B.F. J., L.L. McDonald, D.L. Thomas, T.L. McDonald, and W.P. Erickson. 2002. *Resource Selection by Animals: Statistical Analysis and Design for Field Studies.* 2nd ed. Kluwer Press, Boston, MA.

Martínez, I., F. González-Taboada, T. Wiegand, J.J. Camarero, and E. Gutiérrez. 2012. Dispersal limitation and spatial scale affect model based projections of *Pinus uncinata* response to climate change in the Pyrenees. *Global Change Biology* 18: 1714–1724.

Martínez, I., T. Wiegand, J.J. Camarero, E. Batllori, and E. Gutiérrez. 2011. Elucidating demographic processes underlying tree line patterns: A novel approach to model selection for individual-based models using Bayesian methods and MCMC. *American Naturalist* 177: E136–E152.

Mast, J.N. and T.T. Veblen. 1999. Tree spatial patterns and stand development along the pine-grassland ecotone in the Colorado Front Range. *Canadian Journal of Forest Research* 29: 575–584.

May, R.M. 1976. *Theoretical Ecology: Principles and Applications.* Blackwell Scientific Publications, Oxford.

McGarigal, K. and B.J. Marks. 1995. FRAGSTATS: Spatial analysis program for quantifying landscape structure. USDA Forest Service General Technical Report PNW-GTR-351.

McGill, B.J. 2010. Towards a unification of unified theories of biodiversity. *Ecology Letters* 13: 627–642.

McIntire, E.J.B. and A. Fajardo. 2009. Beyond description: The active and effective way to infer processes from spatial patterns. *Ecology* 90: 46–56.

Mecke, K.R. and D. Stoyan. 2005. Morphological characterization of point patterns. *Biometrical Journal* 47: 473–88.

Metropolis, N., A.W. Rosenbluth, M.N. Rosenbluth, A.H. Teller, and E. Teller. 1953. Equation of state calculations by fast computing machines. *Journal of Chemical Physics* 21: 1087–1092.

Møller, J. 2010. Parametric methods. In *A Handbook of Spatial Statistics*, eds. A.E. Gelfand, P. Diggle, M. Fuentes, and P. Guttorp, pp. 317–338. Chapman & Hall/ CRC Press, Boca Raton, FL.

Møller, J. and R. Waagepetersen. 2004. *Statistical Inference and Simulation for Spatial Point Processes.* Chapman & Hall/CRC Press, Boca Raton, FL.

Møller J. and R. Waagepetersen. 2007. Modern statistics for spatial point processes. *Scand. J. Statist.* 34: 643–684.

Molofsky, J., J.D. Bever, J. Antonovics, and T.J. Newman. 2002. Negative frequency dependence and the importance of spatial scale. *Ecology* 83: 21–27.

Moloney, K.A. 1993. Determining process through pattern: Reality or fantasy? In *Patch Dynamics*, eds. S.A., Levin, T. Powell, and J. Steele, 61–69. Lecture Notes in Biomathematics Vol. 96. Springer-Verlag, Berlin.

Morlon H., G. Chuyong, R. Condit, S.P. Hubbell, D. Kenfack, D. Thomas, R. Valencia, and J.L. Green. 2008. A general framework for the distance-decay of similarity in ecological communities. *Ecology Letters* 11: 904–917.

Moustakas A., M. Günther, K. Wiegand, K.-H. Müller, D. Ward, and K.M. Meyer. 2006. Mortality of *Acacia erioloba*: Influence of climate and intra-specific competition. The middle class shall die! *Journal of Vegetation Science* 17: 473–480.

Murrell, D.J., D.W. Purves, and R. Law. 2001. Uniting pattern and process in plant ecology. *Trends in Ecology and Evolution* 16: 529–530.

Nekola, J.C. and P.S. White. 1999. Distance decay of similarity in biogeography and ecology. *Journal of Biogeography* 26: 867–878.

Nelson, T., K.O. Niemann, and M.A. Wulder. 2002. Spatial statistical techniques for aggregating point objects extracted from high spatial resolution remotely sensed imagery. *Journal of Geographical Systems* 4: 423–433.

Neyman, J. and E.L. Scott. 1958. Statistical approach to problems of cosmology (with discussion). *Journal of the Royal Statistical Society B* 20: 1–43.

Niggemann, M., T. Wiegand, J.J. Robledo-Arnuncio, and R. Bialozyt. 2012. Marked point pattern analysis on genetic paternity data for estimation and uncertainty assessment of pollen dispersal kernels. *Journal of Ecology* 100: 264–276.

Nuske, R.S., S. Sprauer, and J. Saborowski. 2009. Adapting the pair-correlation function for analysing the spatial distribution of canopy gaps. *Forest Ecology and Management* 259: 107–116.

Ogata, Y. and M. Tanemura. 1985. Estimation of interaction potentials of marked spatial point patterns through the maximum likelihood method. *Biometrics* 41: 421–433.

Ohser J. and F. Mücklich. 2000. *Statistical Analysis of Microstructures in Materials Science.* J. Wiley & Sons, Chichester.

Pacala, S.W. 1997. Dynamics of plant communities. In *Plant Ecology*, ed. M.J. Crawley, pp. 532–555. 2nd ed. Blackwell Scientific, Oxford.

Pacala, S.W. and S.A. Levin. 1997. Biologically generated spatial pattern and the coexistence of competing species. In: *Spatial Ecology: The Role of Space in Population Dynamics and Interspecific Interactions*, eds. D. Tilman and P. Kareiva, pp. 204–232. Princeton University Press, Princeton, NJ.

Passos, L. and P.S. Oliveira. 2002. Ants affect the distribution and performance of seedlings of *Clusia criuva*, a primarily birddispersed rain forest tree. *Journal of Ecology* 90: 517–528.

Pélissier, R. and F. Goreaud. 2001. A practical approach to the study of spatial structure in simple cases of heterogeneous vegetation. *Journal of Vegetation Science* 12: 99–108.

Penttinen, A. 1984. Modelling interaction in spatial point patterns: Parameter estimation by the maximum likelihood method. Jyväskylä Studies in Computer Science, Economics and Statistics 7.

Perry, G.L.W., N.J. Enright, B.P. Miller, and B.B. Lamont. 2009. Nearest-neighbour interactions in species-rich shrublands: The roles of abundance, spatial patterns and resources. *Oikos* 118: 161–174.

Perry, G.L.W., B.P. Miller, and N.J. Enright. 2006. A comparison of methods for the statistical analysis of spatial point patterns. *Plant Ecology* 187: 59–82.

Petchey, O.L. and K.J. Gaston. 2006. Functional diversity: Back to basics and looking forward. *Ecology Letters* 9: 741–758.

Peters, H.A. 2003. Neighbour-regulated mortality: The influence of positive and negative density dependence on tree populations in species-rich tropical forests. *Ecology Letters* 6: 757–765.

Peterson, C.J. and E.R. Squiers. 1995. Competition and succession in an aspen-white-pine forest. *Journal of Ecology* 83: 449–457.

Picard, N., A. Bar-Hen, F. Mortier, and J. Chadoeuf. 2009. Understanding the dynamics of an undisturbed tropical rain forest from the spatial pattern of trees. *Journal of Ecology* 97: 97–108.

Pielou, E.C. 1959. The use of point-to-point distances in the study of the pattern of plant populations. *Journal of Ecology* 47: 607–613.

Pielou, E.C. 1977. *Mathematical Ecology*. John Wiley & Sons, New York.

Platt, W.J., G.W. Evans, and S.L. Rathbun. 1988. The population dynamics of a long-lived conifer (*Pinus palustris*). *The American Naturalist* 131:491–525.

Plotkin, J.B., J. Chave, and P.S. Ashton. 2002. Cluster analysis of spatial patterns in Malaysian tree species. *American Naturalist* 160: 629–644.

Plotkin, J.B., M.D. Potts, N. Leslie, N. Manokaran, J.V. LaFrankie, and P.S. Ashton. 2000. Species–area curves, spatial aggregation, and habitat specialization in tropical forests. *Journal of Theoretical Biology* 207: 81–99.

Podani, J. and T. Czaran. 1997. Individual-centered analysis of mapped point patterns representing multispecies assemblages. *Journal of Vegetation Science* 8: 259–270.

Pommerening, A., A.C. Gonçalves, and R. Rodriquez-Soalleiro. 2011. Species mingling and diameter differentiation as second-order characteristics. *Allgemeine Forst- und Jagdzeitung* 182: 115–128.

Pommerening, A. and D. Stoyan. 2006. Edge-correction needs in estimating indices of spatial forest structure. *Canadian Journal of Forest Research* 36: 1723–1739.

Potts, M.D., S.J. Davies, W.H. Bossert, S. Tan, and M.N. Supardi. 2004. Habitat heterogeneity and niche structure of trees in two tropical rain forests. *Oecologia* 139: 446–453.

Prentice, I.C. and M.J.A. Werger. 1985. Clump spacing in a desert dwarf shrub community. *Vegetatio* 63: 133–139.

Punchi-Manage, R., S. Getzin, T. Wiegand, R. Kanagaraj, C.V.S. Gunatilleke, I.A.U.N. Gunatilleke, K. Wiegand, and A. Huth. 2013. Effects of topography on structuring local species assemblages in a Sri Lankan mixed dipterocarp forest. *Journal of Ecology* 101: 149–160.

Purves, D.W. and R. Law. 2002. Fine-scale spatial structure in a grassland community: Quantifying the plant's-eye view. *Journal of Ecology* 90: 121–129.

Railsback, S.F. and V. Grimm. 2012. *Agent-Based and Individual-Based Modeling: A Practical Introduction*. Princeton University Press, Princeton, NJ.

Rajala, T. and J.B. Illian. 2012. A family of spatial biodiversity measures based on graphs. *Environmental and Ecological Statistics* 19: 545–572.

Raventós, J., E. Mujica, T. Wiegand, and A. Bonet. 2011. Analyzing the spatial structure of *Broughtonia cubensis* (Orchidaceae) populations in the dry forests of Guanahacabibes, Cuba. *Biotropica* 43: 173–182.

Raventós, J., T. Wiegand, and M. De Luis. 2010. Evidence for the spatial segregation hypothesis: A test with nine-year survivorship data in a Mediterranean fire-prone shrubland show that interspecific and density-dependent spatial interactions dominate. *Ecology* 91: 2110–2120.

Raventós, J., T. Wiegand, F.T. Maestre, and M. De Luis. 2012. A resprouter herb reduces negative density-dependent effects among neighboring seeders after fire. *Acta Oecologica* 38: 17–23.

Reardon, S. and D. O'Sullivan. 2004. Measures of spatial segregation. *Sociological Methodology* 34: 121–162.

Renshaw, E. and A. Särkkä. 2001. Gibbs point processes for studying the development of spatial–temporal stochastic processes. *Computational Statistics and Data Analysis* 36: 85–105.

Riginos, C., S.J. Milton, and T. Wiegand. 2005. Context-dependent negative and positive interactions between adult shrubs and seedlings in a semi-arid shrubland. *Journal of Vegetation Science* 16: 331–340.

Rintoul, M.D. and S. Torquato. 1997. Reconstruction of the structure of dispersions. *Journal of Colloid and Interface Science* 186: 467–476.

Ripley, B.D. 1976. The second-order analysis of stationary point processes. *Journal of Applied Probability* 13: 255–266.

Ripley, B.D. 1977. Modelling spatial patterns (with discussion). *Journal of the Royal Statistical Society Series B* 39: 172–212.

Ripley, B.D. 1981. *Spatial Statistics*. John Wiley Sons, New York.

Ripley, B.D. 1988. *Statistical Inference for Spatial Processes*. Cambridge University Press, Cambridge, England.

Robledo-Arnuncio, J.J. and C. García. 2007. Estimation of the seed dispersal kernel from exact identification of source plants. *Molecular Ecology* 16: 5098–5109.

Rodríguez-Pérez, J., T. Wiegand, and A. Traveset. 2012. Adult proximity and frugivore activity structure plant populations—Spatial patterns after the disperser's loss. *Functional Ecology* 26: 1221–1229.

Russo, S.E. and C.K. Augspurger. 2004. Aggregated seed dispersal by spider monkeys limits recruitment to clumped patterns in *Virola calophylla*. *Ecology Letters* 7: 1058–1067.

Schiffers, K., F.M. Schurr, K. Tielbörger, C. Urbach, K.A. Moloney, and F. Jeltsch. 2008. Dealing with virtual aggregation—A new index for analysing heterogeneous point pattern. *Ecography* 31: 545–555.

Schladitz, K. and A.J. Baddeley. 2000. A third order point process characteristic. *Scandinavian Journal of Statistics* 27: 657–671.

Schlather, M., P. Ribeiro, and P. Diggle, 2004. Detecting dependence between marks and locations of marked point processes. *Journal of the Royal Statistical Society*, Series B 66: 79–83.

Schurr, F.M., O. Bossdorf, S.J. Milton, and J. Schumacher. 2004. Spatial pattern formation in semi-arid shrubland: A priori predicted versus observed pattern characteristics. *Plant Ecology* 173: 271–282.

Seidler, T.G. and J.B. Plotkin. 2006. Seed dispersal and spatial pattern in tropical trees. *PLoS Biology* 4: 2132–2137.

Shannon, C.E. 1948. A mathematical theory of communication. *Bell System Technical Journal* 27: 379–423 and 623–656.

Shen, G., M. Yu, X. Hu, X. Mi, H. Ren, I. Sun, and K. Ma. 2009. Species–area relationships explained by the joint effects of dispersal limitation and habitat heterogeneity. *Ecology* 90: 3033–3041.

Shen, G., T. Wiegand, X. Mi, and F. He. in press. Quantifying spatial phylogenetic structures of fully mapped plant communities. *Methods in Ecology and Evolution*. DOI: 10.1111/2041-210X.12119.

Shimatani, K. 2001. Multivariate point processes and spatial variation of species diversity. *Forest Ecology and Management* 142: 215–229.

Shimatani K. and Y. Kubota. 2004. Quantitative assessment of multi-species spatial pattern with high species diversity. *Ecological Research* 19: 149–163.

Simberloff, D. 1979. Nearest-neighbor assessment of spatial configurations of circles rather than points. *Ecology* 60: 679–685.

Simpson, E.H. 1949. Measurement of diversity. *Nature* 688: 163.

Sneath, P.H.A. and R.R. Sokal. 1973. *Numerical Taxonomy: The Principles and Practice of Numerical Classification*. Freeman, San Francisco.

Snedecor, G.W. and W.G. Cochran. 1989. *Statistical Methods*, 7th ed. Iowa State University Press, Ames, IA.

Sørensen, R., U. Zinko, and J. Seibert. 2006. On the calculation of the topographic wetness index: Evaluation of different methods based on field observations. *Hydrology and Earth System Sciences* 10: 101–112.

Sterner, F.J., C.A. Ribic, and G.E. Schatz. 1986. Testing for life history changes in spatial patterns of tropical tree species. *Journal of Ecology* 74: 621–633.

Stoll, P. and D.M. Newbery. 2005. Evidence of species-specific neighborhood effects in the Dipterocarpaceae of a Bornean rain forest. *Ecology* 86: 3048–3062.

Stoll, P. and D. Prati. 2001. Intraspecific aggregation alters competitive interactions in experimental plant communities. *Ecology* 82: 319–327.

Stoll, P., J. Weiner, and B. Schmid. 1994. Growth variation in a natural established population of *Pinus sylvestris*. *Ecology* 75: 660–670.

Stoyan, D. 1984. On correlations of marked point processes. *Mathematische Nachrichten* 116: 197–207.

Stoyan, D. 1987. Statistical analysis of spatial point processes: A soft-core model and cross-correlations of marks. *Biometrical Journal* 29: 971–980.

Stoyan, D. 2006a. Fundamentals of point process statistics. In *Case Studies in Spatial Point Pattern Modelling*, eds. A. Baddeley, P. Gregori, J. Mateu, R. Stoica, and D. Stoyan, pp. 3–22. Lecture Notes in Statistics 185. Springer-Verlag, New York.

Stoyan, D. 2006b. On estimators of the nearest neighbour distance distribution function for stationary point processes. *Metrika* 64: 139–150.

Stoyan, D., W.S., Kendall, and J. Mecke. 1995. *Stochastic Geometry and Its Applications*. J. Wiley and Sons, Chichester.

Stoyan, D. and H. Stoyan. 1994. *Fractals, Random Shapes and Point Fields. Methods of Geometrical Statistics*. John Wiley & Sons, Chichester.

Stoyan, D. and H. Stoyan. 1996. Estimating pair correlation functions of planar cluster processes. *Biometrical Journal* 38: 259–271.

Stoyan, D. and H. Stoyan. 2000. Improving ratio estimators of second order point process characteristics. *Scandinavian Journal of Statistics* 27: 641–656.

Stoyan, D., H. Stoyan, A. Tscheschel, and T. Mattfeld. 2001. On the estimation of distance distribution functions for point processes and random sets. *Image Analysis and Stereology* 20: 65–66.

Swenson, N.G., B.J. Enquist, J. Pither, J. Thompson, and J.K. Zimmerman. 2006. The problem and promise of scale dependency in community phylogenetics. *Ecology* 87: 2418–2424.

Tanaka, U., Y. Ogata, and D. Stoyan. 2008. Parameter estimation and model selection for Neyman–Scott point processes. *Biometrical Journal* 50: 43–57.

Tarboton, D.G. 1997. A new method for determination of flow directions and upslope areas in grid digital elevation models. *Water Resource Research* 33: 309–319.

Thomas, M. 1949. A generalization of Poisson's binomial limit for use in ecology. *Biometrika* 36: 18–25.

Thompson, H.R. 1956. Distribution of distance to nth neighbour in a population of randomly distributed individuals. *Ecology* 37: 391–394.

Tilman, D. and P. Kareiva (eds.) 1997. *Spatial Ecology: The Role of Space in Population Dynamics and Interspecific Interactions*. Princeton University Press, Princeton.

Torquato, S. 1998. Morphology and effective properties of disordered heterogeneous media. *International Journal of Solids and Structures* 35: 2385–2406.

Tscheschel, A. and D. Stoyan. 2006. Statistical reconstruction of random point patterns. *Computational Statistics and Data Analysis* 51: 859–871.

Turkington, R.A. and J.L. Harper. 1979. The growth, distribution and neighbor relationships of *Trifolium repens* in a permanent pasture. I. Ordination, pattern and contact. *Journal of Ecology* 67: 201–208.

Upton, G. and B. Fingleton. 1985. *Spatial Data Analysis by Example. Volume 1: Point Pattern and Quantitative Data.* J. Wiley & Sons, Chichester.

Uriarte, M., R. Condit, C.D. Canham, and S.P. Hubbell. 2004. A spatially explicit model of sapling growth in a tropical forest: Does the identity of neighbours matter? *Journal of Ecology* 92: 348–360.

Uriarte, M., S.P. Hubbell, R. John, R. Condit, and C.D. Canham. 2005. Neighborhood effects on sapling growth and survival in a neotropical forest and the ecological equivalence hypothesis. In *Biotic Interactions in the Tropics: Their Role in the Maintenance of Species diversity*, eds. D.F.R.P. Burslem, M. Pinard, and S. Hartley, pp. 89–106. Cambridge University Press, Cambridge, UK.

Valencia, R., R.B. Foster, G. Villa, R. Condit, J.-C. Svenning, C. Hernandez, K. Romoleroux, E. Losos, E. Magard, and H. Balslev. 2004. Tree species distributions and local habitat variation in the Amazon: A large plot in eastern Ecuador. *Journal of Ecology* 92: 214–229.

Volkov, I., J.R. Banavar, S.P. Hubbell, and A. Maritan. 2009. Inferring species interactions in tropical forests. *Proceedings of the National Academy of Sciences, USA* 106: 13854–13859.

Waagepetersen, R. 2007. An estimating function approach to inference for inhomogeneous Neyman–Scott processes. *Biometrics* 63: 252–258.

Waagepetersen, R. and Y. Guan. 2009. Two-step estimation for inhomogeneous spatial point processes and a simulation study. *Journal of the Royal Statistical Society, Series B* 71: 685–702.

Wang, X., N.G. Swenson, T. Wiegand, A. Wolf, R. Howe, Y. Zhao, X. Bai, D. Xing, and Z. Hao. 2013. Phylogenetic and functional area relationships in two temperate forests. *Ecography* 36: 883–893.

Wang, X., T. Wiegand, Z. Hao, B. Li, J. Ye, and J. Zhang. 2010. Species associations in an old-growth temperate forest in north-eastern China. *Journal of Ecology* 98: 674–686.

Wang, X., T. Wiegand, A. Wolf, R. Howe, S. Davis, and Z. Hao. 2011. Spatial patterns of tree species richness in two temperate forests. *Journal of Ecology* 99: 1382–1393.

Ward, J.S. and F.J. Ferrandino. 1999. New derivation reduces bias and increases power of Ripley's L index. *Ecological Modeling* 116: 225–236.

Watson, D.M., D.A. Roshier, and T. Wiegand. 2007. Spatial ecology of a parasitic shrub: Patterns and predictions. *Austral Ecology* 32: 359–369.

Watt, A.S. 1947. Pattern and process in the plant community. *Journal of Ecology* 35: 1–22.

Webb, C.O., D.D. Ackerly, and S.W. Kembel. 2008. Phylocom: Software for the analysis of phylogenetic community structure and trait evolution. *Bioinformatics* 24: 2098–2100.

Webster, R. and A.B. McBratney. 1989. On the Akaike information criterion for choosing models for variograms of soil properties. *Journal of Soil Science* 40: 493–496.

Webster, R. and M.A. Oliver. 2007. *Geostatistics for Environmental Scientists.* 2nd ed. John Wiley & Sons, Chichester.

Weiher, E. and P.A. Keddy. 1995. Assembly rules, null models and trait dispersion—New questions from old patterns. *Oikos* 74: 159–164.

Weiner, J. and C. Damgaard. 2006. Size-asymmetric competition and size-asymmetric growth in a spatially-explicit zone-of-influence model. *Ecological Research* 21: 707–712.

Weiner J., P. Stoll, H. Muller-Landau, and A. Jasentuliyana. 2001. The effects of density, spatial pattern and competitive symmetry on size variation in simulated plant populations. *American Naturalist* 158: 438–450.

Wiegand, T., C.V.S. Gunatilleke, and I.A.U.N. Gunatilleke. 2007a. Species associations in a heterogeneous Sri Lankan Dipterocarp forest. *The American Naturalist* 170: E77–E95.

Wiegand, T., C.V.S. Gunatilleke, I.A.U.N. Gunatilleke, and A. Huth. 2007b. How single species increase local diversity in tropical forests. *Proceedings of the National Academy of Sciences, USA* 104: 19029–19033.

Wiegand, T., C.V.S. Gunatilleke, I.A.U.N. Gunatilleke, and T. Okuda. 2007c. Analyzing the spatial structure of a Sri Lankan tree species with multiple scales of clustering. *Ecology* 88: 3088–3102.

Wiegand, T., F. He, and S.P. Hubbell. 2013a. A systematic comparison of summary characteristics for quantifying point patterns in ecology. *Ecography* 36: 92–103.

Wiegand, T., A. Huth, S. Getzin, X. Wang, Z. Hao, C.V.S. Gunatilleke, and I.A.U.N. Gunatilleke. 2012. Testing the independent species arrangement assertion made by theories of stochastic geometry of biodiversity. *Proceedings of the Royal Society B* 279: 3312–3320.

Wiegand, T., A. Huth, and I. Martínez. 2009. Recruitment in tropical tree species: Revealing complex spatial patterns. *The American Naturalist* 174: E106–E140.

Wiegand, T., F. Jeltsch, I. Hanski, and V. Grimm. 2003. Using pattern-oriented modeling for revealing hidden information: A key for reconciling ecological theory and application. *Oikos* 100: 209–222.

Wiegand, T., W.D. Kissling, P.A. Cipriotti, and M.R. Aguiar. 2006. Extending point pattern analysis to objects of finite size and irregular shape. *Journal of Ecology* 94: 825–837.

Wiegand, T. and K.A. Moloney. 2004. Rings, circles and null-models for point pattern analysis in ecology. *Oikos* 104: 209–229.

Wiegand, T., K.A. Moloney, and S.J. Milton. 1998. Population dynamics, disturbance, and pattern evolution: Identifying the fundamental scales of organization in a model ecosystem. *American Naturalist* 152: 321–337.

Wiegand, T., K.A. Moloney, J. Naves, and F. Knauer. 1999. Finding the missing link between landscape structure and population dynamics: A spatially explicit perspective. *American Naturalist* 154: 605–627.

Wiegand, T., J. Naves and M. Garbulsky, and N. Fernández. 2008. Animal habitat quality and ecosystem functioning: Exploring seasonal patterns using NDVI. *Ecological Monographs* 78: 87–103.

Wiegand, T., J. Raventós, E. Mujica, E. González, and A. Bonet. 2013b. Spatio-temporal analysis of the effects of hurricane Ivan on two contrasting epiphytic orchid species in Guanahacabibes, Cuba. *Biotropica* 45: 441–449.

Wiegand, T., E. Revilla, and F. Knauer. 2004. Reducing uncertainty in spatially explicit population models. *Biodiversity and Conservation* 13: 53–78.

Wood, S.N. 2010. Statistical inference for noisy nonlinear ecological dynamic systems. *Nature* 466: 1102–1104.

Yamada, I. and P. Rogerson, 2003. An empirical comparison of edge effect correction methods applied to K-function analysis. *Geographical Analysis* 35: 97–109.

Yeong, C.L.Y. and S. Torquato. 1998. Reconstructing random media. *Physical Review E* 57: 495–506.

Yodzis, P. 1989. *Introduction to Theoretical Ecology*. Harper & Row, New York.

Yu, H., T. Wiegand, X. Yang, and L. Ci. 2009. The impact of fire and density-dependent mortality on the spatial patterns of a pine forest in the Hulun Buir sandland, Inner Mongolia, China. *Forest Ecology and Management* 257: 2098–2107.

Index

Printed and bound by CPI Group (UK) Ltd, Croydon, CR0 4YY

24/10/2024

01778302-0016